Rebecca J. E

Student Solutions Manual for

General Chemistry

by Clifford W. Hand

Leslie N. Kinsland
The University of Southwestern Louisiana

Saunders Golden Sunburst Series
SAUNDERS COLLEGE PUBLISHING
Fort Worth Philadelphia San Diego New York Orlando San Antonio
Toronto Montreal London Sydney Tokyo

Printed in the United States of America.

Kinsland: Student Solutions Manual for General Chemistry by Hand

ISBN 0-03-076831-4

345 021 987654321

Foreword

This book consists of the answers to all odd-numbered questions in Clifford W. Hand's **General Chemistry** text. To reinforce the problem solving scheme presented by Dr. Hand in the first chapter and used throughout his text, all solutions are presented in the same format as that in the text. In the first chapters of this book, almost all solutions are worked in complete detail with **Target**, **Knowns**, **Relationships**, **Plan**, and **Work**. As your chemical background and expertise increase, this format will often be abridged, as in the main text, to **Analysis/Plan** and **Work**. Throughout the book, more difficult problems and those involving many steps are always shown in complete detail.

All Table and Figure references are to Hand's **General Chemistry**. Physical constants (molar masses, Avogadro's number, the speed of light, etc.) are used to their maximum number of significant digits as presented in the main text. When a problem has several steps, intermediate answers are shown rounded to the correct number of significant digits, however, **unrounded** numbers are used in successive calculations. If you use rounded numbers for successive calculations or for molar masses or other constants, your answers may differ in the last significant digit. Exponential functions (10^x and e^x) are particularly sensitive to rounding and you should not worry about discrepancies unless your numbers differ by more than 20 - 30%.

To use this book wisely, you should not use it until you have tried to work the exercises on your own. To prepare to work the end-of-chapter problems, you should, first and foremost, go to class to find out which problems your instructor considers to be most important by assigning them for homework. Before you begin the assignment, go over your class notes, read the discussion in the text, study the examples and work the sample problems after the examples. Only after you have completed these steps are you really prepared to tackle the exercises. Even after all that groundwork, you should **not** do homework with this book open. Being able to "follow along" with the printed solution is not the same as compiling your own list of targets, knowns, and relationships and devising your own plan of action. You will learn the material more thoroughly and retain the information longer if you can finally determine the solution as presented in Appendix K of the text on your own.

If we discourage you from using this book, then why do so many instructors recommend that their students buy a solutions manual? This book provides a helping hand when you've tried the problem 10 times and can't get the answer or when you have no idea of where to begin. When you turn to this book at those times, don't immediately read the entire solution. First, look at the analysis. Did you find a molarity when the book asked for molality? Did you give an answer in a different form, such as calories instead of joules? Next, survey the knowns and relationships. Did you forget that a simplifying relationship exists? Did you copy a number down wrong or write an equivalence statement backwards? Did you find the molar mass of CO instead of Co? Often this little boost will be enough for you to complete the problem solving process on your own. If it is not, then, finally, look at the rest of the solution, with your work in hand for comparison, to see where you went wrong. Remember your goal is to master the material in this course. The book and these solutions are aids to help you achieve success in your study of chemistry.

All solutions in this manual have been cross checked in consultation with Dr. Hand. I hope that you see in this text his enthusiasm for chemistry and his dedication to teaching. It was these qualities that I saw when I agreed to work on this project. He has made the long process of working on this ancillary text a satisfying experience.

My special thanks go to USL senior chemistry major Frank D. Verret who proofread the manual and double checked the calculations for all steps for all answers in this text. My daughter, Cynthia L. Kinsland, also a senior chemistry major at USL, proofread the entire book, offered many suggestions, caught additional errors, both typographical and chemical, and provided moral support throughout the duration of this project. Saunders College Publishing provided a grant for my original chemical structure drawing program (ChemDraw™). Editor Sandi Kiselica of Saunders made suggestions which were incorporated into the formatting of this manuscript.

This entire manuscript was prepared on a Macintosh 512 computer, enhanced with a Novy Quik 30 accelerator and upgraded to 16 Meg RAM. The word processing program used was Microsoft® Word 5.1. Headings are in Krone Extra Bold, while the main text is in Schoolbook. Both fonts are from KeyFonts® 3.0 from SoftKey. Delta Point's DeltaGraph® Professional 2.0 was used for graphs. SoftShell's Chemintosh® 2.1 was the chemical structure drawing program used for the production of this book. Printing was done on an Apple® Personal Laser Writer® LS printer.

Table of Contents

1 Scope and Vocabulary of Chemistry

1.1 Give a concise definition of "matter". Does the term "substance" have a different meaning?

Matter is anything which has mass and occupies space. A substance is any type of matter with constant composition and invariant properties under a given set of conditions. Chemically, a substance is either an element or a compound.

1.3 Describe the components of the scientific method. Name some areas of human concern, outside of science, for which the scientific method is a valuable approach. Name some areas of human concern in which the scientific method is not useful.

Observations consist of gathering data about some aspect of matter. **Facts** are observations which have been duplicated by numerous observers. A **hypothesis** is a model or tentative explanation of a group of facts. **Testing** is done to see if any data can be found which modifies, reinforces or disproves the hypothesis. A **theory** is a hypothesis which has survived many rounds of testing without being disproved. The scientific method is good for fixing broken vehicles, growing gardens and training athletes. It is not good for religion, love or predicting fashion trends.

1.5 Comment on the statement, "You needn't take it too seriously; after all, it's only a theory."

This statement is using the nonscientific interpretation of the word "theory", which implies a guess. Scientifically, theories are well-tested and should be respected.

1.7 Describe each of the following as observation, fact, hypothesis, theory, and/or law. There may be some differences of opinion as to the correct descriptions, so be prepared to defend your choices.

a. Objects are attracted to the earth with a force proportional to their masses.
b. The force of gravity depends on the color of an object
c. Some objects, like hot-air balloons, rise rather than fall when released
d. The weight of an astronaut on the moon is less than on the earth

a. This is a well-tested observation, so it is a **fact**.
b. This is a rather sketchy proposal, so it is a **hypothesis**.
c. This is a well-tested observation, so it is a **fact**.
d. This is a well-tested observation, so it is a **fact**.

1.9 Name, describe, and give examples of the three most common physical states of matter.

Solids have invariant volume and shape at a given temperature and pressure. Ice cubes, nails, books and baseball bats are solids. **Liquids** have invariant volume, but a variable shape at a given temperature and pressure. Rubbing alcohol, gasoline and drinking water are liquids. **Gases** have variable shape and volume at a given temperature and pressure. The air around us, the carbon dioxide we exhale, and the neon in neon lights are gases.

1.11 Identify each of the following as a pure substance or a mixture.

a. milk   b. air   c. compressed oxygen   d. salt   e. sugar

a. Milk is **mixture** of many substances dissolved in water, plus small globules of butterfat.
b. Air is a **mixture** of oxygen, water vapor, carbon dioxide, nitrogen, etc.
c. Compressed oxygen is a **pure substance** (an element).
d. Salt is a **pure substance** (a compound).
e. Sugar is a **pure substance** (a compound).

1.13 What are the two types of impure substances?

Impure substances consist of homogeneous mixtures (solutions) and heterogeneous mixtures.

1.15 Identify each of the following mixtures as homogeneous or heterogeneous.

a. gasoline b. salad oil c. mayonnaise d. chicken noodle soup e. milk

a. Gasoline is uniform in appearance. It is **homogeneous**.
b. Salad oil is uniform in appearance. It is **homogeneous.**
c. Mayonnaise is not uniform in appearance because it has small fat globules. It is **heterogeneous**.
d. Chicken noodle soup is not uniform in appearance. It is **heterogeneous**.
e. Milk is not uniform in appearance because it has small fat globules. It is **heterogeneous**.

1.17 Define and give some examples of physical change.

A physical change is a change in appearance or form without a composition change. Examples are melting ice, bending a wire, powdering sugar and chopping wood.

1.19 Describe each of the following as either a chemical or a physical change, or both.
a. A wet towel dries in the sun.
b. Lemon juice is added to tea, causing a color change.
c. Hot air rises over a radiator.
d. Coffee is brewed by passing hot water through ground coffee.

a. This is a **physical change** as the water evaporates from the towel.
b. This is a **chemical change** as the lemon reacts with some component of the tea.
c. This is a **physical change** as the air rises because it expands, giving it a lower density.
d. This is a **chemical change** as water reacts with components in the grounds.

1.21 Which of the following are chemical properties, and which are physical properties?
a. Baking powder gives off bubbles of carbon dioxide when added to water.
b. A particular type of steel consists of 95% iron, 4% carbon, and 1% miscellaneous other elements.
c. The density of gold is 19.3 g mL^{-1}.
d. Iron dissolves in hydrochloric acid with the evolution of hydrogen gas.
e. Fine steel wool burns in air.

a. This is a **chemical property** since a new substance is formed.
b. This is a **physical property** which can be determined without a composition change.
c. This is a **physical property** which can be determined without a composition change.
d. This is a **chemical property** since a new substance is formed.
e. This is a **chemical property** since a new substance is formed.

1.23 Which of the following pure substances are elements, and which are compounds?
a. iron b. water c. carbon dioxide d. hydrogen e. mercury

To be an element, the substance must be on the periodic table of the elements. Iron (a), hydrogen (d), and mercury (e) are elements. Water (b) and carbon dioxide (c) are compounds.

1.25 Identify each of the following as a pure substance or as a mixture. Further identify each substance as an element or compound if pure, or as heterogeneous or homogeneous if a mixture.
a. dew b. maple syrup c. concrete d. paint e. table sugar

a. Dew is condensed water vapor; a **compound**.
b. Maple syrup is a **homogenous mixture** of sugar in water.
c. Concrete is a **heterogeneous mixture** of cement and stones.
d. Paint is a **heterogeneous mixture** of solids suspended in a liquid base.
e. Table sugar is a **compound** (sucrose).

1.27 Identify each of the following units as SI or other. If SI, tell whether it is a base or derived unit.
a. cubic centimeter b. liter c. gallon d. meter e. gram

a. A cubic centimeter, which is a derived unit for volume, is an SI unit.
b. This is not an SI unit for volume.
c. This is not an SI unit for volume.
d. A meter is the base SI unit for length.
e. A gram is an SI unit which is a derived unit for mass. (A kilogram is the base unit.)

1.29 Identify the numerical multiplier corresponding to each of the following metric prefixes, or vice versa. Do as many as you can from memory.
a. nano- b. 10^3 c. mega- d. centi- e. tera-

a. nano = $\mathbf{10^{-9}}$ b. 10^3 = **kilo** c. mega = $\mathbf{10^6}$
d. centi = $\mathbf{10^{-2}}$ e. tera = $\mathbf{10^{12}}$

1.31 Convert each of the following from arithmetic to scientific notation, or vice versa.
a. 7325 b. 0.83 c. 18,236 d. 7.5×10^6 e. 0.000972

The power of ten is increased by one for each place the decimal point is shifted to the left and decreased by one for each place the decimal point is shifted to the right.

a. $\mathbf{7.325 \times 10^3}$ b. $\mathbf{8.3 \times 10^{-1}}$ c. $\mathbf{1.8326 \times 10^4}$ d. **7,500,000** e. $\mathbf{9.72 \times 10^{-4}}$

1.33 Convert each of the following from arithmetic to scientific notation, or vice versa. (Do not change the units.)
a. 692 m b. 0.003151596 g c. 2.54 cm $inch^{-1}$ d. 3.00×10^8 m s^{-1} e. 0.000972 m^3

a. $\mathbf{6.92 \times 10^2}$ **m** b. $\mathbf{3.151596 \times 10^{-3}}$ **g** c. **2.54 cm** $\mathbf{inch^{-1}}$
d. **300,000,000 m** $\mathbf{s^{-1}}$ e. $\mathbf{9.72 \times 10^{-4}\ m^3}$

1.35 Express the following quantities using a more appropriate SI prefix. Choose a prefix corresponding to an exponent that is divisible by three, and such that the value of the quantity is in the range 1 to 999. For example, 0.0050 L is better expressed as 5.0 mL.
a. 0.083 kg b. 1.5×10^8 m c. 15,626 K d. 8372 mg e. 0.946 L

a. 0.083 kg = 0.083×10^3 g = **83 g**
b. 1.5×10^8 m = 150×10^6 m = **150 Mm**
c. 15,626 K = 15.626×10^3 K = **15.626 kK**
d. 8372 mg = 8372×10^{-3} g = **8.372 g**
e. 0.946 L = 946×10^{-3} L = **946 mL**

1.37 Express the following quantities using a more appropriate SI prefix.
a. 7,853 nm b. 0.000285 kPa c. 0.00183 °C d. 4.00×10^5 J e. 4.72×10^{-11} g

a. 7,853 nm = $7{,}853 \times 10^{-9}$ m = 7.853×10^{-6} m = **7.853 μm**
b. 0.000285 kPa = 0.000285×10^3 Pa = 0.285 Pa = 285×10^{-3} Pa = **285 mPa**

c. 0.00183 °C = 1.83×10^{-3} °C = **1.83 m°C**
d. 4.00×10^{5} J = 400×10^{3} J = **400 kJ**
e. 4.72×10^{-11} g = 47.2×10^{-12} g = **47.2 pg**

1.39 Express the following derived quantities using the most appropriate SI prefix. It is customary to express the denominator without a prefix. For example, the speed 3,458 m s^{-1} is normally expressed as 3.458 km s^{-1} rather than 3.458 m ms^{-1} although the latter is not incorrect.
a. 4.82×10^{3} cm s^{-1} b. 0.00328 mol L^{-1} c. 50.3×10^{-6} L s^{-1} d. 3803 J m^{-3}

a. 4.82×10^{3} cm s^{-1} = $4.82 \times 10^{3} \times 10^{-2}$ m s^{-1} = 4.82×10^{1} m s^{-1} = **48.2 m s^{-1}**
b. 0.00328 mol L^{-1} = 3.28×10^{-3} mol L^{-1} = **3.28 mmol L^{-1}**
c. 50.3×10^{-6} L s^{-1} = **50.3 μL s^{-1}**
d. 3803 J m^{-3} = 3.803×10^{3} J m^{-3} = **3.803 kJ m^{-3}**

1.41 Express the following derived quantities using the indicated SI prefix(es). These conversions are more difficult than those in the preceding problems, and they are very important (see Example 1.3).
a. 1.5×10^{4} cm^{2} (express in m^{2}) b. 9.03×10^{2} kg m^{-3} (express in g cm^{-3})

a. **Analysis:**
Target: An area in m^2. (? m^2) =
Known: An area in cm^2.
Relationship: 1 cm ↔ 10^{-2} m.
Plan: The conversion must be applied twice (squared) for the square term.
Work:

$$? \text{ m}^2 = 1.5 \times 10^{4} \text{ cm}^2 \times \left(\frac{10^{-2}\text{ m}}{1\text{ cm}}\right)^2 = \mathbf{1.5\ m^2}$$

b. **Analysis:**
Target: A density in g cm^{-3}. (? g cm^{-3}) =
Known: A density in kg m^{-3}.
Relationships: 1 kg ↔ 10^{3} g and 1 cm ↔ 10^{-2} m.
Plan: The mass must be converted to kg and the volume to cm^{-3}. The distance conversion must be applied three times (cubed) for the cubic term.
Work:

$$? \text{ g cm}^{-3} = \frac{9.03 \times 10^{2}\text{ kg}}{\text{m}^3} \times \frac{10^{3}\text{ g}}{1\text{ kg}} \times \left(\frac{10^{-2}\text{ m}}{1\text{ cm}}\right)^3 = \mathbf{9.03 \times 10^{-1}\ g\ cm^{-3}}$$

1.43 How many significant figures are there in each of the following numbers?
a. 105 b. 0.82 c. 0.0082 d. 1500.

The significant digits are shown in boldface.
a. **105** (3) b. 0.**82** (2) c. 0.00**82** (2) d. **1500.** (4)

1.45 How many significant figures are there in each of the following numbers?
a. 108 nm b. 180 nL c. 018 Gg d. 180. pg

The significant digits are shown in boldface.
a. **108** nm (3) b. **18**0 or **180** nL (2 or 3) c. 0**18** Gg (2) d. **180.** pg (3)

1.47 How many significant figures are there in the quantities in each of the following statements?
a. The length of a meter is 100 centimeters.
b. The flask holds 125 mL.

c. The flask holds 500 mL.
d. An English pound is equivalent to 454 grams.
e. Most dogs have 2 ears.

a. This is an **exact** defined term, so it has an infinite number of significant figures.
b. This is measured to **three** significant figures.
c. This is ambiguous; it could be **one, two** or **three** significant figures.
d. This is not exact, but is presented to **three** significant figures.
e. This is an **exact**, counted term, so it has an infinite number of significant figures.

1.49 Criticize the following statement: "According to the weather report, the high tomorrow will be 75 °F, or 23.89 °C."

The temperature is being estimated to be about 75 °F, which has two significant figures. There is no reason to report the Celsius temperature to four significant figures, which would indicate a much higher degree of certainty. The temperature should be reported as 24 °C.

1.51 Perform the following calculations, and use the proper number of significant digits to express the answer.
a. $(185) \times (26)$ b. $(0.032) \times (6.4 \times 10^2)$ c. $(8.02 \times 10^{-3})/(2.218 \times 10^{-7})$

Analysis/Plan: Remember that when we add or subtract, the final answer must have the same number of **decimal places** as the component with the fewest. When we multiply or divide, the final answer must have the same number of **significant digits** as the component with the fewest.
Work:
a. $(185) \times (26) = 4810 = \mathbf{4.8 \times 10^3}$
b. $(0.032) \times (6.4 \times 10^2) = 20.48 = \mathbf{20.}$ or $\mathbf{2.0 \times 10^1}$
c. $(8.02 \times 10^{-3})/(2.218 \times 10^{-7}) = 36158.70153 = \mathbf{3.62 \times 10^4}$

1.53 Perform the following calculations, and use the proper number of significant digits to express the answer.
a. 75 + 103 b. 752 + 103 c. 752 + 10.3 d. 752 - 10.3 e. 75.2 - 10.3 f. 0.752 + 10.3

Analysis/Plan: Remember that when we add or subtract, the final answer must have the same number of **decimal places** as the component with the fewest. When we multiply or divide, the final answer must have the same number of **significant digits** as the component with the fewest.
Work:
a. 75 + 103 = **178**
b. 752 + 103 = **855**
c. 752 + 10.3 = 762.3 = **762**
d. 752 - 10.3 = 741.7 = **742**
e. 75.2 - 10.3 = **64.9**
f. 0.752 + 10.3 = 11.052 = **11.1**

1.55 Perform the following calculations, and use the proper number of significant digits to express the answer.
a. $\dfrac{(6.91 \times 10^3) + 276}{185}$ b. $(2.54 - 454) + (9.46 - 0.70)$ c. $72.53 - (0.0322 \cdot 213)$

Analysis/Plan: Remember that when we add or subtract, the final answer must have the same number of **decimal places** as the component with the fewest. When we multiply or divide, the final answer must have the same number of **significant digits** as the component with the fewest.

Work: In the intermediate steps, the significant digits are shown in boldface.

a. $\frac{(6.91 \times 10^3) + 276}{185} = \frac{(\mathbf{6.91} \times 10^3) + (\mathbf{0.276} \times 10^3)}{\mathbf{185}} = \frac{\mathbf{7.186} \times 10^3}{\mathbf{6}}$ = 38.84324324 = **38.8**

b. (2.54 - 454) + (9.46 - 0.70) = **-451.46** + **8.76** = **-442**.70 = **-443**

c. 72.53 - (0.0322) x (213) = **72.53** - **6.8586** = **65.67**14 = **65.67**

1.57 The masses of substances used in the laboratory are often determined by weighing the substance in a container of some sort, then subtracting the mass of the empty container. If a chemist finds the mass of an empty beaker to be 37.603 g, and to be 37.682 g after a sample of titanium is placed in it, to how many significant digits does he know the mass of the titanium?

Analysis/Plan: Remember that when we add or subtract, the final answer must have the same number of **decimal places** as the component with the fewest.

Work:

37.682 g - 37.603 g = 0.0**79 g** (**two** significant digits)

1.59 Perform the indicated calculations and express the result to the correct number of significant figures. Note that the number of significant digits in the answers to (a) and (b) are different, and that (a) is therefore a much more reliable number. Often there are several different ways to go about making a measurement, and you should, if possible, avoid ways that require taking the difference between two large numbers.

a. 60725 + 60703 b. 60725 - 60703

Analysis/Plan: Remember that when we add or subtract, the final answer must have the same number of **decimal places** as the component with the fewest.

Work:

a. 60725 + 60703 = **121,428** (**six** significant digits)

b. 60725 - 60703 = **22** (**two** significant digits)

1.61 Convert the following Celsius temperatures to the Kelvin scale.

a. 25 °C b. -78 °C c. 500 °C d. 1234 °C

Analysis:

Target: A temperature in Kelvin (K). (? K) =

Known: A temperature in °C.

Relationship: °C + 273.15 = K.

Plan: Use the relationship directly and solve.

Work:

a. ? K = 25 °C + 273.15 = 298.15 K= **298 K**

b. ? K = -78 °C + 273.15 = 195.15 K = **195 K**

c. ? K = 500 °C + 273.15 = 773.15 K = **773 K**

d. ? K = 1234 °C + 273.15 = 1507.15 K = **1507 K**

1.63 Convert the following Fahrenheit temperatures to the Celsius scale.

a. 68 °F b. 40 °F c. 1234 °F d. 805°F

Analysis:

Target: A temperature in Celsius (°C). (?°C) =

Known: A temperature in °F.

Relationship: °C = (°F - 32)/1.8.

Plan: Use the relationship directly and solve.

Work: Remember that the 32 and 1.8 are the defined differences between the scales and will not limit significant digits.

a. ? °C = (68 - 32)/1.8 = 36/1.8 = **20 °C**

b. ? °C = (40 - 32)/1.8 = 8/1.8 = **4 °C**

c. ? °C = (1234 - 32)/1.8 = 1202/1.8 = **668 °C**

d. ? °C = (805 - 32)/1.8 = 773/1.8 = **429 °C**

1.65 Convert the following Celsius temperatures to temperatures on the Fahrenheit scale.
a. 10 °C b. -30 °C c. -196 °C d. 500 °C

Analysis:
Target: A temperature in Fahrenheit (°F). (? °F) =
Known: A temperature in °C.
Relationship: °F = 1.8(°C) + 32.

Plan: Use the relationship directly and solve.

Work: Remember that the 32 and 1.8 are the defined differences between the scales and will not limit significant digits.

a. ? °F = 1.8(10) + 32 = 18 + 32 = **50 °F**
b. ? °F = 1.8(-30) + 32 = -54 + 32 = **-22 °F**
c. ? °F = 1.8(-196) + 32 = -353 + 32 = **-321 °F**
d. ? °F = 1.8(500) + 32 = 900 + 32 = **932 °F**

1.67 Liquified gases have boiling points well below room temperature. On the Kelvin scale the boiling points of the following gases are: He, 4.2 K; H_2, 20.4 K; N_2, 77.4 K; O_2, 90.2 K. Convert these temperatures to the Celsius and the Fahrenheit scales.

Analysis:
Target: A temperature in Celsius (°C) and Fahrenheit (°F). (? °C) = and (? °F) =
Known: A temperature in Kelvin.
Relationships: °C = K - 273.15 and °F = 1.8(°C) + 32.

Plan: Use the known relationships directly to convert each temperature to Celsius and then to Fahrenheit.

Work:

He: ? °C = 4.2 - 273.15 = -268.95 °C = **-269.0 °C**
? °F = 1.8(-269.0) + 32 = -484.1 + 32 = **-452.1 °F**

H_2: ? °C = 20.4 - 273.15 = -252.75 = **-252.8 °C**
? °F = 1.8(-252.8) + 32 = -455.0 + 32 = **-423.0 °F**

N_2: ? °C = 77.4 - 273.15 = -195.75 = **-195.8 °C**
? °F = 1.8(-195.8) + 32 = -352.4 + 32 = **-320.4 °F**

O_2: ? °C = 90.2 - 273.15 = -182.95 = **-183.0 °C**
? °F = 1.8(-183.0) + 32 = -329.3 + 32 = **-297.3°F**

1.69 Bored with sightseeing on a trip to England, you wish to bake some cookies. You remember the recipe, and that the baking temperature should be 375 °F. Unfortunately, the oven in your flat is calibrated in °C. To what temperature should it be set? (The kitchen is not equipped with a calculator, so do this one in your head, or with a paper and pencil.)

Analysis:
Target: A temperature in Celsius(°C). (? °C) =
Known: A temperature in °F.
Relationship: °C = (°F - 32)/1.8.

Plan: Use the known relationship directly.

Work:
? °C = (375 - 32)/1.8 = 343/1.8 = **190 °C**

1.71 The density of cork varies somewhat from one sample to the next, but is usually about 0.21 g mL^{-1} at 30 °C. What is the specific gravity of cork? The density of water at 30 °C is 0.9956 g mL^{-1}.

Analysis:

Target: Specific gravity. (? Sp. Gr.) =

Knowns: Density of cork = 0.21 g mL^{-1} and density of water = 0.9956 g mL^{-1} at 30 °C.

Relationship: The definition of specific gravity, Sp. Gr. = $\frac{D_{substance}}{D_{water}}$.

Plan: The definition of specific gravity can be used directly.

Work:

$$\text{Sp. Gr.} = \frac{0.21 \text{ g mL}^{-1}}{0.9956 \text{ g mL}^{-1}} = \mathbf{0.21}$$

1.73 What is the density of copper at 30 °C, if its specific gravity is 8.92? The density of water at 30 °C is 0.9956 g mL^{-1}.

Analysis:

Target: The density of copper. (? g mL^{-1}) =

Knowns: Specific gravity of copper = 8.92 and density of water = 0.9956 g mL^{-1} at 30 °C.

Relationship: The definition of specific gravity, Sp. Gr. = $\frac{D_{substance}}{D_{water}}$.

Plan: The definition of specific gravity can be rearranged to yield the density of copper.

$$\text{Sp. Gr.} = \frac{D_{copper}}{D_{water}} \text{ so } D_{copper} = (\text{Sp.Gr.}) \times (D_{water})$$

Work:

$$? \text{ g mL}^{-1} = D_{copper} = (\text{Sp. Gr.})(D_{water}) = 8.92(0.9956 \text{ g mL}^{-1}) = \mathbf{8.88 \text{ g mL}^{-1}}$$

1.75 Does the specific gravity of water vary as the temperature is changed? Explain.

The specific gravity of water is independent of the temperature because of the way specific gravity is defined:

$$\text{Sp. Gr.} = \frac{D_{substance}}{D_{water}}. \text{ For water Sp. Gr.} = \frac{D_{water}}{D_{water}} = 1 \text{ always.}$$

1.77 What is the density of cadmium metal, if 10.0 g occupies exactly 1.16 mL?

Analysis:

Target: The density of cadmium. (? g mL^{-1}) =

Knowns: The mass of cadmium and the volume of cadmium.

Relationship: The definition of density is $\frac{\text{mass}}{\text{volume}}$.

Plan: The definition of density can be used directly.

Work:

$$? \text{ g mL}^{-1} = \frac{10.0 \text{ g}}{1.16 \text{ mL}} = \mathbf{8.62 \text{ g mL}^{-1}}$$

1.79 A convenient method of determining the density of an object is to submerge the object in water and measure the volume of displaced water. If a small bar of zinc weighing 24.3 g is dropped into a partially filled graduated cylinder and the water level rises from 40.8 mL to 44.2 mL, what is the density of zinc?

Analysis:

Target: The density of zinc. (? g mL^{-1}) =

Known: The mass of zinc, the volume of water before the zinc is added, and the volume of the zinc water mixture.

Relationships: The definition of density is $\frac{\text{mass}}{\text{volume}}$. When a metal is added to water, the volumes are additive.

Plan: The volume of zinc can be found by subtracting the volume of water from the volume of zinc plus water. The density relationship can then be used directly.

Work:

? mL zinc = 44.2 mL - 40.8 mL = 3.4 mL

$$? \text{ g mL}^{-1} = \frac{24.3 \text{ g}}{3.4 \text{ mL}} = \mathbf{7.1 \text{ g mL}^{-1}}$$

1.81 What is the volume of a kilogram of nickel, whose density is 8.90 g mL^{-1}?

Analysis:

Target: The volume of nickel. (? mL) =

Knowns: The mass of nickel and the density of nickel.

Relationships: The definition of density is $\frac{\text{mass}}{\text{volume}}$ and 1 kg ↔ 10^3 g.

Plan: First, the mass of nickel in grams can be found from the relationship between kg and g. The mass can then be used with the density relationship rearranged to solve for volume.

$$D = \frac{m}{V} \text{ so } DV = m \text{ and } V = \frac{m}{D}$$

This could also be solved by the factor-label method, in this case, multiplying by $\frac{1 \text{ mL}}{8.90 \text{ g}}$.

Work:

$$1 \text{ kg} \times \frac{10^3 \text{ g}}{1 \text{ kg}} = 1000 \text{ g} \quad \text{and } ? \text{ mL} = \frac{1000 \text{ g}}{8.90 \text{ g mL}^{-1}} = \mathbf{112 \text{ mL}}$$

$$\text{OR } ? \text{ mL} = 1 \text{ kg} \times \frac{10^3 \text{ g}}{1 \text{ kg}} \times \frac{1 \text{ mL}}{8.90 \text{ g}} = \mathbf{112 \text{ mL}}$$

1.83 The density of dry air at sea level and at 22 °C is 1.12 g L^{-1}. What is the volume of one metric ton (1000 kg ↔ 1 metric ton) of air? Express your answer in (a) liters, (b) cubic meters, and (c) cubic kilometers.

Analysis:

Target: The volume of air with units of L, m^3 and km^3. (? L) = and (? m^3) = and (? km^3) =

Knowns: The mass of air and the density of air.

Relationships: Density = $\frac{\text{mass}}{\text{volume}}$, 1 kg ↔ 10^3 g, 1 L ↔ 1000 cm^3, 10^{-2} m ↔ 1 cm, 1000 kg ↔ 1 metric ton, 10^{-2} m ↔ 1 cm, and 1 km ↔ 10^3 m.

Plan: The mass must be converted to grams and the density relationship rearranged to solve for volume as in Problem 1.81. Alternatively, the factor-label method can be used to find the volume in liters. The factor-label method can be used to convert to the other volume units.

Work:

$$? \text{ g} = 1000 \text{ kg} \times \frac{10^3 \text{ g}}{1 \text{ kg}} = 1.00 \times 10^6 \text{ g} \quad \text{and } ? \text{ L} = V = \frac{m}{D} = \frac{1.00 \times 10^6 \text{ g}}{1.12 \text{ g L}^{-1}} = \mathbf{8.93 \times 10^5 \text{ L}}$$

$$\text{OR } 1.00 \times 10^6 \text{ g} \times \frac{1 \text{ L}}{1.12 \text{ g}} = \mathbf{8.93 \times 10^5 \text{ L}}$$

$$? \text{ m}^3 = 8.93 \times 10^5 \text{ L} \times \frac{1000 \text{ cm}^3}{\text{L}} \times \left(\frac{10^{-2} \text{ m}}{1 \text{ cm}}\right)^3 = \mathbf{8.93 \times 10^2 \text{ m}^3} \text{ or } \mathbf{893 \text{ m}^3}$$

$$? \text{ km}^3 = 8.93 \times 10^2 \text{ m}^3 \times \left(\frac{1 \text{ km}}{10^3 \text{m}}\right)^3 = \mathbf{8.93 \times 10^{-7} \text{ km}^3}$$

1.85 Calculate the mass of a cubic meter (a "stere") of seawater, whose density is 1.025 g mL^{-1}. Owing to dissolved salt, the density of seawater is greater than that of fresh water. [Since the amount of dissolved salt (the "salinity") varies from place to place in the earth's oceans, so does the density.]

Analysis:
Target: The mass of water. (? Mg) =
Knowns: The volume of water, 1 m^3, defined by the name cubic meter, and the density of sea water.
Relationships: Density = $\frac{\text{mass}}{\text{volume}}$, 10^{-2} m ↔ 1 cm, 1 Mg ↔ 10^6 g, and 1 cm^3 ↔ 1 mL.
Plan: The volume must be converted to mL and the density relationship rearranged to solve for mass. The factor-label method can also be used to solve for the mass.
Work:

$$? \text{ mL} = 1 \text{ m}^3 \times \left(\frac{1 \text{ cm}}{10^{-2} \text{ m}}\right)^3 \times \frac{1 \text{ mL}}{1 \text{ cm}^3} = 1.000 \times 10^6 \text{ mL}$$

$$? \text{ g} = 1.025 \text{ g mL}^{-1} \times 1.000 \times 10^6 \text{ mL} = \mathbf{1.025 \times 10^6 \text{ g} \text{ or } 1.025 \text{ Mg}}, \quad \text{since } 10^6 \text{ g} = 1 \text{ Mg}$$

1.87 One cubic foot of lead weighs 708 pounds. Calculate its density in g mL^{-1}. You will need the equivalence statements 1 lb ↔ 454 g, 1 inch ↔ 2.54 cm, and 1 foot ↔ 12 inches.

Analysis:
Target: Density in g mL^{-1}. (? g mL^{-1}) =
Known: Density in lb ft^{-3}.
Relationships: 1 cm^3 ↔ 1 mL, 1 lb ↔ 454 g, 1 inch ↔ 2.54 cm, and 1 foot ↔ 12 inches.
Plan: This problem can be done by the factor-label method.
Work:

$$? \text{ g mL}^{-1} = \frac{708 \text{ lb}}{\text{ft}^3} \times \left(\frac{1 \text{ ft}}{12 \text{ in}}\right)^3 \times \left(\frac{1 \text{ in}}{2.54 \text{ cm}}\right)^3 \times \frac{1 \text{ cm}^3}{1 \text{ mL}} \times \frac{454 \text{ g}}{1 \text{ lb}} = \mathbf{11.4 \text{ g mL}^{-1}}$$

1.89 The average density of dry balsa wood is 0.13 g mL^{-1}, that of Douglas fir is about 0.48 g mL^{-1}, and that of a typical hardwood such as white oak is 0.71 g mL^{-1}. The most dense of the common woods is ironwood, whose density when dry is 1.08 g mL^{-1}.
a. Calculate the volume occupied by 5.00 lb of each of these woods.
b. Why did Thor Heyerdahl choose balsa rather than ironwood (a far stronger material) for his raft Kon-Tiki?

a. **Analysis:**
Target: The volume of wood with convenient units of L. (? L) =
Knowns: The mass of each wood and the density of each wood.
Relationships: Density = $\frac{\text{mass}}{\text{volume}}$, 1 lb ↔ 454 g; and 1 mL ↔ 10^{-3} L.
Plan: The mass must be converted to grams and the density relationship rearranged to solve for volume as in Problem 1.81. The factor-label method can also be used to find the volume in mL.
Work:
The mass in grams of each wood sample is found first.

$$? \text{ g} = 5.00 \text{ lb} \times \frac{454 \text{ g}}{1 \text{ lb}} = 2.27 \times 10^3 \text{ g}$$

$$\text{For balsa, } ? \text{ L} = \frac{2.27 \times 10^3 \text{ g}}{0.13 \text{ g mL}^{-1}} \times \frac{10^{-3} \text{ L}}{1 \text{ mL}} = \mathbf{17 \text{ L}}$$

$$\text{For fir, } ? \text{ L} = \frac{2.27 \times 10^3 \text{ g}}{0.48 \text{ g mL}^{-1}} \times \frac{10^{-3} \text{ L}}{1 \text{ mL}} = \mathbf{4.7 \text{ L}}$$

For white oak, $? \text{ L} = \frac{2.27 \times 10^3 \text{ g}}{0.71 \text{ g mL}^{-1}} \times \frac{10^{-3} \text{ L}}{1 \text{ mL}} = \mathbf{3.2 \text{ L}}$

For ironwood, $? \text{ L} = \frac{2.27 \times 10^3 \text{ g}}{1.08 \text{ g mL}^{-1}} \times \frac{10^{-3} \text{ L}}{1 \text{ mL}} = \mathbf{2.10 \text{ L}}$

b. Even though the ironwood is stronger, it will not float on water because of its density is higher than that of seawater. If it cannot float, it would not make a good raft.

1.91 Oil floats on water because it is less dense and it does not dissolve. Oil also does not dissolve in alcohol, but does it float? The density of salad oil is typically about 0.93 g mL^{-1}, and a 5.00 mL sample of ethanol (grain alcohol) weighs 3.95 g.

Analysis:
Target: Floating is density dependent. The density of alcohol is needed. (? g mL^{-1}) =
Known: The density of oil, the volume of ethanol, and the mass of ethanol.
Relationship: Density $= \frac{\text{mass}}{\text{volume}}$.
Plan: The density formula can be used directly to determine the density of alcohol. The densities can then be compared.
Work:

$$? \text{ g mL}^{-1} = \frac{3.95 \text{ g}}{5.00 \text{ mL}} = 0.790 \text{ g mL}^{-1}$$

The density of the ethanol is less than that of oil. Therefore, alcohol will **float** on oil.

1.93 Aluminum foil is sold in supermarkets in long rolls measuring 66 2/3 yards by 12 inches, in a thickness of 0.00065 inch. If the specific gravity of aluminum at 22 °C is 2.70, calculate the mass of a roll. The density of water at 22 °C is 0.998 g mL^{-1}.

Analysis:
Target: The mass of a roll of aluminum. (? kg) =
Knowns: The length, the width, and the thickness of the foil, the specific gravity of the foil and the density of water at the same temperature.
Relationships: The definition of specific gravity, Sp. Gr. $= \frac{D_{substance}}{D_{water}}$, the definition of density, $D = \frac{\text{mass}}{\text{volume}}$, and the relationship for the volume is length x thickness x width.
Several equivalence statements are also needed to convert volume to the same units as in the density: 12 inches ↔ 1 foot, 1 yard ↔ 36 inches, 1 inch ↔ 2.54 cm,

1 kg ↔ 10^3 g, and 1 cm^3 ↔ 1 mL.

Plan: **Step 4.** If the volume and density are known, the density relationship can be rearranged to yield mass. The factor-label method is used to convert the mass to kg.
Step 3. The density of the aluminum foil can be found from the specific gravity.
Step 2. The volume can be found from the formula above.
Step 1. Measurements can be converted to appropriate units (cm) to find the volume.

Work:
Step 1. Convert the measurements to cm.

length: $? \text{ cm} = 66\ 2/3 \text{ yd} \times \frac{36 \text{ in}}{1 \text{ yd}} \times \frac{2.54 \text{ cm}}{1 \text{ in}} = 6.10 \times 10^3 \text{ cm}$

width: $? \text{ cm} = 12 \text{ in} \times \frac{2.54 \text{ cm}}{1 \text{ in}} = 30. \text{ cm}$

thickness: $? \text{ cm} = 0.00065 \text{ in} \times \frac{2.54 \text{ cm}}{1 \text{ in}} = 1.7 \times 10^{-3} \text{ cm}$

Step 2. Find the volume and convert it to mL.

$$? \text{ mL} = 6.10 \times 10^{3} \text{ cm} \times 30. \text{ cm} \times 1.7 \times 10^{-3} \text{ cm} = 3.1 \times 10^{2} \text{ cm}^{3} \times \frac{1 \text{ mL}}{1 \text{ cm}^{3}} = 3.1 \times 10^{2} \text{ mL}$$

Step 3. Find the density.

$$? \text{ g mL}^{-1} = D_{aluminum} = (\text{Sp. Gr.})(D_{water}) = 2.70(0.998 \text{ g mL}^{-1}) = 2.69 \text{ g mL}^{-1}$$

Step 4. Find the mass in kg.

$$D = \frac{m}{V} \quad \text{so} \quad m = DV$$

$$? \text{ g} = \frac{2.69 \text{ g}}{\text{mL}} \times (3.1 \times 10^{2} \text{ mL}) \times \frac{1 \text{ kg}}{10^{3} \text{ g}} = \mathbf{0.83 \text{ kg}}$$

(Remember that, even though all intermediate answers are shown rounded to the proper number of significant digits, the final answer is computed from the unrounded numbers.)

2 Atoms and Molecules

2.1 Describe, in your own words, what the word "atom" meant to each of the following: Democritus, Dalton, Thomson, Rutherford.

Democritus considered an atom to be a tiny, indivisible particle which makes up matter.
Dalton also considered an atom to be a tiny, indivisible particle but he also said that the atoms of an element are identical and different from those of all other elements.
Thomson's view of an atom was basically the same as Dalton's, however, he added the idea that an atom consisted of positive regions buried in a negative sphere.
Rutherford's view of the atom was similar to Thomson's, however, he added the idea that the positive center was a dense "nucleus" at the center of the negative region.

2.3 What is a nucleon?

A nucleon is any particle found in the nucleus, therefore, either a proton or a neutron.

2.5 What are isotopes?

Isotopes of an element have the same atomic number (which is equal to the number of protons), however, they differ in the number of neutrons. Therefore, they differ in mass number, which is the sum of the number of neutrons and the number of protons.

2.7 Distinguish between an empirical formula and a molecular formula.

An empirical formula is the lowest whole number ratio of the different atoms (or ions) in a compound. A molecular formula is the actual number of each type of atom found in one molecule of the compound.

2.9 What information is given by a structural formula and not given by a molecular formula?

The structural formula not only tells which atoms make up the compound, but it tells how those atoms are attached to each other in space.

2.11 What is the definition of Avogadro's number?

Avogadro's number is the number of carbon atoms in a pure sample of carbon-12 with a mass of exactly 12 g. It is equal to 6.022137×10^{23} atoms.

2.13 What experimental technique can be used to determine the precise masses of isotopes?

Mass spectrometry separates the atoms of an element on the basis of their masses and precisely compares the masses to a standard.

2.15 What is meant by the term "natural abundance"? How is it measured?

The "natural abundance" of an isotope is the relative amount of that isotope in naturally occurring samples on Earth. It is usually expressed as a percentage of all atoms of the element. Relative numbers of the isotopes of an element are determined by mass spectrometry.

2.17 What is meant by a monoisotopic element? Name a common one.

A monoisotopic element is one which exists on Earth with all of its non-radioactive atoms having the same mass number. In other words, all atoms of a monoisotopic element have both

A monoisotopic element is one which exists on Earth with all of its non-radioactive atoms having the same mass number. In other words, all atoms of a monoisotopic element have both the same number of protons and the same number of neutrons. Helium, beryllium, fluorine, sodium and gold are among the monoisotopic elements.

2.19 In what order are the components listed in the name of an ionic compound?

The cation (positive ion) is named first, followed by the anion (negative ion).

2.21 Give the name of the following monoisotopic elements. How many neutrons are in the nuclei of each?

a. $^{23}_{11}Na$ b. $^{27}_{13}Al$ c. $^{59}_{27}Co$ d. $^{19}_{9}F$ e. $^{55}_{25}Mn$ f. $^{75}_{33}As$

Analysis:

Target: The name of the element and the number of neutrons in that isotope.
Known: The isotopic symbol. The names which agree with the symbols must be memorized or looked up inside the front cover of the text.
Relationships: All of the information needed to determine the number of neutrons is contained in the symbols of the elements. The subscripted number is the atomic number, which is the number of protons. The superscripted number is the mass number, which is the sum of the number of protons and the number of neutrons. The number of neutrons is the mass number minus the number of protons.
Plan: Use the known relationships directly.
Work:

a. The element is sodium. $^{23}_{11}Na$ has 23 - 11 = **12 neutrons**

b. The element is aluminum. $^{27}_{13}Al$ has 27 - 13 = **14 neutrons**

c. The element is cobalt. $^{59}_{27}Co$ has 59 - 27 = **32 neutrons**

d. The element is fluorine. $^{19}_{9}F$ has 19 - 9 = **10 neutrons**

e. The element is manganese. $^{55}_{25}Mn$ has 55 - 25 = **30 neutrons**

f. The element is arsenic. $^{75}_{33}As$ has 75 - 33 = **42 neutrons**

2.23 Identify each of the isotopes in Problem 2.21 using the other abbreviated form, as, for example C-12.

Analysis:

Target: The abbreviated form of the isotope.
Known: The isotopic symbol of the isotope.
Relationship: The abbreviated form consists of the symbol and the mass number, which is the superscripted number.
Plan: Use the known relationships directly.
Work:

a. Na-23 b. Al-27 c. Co-59 d. F-19 e. Mn-55 f. As-75

2.25 Give the name and symbol of the element represented by x in each of the following isotopes. Each, incidentally, is the most plentiful of its element.

a. $^{16}_{8}X$ b. $^{24}_{12}X$ c. $^{39}_{19}X$ d. $^{40}_{18}X$ e. $^{28}_{14}X$ f. $^{138}_{56}X$

Analysis:

Target: The name and symbol of each element.
Known: The symbol of the isotope.

agrees with the symbol should be memorized or looked up inside the front cover of the text.

Plan: Use the known relationships directly.

Work:

a. 8 protons; oxygen (O) b. 12 protons; magnesium (Mg)
c. 19 protons; potassium (K) d. 18 protons; argon (Ar)
e. 14 protons; silicon (Si) f. 56 protons; barium (Ba)

2.27 Determine the number of protons, neutrons, and electrons in each of the following species.

a. $^{24}_{12}Mg$ b. $^{45}_{21}Sc$ c. $^{91}_{40}Zr$ d. $^{27}_{13}Al^{3+}$ e. $^{65}_{30}Zn^{2+}$ f. $^{108}_{47}Ag^{+}$

Analysis:

Target: The number of protons, neutrons and neutrons electrons in each species.

Known: The isotopic symbol of the element.

Relationships: The number of protons is equal to the atomic number, which is the subscripted number on the left hand side. The mass number, which is the sum of the protons and neutrons, is the superscripted number on the left hand side. The charge of the ion is the superscripted number on the right hand side. In a neutral atom, the number of protons equals the number of electrons. A positive charge means that a number of electrons equal to the charge has been removed, while a negative charge means that a number of electrons equal to the charge has been added.

Plan: Use the known relationships directly.

Work:

a. **12 protons**; neutral atom, **12 electrons;** 24 - 12 = **12 neutrons**
b. **21 protons**; neutral atom, **21 electrons**; 45 - 21 = **24 neutrons**
c. **40 protons**; neutral atom, **40 electrons**; 91 - 40 = **51 neutrons**
d. **13 protons**; 3+ charge, 13 - 3 = **10 electrons**; 27 - 13 = **14 neutrons**
e. **30 protons**; 2+ charge, 30 - 2 = **28 electrons**; 65 - 35 = **35 neutrons**
f. **47 protons**; + charge, 47 - 1 = **46 electrons**; 108 - 47 = **61 neutrons**

2.29 Determine the number of protons, neutrons, and electrons in each of the following species.

a. N-15 b. Os-188 c. W-184 d. $^{37}_{17}Cl$ e. S-34, doubly charged anion
f. Rb-85, singly charged cation

Analysis:

Target: The number of protons, neutrons and electrons in each of the species.

Known: The isotopic symbol or the abbreviated form of the symbol.

Relationships: All necessary information can be found using either the symbol or the periodic table with the abbreviated form of the symbol to determine the atomic number. All of the relationships in Problem 2.27 apply.

Plan: Use the known relationships directly.

Work:

a. N-15 has **7 protons** (table); neutral atom, **7 electrons**; 15 - 7 = **8 neutrons**
b. Os-188 has **76 protons** (table); neutral atom, **76 electrons**; 188 - 76 = **112 neutrons**
c. W-184 has **74 protons** (table); neutral atom, **74 electrons**; 184 - 74 = **110 neutrons**
d. $^{37}_{17}Cl$ has **17 protons**; neutral atom, **17 electrons**; 37 - 17 = **20 neutrons**
e. S-34 has **16 protons** (table); 2- charge, 16 + 2 = **18 electrons**; 34 - 16 = **18 neutrons**
f. Rb-85 has **37 protons** (table); + charge, 37 - 1 = **36 electrons**; 85 - 37 = **48 neutrons**

2.31 What is the symbol of the species composed of each of the following sets of subatomic particles?

a. 25p, 30n, 25e b. 20p, 20n, 18e c. 33p, 42n, 33e d. 53p, 74n, 54e

Analysis:

Target: The symbol of the species in the form $^{A}_{Z}X^{n}$, where A is the mass number, Z is the atomic number, X is the symbol of the element and n is the charge, if any.

Known: The number of protons, neutrons and electrons in each species.

Relationships: All necessary information can be found using the relationships between the symbol and the subatomic particles. All of the relationships in Problem 2.27 apply.
Plan: use the known relationships to write the symbol directly.
Work:

a. 25p; element = Mn; mass number = 25p + 30n = 55; 25p - 25e = 0, (neutral); $^{55}_{25}Mn$

b. 20p; element = Ca; mass number = 20p + 20n = 40; 20p - 18 e = 2+ charge; $^{40}_{20}Ca^{2+}$

c. 33p; element = As; mass number = 33p + 42n = 75; 33p - 33e = 0, (neutral); $^{75}_{33}As$

d. 53p; element = I; mass number = 53p + 74n = 127; 53p - 54e = - charge ; $^{127}_{53}I^-$

2.33 How many oxygen atoms are there in one molecule of each of the following?
a. O_2 b. P_4O_{10} c. $C_{12}H_{22}O_{11}$ d. $HClO_2$

Analysis:
The number of atoms is available directly from the formula. The conventions for subscripts, parentheses and dots in hydrates must be learned.
Work:
a. O_2; **2 oxygen atoms**
b. P_4O_{10}; **10 oxygen atoms**
c. $C_{12}H_{22}O_{11}$; **11 oxygen atoms**
d. $HClO_2$; **2 oxygen atoms**

2.35 How many hydrogen atoms are there in one molecule of each of the following?
a. $(CH_3)_2O$ b. CH_3CH_2Cl c. CH_3CH_2OH d. $CH_3(CH_2)_6CH_3$

Analysis:
The number of atoms is available directly from the formula. Remember that when there are parentheses, everything inside the parentheses is multiplied by the subscript to the right.
Work:
a. $(CH_3)_2O$; **6 hydrogen atoms**
b. CH_3CH_2Cl ; **5 hydrogen atoms**
c. CH_3CH_2OH; **6 hydrogen atoms**
d. $CH_3(CH_2)_6CH_3$; **18 hydrogen atoms**

2.37 How many oxygen atoms are there in each of the following formula units?
a. K_2SO_4 b. $(NH_4)_2CO_3$ c. $FeSO_4$ d. $Ca(OH)_2$ e. $Al_2(CO_3)_3$ f. $Fe_2(SO_4)_3$

Analysis:
The number of atoms is available directly from the formula. Remember that when there are parentheses, everything inside the parentheses is multiplied by the subscript to the right.
Work:
a. K_2SO_4; **4 oxygen atoms**
b. $(NH_4)_2CO_3$; **3 oxygen atoms**
c. $FeSO_4$; **4 oxygen atoms**
d. $Ca(OH)_2$; **2 oxygen atoms**
e. $Al_2(CO_3)_3$; **9 oxygen atoms**
f. $Fe_2(SO_4)_3$; **12 oxygen atoms**

2.39 How many hydrogen atoms are there in each of the following formula units?
a. $(NH_4)_2S$ b. NH_4HCO_3 c. $BaCl_2 \cdot 2H_2O$ d. $CuSO_4 \cdot 5H_2O$

Analysis:
The number of atoms is available directly from the formula. Remember that when there is a hydrate, the number of water molecules is multiplied by the coefficient to the right of the dot. The number of oxygen atoms is equal to that coefficient. The number of hydrogen atoms is twice that coefficient.
Work:
a. $(NH_4)_2S$; **8 hydrogen atoms**
b. NH_4HCO_3; **5 hydrogen atoms**
c. $BaCl_2 \cdot 2H_2O$; **4 hydrogen atoms**
d. $CuSO_4 \cdot 5H_2O$; **10 hydrogen atoms**

2.41 Given that the masses and natural abundances of the isotopes of magnesium are as follows, calculate the molar mass.

$^{24}_{12}Mg$	23.985042 amu	78.99%	$^{25}_{12}Mg$	24.985837 amu	10.00%
$^{26}_{12}Mg$	25.982593 amu	11.01%			

Analysis:

Target: The molar mass of the element magnesium. (? g) =

Knowns: Isotopic masses and their percent abundances.

Relationships The molar mass is the weighted average of all naturally occurring isotopes of the element. The mass of one atom in atomic mass units is numerically equivalent to the mass of one mole of that element in grams.

Plan: Multiply the mass of the isotope by its decimal percent abundance and sum the results.

Work:

(23.985042 g) x (0.7899)	=	18.95 g	(mass of 78.99% of one mole of Mg-24)
(24.985837 g) x (0.1000)	=	2.499 g	(mass of 10.00% of one mole of Mg-25)
(25.982593 g) x (0.1101)	=	2.861 g	(mass of 11.01% of one mole of Mg-26)
		24.31 g	(mass of one mole of natural Mg)

2.43 Approximately 50% of naturally occurring bromine is composed of the isotope Br-79, while the other half is Br-81. Without looking up the exact masses of the isotopes, estimate the molar mass of bromine.

Analysis:

Target: The approximate molar mass of the element bromine. (? g) =

Knowns: Approximate isotopic masses and their percent abundances.

Relationship: The molar mass is the weighted average of all naturally occurring isotopes of the element.

Plan: Multiply the mass of the isotope by its decimal percent abundance and sum the results.

Work:

(79 g) x (0.50)	=	40 g	(mass of 50% of one mole of Br-79)
(81 g) x (0.50)	=	40 g	(mass of 50% of one mole of Br-81)
		80 g	(mass of one mole of natural Br)

2.45 There are two stable isotopes of bromine, having masses of 78.918336 and 80.916289 amu, respectively. The molar mass of the naturally occurring mixture of these isotopes is 79.904 g mol^{-1}. Determine the natural abundance of the isotopes, and report the answer to three significant digits.

Analysis:

Target: The natural abundances of the two isotopes of bromine. (? %) =

Knowns: The relative masses of the isotopes and the molar mass of the element.

Relationship: The molar mass is the weighted average of the masses of the individual isotopes. The percentages of the isotopes must add to 100%; therefore, the decimal percentages must add to **1**.

Plan: Let one isotope have a decimal percent abundance of **x**, then the other would be **1 - x**. Multiply the masses by the decimal percent abundances and set the result equal to the molar mass. Solve algebraically.

Work:

$$(x)(78.918336) + (1 - x)(80.916829) = 79.904$$

$$78.918336x + 80.916829 - 80.916829x = 79.904$$

$$-1.998493x = 79.904 - 80.916829 = -1.013$$

$$x = \frac{-1.013}{-1.998493} = 0.507; \quad 1 - x = 1 - 0.507 = 0.493$$

Natural bromine is **50.7%** mass 78.918336 and **49.3%** mass 80.916829.

2.47 Express the following amounts in moles or (millimoles).
a. 3.01×10^{23} F atoms b. 999 Ca atoms c. 1.59×10^{22} NO_3^- ions
d. 7.50×10^{22} electrons

Analysis:
Target: The number of moles in each of the amounts given. (? mol) =
Knowns: The number of particles is given and Avogadro's number is known.
Relationships: The number of moles is related to the number of particles (atoms, molecules or formula units) by Avogadro's number, 1 mol ↔ 6.022137×10^{23} particles.
1 mmol ↔ 10^{-3} mol.
Plan: Use the factor-label method to solve the problem.
Work:

a. $? \text{ mol F} = 3.01 \times 10^{23} \text{ F atoms} \times \frac{1 \text{ mol F}}{6.022137 \times 10^{23} \text{ F atoms}} = \mathbf{0.500 \text{ mol F}}$

b. $? \text{ mol Ca} = 999 \text{ Ca atoms} \times \frac{1 \text{ mol Ca}}{6.022137 \times 10^{23} \text{ Ca atoms}} = \mathbf{1.66 \times 10^{-21} \text{ mol Ca}}$

c. $? \text{ mmol } NO_3^- \text{ ions} = 1.59 \times 10^{22} \text{ ions} \times \frac{1 \text{ mol ions}}{6.022137 \times 10^{23} \text{ ions}} \times \frac{1 \text{ mmol ions}}{10^{-3} \text{ mol ions}} =$

26.4 mmol NO_3^-

d. $? \text{ mmol electrons} = 7.50 \times 10^{22} \text{ e}^- \times \frac{1 \text{ mol e}^-}{6.022137 \times 10^{23} \text{ e}^-} \times \frac{1 \text{ mmol e}^-}{10^{-3} \text{ mol e}^-} =$

125 mmol electrons

2.49 How many atoms are there in each of the following amounts?
a. 3.98 mmol He b. 6.02 mol tungsten c. 8.83×10^{-20} mol Sr-90 d. 1.234 mol In

Analysis:
Target: A number of atoms. (? atoms) =
Knowns: The number of moles of each type of atom is given and Avogadro's number is known.
Relationships: The number of moles is related to the number of particles (atoms, molecules or formula units) by Avogadro's number, 1 mol ↔ 6.022137×10^{23} particles.
1 mmol ↔ 10^{-3}mol.
Plan: Use the factor-label method to solve the problem, converting to moles when appropriate.
Work:

a. $? \text{ He atoms} = 3.98 \text{ mmol He} \times \frac{10^{-3} \text{ mol He}}{1 \text{ mmol He}} \times \frac{6.022137 \times 10^{23} \text{ atoms He}}{1 \text{ mol He}} =$

2.40×10^{21} atoms He

b. $? \text{ W atoms} = 6.02 \text{ mol W} \times \frac{6.022137 \times 10^{23} \text{ atoms W}}{1 \text{ mol W}} = \mathbf{3.63 \times 10^{24} \text{ atoms W}}$

c. $? \text{ Sr-90 atoms} = 8.83 \times 10^{-20} \text{ mol Sr-90} \times \frac{6.022137 \times 10^{23} \text{ atoms Sr-90}}{1 \text{ mol Sr-90}} =$

5.32×10^{4} atoms Sr-90

d. ? In atoms = 1.234 mol In x $\frac{6.022137 \times 10^{23} \text{ atoms In}}{1 \text{ mol In}}$ = **7.431 x 10^{23} atoms In**

2.51 How many molecules or ions are there in each of the following amounts?
a. 5.00 x 10^{-6} mol water b. 1.9 x 10^{-23} mol C_2H_4 c. 165 mmol HCO_3^- ions
d. 25 mol NH_3

Analysis:
Target: The number of molecules or ions in each of the amounts given. (? particles) =
Known: The number of moles or millimoles is given and Avogadro's number is known.
Relationship: The number of moles is related to the number of particles (atoms, molecules or formula units) by Avogadro's number, 1 mol ↔ 6.022137 x 10^{23} particles.
1 mmol = 10^{-3} mol.
Plan: Use the factor-label method to solve the problem.
Work:

a. ? molecules H_2O = 5.00 x 10^{-6} mol H_2O x $\frac{6.022137 \times 10^{23} \text{ molecules } H_2O}{1 \text{ mol } H_2O}$ =

3.01 x 10^{18} molecules H_2O

b. ? molecules C_2H_4 = 1.9 x 10^{-23} mol C_2H_4 x $\frac{6.022137 \times 10^{23} \text{ molecules } C_2H_4}{1 \text{ mol } C_2H_4}$ =

11 molecules C_2H_4

c. ? ions = 165 mmol HCO_3^- x $\frac{10^{-3} \text{ mol ions}}{1 \text{ mmol ions}}$ x $\frac{6.022137 \times 10^{23}\ HCO_3^- \text{ ions}}{1 \text{ mol } HCO_3^- \text{ions}}$ =

9.94 x 10^{22} ions HCO_3^-

d. ? molecules NH_3 = 25 mol NH_3 x $\frac{6.022137 \times 10^{23} \text{ molecules } NH_3}{1 \text{ mol } NH_3}$ =

1.5 x 10^{25} molecules NH_3

2.53 Calculate the molar mass of the following compounds:
a. CaH_2 b. N_2O_5 c. BaF_2 d. CuCl

Analysis:
Target: A molar mass. (? g mol^{-1}) =
Known: The formula of the compound.
Relationships: The molar mass is the sum of the molar masses of the constituent elements. The subscripts give the number of atoms in each unit of the compound and, therefore, the number of moles of each element in one mole of the compound.
Plan: Multiply the molar mass of each element by its number of moles in the compound and add the results.
Work:

a. Ca: 1 mol x 40.078 g mol^{-1}	=	40.078 g Ca	
H: 2 mol x 1.0079 g mol^{-1}	=	2.0158 g H	
CaH_2: 1 mol	=	42.094 g CaH_2	(The molar mass is **42.094 g mol^{-1}**.)

b. N $2\ mol \times 14.0067\ g\ mol^{-1}$ = $28.0134\ g\ N$
O $5\ mol \times 15.9994\ g\ mol^{-1}$ = $79.9970\ g\ O$

N_2O_5: 1 mol = $108.0104\ g\ N_2O_5$ (The molar mass is **108.0104 g mol^{-1}**.)

c. Ba: $1\ mol \times 137.33\ g\ mol^{-1}$ = $137.33\ g\ Ba$
F: $2\ mol \times 18.998403\ g\ mol^{-1}$ = $37.996806\ g\ F$

BaF_2: 1 mol = $175.33\ g\ BaF_2$ (The molar mass is **175.33 g mol^{-1}**.)

d. Cu: $1\ mol \times 63.546\ g\ mol^{-1}$ = $63.546\ g\ Cu$
Cl: $1\ mol \times 35.453\ g\ mol^{-1}$ = $35.453\ g\ Cl$

CuCl: 1 mol = 98.999 g CuCl (The molar mass is **98.999 g mol^{-1}**.)

2.55 Calculate the molar mass of the following compounds:
a. carbon dioxide b. dinitrogen tetroxide c. sulfur hexafluoride

Analysis:
Target: A molar mass. (? g mol^{-1}) =
Known: The name of the compound.
Relationships: The molar mass is the sum of the molar masses of the constituent elements. The subscripts in the formula give the number of atoms in each unit of the compound, and, therefore, the number of moles of each element in one mole of the compound. The formula can be determined by applying nomenclature rules.
Plan: Determine the formula of each compound. Multiply the molar mass of each element by its number of moles in the compound and then add the results.
Work:
a. Carbon dioxide means 1 C and 2 O, therefore, the formula is CO_2.

C: $1\ mol \times 12.011\ g\ mol^{-1}$ = $12.011\ g\ C$
O: $2\ mol \times 15.9994\ g\ mol^{-1}$ = $31.9988\ g\ O$

CO_2: 1 mol = $44.010\ g\ CO_2$ (The molar mass is **44.010 g mol^{-1}**.)

b. Dinitrogen tetroxide means 2 N and 4 O, therefore, the formula is N_2O_4.

N: $2\ mol \times 14.0067\ g\ mol^{-1}$ = $28.0134\ g\ N$
O: $4\ mol \times 15.9994\ g\ mol^{-1}$ = $63.9976\ g\ O$

N_2O_4: 1 mol = $92.0110\ g\ N_2O_4$ (The molar mass is **92.0110 g mol^{-1}**.)

c. Sulfur hexafluoride means 1 S and 6 F, therefore, the formula is SF_6.

S: $1\ mol \times 32.06\ g\ mol^{-1}$ = $32.06\ g\ S$
F: $6\ mol \times 18.998403\ g\ mol^{-1}$ = $113.990418\ g\ F$

SF_6: 1 mol = $146.05\ g\ SF_6$ (The molar mass is **146.05 g mol^{-1}**.)

2.57 Calculate the mass of each of the following.
a. 0.500 mol $NiSO_4$ b. 605 mmol $CuSO_4 \cdot 5H_2O$ c. 0.875 mol H_2CO_3
d. 6.02×10^{-6} mol XeF_4

Analysis:
Target: The mass of the given amount of substance. (? g) =
Known: The number of moles or millimoles of each substance.

Relationships: 1 mmol ↔ 10^{-3} mol. The equivalence statement 1 mol ↔ x g can be determined from the molar mass.

Plan: If we knew the molar mass of the substance, we could use the factor-label method to find the number of grams.

Step 2. Use the factor-label method to determine mass from the number of moles and the molar mass.

Step 1. Determine the molar mass by multiplying the number of atoms of each element in the formula by the molar mass of the element.

Work:

a. **Step 1.** Find the molar mass of $NiSO_4$.

Ni: 1 mol x 58.69 g mol^{-1} = 58.69 g Ni
S: 1 mol x 32.06 g mol^{-1} = 32.06 g S
O: 4 mol x 15.9994 g mol^{-1} = 63.9976 g O

$NiSO_4$: 1 mol = 154.75 g $NiSO_4$ (The molar mass is 154.75 g mol^{-1}.)

Step 2. Use the factor-label method to find the mass in grams.

$$? \text{ g } NiSO_4 = 0.500 \text{ mol } NiSO_4 \times \frac{154.75 \text{ g } NiSO_4}{1 \text{ mol } NiSO_4} = \mathbf{77.4 \text{ g } NiSO_4}$$

b. **Step 1.** Find the molar mass of $CuSO_4{\bullet}5H_2O$.

Cu: 1 mol x 63.546 g mol^{-1} = 63.546 g Cu
S: 1 mol x 32.06 g mol^{-1} = 32.06 g S
O: 9 mol x 15.9994 g mol^{-1} = 143.9946 g O
H: 10 mol x 1.0079 g mol^{-1} = 10.079 g H

$CuSO_4{\bullet}5H_2O$: 1 mol = 249.68 g $CuSO_4{\bullet}5H_2O$ (Molar mass is 249.68 g mol^{-1}.)

Step 2. Use the factor-label method to find the mass in grams.

$$? \text{ g } CuSO_4{\bullet}5H_2O = 605 \text{ mmol} \times \frac{10^{-3} \text{ mol}}{1 \text{ mmol}} \times \frac{249.68 \text{ g } CuSO_4{\bullet}5H_2O}{1 \text{ mol } CuSO_4{\bullet}5H_2O} =$$

151 g $CuSO_4{\bullet}5H_2O$

c. **Step 1.** Find the molar mass of H_2CO_3.

H: 2 mol x 1.0079 g mol^{-1} = 2.0158 g H
C: 1 mol x 12.011 g mol^{-1} = 12.011 g C
O: 3 mol x 15.9994 g mol^{-1} = 47.9982 g O

H_2CO_3: 1 mol = 62.025 g H_2CO_3 (The molar mass is 62.025 g mol^{-1}.)

Step 2. Use the factor-label method to find the mass in grams.

$$? \text{ g } H_2CO_3 = 0.875 \text{ mol } H_2CO_3 \times \frac{62.025 \text{ g } H_2CO_3}{1 \text{ mol } H_2CO_3} = \mathbf{54.3 \text{ g } H_2CO_3}$$

d. **Step 1.** Find the molar mass of XeF_4.

Xe: 1 mol x 131.29 g mol^{-1} = 131.29 g Xe
F: 4 mol x 18.998403 g mol^{-1} = 75.993612 g F

XeF_4: 1 mol = 207.28 g XeF_4 (The molar mass is 207.28 g mol^{-1}.)

Step 2. Use the factor-label method to find the mass in milligrams.

$$? \text{ g } XeF_4 = 6.02 \times 10^{-6} \text{ mol } XeF_4 \times \frac{207.28 \text{ g } XeF_4}{1 \text{ mol } XeF_4} \times \frac{1 \text{ mg}}{10^{-3} \text{ g}} = \mathbf{1.25 \text{ mg } XeF_4}$$

2.59 What amount (how many moles) of substance is in each of the following quantities?
a. 7.5 g $Sr(OH)_2$ b. 100. g SiO_2 c. 23 mg UF_6 d. 10.0 g N_2O_4

Analysis:

Target: The number of moles of a given substance. (? mol) =

Known: The number of grams or milligrams of each substance.

Relationships: 1 mg ↔ 10^{-3} g. The equivalence statement g ↔ mol can be determined from the molar mass of the compound.

Plan: If we knew the molar mass of the substance, we could use the factor-label method to find the number of moles of each substance.

Step 2. Use the factor-label method to determine the number of moles from the mass and the molar mass.

Step 1. Determine the molar mass by multiplying the number of atoms of each element in the formula by the molar mass of the element.

Work:

a. **Step 1.** Find the molar mass of $Sr(OH)_2$.

Sr:	1 mol x 87.62 g mol^{-1}	=	87.62 g Sr
O:	2 mol x 15.9994 g mol^{-1}	=	31.9988 g O
H:	2 mol x 1.0079 g mol^{-1}	=	2.0158 g H
$Sr(OH)_2$:	1 mol	=	121.63 g $Sr(OH)_2$ (The molar mass is 121.63 g mol^{-1}.)

Step 2. Use the factor-label method to find the number of moles.

$$? \text{ mol } Sr(OH)_2 = 7.5 \text{ g } Sr(OH)_2 \times \frac{1 \text{ mol } Sr(OH)_2}{121.63 \text{ g } Sr(OH)_2} = \mathbf{0.062 \text{ mol } Sr(OH)_2}$$

b. **Step 1.** Find the molar mass of SiO_2.

Si:	1 mol x 28.0855 g mol^{-1}	=	28.0855 g Si
O:	2 mol x 15.9994 g mol^{-1}	=	31.9988 g O
SiO_2:	1 mol	=	60.0843 g SiO_2 (The molar mass is 60.0843 g mol^{-1}.)

Step 2. Use the factor-label method to find the number of moles.

$$? \text{ mol } SiO_2 = 100. \text{ g } SiO_2 \times \frac{1 \text{ mol } SiO_2}{60.0843 \text{ g } SiO_2} = \mathbf{1.66 \text{ mol } SiO_2}$$

c. **Step 1.** Find the molar mass of UF_6.

U:	1 mol x 238.0289 g mol^{-1}	=	238.0289 g U
F:	6 mol x 18.998403 g mol^{-1}	=	113.99042 g F
UF_6:	1 mol	=	352.0193 g UF_6 (The molar mass is 352.0193 g mol^{-1}.)

Step 2. Use the factor-label method to find the number of moles.

$$? \text{ mol } UF_6 = 23 \text{ mg } UF_6 \times \frac{10^{-3} \text{ g}}{1 \text{ mg}} \times \frac{1 \text{ mol } UF_6}{352.0193 \text{ g } UF_6} = \mathbf{6.5 \times 10^{-5} \text{ mol } UF_6}$$

d. **Step 1.** Find the molar mass of N_2O_4.

N: 2 mol x 14.0067 g mol^{-1} = 28.0134 g N
O: 4 mol x 15.9994 g mol^{-1} = 63.9976 g O

N_2O_4: 1 mol = 92.0110 g N_2O_4 (The molar mass is 92.0110 g mol^{-1}.)

Step 2. Use the factor-label method to find the number of moles.

$$? \text{ mmol } N_2O_4 = 10.0 \text{ g } N_2O_4 \times \frac{1 \text{ mol } N_2O_4}{92.0110 \text{ g } N_2O_4} = \mathbf{0.109 \text{ mol } N_2O_4}$$

2.61 Calculate the number of moles in a half-pound of each of the noble gases.

Analysis:
Target: The number of moles of a given substance. (? mol) =
Known: The number of pounds of each noble gas.
Relationships: 1 lb ↔ 0.45359 kg. 1 kg ↔ 10^3 g. The equivalence statement mol ↔ g can be determined from the molar mass, which is read directly from the periodic table.
Plan: Find the mass in grams of a half-pound first (assuming three significant digits) and then use the factor-label method to find the number of moles of each noble gas.
Work:

$$? \text{ g} = 0.500 \text{ lb} \times \frac{0.45359 \text{ kg}}{1 \text{ lb}} \times \frac{10^3 \text{ g}}{1 \text{ kg}} = 227 \text{ g}$$

$$? \text{ mol He} = 227 \text{ g He} \times \frac{1 \text{ mol He}}{4.00260 \text{ g He}} = \mathbf{56.7 \text{ mol He}}$$

$$? \text{ mol Ne} = 227 \text{ g Ne} \times \frac{1 \text{ mol Ne}}{20.180 \text{ g Ne}} = \mathbf{11.2 \text{ mol Ne}}$$

$$? \text{ mol Ar} = 227 \text{ g Ar} \times \frac{1 \text{ mol Ar}}{39.948 \text{ g Ar}} = \mathbf{5.68 \text{ mol Ar}}$$

$$? \text{ mol Kr} = 227 \text{ g Kr} \times \frac{1 \text{ mol Kr}}{83.80 \text{ g Kr}} = \mathbf{2.71 \text{ mol Kr}}$$

$$? \text{ mol Xe} = 227 \text{ g Xe} \times \frac{1 \text{ mol Xe}}{131.29 \text{ g Xe}} = \mathbf{1.73 \text{ mol Xe}}$$

$$? \text{ mol Rn} = 227 \text{ g Rn} \times \frac{1 \text{ mol Rn}}{222 \text{ g Rn}} = \mathbf{1.02 \text{ mol Rn}}$$

2.63 A laboratory recipe calls for 0.15 mol $MgSO_4$ (molar mass = 120.4 g mol^{-1}). How many grams should be weighed out?

Analysis/Plan: The mass of the sample is needed and the molar mass is provided (1 mol ↔ 120.4 g). The factor-label method can be used to find the mass.
Work:

$$? \text{ g } MgSO_4 = 0.15 \text{ mol } MgSO_4 \times \frac{120.4 \text{ g } MgSO_4}{1 \text{ mol } MgSO_4} = \mathbf{18 \text{ g } MgSO_4}$$

2.65 If each mole of mercury(II) oxide (HgO) yields one-half mole of oxygen (O_2) on decomposition, how many moles of oxygen are produced by the decomposition of 100 g HgO?

Analysis:
Target: The number of moles of oxygen. (? mol O_2) =
Knowns: The mass of HgO and the number of moles of oxygen formed when one mole of HgO decomposes.
Relationships: The equivalence statement mol ↔ g can be determined from the molar mass of HgO. One mole of HgO produces 1/2 mole O_2.
Plan: If we knew the number of moles of HgO, we could find the number of moles of O_2. The number of moles of HgO can be found from its mass and molar mass.
Step 3. Use the relationship between the number of moles of HgO and the number of moles of O_2 to find the number of moles of O_2.
Step 2. Use the mass of HgO and its molar mass to find the number of moles of HgO.
Step 1. Find the molar mass of HgO.
Work:
Step 1. Find the molar mass of HgO.

Hg: 1 mol x 200.59 g mol^{-1} = 200.59 g Hg
O: 1 mol x 15.9994 g mol^{-1} = 15.9994 g O

HgO: 1 mol = 216.59 g HgO (The molar mass is 216.59 g mol^{-1}.)

Step 2. Use the factor-label method to find the number of moles of HgO.

$$? \text{ mol HgO} = 100 \text{ g HgO} \times \frac{1 \text{ mol HgO}}{216.59 \text{ g HgO}} = 0.462 \text{ mol HgO}$$

Step 3. Use the relationship between moles of O_2 and moles of HgO.

$$? \text{ mol } O_2 = 0.462 \text{ mol HgO} \times \frac{0.5 \text{ mol } O_2}{1 \text{ mol HgO}} = \mathbf{0.231 \text{ mol } O_2}$$

2.67 How many hydrogen atoms are there in each of the following?
a. 1.5 mol $HClO_2$ b. 65 mmol H_2SO_4 c. 0.094 mol $(NH_4)_2CO_3$
d. 0.750 mol $BaCl_2 \cdot 2H_2O$

Analysis:
Target: The number of hydrogen atoms in each amount of a substance. (? H atoms)=
Knowns: The number of moles of each substance and the formula of each substance.
Relationships: The number of moles of hydrogen atoms in each mole of the substance is known from the formula of the substance. The number of moles is related to the number of particles (atoms, molecules or formula units) by Avogadro's number, 1 mol ↔ 6.022137 x 10^{23} particles. 1 mmol ↔ 10^{-3} mol.
Plan: Use the known relationships and the factor-label method.
Work:

a. $$? \text{ atoms H} = 1.5 \text{ mol } HClO_4 \times \frac{1 \text{ mol H}}{1 \text{ mol } HClO_4} \times \frac{6.022137 \times 10^{23} \text{ H atoms}}{1 \text{ mol H}} =$$

9.0 x 10^{23} atoms H

b. $$? \text{ atoms H} = 65 \text{ mmol } H_2SO_4 \times \frac{1 \text{ mol}}{10^{-3} \text{ mmol}} \times \frac{2 \text{ mol H}}{1 \text{ mol } H_2SO_4} \times \frac{6.022137 \times 10^{23} \text{ H atoms}}{1 \text{ mol H}} =$$

7.8 x 10^{22} atoms H

c. $$? \text{ atoms H} = 0.094 \text{ mol } (NH_4)_2SO_4 \times \frac{8 \text{ mol H}}{1 \text{ mol } (NH_4)_2SO_4} \times \frac{6.022137 \times 10^{23} \text{ H atoms}}{1 \text{ mol H}} =$$

4.5 x 10^{23} atoms H

d. $? \text{ atoms H} = 0.750 \text{ mol } BaCl_2{\bullet}2H_2O \times \frac{4 \text{ mol H}}{1 \text{ mol } BaCl_2{\bullet}2H_2O} \times \frac{6.022137 \times 10^{23} \text{ H atoms}}{1 \text{ mol H}} =$

1.81×10^{24} atoms H

2.69 How many nitrogen atoms are there in 50 g (two significant digits) of each of the following?
a. NH_3 b. $NaNO_2$ c. C_2N_2 d. $(NH_4)_2CO_3$

Analysis:
Target: The number of N atoms in a given mass of a compound. (? N atoms) =
Knowns: The mass of each substance and the formula of the substance.
Relationships: The number of moles of nitrogen atoms in one mole of the substance is known from the formula of the substance. The equivalence statement mol ↔ g can be determined from the molar mass of the compound. The number of moles is related to the number of particles (atoms, molecules or formula units) by Avogadro's number, 1 mol ↔ 6.022137×10^{23} particles.
Plan: If we knew the number of moles of nitrogen, we could find the number of nitrogen atoms. If we knew the number of moles of the compound, we could find the number of moles of nitrogen. If we knew the molar mass of the compound, we could find the number of moles of the compound.
Step 3. Use the factor-label method to find the number of nitrogen atoms.
Step 2. Determine the number of moles of the compound.
Step 1. Find the molar mass of the compound.
Work:
a. **Step 1.** Find the molar mass of NH_3.

N: 1 mol x 14.0067 g mol^{-1} = 14.0067 g N
H: 3 mol x 1.0079 g mol^{-1} = 3.0237 g H

NH_3: 1 mol = 17.0304 g NH_3 (The molar mass is 17.0304 g mol^{-1}.)

Step 2. Use the factor-label method to find the number of moles of the compound.

$? \text{ mol } NH_3 = 50 \text{ g } NH_3 \times \frac{1 \text{ mol } NH_3}{17.0304 \text{ g } NH_3} = 2.9 \text{ mol } NH_3$

Step 3. Use the factor-label method to find the number of atoms of nitrogen.

$? \text{ atoms N} = 2.9 \text{ mol } NH_3 \times \frac{1 \text{ mol N}}{1 \text{ mol } NH_3} \times \frac{6.022137 \times 10^{23} \text{ atoms N}}{1 \text{ mol N}} =$

1.8×10^{24} atoms N

b. **Step 1.** Find the molar mass of $NaNO_2$.

Na: 1 mol x 22.98977 g mol^{-1} = 22.98977 g Na
N: 1 mol x 14.0067 g mol^{-1} = 14.0067 g N
O: 2 mol x 15.9994 g mol^{-1} = 31.9988 g O

$NaNO_2$: 1 mol = 68.9953 g $NaNO_2$ (The molar mass is 68.9953 g mol^{-1}.)

Step 2. Use the factor-label method to find the number of moles of the compound.

$? \text{ mol } NaNO_2 = 50 \text{ g } NaNO_2 \times \frac{1 \text{ mol } NaNO_2}{68.9953 \text{ g } NaNO_2} = 0.72 \text{ mol } NaNO_2$

Step 3. Use the factor-label method to find the number of atoms of nitrogen.

$$? \text{ atoms N} = 0.72 \text{ mol NaNO}_2 \times \frac{1 \text{ mol N}}{1 \text{ mol NaNO}_2} \times \frac{6.022 \times 10^{23} \text{ atoms N}}{1 \text{ mol N}} =$$

4.4×10^{23} atoms N

c. **Step 1.** Find the molar mass of C_2N_2.

C: 2 mol x 12.011 g mol^{-1} = 24.022 g C
N: 2 mol x 14.0067 g mol^{-1} = 28.0134 g N

C_2N_2: 1 mol = 52.035 g C_2N_2 (The molar mass is 52.035 g mol^{-1}.)

Step 2. Use the factor-label method to find the number of moles of the compound.

$$? \text{ mol C}_2\text{N}_2 = 50 \text{ g C}_2\text{N}_2 \times \frac{1 \text{ mol C}_2\text{N}_2}{52.035 \text{ g C}_2\text{N}_2} = 0.96 \text{ mol C}_2\text{N}_2$$

Step 3. Use the factor-label method to find the number of atoms of nitrogen.

$$? \text{ atoms N} = 0.96 \text{ mol C}_2\text{N}_2 \times \frac{2 \text{ mol N}}{1 \text{ mol C}_2\text{N}_2} \times \frac{6.022137 \times 10^{23} \text{ atoms N}}{1 \text{ mol N}} =$$

1.2×10^{24} atoms N

d. **Step 1.** Find the molar mass of $(NH_4)_2CO_3$.

N: 2 mol x 14.0067 g mol^{-1} = 28.0134 g N
H: 8 mol x 1.0079 g mol^{-1} = 8.0632 g H
C: 1 mol x 12.011 g mol^{-1} = 12.011 g C
O: 3 mol x 15.9994 g mol^{-1} = 47.9982 g O

$(NH_4)_2CO_3$: 1 mol = 96.086 g $(NH_4)_2CO_3$ (The molar mass is 96.086 g mol^{-1}.)

Step 2. Use the factor-label method to find the number of moles of the compound.

$$? \text{ mol (NH}_4)_2\text{CO}_3 = 50 \text{ g (NH}_4)_2\text{CO}_3 \times \frac{1 \text{ mol (NH}_4)_2\text{CO}_3}{96.086 \text{ g (NH}_4)_2\text{CO}_3} = 0.52 \text{ mol (NH}_4)_2\text{CO}_3$$

Step 3. Use the factor-label method to find the number of atoms of nitrogen.

$$? \text{ atoms N} = 0.52 \text{ mol (NH}_4)_2\text{CO}_3 \times \frac{2 \text{ mol N}}{1 \text{ mol (NH}_4)_2\text{CO}_3} \times \frac{6.022137 \times 10^{23} \text{ atoms N}}{1 \text{ mol N}} =$$

6.3×10^{23} atoms N

2.71 Calculate the mass percent of the Group 1A element in each of the following: LiOH, NaOH, KOH.

Analysis:
Target: The mass percent of the elèment in a compound. (? % x) =
Known: The formula of the compound is given and the molar masses are implicit knowns.
Relationship: The percent of an element in a compound is found from the relationship:

$$\frac{\text{mass of the element in one mole of the compound}}{\text{mass of one mole of the compound}} \times 100$$

Plan: Calculate the molar mass of the compound and then apply the above relationship.

Work:
Calculate the molar mass of LiOH.

Li: 1 mol x 6.941 g mol^{-1}	=	6.941 g Li	
O: 1 mol x 15.9994 g mol^{-1}	=	15.9994 g O	
H: 1 mol x 1.0079 g mol^{-1}	=	1.0079 g H	
LiOH: 1 mol	=	23.948 g LiOH	(The molar mass is 23.948 g mol^{-1}.)

Find the percent Li: $\frac{6.941 \text{ g Li}}{23.948 \text{ g LiOH}} \times 100 =$ **28.98% Li**

Calculate the molar mass of NaOH.

Na: 1 mol x 22.98977 g mol^{-1}	=	22.98977 g Na	
O: 1 mol x 15.9994 g mol^{-1}	=	15.9994 g O	
H: 1 mol x 1.0079 g mol^{-1}	=	1.0079 g H	
NaOH: 1 mol	=	39.9971 g NaOH	(The molar mass is 39.9971 g mol^{-1}.)

Find the percent Na: $\frac{22.98977 \text{ g Na}}{39.9971 \text{ g NaOH}} \times 100 =$ **57.4786% Na**

Calculate the molar mass of KOH.

K: 1 mol x 39.0983 g mol^{-1}	=	39.0983 g K	
O: 1 mol x 15.9994 g mol^{-1}	=	15.9994 g O	
H: 1 mol x 1.0079 g mol^{-1}	=	1.0079 g H	
KOH: 1 mol	=	56.1056 g KOH	(The molar mass is 56.1056 g mol^{-1}.)

Find the percent K: $\frac{39.0983 \text{ g K}}{56.1056 \text{ g KOH}} \times 100 =$ **69.6870% K**

2.73 Which compound has the higher percent carbon: carbon monoxide or carbon dioxide?

Analysis:
Target: The higher mass percent of the carbon in a compound. (? % x) =
Knowns: The formula of the compound can be determined from the name and the molar masses are implicit knowns.
Relationship: The percent of an element in a compound is found from the relationship:

$$\frac{\text{mass of the element in one mole of the compound}}{\text{mass of one mole of the compound}} \times 100$$

Plan: Calculate the molar mass of each compound and then apply the above relationship. Compare the two percentages to determine which is highest.
Work:
Calculate the molar mass of carbon monoxide, CO.

C: 1 mol x 12.011 g mol^{-1}	=	12.011 g C	
O: 1 mol x 15.9994 g mol^{-1}	=	15.9994 g O	
CO: 1 mol	=	28.010 g CO	(The molar mass is 28.010 g mol^{-1}.)

Find the percent C: $\frac{12.011 \text{ g C}}{28.010 \text{ g CO}} \times 100 = 42.881\%$ C

Calculate the molar mass of carbon dioxide, CO_2.

C: 1 mol x 12.011 g mol^{-1}		= 12.011 g C
O: 2 mol x 15.9994 g mol^{-1}		= 31.9988 g O
CO_2:	1 mol	= 44.010 g CO_2

(The molar mass is 44.010 g mol^{-1}.)

Find the percent C: $\frac{12.011 \text{ g C}}{44.010 \text{ g } CO_2} \times 100 = 27.292\%$ C

CO has the higher percent carbon.

2.75 A xenon oxofluoride contains 6.12% O, 43.6% F, and the remainder is Xe. What is the empirical formula?

Analysis:

Target: The empirical formula, which is the relative number of atoms of each element in the compound expressed as the smallest whole numbers that indicate the correct atom ratios.

Knowns: The identity of the elements and the mass percent of two of them. The percent of the other element is an implicit known, as are the molar masses of the elements.

Relationships: Because the mole is a counting unit, the relative number of *atoms* is the same as the relative number of *moles.* If we knew the relative number of grams, we could find the relative number of moles. If we consider a ***100.0 g sample of the compound, the number of grams of each element is equal to its mass percent.***

Plan: The stepwise plan is:

Step 3. Find the ratio of the molar amounts. This is the same as the ratio of the atoms in the molecule, and when, converted to a whole-number ratio, gives the subscripts in the empirical formula.

Step 2. Find the number of moles of each element from its mass and its molar mass.

Step 1. Find the number of grams of each element in 100 g of the compound.

Work:

Step 1. Find the mass of each element in 100.0 g of the compound, using the percentages.

$$100.0 \text{ g compound} \times \frac{6.18 \text{ g O}}{100 \text{ g compound}} = 6.18 \text{ g O}$$

$$100.0 \text{ g compound} \times \frac{43.6 \text{ g F}}{100 \text{ g compound}} = 43.6 \text{ g F}$$

The rest of the compound is xenon.

$$100.0 \text{ g compound} - 6.18 \text{ g O} - 43.6 \text{ g F} = 50.2 \text{ g Xe}$$

Step 2. Convert the mass of each element to number of moles using its molar mass.

$$? \text{ mol O} = 6.18 \text{ g O} \times \frac{1 \text{ mol O}}{15.9994 \text{ g O}} = 0.386 \text{ mol O}$$

$$? \text{ mol F} = 43.6 \text{ g F} \times \frac{1 \text{ mol F}}{18.998403 \text{ g F}} = 2.29 \text{ mol F}$$

$$? \text{ mol Xe} = 50.2 \text{ g Xe} \times \frac{1 \text{ mol Xe}}{131.29 \text{ g Xe}} = 0.382 \text{ mol Xe}$$

Step 3. Try to convert the mole ratios to whole numbers by dividing each by the smallest.

$$\text{O:} = \frac{0.386}{0.382} = 1.01 \qquad \text{F:} \frac{2.29}{0.382} = 5.99 \qquad \text{Xe:} \frac{0.382}{0.382} = 1.00$$

Because the number of atoms must be a whole number, the mole ratios must also be whole numbers. The results are rounded to give the empirical formula of the compound: **$XeOF_6$**.

2.77 A 5.682 g sample of an oxide of phosphorus contains 2.480 g P. What is its empirical formula?

Analysis:

Target: The empirical formula, which is the relative number of atoms of each element in the compound, expressed as the smallest whole numbers that indicate the correct atom ratios.

Knowns: The identity of the elements, the mass of the compound and the mass of one of the elements in the compound. The mass of the other element is an implicit known, as are the molar masses of the elements.

Relationship: The relative number of moles of each element can be found from its mass and molar mass.

Plan: The stepwise plan is:

Step 3. Find the ratio of the molar amounts. This is the same as the ratio of the atoms in the molecule, and when, converted to a whole-number ratio, gives the subscripts in the empirical formula.

Step 2. Find the number of moles of each element from its mass and its molar mass.

Step 1. Find the number of grams of oxygen from the mass of the sample and the mass of phosphorus.

Work:

Step 1. Find the mass of oxygen in the sample.

$$5.682 \text{ g sample} - 2.480 \text{ g P} = 3.202 \text{ g O}$$

Step 2. Convert the mass of each element to number of moles using the molar mass.

$$? \text{ mol P} = 2.480 \text{ g P} \times \frac{1 \text{ mol P}}{30.97376 \text{ g P}} = 0.08007 \text{ mol P}$$

$$? \text{ mol O} = 3.202 \text{ g O} \times \frac{1 \text{ mol O}}{15.9994 \text{ g O}} = 0.2001 \text{ mol O}$$

Step 3. Try to convert the mole ratios to whole numbers by dividing by the smaller.

$$\text{P: } \frac{0.08077}{0.08077} = 1.000 \qquad \text{O: } = \frac{0.2001}{0.08007} = 2.499$$

Because the number of atoms must be a whole number, the mole ratios must also be whole numbers. The ratio of oxygen is too different from a whole number to simply round off. If we multiply each mole ratio by two, the ratios become:

P: 2.000 O: 4.998

The results are rounded to give the empirical formula of the compound: **P_2O_5**.

2.79 Analysis of a 6.3200 g sample yields the following amounts of nitrogen and oxygen: N, 1.923 g; O, 4.397 g. No other elements are present. What is the empirical formula of the compound? If the molar mass is known to be in the range 80 - 100 g mol^{-1}, what is the molecular formula?

Analysis:

Target: The empirical formula, which is the lowest whole number ratio of atoms in the compound and the molecular formula, which is the actual number of atoms of each element in a molecule of the compound.

Knowns: The identity of the elements, the mass of each element in the compound, the molar mass of each element and the approximate mass of one mole of the compound.

Relationships: The relative number of moles of each element can be found from the mass and molar mass. The number of atoms of each element in the molecule is a whole-number multiple of the subscript in the empirical formula.

Plan: If we knew the empirical formula, we could find its molar mass. The molar mass of the compound divided by the mass of the empirical unit gives the number of empirical units in one molecule. The empirical formula can be found from the masses and molar masses as in previous examples. The stepwise plan is:

Step 5. Multiply each subscript in the empirical formula by the number of empirical units in one molecule.

Step 4. Divide the molar mass of the compound by the molar mass of the empirical unit and round to a whole number.

Step 3. Find the molar mass of the empirical unit.

Step 2. Find the ratio of the molar amounts. This is the same as the ratio of the atoms in the molecule, and, when converted to a whole-number ratio, gives the subscripts in the empirical formula.

Step 1. Find the number of moles of each element from its mass and its molar mass.

Work:

Step 1. Find the number of moles of each element.

$$? \text{ mol N} = 1.923 \text{ g N} \times \frac{1 \text{ mol N}}{14.0067 \text{ g N}} = 0.1373 \text{ mol N}$$

$$? \text{ mol O} = 4.397 \text{ g O} \times \frac{1 \text{ mol O}}{15.9994 \text{ g O}} = 0.2748 \text{ mol O}$$

Step 2. Try to convert the mole ratios to whole numbers by dividing by the smaller.

$$\text{N: } \frac{0.1373}{0.1373} = 1.000 \qquad \text{O: } \frac{0.2748}{0.1373} = 2.002$$

Because the ratio must be a whole number, the empirical formula is $\mathbf{NO_2}$.

Step 3. Find the molar mass of NO_2.

N: 1 mol x	$14.0067 \text{ g mol}^{-1}$	=	14.0067 g N
O: 2 mol x	$15.9994 \text{ g mol}^{-1}$	=	31.9988 g O
NO_2:	1 mol	=	46.0055 g NO_2

(The molar mass is $46.0055 \text{ g mol}^{-1}$.)

Step 4. Divide the molar mass of the compound by the molar mass of the empirical unit and round to a whole number.

$$\frac{\approx 90 \text{ g mol}^{-1}}{46.0055 \text{ g mol}^{-1}} \approx 2$$

Step 5. Multiply each subscript in the empirical formula by 2 to yield the molecular formula.

The formula is $\mathbf{N_2O_4}$.

2.81 Zinc chloride is an ionic compound that contains 48.0% Zn. What is the charge of the zinc cation?

Analysis:

Target: The charge of a zinc cation.

Knowns: The mass percent of zinc in the compound. The mass percent of chlorine is an implicit known, as is the molar mass of each element. Chlorine has a 1- charge in its ionic compounds.

Relationship: The compound must be neutral.

Plan: If we knew the empirical formula of the compound, we could determine the charge that the zinc ion must have to make the compound neutral. We can find the empirical formula of the compound from the percent of each element and their molar masses as in previous examples.

Step 4. Determine the charge on the zinc cation from the formula and the charge of the chloride ion.

Step 3. Find the ratio of the molar amounts to determine the empirical formula.

Step 2. Find the number of moles of each element from its mass and its molar mass.

Step 1. Find the number of grams of each element in 100 g of the compound.

Work:

Step 1. Find the mass of each element in 100.0 g of the compound by using the percentages.

$$100.0 \text{ g compound} \times \frac{48.0 \text{ g Zn}}{100 \text{ g compound}} = 48.0 \text{ g Zn}$$

The rest of the compound is chlorine.

$$100.0 \text{ g compound} - 48.0 \text{ g Zn} = 52.0 \text{ g Cl}$$

Step 2. Convert the mass of each element to the number of moles using the molar mass.

$$? \text{ mol Zn} = 48.0 \text{ g Zn} \times \frac{1 \text{ mol Zn}}{65.38 \text{ g Zn}} = 0.734 \text{ mol Zn}$$

$$? \text{ mol Cl} = 52.0 \text{ g Cl} \times \frac{1 \text{ mol Cl}}{35.453 \text{ g Cl}} = 1.47 \text{ mol Cl}$$

Step 3. Try to convert the mole ratios to whole numbers by dividing by the smaller.

$$\text{Zn: } \frac{0.734}{0.734} = 1.00 \qquad \text{Cl : } \frac{1.47}{0.734} = 2.00$$

The empirical formula is $ZnCl_2$.

Step 4. Determine the charge on the zinc cation from the formula and the charge of the chloride ion.

Because the formula is $ZnCl_2$ and the chloride ion has a charge of 1-, the ion is $\mathbf{Zn^{2+}}$.

2.83 One of many compounds containing only carbon and hydrogen (hydrocarbons) is cyclohexane, whose empirical formula is CH_2. If its molar mass is in the range 80 - 90 g mol^{-1}, what is its molecular formula?

Analysis:

Target: The molecular formula, which is the actual number of atoms of each element in a molecule of the compound.

Knowns: The empirical formula of the compound and the approximate mass of one mole of the compound.

Relationship: The number of atoms of each element in the molecule is a whole-number multiple of the numbers in the empirical formula.

Plan: The molar mass of the compound divided by the mass of the empirical unit gives the number of empirical units in one molecule.

Step 3. Multiply each subscript in the empirical formula by the number of empirical units in one molecule.

Step 2. Divide the molar mass of the compound by the molar mass of the empirical unit and round to a whole number.

Step 1. Find the molar mass of the empirical unit.

Work:

Step 1. Find the molar mass of the empirical unit from the empirical formula and the molar masses.

C: 1 mol x 12.011 g mol^{-1}		=	12.011 g C
H: 2 mol x 1.0079 g mol^{-1}		=	2.0158 g H
CH_2:	1 mol	=	14.027 g CH_2 (The molar mass is 14.027 g mol^{-1}.)

Step 2. Divide the molar mass of the compound by the molar mass of the empirical unit.

$$\frac{\sim 85 \text{ g mol}^{-1}}{14.027 \text{ g mol}^{-1}} \sim 6$$

Step 3. Multiply each subscript in the empirical unit by 6 to yield the molecular formula.

The formula is C_6H_{12}.

2.85 When 6.25 mg of the anesthetic ether, which contains C, H, and O, is subjected to combustion analysis, 14.84 mg CO_2 and 7.56 mg H_2O are obtained. What is the empirical formula? If the molar mass is known to be between 50 and 100 g mol^{-1}, what is the molecular formula?

Analysis:

Target: The empirical formula of the compound and the molecular formula of the compound.

Knowns: The original mass of the compound and the masses of CO_2 and H_2O produced. The molar masses of the elements are implicit knowns.

Relationships: The carbon in the sample is related to the amount of CO_2 produced and the hydrogen in the sample is related to the amount of H_2O produced. The amount of oxygen can be found from by subtraction. The relative number of moles of each element in the formula can be found from the mass and molar mass. The molecular formula can be found from the molar mass of the compound and the molar mass of the empirical unit.

Plan: We can use known mass relationships to find the masses of carbon and hydrogen in the original compound and can find the mass of oxygen by subtraction. From those masses, we can proceed as in previous examples to find the empirical and molecular formulas.

Step 8. Multiply each subscript in the empirical formula by the number of empirical units in one molecule.

Step 7. Divide the molar mass of the compound by the molar mass of the empirical unit and round to a whole number.

Step 6. Find the molar mass of the empirical unit.

Step 5. Find the ratio of the molar amounts. This is the same as the ratio of the atoms in the molecule, and, when converted to a whole-number ratio, gives the subscripts in the empirical formula.

Step 4. Find the number of moles of oxygen from its mass and its molar mass.

Step 3. Find the mass of oxygen in the original sample by subtracting the mass of H and C from the original sample mass.

Step 2. Find the mass of hydrogen in the original compound from the mass of H_2O collected by:

a. Converting the given mass of water to moles using the molar mass (18.0152 g mol^{-1}).
b. Converting from moles of H_2O to moles of hydrogen in the original sample.
c. Converting from moles of hydrogen to mass of hydrogen in the original sample using the molar mass (1.0079 g mol^{-1}).

Step 1. Find the mass of carbon in the original compound from the mass of CO_2 collected by:

a. Converting the given mass of CO_2 to moles using the molar mass (44.010 g mol^{-1}).
b. Converting from moles of CO_2 to moles of carbon in the original sample.
c. Converting from moles of carbon to mass of carbon in the original sample using the the molar mass (12.011 g mol^{-1}).

Note that since the masses are in mg, the problem can be worked in terms of *millimoles*.

Work:

Step 1. a. Find the number of mmol CO_2 in the CO_2 produced.

$$? \text{ mmol } CO_2 = 14.84 \text{ mg } CO_2 \times \frac{1 \text{ mmol } CO_2}{44.010 \text{ mg } CO_2} = 0.3372 \text{ mmol } CO_2$$

b. Find the number of mmol C in the CO_2 produced.

$$? \text{ mmol C} = 0.3372 \text{ mmol } CO_2 \times \frac{1 \text{ mmol C}}{1 \text{ mmol } CO_2} = 0.3372 \text{ mmol C}$$

c. Find the mass of C in the sample.

$$? \text{ mg C} = 0.3372 \text{ mmol C} \times \frac{12.011 \text{ mg C}}{1 \text{ mmol C}} = 4.050 \text{ mg C}$$

Step 2. a. Find the number of mmol H_2O in the H_2O produced.

$$? \text{ mmol } H_2O = 7.56 \text{ mg } H_2O \times \frac{1 \text{ mmol } H_2O}{18.0152 \text{ mg } H_2O} = 0.420 \text{ mmol } H_2O$$

b. Find the number of mmol of H in the H_2O produced.

$$? \text{ mmol H} = 0.420 \text{ mmol } H_2O \times \frac{2 \text{ mmol H}}{1 \text{ mmol } H_2O} = 0.839 \text{ mmol H}$$

c. Find the mass of H in the sample.

$$? \text{ mg H} = 0.839 \text{ mmol H} \times \frac{1.0079 \text{ mg H}}{1 \text{ mmol H}} = 0.846 \text{ mg H}$$

Step 3. Find the mass of O in the sample.

$$? \text{ g O} = 6.25 \text{ mg} - 4.050 \text{ mg C} - 0.846 \text{ mg H} = 1.35 \text{ mg O}$$

Step 4. Find the number of mmol of O in the sample.

$$? \text{ mmol O} = 1.35 \text{ mg O} \times \frac{1 \text{ mmol O}}{15.9994 \text{ mg O}} = 0.0844 \text{ mmol O}$$

Step 5. Try to convert the mmole ratios to whole numbers by dividing by the smallest.

$$\text{C: } \frac{0.3372}{0.0844} = 4.00 \qquad \text{H: } \frac{0.839}{0.0844} = 9.94 \qquad \text{O: } \frac{0.0844}{0.0844} = 1.00$$

These are rounded to whole numbers to yield an empirical formula of $\mathbf{C_4H_{10}O}$.

Step 6. Find the mass of the empirical unit.

C: 4 mol x 12.011 g mol^{-1}		=	48.044 g C
H: 10 mol x 1.0079 g mol^{-1}		=	10.008 g H
O: 1 mol x 15.9994 g mol^{-1}		=	15.9994 g O
$C_4H_{10}O$:	1 mol	=	74.051 g (The molar mass is 74.051 g mol^{-1}.)

Step 7. Divide the molar mass of the compound by the molar mass of the empirical unit.

$$\frac{\sim 75 \text{ g mol}^{-1}}{74.051 \text{ g mol}^{-1}} \sim 1$$

Step 8. Multiply each subscript in the empirical formula by 1.

The molecular formula is also $\mathbf{C_4H_{10}O}$.

2.87 Write the formula of the following ionic compounds.

a. rubidium fluoride b. ammonium sulfate c. potassium sulfide
d. barium sulfide e. magnesium hydrogen carbonate f. sodium nitrite

Analysis:

Target: The formula of the compound.
Known: The name of the compound. The formulas of ions must be memorized. See Tables 2.3, 2.6 and 2.7 in the text.
Relationship: The compound must be electrically neutral.

Plan: Determine the charges of the constituent ions and use subscripts and parentheses if necessary to make the compound neutral.

Work:

a. Rubidium is Rb^+ (Table 2.6) and fluoride is F^- (Table 2.6). The compound is **RbF**.

b. Ammonium is NH_4^+ (Table 2.3) and sulfate is SO_4^{2-} (Tables 2.3 and 2.7). The compound is $\mathbf{(NH_4)_2SO_4}$.

c. Potassium is K^+ (Table 2.6) and sulfide is S^{2-} (Table 2.6). The compound is $\mathbf{K_2S}$.

d. Barium is Ba^{2+} (Table 2.6) and sulfide is S^{2-} (Table 26). The compound is **BaS**.

e. Magnesium is Mg^{2+} (Table 2.6) and hydrogen carbonate is HCO_3^- (Table 2.3). The compound is $\mathbf{Mg(HCO_3)_2}$.

f. Sodium is Na^+ (Table 2.6) and nitrite is NO_2^- (Tables 2.3 and 2.6). The compound is $\mathbf{NaNO_2}$.

2.89 Write the formula of each of the following molecular compounds.

a. sulfur trioxide b. chlorous acid c. carbon tetrachloride
d. phosphorus pentachloride e. ammonia f. hydrogen iodide

Analysis:

Target: The formula of a molecular compound.
Known: The name of the compound. Names of elements, common names and values of the prefixes (Section 2.8) must be memorized. Names of acids and anions derived from them are in Table 2.7.
Relationship: The prefix on the element name gives the number of atoms of that element in a molecule of the compound.

Plan: Use the prefixes to write the formula from the name and learn the common names and acid names as provided in the text.

Work:

a. Sulfur is S, oxide is O, the prefix tri means 3. The compound is $\mathbf{SO_3}$.

b. Chlorous acid is $\mathbf{HClO_2}$ (Table 2.7).

c. Carbon is C, chloride is Cl, tetra means 4. The compound is $\mathbf{CCl_4}$.

d. Phosphorus is P, chloride is Cl, penta means 5. The compound is $\mathbf{PCl_5}$.

e. Ammonia is $\mathbf{NH_3}$.

f. Hydrogen is H, iodide is I. The compound is **HI**.

2.91 Give the formula of each of the following compounds, and label each as a molecular or empirical formula. If the empirical and molecular formulas are different, give both. If you can, identify each compound as ionic or molecular.

a. manganese(IV) oxide b. sulfur dioxide c. ferric chloride
d. dinitrogen tetroxide e. cobalt(III) dichromate f. hydrogen peroxide

Analysis:

Targets: The formula of the compound, whether it is an empirical or molecular formula, and whether it is ionic or covalent.

Known: The name of the compound, the naming rules and the charges of characteristic ions (Consult Tables 2.3, 2.6 and 2.7 and Sections 2.8.) An empirical formula shows the lowest whole number ratio of atoms, while the molecular formula gives the actual number of atoms in a molecule of the compound.
Relationships: If ionic, the compound must be electrically neutral. If molecular, the prefixes give the number of atoms of each element in the molecule. Presence of a metal often indicates that the compound is ionic. Compounds with two nonmetals are always molecular.
Plan: Study the name and apply the rules above, consulting tables and sections as needed.
Work:

a. Manganese(IV) is Mn^{4+} and oxide is O^{2-}. The compound is **MnO_2 (empirical formula)**. The compound could be either **molecular or ionic**.
b. Sulfur is S, oxide is O and di means 2. The compound is **SO_2 (empirical** and **molecular formula)**. The compound is **molecular**.
c. Ferric is Fe^{3+} (Section 2.8) and chloride is Cl^- (Table 2.6). The compound is **$FeCl_3$ (empirical formula)**. The compound is **ionic**.
d. Nitrogen is N, oxide is O, di means 2 and tetra means 4. The compound is **N_2O_4 (molecular formula)**. The empirical formula is **NO_2**. The compound is **molecular**.
e. Cobalt(III) is Co^{3+} and dichromate is $Cr_2O_7^{2-}$(Table 2.3). The compound is **$Co_2(Cr_2O_7)_3$ (empirical formula)**. The compound is **ionic**.
f. Hydrogen is H^+ with nonmetals and peroxide is O_2^{2-} (Table 2.3). The compound is **H_2O_2 (molecular formula)**. The empirical formula is **HO**. The compound is **molecular**.

2.93 Name the following ionic compounds.

a. NH_4HCO_3	b. $Na_2C_2O_4$	c. Mg_3N_2
d. $K_2Cr_2O_7$	e. CaH_2	f. $Ca(HSO_4)_2$

Analysis:
Target: The name of the compounds.
Known: The formula of the compounds. Ion names are tabulated in Tables 2.3, 2.6 and 2.7.
Relationship: The ions are named in the order they appear in the compound.
Plan: Name the ions in order.
Work:

a. NH_4^+ is ammonium and HCO_3^- is hydrogen carbonate. The compound is **ammonium hydrogen carbonate**.
b. Na^+ is sodium and $C_2O_4^{2-}$ is oxalate. The compound is **sodium oxalate**.
c. Mg^{2+} is magnesium and N^{3-} is nitride. The compound is **magnesium nitride**.
d. K^+ is potassium and $Cr_2O_7^{2-}$ is dichromate. The compound is **potassium dichromate**.
e. Ca^{2+} is calcium and H^- is hydride. The compound is **calcium hydride**.
f. Ca^{2+} is calcium and HSO_4^- is hydrogen sulfate. The compound is **calcium hydrogen sulfate**.

2.95 Name the following molecular compounds.

a. $HClO_3$	b. Cl_2O_7	c. N_2O_5
d. CO	e. SO_3	f. CI_4

Analysis:
Target: The name of a molecular compound.
Known: The formula of the compound. Names of elements, common names and values of the prefixes (Section 2.8) must be memorized. Names of acids and anions derived from them are in Table 2.7.
Relationship: The prefix of the element name gives the number of atoms of that element in a molecule of the compound.

Plan: Use the prefixes to write the name and learn the common names and acid names as provided in the text.
Work:
a. $HClO_3$ is **chloric acid** (Table 2.7).
b. Cl is chlorine, O is oxide, 2 is di and 7 is hepta. The compound is **dichlorine heptoxide**.
c. N is nitrogen, O is oxide, 2 is di and 5 is penta. The compound is **dinitrogen pentoxide**
d. C is carbon, O is oxide and 1 is mono. The compound is **carbon monoxide**.
e. S is sulfur, O is oxide and 3 is tri. The compound is **sulfur trioxide.**
f. C is carbon, I is iodide and 4 is tetra. The compound is **carbon tetraiodide**.

2.97 Name the following compounds.
a. $FeCl_2$ b. HClO c. $GeCl_4$
d. CaH_2 e. Na_2O f. $MgCl_2$

Analysis:
Target: The name of the compound.
Knowns: The formula of the compound, the naming rules and the charges of characteristic ions (Consult Tables 2.3, 2.6 and 2.7 and Sections 2.8 and 2.8.
Relationships: If ionic the compound must be electrically neutral. If molecular, the prefixes give the number of atoms of each element in the molecule.
Plan: Study the formula and apply the rules above, consulting tables and sections as needed.
Work:
a. Cl^- is chloride. To be neutral, the iron must be Fe^{2+}, iron(II). The compound is **iron(II) chloride.**
b. HClO is **hypochlorous acid** (Table 2.7).
c. Ge is germanium and Cl is chloride. The compound is **germanium tetrachloride.**
d. Ca^{2+} is calcium and H^- is hydride. The compound is **calcium hydride**.
e. Na^+ is sodium and O^{2-} is oxide. The compound is **sodium oxide.**
f. Mg^{2+} is magnesium and Cl^- is chloride. The compound is **magnesium chloride.**

2.99 Give the names and elemental symbols of all the noble gases, in order of increasing molar mass.

Analysis:
Target: The names of the noble gases, in order of increasing molar mass.
Knowns: The noble gases are in Group 8A. Molar masses can be determined by consulting the periodic table.
Relationship: Molar masses increase down a group.
Plan: Consult the periodic table.
Work: **Helium** (He), **Neon** (Ne), **Argon** (Ar), **Krypton** (Kr), **Xenon** (Xe) and **Radon** (Rn).

2.101 Identify the noble gas that is
a. the most common (most abundant)
b. the least abundant
c. the most reactive

Analysis: This must be determined by learning the information in Section 2.9.
Work:
a. **Argon** (Ar) is most abundant.
b. **Radon** (Rn) is least abundant.
c. **Xenon** (Xe) is most reactive.

3 Chemical Equations and Stoichiometry

3.1 Explain in your own words what information is carried by an unbalanced chemical equation.

The unbalanced equation gives the proper formulas for what reacts (the reactants) and what is produced (the products). It may also tell about reaction conditions or the physical states of the substances involved.

3.3 How can a balanced equation be distinguished from an unbalanced one?

In a balanced equation, the same number of atoms of each element appears on both sides of the equation.

3.5 What is the relationship between the law of conservation of mass and a balanced chemical equation?

Because the number and kind of atoms is the same on each side of the equation, the total mass must be the same. This means that mass is conserved.

3.7 What is a stoichiometric ratio?

A stoichiometric ratio is the molar ratio between two reactants, between two products or between a reactant and a product in the balanced chemical equation.

3.9 Give some reasons why the actual yield might be less than the theoretical yield.

Actual yield may be less than theoretical yield if there are competing reactions leading to other products, if the reaction does not go to completion or if product is lost during the processes of isolation and purification.

3.11 What is a limiting reactant:?

A limiting reactant is the reactant that is present in the smallest relative stoichiometric amount. That means that it is used up before any other reactant is totally consumed. Because the limiting reactant is consumed first, it limits the amount of product that can be made.

3.13 Distinguish between solvent and solute.

In a solution, the solvent is the component that, when pure, has the same physical state as the solution. If more than one component has the same physical state as the solution, the one present in greater amount is the solvent. A solute is anything that is dissolved in the solvent.

3.15 Distinguish between the terms "concentration" and "molarity".

Concentration is the general term for the amount of solute dissolved in a specified amount of solvent or solution. Molarity is defined as the number of moles of solute dissolved in one liter of solution.

3.17 Which of the following equations are balanced, and which are not?

a. $Zn + 2\ HCl \longrightarrow ZnCl_2 + H_2$

b. $HNO_3 + 2\ HCl \longrightarrow NOCl + Cl_2 + 2\ H_2O$

c. $H_2O + Cl_2 \longrightarrow 2\ HCl + O_2$

Analysis/Plan: A balanced equation must have the same number and kind of atoms on each side. The number of atoms on each side of the equation must be counted and compared.
Work:

a. **Left Side**

There is 1 Zn atom. There are 2 HCl molecules, each of which has 1 atom of H and 1 atom of Cl.
Zn = 1(1) = 1 atom
H = 2(1) = 2 atoms
Cl = 2(1) = 2 atoms

Right Side

There is 1 $ZnCl_2$ formula unit which has 1 Zn atom and 2 Cl atoms. There is 1 molecule of H_2 which has 2 H atoms.
Zn = 1(1) = 1 atom
H = 2(1) = 2 atoms
Cl = 2(1) = 2 atoms

There are the same number and kind of atoms on each side so the equation is **balanced.**

b. There is 1 HNO_3 molecule which contains 1 H atom, 1 N atom, and 3 O atoms. Each of the 2 HCl molecules contains 1 H atom and 1 Cl atom.
H = 1(1) + 2(1) = 3 atoms
N = 1(1) = 1 atom
Cl = 2(1) = 2 atoms
O = 1(3) = 3 atoms

There is 1 NOCl molecule which contains 1 N atom, 1 O atom, and 1 Cl atom. The Cl_2 molecule contains 2 Cl atoms. Each of the 2 H_2O molecules contains 2 H atoms and 1 O atom.
H = 2(2) = 4 atoms
N = 1(1) = 1 atom
Cl = 1(1) + 1(2) = 3 atoms
O = 1(1) + 2(1) = 3 atoms

The number of H atoms and the number of Cl atoms are not the same on each side, so the equation is **not balanced.**

c. There is 1 H_2O molecule which contains 2 H atoms and 1 O atom. The Cl_2 molecule contains 2 Cl atoms.
H = 1(2) = 2 atoms
Cl = 1(2) = 2 atoms
O = 1(1) = 1 atom

There are 2 HCl molecules, each of which contains 1 H atom and 1 Cl atom. The O_2 molecule contains 2 O atoms.
H= 2(1) = 2 atoms
Cl = 1(2) = 2 atoms
O = 1(2)= 2 atoms

The number of O atoms is not the same on each side so the equation is **not balanced.**

3.19 Balance those equations in Problem 3.17 that are unbalanced.

Analysis/Plan: Remove all current coefficients, then follow the systematic guidelines presented in the text, beginning with the most complicated species and an element that appears in the fewest substances on each side.
Work:

b. HNO_3 is a more complicated species than HCl. N is already balanced with 1 atom on each side. Because O is in only one compound on the left, balance it next, by using a coefficient of 2 for H_2O because that will not unbalance the N.

$HNO_3 + HCl \longrightarrow NOCl + Cl_2 + 2\ H_2O$ 3(1) O left = 1(1) +2(1) O right

Now balance H by using a coefficient of 3 for HCl because that will not unbalance N.

$\mathbf{HNO_3 + 3\ HCl \longrightarrow NOCl + Cl_2 + 2\ H_2O}$ 1(1) + 3(1) H left = 2(2) H right

Notice that this also balances the Cl and is the completed balanced equation. 3(1) Cl right = 1(1) + 2(1) Cl left

c. H_2O is a more complicated species than Cl_2. Balance H, before balancing O, because all of the O goes to make the element O_2. H is balanced by using a coefficient of 2 for HCl.

$H_2O + Cl_2 \longrightarrow 2\ HCl + O_2$ 1(2) H left = 2(1) H right

This step also balances the Cl_2. 1(2) Cl left = 2(1) Cl right

The oxygen can be balanced using the coefficient 1/2 for O_2.

$H_2O + Cl_2 \longrightarrow 2\ HCl + 1/2\ O_2$ 1(1) O left = 1/2(1) O right

To convert to whole number coefficients, multiply all coefficients by 2, giving the balanced equation: **$2\ H_2O + 2\ Cl_2 \longrightarrow 4\ HCl + O_2$**

3.21 Which of the following equations are balanced, and which are not?

a. $K_4Fe(CN)_6 \longrightarrow 4\ KCN + Fe + 2\ C + N_2$

b. $Na + H_2O \longrightarrow NaOH + H_2$

c. $2\ KNO_3 + S + 2\ C \longrightarrow K_2S + N_2 + 2\ CO_2$

Analysis/Plan: A balanced equation must have the same number and kind of atoms on each side. The number of atoms on each side of the equation must be counted and compared.

Work:

	Left Side	Right Side
a.	There is 1 $K_4Fe(CN)_6$ formula unit which is composed of 4 K atoms, 1 Fe atom, 6 C atoms and 6 N atoms. K = 1(4) = 4 atoms Fe= 1(1) = 1 atom C = 1(6) = 6 atoms N = 1(6) = 6 atoms	There are 4 KCN formula units , each of which has 1 K atom, 1 C atom and 1 N atom. There are 2 atoms of C and 1 atom of Fe in the elements. The 1 N_2 molecule has 2 N atoms. K = 4(1) = 4 atoms Fe = 1(1) = 1 atom C = 4(1) + 2(1) = 6 atoms N = 4(1) + 1(2) = 6 atoms

There are the same number and kind of atoms on each side so the equation is **balanced.**

	Left Side	Right Side
b.	There is 1 Na atom. There is 1 H_2O molecule that contains 2 H atoms and 1 O atom Na = 1(1) = 1 atom H =2(1) = 2 atoms O = 1(1) = 1 atom	There is 1 NaOH formula unit which has 1 Na atom, 1 O atom and 1 H atom. There is 1 H_2 molecule that contains 2 H atoms. Na = 1(1)= 1 atom H = 1(1) + 1(2) = 3 atoms O = 1(1) = 1 atom

The number of H atoms is not the same on each side so the equation is **not balanced.**

	Left Side	Right Side
c.	There are 2 KNO_3 formula units, each of which contains 1 K atom, 1 N atom and 3 O atoms. There are 2 C atoms and 1 S atom. K = 2(1) = 2 atoms N = 2(1) = 2 atoms O = 2(3) = 6 atoms S = 1(1) = 1 atom C = 2(1) = 2 atoms	There is 1 K_2S formula unit which contains 2 K atoms and 1 S atom. There is 1 N_2 molecule that contains 2 N atoms. There are 2 CO_2 molecules, each of which contains 1 C atom and 2 O atoms. K = 2(1) = 2 atoms N = 1(2) = 2 atoms O = 2(2) = 4 atoms S = 1(1) = 1 atom C = 2(1) = 2 atoms

The number of O atoms is not the same on each side so the equation is **not balanced.**

3.23 Balance those equations in Problem 3.21 that are unbalanced.

Analysis/Plan: Remove all current coefficients, then follow the systematic guidelines presented in the text, beginning with the most complicated species and the element that appears in the fewest substances on each side.

Work:

b. H_2O is a more complicated species than Na. O is already balanced, as is Na, with 1 atom on each side. H can be balanced without disturbing the balance of O and Na by using a coefficient of 1/2 for H_2.

$Na + H_2O \longrightarrow NaOH + 1/2\ H_2$ 2(1) H right = 1(1) + 1/2(2) H left

To convert to whole number coefficients, multiply all coefficients by 2, giving the balanced equation: $\mathbf{2\ Na + 2\ H_2O \longrightarrow 2\ NaOH + H_2}$

c. KNO_3 is the most complicated species. K can be balanced by a coefficient of 2 for KNO_3.

$2\ KNO_3 + S + C \longrightarrow K_2S + N_2 + CO_2$ 2(1) K left = 1(2) K right

That also balances N with 2 atoms on each side. Balance O by using a coefficient of 3 for CO_2.

$2\ KNO_3 + S + C \longrightarrow K_2S + N_2 + 3\ CO_2$ 2(3) O left = 3(2) O right

The S is already balanced with 1 atom on each side. The process can be completed by using a coefficient of 3 for C.

$\mathbf{2\ KNO_3 + S + 3\ C \longrightarrow K_2S + N_2 + 3\ CO_2}$ 3(1) C left = 3(1) C right

3.25 Balance the following equations.

a. $Xe + F_2 \longrightarrow XeF_4$

b. $Mg + O_2 \longrightarrow MgO$

c. $N_2 + H_2 \longrightarrow NH_3$

d. $Al + HCl \longrightarrow AlCl_3 + H_2$

e. $Al + Cr_2O_3 \longrightarrow Al_2O_3 + Cr$

f. $Zn + HNO_3 \longrightarrow Zn(NO_3)_2 + H_2$

Analysis/Plan: Follow the systematic guidelines presented in the text, beginning with the most complicated species and the element that appears in the fewest substances on each side.

Work:

a. Xe is already balanced with 1 atom on each side. The process can be completed by using a coefficient of 2 for F_2.

$\mathbf{Xe + 2\ F_2 \longrightarrow XeF_4}$ 2(2) F left = 1(4) F right

b. Mg is balanced with 1 atom on each side. O can be balanced without unbalancing the Mg by using the coefficient 1/2 for O_2.

$Mg + 1/2\ O_2 \longrightarrow MgO$ 1/2(2) O left = 1(1) O right

To convert to whole number coefficients, multiply all coefficients by 2, giving the balanced equation: $\mathbf{2\ Mg + O_2 \longrightarrow 2\ MgO}$

c. Both N_2 and H_2 are elements and neither is balanced initially. N can be balanced by using a coefficient of 2 for NH_3.

$N_2 + H_2 \longrightarrow 2\ NH_3$ 1(2) N left = 2(1) N right

H can now be balanced by using a coefficient of 3 for H_2.

$N_2 + 3\ H_2 \rightarrow 2\ NH_3$ 3(2) H left = 2(3) H right

d. HCl is a more complicated species than Al. Al is already balanced with 1 atom on each side. Cl can be balanced by using a coefficient of 3 for HCl.
$Al + 3\ HCl \rightarrow AlCl_3 + H_2$ 3(1) Cl left = 1(3) Cl right
H can now be balanced by using a coefficient of 3/2 for H_2.
$Al + 3\ HCl \rightarrow AlCl_3 + 3/2\ H_2$ 3(1) H left = 3/2(2) H right
To convert to whole number coefficients, multiply all coefficients by 2, giving the balanced equation: **$2\ Al + 6\ HCl \rightarrow 2\ AlCl_3 + 3\ H_2$**

e. Cr_2O_3 is a more complicated species than Al. O is already balanced with 3 atoms on each side. Cr can be balanced by using a coefficient of 2 for Cr.
$Al + Cr_2O_3 \rightarrow Al_2O_3 + 2\ Cr$ 1(2) Cr left = 2(1) Cr right
The balancing is completed by using a coefficient of 2 for Al.
$2\ Al + Cr_2O_3 \rightarrow Al_2O_3 + 2\ Cr$ 2(1) Al left = 1(2) Al right

f. HNO_3 is a more complicated species than Zn. Zn is already balanced with 1 atom on each side. All species can be balanced by using a coefficient of 2 for HNO_3.
$Zn + 2\ HNO_3 \rightarrow Zn(NO_3)_2 + H_2$ 2(1) H left = 1(2) H right
2(1) N left = 1(2) N right
2(3) O left = 1(6) O right

3.27 Balance the following equations.
a. $CS_2 + O_2 \rightarrow CO_2 + SO_2$
b. $BiCl_3 + H_2S \rightarrow Bi_2S_3 + HCl$
c. $RbOH + SO_2 \rightarrow Rb_2SO_3 + H_2O$
d. $HBF_4 + H_2O \rightarrow H_3BO_3 + HF$
e. $PCl_3 + Cl_2 \rightarrow PCl_5$
f. $Na_2SO_4 + C \rightarrow Na_2S + CO_2$

Analysis/Plan: Follow the systematic guidelines presented in the text, beginning with the most complicated species and the element that appears in the fewest substances on each side.
Work:
a. CS_2 is a more complicated species than O_2. C is already balanced with 1 atom on each side. S can be balanced by using a coefficient of 2 for SO_2.
$CS_2 + O_2 \rightarrow CO_2 + 2\ SO_2$ 1(2) S left = 2(1) S right
O can then be balanced by using a coefficient of 3 for O_2.
$CS_2 + 3\ O_2 \rightarrow CO_2 + 2\ SO_2$ 3(2) O left = 1(2) + 2(2) O right

b. Both reactants are equally complicated. Begin by balancing the Bi by using a coefficient of 2 for $BiCl_3$.
$2\ BiCl_3 + H_2S \rightarrow Bi_2S_3 + HCl$ 2(1) Bi left = 1(2) Bi right
Now Cl can be balanced by using a coefficient of 6 for HCl.
$2\ BiCl_3 + H_2S \rightarrow Bi_2S_3 + 6\ HCl$ 2(3) Cl left = 6(1) Cl right
A coefficient of 3 for H_2S balances both the H and the S to complete the process.

$2\ BiCl_3 + 3\ H_2S \longrightarrow Bi_2S_3 + 6\ HCl$ 3(2) H left = 6(1) H right
3(1) S left = 1(3) S right

c. RbOH is a more complicated species than SO_2. Balance the Rb by using a coefficient of 2 for RbOH. This also balances the rest of the elements.

$2\ RbOH + SO_2 \longrightarrow Rb_2SO_3 + H_2O$ 2(1) Rb left = 1(2) Rb right
2(1) + 1(2) O left = 1(3) + 1(1) O right
2(1) H left = 1(2) H right
1(1) S left = 1(1) S right

d. HBF_4 is a more complicated species than H_2O. H will be balanced last because it appears in all the compounds. B is already balanced with 1 atom on each side. F can be balanced by using a coefficient of 4 for HF.

$HBF_4 + H_2O \longrightarrow H_3BO_3 + 4\ HF$ 1(4) F left = 4(1) F right

O can now be balanced by using a coefficient of 3 for H_2O. This also balances H.

$HBF_4 + 3\ H_2O \longrightarrow H_3BO_3 + 4\ HF$ 3(1) O left = 1(3) O right
1(1) + 3(2) H left = 1(3) + 4(1) H right

e. This equation is already balanced with 1 P atom on each side. There are 1(3) + 1(2) = 5 Cl atoms on the left and 1(5) Cl atoms on the right.

$PCl_3 + Cl_2 \longrightarrow PCl_5$

f. Na_2SO_4 is a more complicated species than C. Na is already balanced with 2 atoms on each side. S is already balanced with 1 atom on each side. O can be balanced by using a coefficient of 2 for CO_2.

$Na_2SO_4 + C \longrightarrow Na_2S + 2\ CO_2$ 1(4) O left = 2(2) O right

The process is completed by using a coefficient of 2 for C.

$Na_2SO_4 + 2\ C \longrightarrow Na_2S + 2\ CO_2$ 2(1) C left = 2(1) C right

3.29 Balance the following equations.

a. $P_4O_{10} + H_2O \longrightarrow H_3PO_4$

b. $P_2H_4 \longrightarrow PH_3 + P_4$

c. $PbO + NH_3 \longrightarrow Pb + N_2 + H_2O$

d. $Mg_3N_2 + H_2O \longrightarrow Mg(OH)_2 + NH_3$

Analysis/Plan: Follow the systematic guidelines presented in the text, beginning with the most complicated species and the element that appears in the fewest substances on each side.

Work:

a. P_4O_{10} and H_2O are equally complicated. Begin with the P because it can be balanced by using the whole number coefficient of 4 for H_3PO_4.

$P_4O_{10} + H_2O \longrightarrow 4\ H_3PO_4$ 1(4) P left = 4(1) P right

O appears in more compounds than does H. Balance H next by using a coefficient of 6 for H_2O. That coefficient also balances O, completing the process.

$P_4O_{10} + 6\ H_2O \longrightarrow 4\ H_3PO_4$ 6(2) H left = 4(3) H right
1(10) + 6(1) O left = 4(4) O right

b. H occurs in fewer species than does P. Balance H first by recognizing that the lowest common multiple of 4 and 3 is 12. Use a coefficient of 3 for P_2H_4 and a coefficient of 4 for PH_3.

$3\ P_2H_4 \longrightarrow 4\ PH_3 + P_4$ 3(4) H left = 4(3) H right

P can now balanced by using a coefficient of 1/2 for P_4.

$3\ P_2H_4 \longrightarrow 4\ PH_3 + 1/2\ P_4$ 3(2) P left = 4(1) + 1/2(4) P right

To convert to whole number coefficients, multiply all coefficients by 2, giving the balanced equation: $\mathbf{6\ P_2H_4 \longrightarrow 8\ PH_3 + P_4}$

c. PbO and NH_3 are equally complicated. Pb and O are initially balanced with 1 atom on each side. N is in an element on the right, so balance H by recognizing that the lowest common multiple of 2 and 3 is 6. Use a coefficient of 2 for NH_3 and a coefficient of 3 for H_2O. This also balances the N.

$PbO + 2\ NH_3 \longrightarrow Pb + N_2 + 3H_2O$ 2(1) N left = 1(2) N right
2(3) H left = 3(2) H right

That process unbalances O. Pb is an element on the right, so balance O next by using a coefficient of 3 for PbO.

$3\ PbO + 2\ NH_3 \longrightarrow Pb + N_2 + 3\ H_2O$ 3(1) O left = 3(1) O right

The process is completed by using a coefficient of 3 for Pb.

$\mathbf{3\ PbO + 2\ NH_3 \longrightarrow 3\ Pb + N_2 + 3\ H_2O}$ 3(1) Pb left = 3(1) Pb right

d. Mg_3N_2 and H_2O are equally complicated. Balance H last because it occurs in the most different compounds. Balance Mg first by using a coefficient of 3 for $Mg(OH)_2$.

$Mg_3N_2 + H_2O \longrightarrow 3\ Mg(OH)_2 + NH_3$ 1(3) Mg left = 3(1) Mg right

Now balance N by using a coefficient of 2 for NH_3.

$Mg_3N_2 + H_2O \longrightarrow 3\ Mg(OH)_2 + 2\ NH_3$ 1(2) N left = 2(1) N right

O can be balanced by using a coefficient of 6 for H_2O. This also balances H, completing the process.

$\mathbf{Mg_3N_2 + 6\ H_2O \longrightarrow 3\ Mg(OH)_2 + 2\ NH_3}$ 6(1) O left = 3(2) O right
6(2) H left = 3(2) + 2(3) H right

3.31 Balance the following equations.

a. $AsF_3 + PCl_5 \longrightarrow PF_5 + AsCl_3$

b. $NH_3 + O_2 \longrightarrow NO + H_2O$

c. $PbO_2 + Pb + H_2SO_4 \longrightarrow PbSO_4 + H_2O$

Analysis/Plan: Follow the systematic guidelines presented in the text, beginning with the most complicated species and the element that appears in the fewest substances on each side.

Work:

a. AsF_3 and PCl_5 are equally complicated. As and P are each initially balanced, with 1 atom on each side. Balance F by recognizing that the lowest common multiple of 3 and 5 is 15. Use a coefficient of 5 for AsF_3 by 5 and a coefficient of 3 for PF_5.

$5\ AsF_3 + PCl_5 \longrightarrow 3\ PF_5 + AsCl_3$ 5(3) F left = 3(5) F right

This process unbalances both As and P. Rebalance As by using a coefficient of 5 for $AsCl_3$.

$5\ AsF_3 + PCl_5 \longrightarrow 3\ PF_5 + 5\ AsCl_3$ 5(1) As left = 5(1) As right

Rebalance P by using a coefficient of 3 for PCl_5. This balances Cl at the same time and completes the process.

$\mathbf{5\ AsF_3 + 3\ PCl_5 \longrightarrow 3\ PF_5 + 5\ AsCl_3}$ 3(1) P left = 3(1) P right
3(5) Cl left = 5(3) Cl right

b. NH_3 is a more complicated species than O_2, the element that should be balanced last because it is in the most species. N is already balanced with 1 atom on each side. Balance H by recognizing that the lowest common multiple of 2 and 3 is 6. Use a coefficient of 2 for NH_3 and a coefficient of 3 for H_2O.

$2\ NH_3 + O_2 \rightarrow NO + 3\ H_2O$ — 2(3) H left = 3(2) H right

This unbalances N, so rebalance it by using a coefficient of 2 for NO.

$2\ NH_3 + O_2 \rightarrow 2\ NO + 3\ H_2O$ — 2(1) N left = 2(1) N right

There are 2(1) + 3(1) O on the right. O can be balanced by using a coefficient of 5/2 for O_2.

$2\ NH_3 + 5/2\ O_2 \rightarrow 2\ NO + 3\ H_2O$ — 5/2(2) O left = 2(1) + 3(1) O right

To convert to whole number coefficients, multiply all coefficients by 2, giving the balanced equation: $\mathbf{4\ NH_3 + 5\ O_2 \rightarrow 4\ NO + 6\ H_2O}$

c. H_2SO_4 is a more complicated species than PbO_2 or Pb, the element that should be balanced last because it appears in the most species. H and S are already balanced with 1 atom on each side. Balance O without unbalancing the H or S by using a coefficient of 1/2 for PbO_2.

$1/2\ PbO_2 + Pb + H_2SO_4 \rightarrow PbSO_4 + H_2O$ — 1/2(2) + 1(4) O left = 1(4) + 1(1) O right

Pb can be balanced by using a coefficient of 1/2 for Pb.

$1/2\ PbO_2 + 1/2\ Pb + H_2SO_4 \rightarrow PbSO_4 + H_2O$ — 1/2(1) + 1/2(1) Pb left = 1(1) Pb right

To convert to whole number coefficients, multiply all coefficients by 2, giving the balanced equation: $\mathbf{PbO_2 + Pb + 2\ H_2SO_4 \rightarrow 2\ PbSO_4 + 2\ H_2O}$

3.33 Balance the following equation (challenging).

$HMnO_4 + MnCl_2 + H_2O \rightarrow MnO_2 + HCl$

Analysis/Plan: Follow the systematic guidelines presented in the text, beginning with the most complicated species and the element that appears in the fewest substances on each side.

Work:

$HMnO_4$ is the most complicated species, however its elements are also available in the other compounds on that side. This combination forces us to use a more trial and error method. Note that Cl can be balanced by using a coefficient of 2 for HCl.

$HMnO_4 + MnCl_2 + H_2O \rightarrow MnO_2 + 2\ HCl$ — 1(2) Cl left = 2(1) Cl right

By inspection, it is evident that the coefficient of $HMnO_4$ must be even because there must be an even number of H on the right to agree with Cl. Also, the coefficient of H_2O must be even because there is an even number of O atoms in MnO_2. Suppose we try the lowest even number as coefficient for these two compounds.

$2\ HMnO_4 + MnCl_2 + 2\ H_2O \rightarrow MnO_2 + 2\ HCl$

If 2 were the correct coefficient, then the coefficient of MnO_2 would have to be 5.

$2\ HMnO_4 + MnCl_2 + 2\ H_2O \rightarrow 5\ MnO_2 + 2\ HCl$ — 2(4) + 2(1) O left = 5(2) O right

Mn can be balanced without disturbing H or O by using a coefficient of 3 for $MnCl_2$.

$2\ HMnO_4 + 3\ MnCl_2 + 2\ H_2O \rightarrow 5\ MnO_2 + 2\ HCl$ — 2(1) + 3(1) Mn left = 5(2) Mn right

That unbalances Cl. The coefficient of HCl is readjusted to be 6, a change that balances the H at the same time, completing the process.

$\mathbf{2\ HMnO_4 + 3\ MnCl_2 + 2\ H_2O \rightarrow MnO_2 + 6\ HCl}$ — 3(2) Cl left = 6(1) Cl right

2(1) + 2(2) H left = 6(1) H right

(This type of equation is called an oxidation reduction equation. We will balance this type of equation by a more systematic method in Chapter 18.)

3.35 Write balanced chemical equations for each of the following processes.

a. Calcium phosphate reacts with sulfuric acid to produce calcium sulfate and phosphoric acid.

b. Calcium phosphate reacts with water containing dissolved carbon dioxide to produce calcium hydrogen carbonate and calcium hydrogen phosphate.

Analysis/Plan: Nomenclature rules, introduced in Chapter 2, must be used to write the formulas for the reactants and products. Then, follow the systematic guidelines presented in the text, beginning with the most complicated species and the element that appears in the fewest substances on each side.

Work:

a. Calcium makes a Ca^{2+} ion and phosphate is PO_4^{3-}, so calcium phosphate is $Ca_3(PO_4)_2$. Sulfuric acid is H_2SO_4. Sulfate is SO_4^{2-}, so calcium sulfate is $CaSO_4$. Phosphoric acid has the formula H_3PO_4. The skeleton equation is:

$Ca_3(PO_4)_2 + H_2SO_4 \longrightarrow CaSO_4 + H_3PO_4$

O should be balanced last because it is in the most different species. $Ca_3(PO_4)_2$ is the most complicated species, so begin with it. Ca can be balanced by using a coefficient of 3 for $CaSO_4$.

$Ca_3(PO_4)_2 + H_2SO_4 \longrightarrow 3\ CaSO_4 + H_3PO_4$ 1(3) Ca left = 3(1) Ca right

Now, the P can be balanced by using a coefficient of 2 for H_3PO_4.

$Ca_3(PO_4)_2 + H_2SO_4 \longrightarrow 3\ CaSO_4 + 2\ H_3PO_4$ 1(2) P left = 2(1) P right

A coefficient of 3 for H_2SO_4 balances the S and the H. It also balances O, completing the process.

$\mathbf{Ca_3(PO_4)_2 + 3\ H_2SO_4 \longrightarrow 3\ CaSO_4 + 2\ H_3PO_4}$ 3(1) S left = 3(1) S right
3(2) H left = 2(3) H right
2(4) + 3(4) O left = 3(4) + 2(4) O right

b. From part a, calcium phosphate is $Ca_3(PO_4)_2$. Water is H_2O. Carbon dioxide is CO_2. A calcium ion is Ca^{2+} and hydrogen carbonate ion is HCO_3^-, so calcium hydrogen carbonate is $Ca(HCO_3)_2$. A hydrogen phosphate ion is HPO_4^{2-}, so calcium hydrogen phosphate is $CaHPO_4$. The skeleton equation is:

$Ca_3(PO_4)_2 + H_2O + CO_2 \longrightarrow Ca(HCO_3)_2 + CaHPO_4$

$Ca_3(PO_4)_2$ is the most complicated species. P appears in only one compound on the right, so it is balanced first by using a coefficient of 2 for $CaHPO_4$. This also balances the Ca.

$Ca_3(PO_4)_2 + H_2O + CO_2 \longrightarrow Ca(HCO_3)_2 + 2\ CaHPO_4$ 1(3) Ca left = 1(1) + 2(1) Ca right
1(2) P left = 2(1) P right

Balance H next, since it appears in just one compound on each side. This is done by using a coefficient of 2 for H_2O.

$Ca_3(PO_4)_2 + 2\ H_2O + CO_2 \longrightarrow Ca(HCO_3)_2 + 2\ CaHPO_4$ 2(2) H left = 1(2) + 2(1) H right

Balance C next, since it also appears in just one compound on each side. This is done by using a coefficient of 2 for CO_2. Inspection reveals that this also balances the O, completing the process.

$\mathbf{Ca_3(PO_4)_2 + 2\ H_2O + 2\ CO_2 \longrightarrow Ca(HCO_3)_2 + 2\ CaHPO_4}$ 2(1) C left = 1(2) C right
1(8) + 2(1) + 2(2) O left = 2(3) + 2(4) O right

3.37 Write balanced chemical equations for each of the following processes.

a. Copper reacts with water, oxygen, and carbon dioxide to produce $Cu_2(OH)_2CO_3$.

b. When heated, a mixture of potassium hydroxide and ammonium bromide produces potassium bromide, water, and ammonia.

Analysis/Plan: Nomenclature rules, introduced in Chapter 2, must be used to write the formulas for the reactants and products. Then, follow the systematic guidelines presented in the text, beginning with the most complicated species and the element that appears in the fewest substances on each side.

Work:

a. Copper is Cu, water is H_2O, oxygen is a diatomic element, O_2, and carbon dioxide is CO_2. The skeleton equation is:

$Cu + H_2O + O_2 + CO_2 \longrightarrow Cu_2(OH)_2CO_3$

O appears in many species, so will be balanced last. Inspection reveals that H is already balanced (2 atoms on each side), as is C (1 atom on each side). Cu can be balanced by using a coefficient of 2 for Cu. This also balances the O. (It was also balanced in the beginning, with 5 atoms on each side.)

$\mathbf{2\ Cu + H_2O + O_2 + CO_2 \longrightarrow Cu_2(OH)_2CO_3}$ 2(1) Cu left = 1(2) Cu right

1(1) + 1(2) + 1(2) O left = 1(5) O right

b. Potassium ion is K^+ and hydroxide ion is OH^-, so potassium hydroxide is KOH. An ammonium ion is NH_4^+ and a bromide ion is Br^- so ammonium bromide is NH_4Br. Using the ion charges, potassium bromide is KBr. Water is H_2O and ammonia is NH_3. The skeleton equation is:

$\mathbf{KOH + NH_4Br \longrightarrow KBr + H_2O + NH_3}$

Inspection reveals that the equation is already balanced, with 1 K atom, 1 O atom, 1 N atom and 1 Br atom on each side. There are 1(1) + 1(4) = 5 H atoms on the left and 1(2) + 1(3) = 5 H atoms on the right.

3.39 Write balanced chemical equations for each of the following processes.

a. Potassium chlorate reacts with table sugar ($C_{12}H_{22}O_{11}$) to produce potassium chloride, carbon dioxide, and water.

b. Lead(IV) oxide reacts with sulfur dioxide, producing lead(II) sulfate.

Analysis/Plan: Nomenclature rules, introduced in Chapter 2, must be used to write the formulas for the reactants and products. Then, follow the systematic guidelines presented in the text, beginning with the most complicated species and the element that appears in the fewest substances on each side.

Work:

a. Potassium makes a K^+ ion and chlorate is ClO_3^-, so potassium chlorate is $KClO_3$. Chloride is Cl^-, so potassium chloride is KCl. Carbon dioxide is CO_2 and water is H_2O. The skeleton equation is:

$KClO_3 + C_{12}H_{22}O_{11} \longrightarrow KCl + CO_2 + H_2O$

$C_{12}H_{22}O_{11}$ is the more complicated species. C can be balanced by using a coefficient of 12 for CO_2.

$KClO_3 + C_{12}H_{22}O_{11} \longrightarrow KCl + 12\ CO_2 + H_2O$ 1(12) C left = 12(1) C right

Now, H can be balanced by using a coefficient of 11 for H_2O.

$KClO_3 + C_{12}H_{22}O_{11} \longrightarrow KCl + 12\ CO_2 + 11\ H_2O$ 1(22) H left = 11(2) H right

To balance the O without unbalancing the C and H, a coefficient of 8 is used for $KClO_3$.

$8\ KClO_3 + C_{12}H_{22}O_{11} \longrightarrow KCl + 12\ CO_2 + 11\ H_2O$

8(3) + 1(11) O left = 12(2) + 11(1) O right

Finally, a coefficient of 8 for KCl balances both the K and the Cl.

$\mathbf{8\ KClO_3 + C_{12}H_{22}O_{11} \longrightarrow 8\ KCl + 12\ CO_2 + 11\ H_2O}$ 8(1) K left = 8(1) K right

8(1) Cl left = 8(1) Cl right

b. Lead(IV) is Pb^{4+}, and oxide is O^{2-}, so lead(IV) oxide is PbO_2. Sulfur dioxide is SO_2. Lead(II) is Pb^{2+} and sulfate is SO_4^{2-}, so lead(II) sulfate is $PbSO_4$. The skeleton equation is:

$\mathbf{PbO_2 + SO_2 \longrightarrow PbSO_4}$

Inspection reveals that Pb and S are already balanced with 1 atom on each side. O is also balanced with 1(2) + 1(2) = 2 atoms on the left and 1(4) atoms on the right.

3.41 How many moles of Cl_2 are formed when 0.0538 mol HCl reacts with nitric acid according to $HNO_3 + 3\ HCl \longrightarrow NOCl + Cl_2 + 2\ H_2O$?

Analysis:

Target: The number of moles of Cl_2. (? mol Cl_2) =

Knowns: The number of moles of HCl and the balanced chemical equation.

Relationship: The balanced chemical equation gives mole ratios of all species.

Plan: Use the balanced chemical equation to generate a conversion factor and use it to convert a number of moles of HCl to a number of moles of Cl_2.

Work: The balanced equation gives the equivalence statement: 3 mol HCl ↔ 1 mol Cl_2.

The necessary conversion factor generated from this equivalency is: $\frac{1\ mol\ Cl_2}{3\ mol\ HCl}$.

Applying the factor-label method gives:

$$? \text{ mol } Cl_2 = 0.0538 \text{ mol HCl} \times \frac{1 \text{ mol } Cl_2}{3 \text{ mol HCl}} = \mathbf{0.0179\ mol\ Cl_2}$$

3.43 How many moles of water are required to react with 0.0446 mol chlorine dioxide in the reaction $2\ ClO_2 + H_2O \longrightarrow HClO_3 + HClO_2$?

Analysis:

Target: The number of moles of water. (? mol water) =

Knowns: The number of moles of chlorine dioxide and the balanced chemical equation. (The formulas of water, H_2O, and chlorine dioxide, ClO_2, are known from their names.)

Relationship: The balanced chemical equation gives mole ratios of all species.

Plan: Use the balanced chemical equation to generate a conversion factor and use it to convert moles of ClO_2 to moles of H_2O.

Work: The balanced equation gives the equivalence statement: 2 mol ClO_2 ↔ 1 mol H_2O.

The conversion factor generated from this equivalency is: $\frac{1\ mol\ H_2O}{2\ mol\ ClO_2}$.

Applying the factor-label method gives:

$$? \text{ mol } Cl_2 = 0.0446 \text{ mol } ClO_2 \times \frac{1 \text{ mol } H_2O}{2 \text{ mol } ClO_2} = \mathbf{0.0223\ mol\ H_2O}$$

3.45 What mass of H_2 is produced when 5.28 g Zn reacts according to $Zn + 2\ HCl \longrightarrow ZnCl_2 + H_2$?

Analysis:

Target: The mass of H_2. (? g H_2) =

Knowns: The mass of Zn and the balanced chemical equation. The molar masses of the substances are implicit knowns.

Relationships: The balanced chemical equation gives mole ratios of all species. The molar masses provide the equivalence statements g ↔ mol.

Plan: Use the standard stoichiometric procedure (SSP) discussed in the text, with the appropriate conversion factors. The summary of the SSP is:

Step 3. Find the mass of H_2 from its number of moles, using its molar mass. 2.0158 g H_2 ↔ 1 mol H_2.

Step 2. Find the number of moles of H_2 from the number of moles of Zn, using the equivalence statement that 1 mol Zn ↔ 1 mol H_2.

Step 1. Find the number of moles of Zn from its mass, using its molar mass. 65.38 g Zn ↔ 1 mol Zn.

Summary of the SSP: g Zn $\longrightarrow$ mol Zn $\longrightarrow$ mol H_2 $\longrightarrow$ g H_2

Work: The SSP can be treated as a single 3 step conversion process.

$$? \text{ g } H_2 = 5.28 \text{ g Zn} \times \frac{1 \text{ mol Zn}}{65.38 \text{ g Zn}} \times \frac{1 \text{ mol } H_2}{1 \text{ mol Zn}} \times \frac{2.0158 \text{ g } H_2}{1 \text{ mol } H_2} = \mathbf{0.163 \text{ g } H_2}$$

3.47 What mass of ClO_2 is required to produce 8.36 kg $HClO_3$, according to
$2\ ClO_2 + H_2O \longrightarrow HClO_3 + HClO_2$?

Analysis:

Target: The mass of ClO_2. (? kg ClO_2) =

Knowns: The mass of $HClO_3$ and the balanced chemical equation. The molar masses of the substances are implicit knowns.

Relationships: The balanced chemical equation gives mole ratios of all species. The molar masses provide the equivalence statements g $\leftrightarrow$ mol.

Plan: Use the standard stoichiometric procedure (SSP), with the appropriate conversion factors. Because the product mass is given in kg, the reactant mass is also expressed in kg.

kg $HClO_3$ $\longrightarrow$ kmol $HClO_3$ $\longrightarrow$ kmol ClO_2 $\longrightarrow$ kg ClO_2

Work: The SSP can be treated as a single 3 step conversion process.
The molar masses required are:

$HClO_3$ = 1(1.0079 H) + 1(35.453 Cl) + 3(15.9994 O) = 84.459 g mol^{-1}

ClO_2 = 1(35.453 Cl) + 2(15.9994 O) = 67.452 g mol^{-1}

The equivalence statement for moles is 2 mol ClO_2 $\leftrightarrow$ 1 mol $HClO_3$.

$$? \text{ kg } ClO_2 = 8.36 \text{ kg } HClO_3 \times \frac{1 \text{ kmol } HClO_3}{84.459 \text{ kg } HClO_3} \times \frac{2 \text{ kmol } ClO_2}{1 \text{ kmol } HClO_3} \times \frac{67.452 \text{ kg } ClO_2}{1 \text{ kmol } ClO_2} =$$

13.4 kg ClO_2

3.49 What mass of anhydrous copper(II) sulfate remains when 150 g copper(II) sulfate pentahydrate ($CuSO_4 \cdot 5H_2O$) is heated sufficiently to drive off all its water of hydration?

Analysis:

Target: The mass of copper(II) sulfate. (? g copper(II) sulfate) =

Knowns: The mass of copper(II) sulfate pentahydrate. The formulas of the compounds are implicitly known from the names. From the formulas of reactants and products, the molar masses of reactants and products can be determined and the balanced chemical equation can be written.

Relationships: The balanced chemical equation gives mole ratios of all species. The molar masses provide the equivalence statements g $\leftrightarrow$ mol.

Plan: If we knew the balanced chemical equation, we could use the SSP. From the names we can determine the formulas, and from the formulas we can write the balanced equation.

Step 3. Use the SSP to determine the mass of copper(II) sulfate.

g copper(II) sulfate pentahydrate $\longrightarrow$ mol copper(II) sulfate pentahydrate $\longrightarrow$

mol copper(II) sulfate $\longrightarrow$ g copper(II) sulfate

Step 2. Use the formulas to write the balanced chemical equation.

Step 1. Use the names to determine the formulas of the compounds.

Work:

Step 1. A copper(II) ion is Cu^{2+} and a sulfate ion is SO_4^{2-}, so copper(II) sulfate is $CuSO_4$. Water is H_2O.

Step 2. The skeleton equation for the decomposition is:

$CuSO_4 \cdot 5H_2O \longrightarrow CuSO_4 + H_2O$

It is apparent that a coefficient of 5 in front of the H_2O balances the H and the O.
$CuSO_4 \cdot 5H_2O \longrightarrow CuSO_4 + 5\ H_2O$ is the balanced equation.

Step 3. The SSP can be treated as a single 3 step conversion process.
The molar masses required are:
$CuSO_4 = 1(63.546\ Cu) + 1(32.06\ S) + 4(15.9994\ O) = 159.60\ g\ mol^{-1}$
$H_2O = 2(1.0079\ H) + 1(15.9994\ O) = 18.0152\ g\ mol^{-1}$
$CuSO_4 \cdot 5H_2O = 1(159.60) + 5(18.0152) = 249.68\ g\ mol^{-1}$
The equivalence statement for moles is 1 mol $CuSO_4 \cdot 5H_2O \leftrightarrow$ 1 mol $CuSO_4$.

$$? \text{ g } CuSO_4 = 150 \text{ g } CuSO_4 \cdot 5H_2O \times \frac{1 \text{ mol } CuSO_4 \cdot 5H_2O}{249.68 \text{ g } CuSO_4 \cdot 5H_2O} \times \frac{1 \text{ mol } CuSO_4}{1 \text{ mol } CuSO_4 \cdot 5H_2O}$$

$$\times \frac{159.60 \text{ g } CuSO_4}{1 \text{ mol } CuSO_4} = \mathbf{95.9\ g\ CuSO_4}$$

3.51 If 0.328 g O_2 is produced in the reaction of Cl_2 with water, $2\ H_2O + 2\ Cl_2 \longrightarrow 4\ HCl + O_2$, how much HCl is produced at the same time?

Analysis:
Target: The mass of HCl. (? g HCl) =
Knowns: The mass of O_2 and the balanced chemical equation. The molar masses of the substances are implicit knowns.
Relationships: The balanced chemical equation gives mole ratios of all species. The molar masses provide the equivalence statements g $\leftrightarrow$ mol.
Plan: Use the standard stoichiometric procedure (SSP), with the appropriate conversion factors.
g $O_2 \longrightarrow$ mol $O_2 \longrightarrow$ mol HCl $\longrightarrow$ g HCl
Work: The SSP can be treated as a single 3 step conversion process.
The molar masses required are:
$HCl = 1(1.0079\ H) + 1(35.453\ Cl) = 36.461\ g\ mol^{-1}$
$O_2 = 2(15.9994\ O) = 31.9988\ g\ mol^{-1}$
The equivalence statement for moles is 1 mol $O_2 \leftrightarrow$ 4 mol HCl.

$$? \text{ g HCl} = 0.328 \text{ g } O_2 \times \frac{1 \text{ mol } O_2}{31.9988 \text{ g } O_2} \times \frac{4 \text{ mol HCl}}{1 \text{ mol } O_2} \times \frac{36.461 \text{ g HCl}}{1 \text{ mol HCl}} = \mathbf{1.49\ g\ HCl}$$

3.53 The density of N_2 at room temperature and atmospheric pressure is 1.15 g L^{-1}. How much lithium is required to react with 10.0 L N_2 according to $6\ Li + N_2 \longrightarrow 2\ Li_3N$?

Analysis:
Target: The mass of lithium, Li. (? g Li) =
Knowns: The volume of N_2, the density of N_2, and the balanced chemical equation. The molar masses of the substances are implicit knowns.
Relationships: The balanced chemical equation gives mole ratios of all species. The molar masses provide the equivalence statements g $\leftrightarrow$ mol. Density $= \frac{\text{mass}}{\text{volume}}$
Plan: If we knew the mass of N_2, we could use the SSP.
Step 2. Use the SSP with the appropriate conversion factors.
g $N_2 \longrightarrow$ mol $N_2 \longrightarrow$ mol Li $\longrightarrow$ g Li
Step 1. Find the mass of N_2 from its volume and its density.
Work:
Step 1. Use the factor-label method to find the mass of N_2.

$$? \text{ g } N_2 = 10.0 \text{ L } N_2 \times \frac{1.15 \text{ g } N_2}{1 \text{ L } N_2} = 11.5 \text{ g } N_2$$

Step 2. The SSP can be treated as a single 3 step conversion process:
The molar masses required are:
Li = 6.941 g mol^{-1}
N_2 = 2(14.0067 N) = 28.0134 g mol^{-1}
The equivalence statement for moles is 1 mol N_2 ↔ 6 mol Li.

$$? \text{ g Li} = 11.5 \text{ g } N_2 \times \frac{1 \text{ mol } N_2}{28.0134 \text{ g } N_2} \times \frac{6 \text{ mol Li}}{1 \text{ mol } N_2} \times \frac{6.941 \text{ g Li}}{1 \text{ mol Li}} = \mathbf{17.1 \text{ g Li}}$$

3.55 When 50 g each of Zn and S react according to Zn + S ⟶ ZnS, how much ZnS is formed?

Analysis:

Target: The mass of ZnS. (? g ZnS) =

Knowns: The mass of Zn, the mass of S, and the balanced chemical equation. The molar masses of the substances are implicit knowns.

Relationships: The balanced chemical equation gives mole ratios of all species. The molar masses provide the equivalence statements g ↔ mol.

Plan: We recognize that this is a limiting reactant problem because starting amounts of both reactants are given. We can use the SSP to determine the amount of product that could be made if **each** reactant could be totally consumed. The smaller amount of product can actually be made and the reactant that produces it is the **limiting reactant**. The other reactant will be present in excess. Some of it will be left over at the end of the process.
Step 2. Compare the amount of ZnS made if each reactant could be totally consumed.
Step 1. Use the SSP with the appropriate conversion factors to predict the amount of product that could be made from each reactant.

g Zn ⟶ mol Zn ⟶ mol ZnS ⟶ g ZnS

g S ⟶ mol S ⟶ mol ZnS ⟶ g ZnS

Work:

Step 1. The SSP for each reactant can be treated as a single 3 step conversion process:
The molar masses required are:
Zn = 65.38 mol^{-1}
S = 32.06 g mol^{-1}
ZnS = 1(65.38 Zn) + 1(32.06 S) = 97.44 g mol^{-1}
The equivalence statements for moles are:
1 mol Zn ↔ 1 mol ZnS and 1 mol S ↔ 1 mol ZnS.

$$? \text{ g ZnS} = 50 \text{ g Zn} \times \frac{1 \text{ mol Zn}}{65.36 \text{ g Zn}} \times \frac{1 \text{ mol ZnS}}{1 \text{ mol Zn}} \times \frac{97.44 \text{ g ZnS}}{1 \text{ mol ZnS}} = 75 \text{ g ZnS}$$

$$? \text{ g ZnS} = 50 \text{ g S} \times \frac{1 \text{ mol S}}{32.06 \text{ g S}} \times \frac{1 \text{ mol ZnS}}{1 \text{ mol S}} \times \frac{97.44 \text{ g ZnS}}{1 \text{ mol ZnS}} = 1.5 \times 10^2 \text{ g ZnS}$$

Step 2. Compare the amounts of product.
75 g is less than 1.5×10^2 g. Zn is the limiting reactant and **75 g ZnS** can be made.

3.57 How much water is produced when 55 g C_3H_8 reacts with 155 g O_2? (The other product is carbon dioxide.)

Analysis:

Target: The mass of water. (? g water) =

Knowns: The mass of C_3H_8 and the mass of O_2. The formulas of the products are implicitly known from their names. From the formulas of reactants and products, the molar masses of reactants and products can be determined and the balanced chemical equation can be written.

Relationships: The balanced chemical equation gives mole ratios of all species. The molar masses provide the equivalence statements g ↔ mol.

Plan: We recognize that this is a limiting reactant problem because starting amounts of both reactants are given. Consult the discussion in the plan for Problem 3.55. From the names, we can determine the formulas. From the formulas, we can write the balanced equation.

Step 4. Compare the amount of water made if each reactant could be totally consumed.

Step 3. Use the SSP with the appropriate conversion factors to predict the amount of product that could be made from each reactant.

g C_3H_8 → mol C_3H_8 → mol water → g water

g O_2 → mol O_2 → mol water → g water

Step 2. Use the formulas to write the balanced chemical equation.

Step 1. Use the names to determine the formulas of the compounds.

Work:

Step 1. Water is H_2O and carbon dioxide is CO_2.

Step 2. The skeleton equation for the combustion is:

$C_3H_8 + O_2 \rightarrow CO_2 + H_2O$

C_3H_8 is the more complicated molecule. C can be balanced by using a coefficient of 3 for CO_2.

$C_3H_8 + O_2 \rightarrow 3\ CO_2 + H_2O$ 1(3) C left = 3(1) C right

Now the H can be balanced by using a coefficient of 4 for H_2O.

$C_3H_8 + O_2 \rightarrow 3\ CO_2 + 4\ H_2O$ 1(8) H left = 4(2) H left

Finally, the O_2 can be balanced by using a coefficient of 5 for O_2

$C_3H_8 + 5\ O_2 \rightarrow 3\ CO_2 + 4\ H_2O$ 5(2) O left = 3(2) + 4(1) O right

Step 3. The SSP for each reactant can be treated as a single 3 step conversion process. The molar masses required are:

C_3H_8 = 3(12.011 C) + 8(1.0079 H) = 44.096 g mol^{-1}

O_2 = 2(15.9994 O) = 31.9988 g mol^{-1}

H_2O = 2(1.0079 H) + 1(15.9994 O) = 18.0152 g mol^{-1}

The equivalence statements for moles are:

1 mol C_3H_8 ↔ 4 mol H_2O and 5 mol O_2 ↔ 4 mol H_2O.

$$? \text{ g } H_2O = 55 \text{ g } C_3H_8 \times \frac{1 \text{ mol } C_3H_8}{44.096 \text{ g } C_3H_8} \times \frac{4 \text{ mol } H_2O}{1 \text{ mol } C_3H_8} \times \frac{18.0152 \text{ g } H_2O}{1 \text{ mol } H_2O} = 90 \text{ g } H_2O$$

$$? \text{ g } H_2O = 155 \text{ g } O_2 \times \frac{1 \text{ mol } O_2}{31.9988 \text{ g } O_2} \times \frac{4 \text{ mol } H_2O}{5 \text{ mol } O_2} \times \frac{18.0152 \text{ g } H_2O}{1 \text{ mol } H_2O} = 69.8 \text{ g } H_2O$$

Step 4. Compare amounts of product.

69.8 g is less than 90 g. The O_2 is the limiting reactant and **69.8 g H_2O** can be made.

3.59 Dinitrogen pentoxide reacts with water to produce nitric acid.

a. How much HNO_3 is formed when 100.0 g N_2O_5 reacts with 20.0 g H_2O?

b. Which is the limiting reactant?

Analysis:

Target: The mass of HNO_3. (? g HNO_3) =

Knowns: The mass of N_2O_5 and the mass of H_2O. From the formulas of reactants and products, the molar masses of reactants and products can be determined and the balanced chemical equation can be written.

Relationships: The balanced chemical equation gives mole ratios of all species. The molar masses provide the equivalence statements g ↔ mol.

Plan: We recognize that this is a limiting reactant problem because starting amounts of both reactants are given. Consult the discussion in the plan for Problem 3.55.

Step 3. Compare the amount of HNO_3 made if each reactant could be totally consumed.

Step 2. Use the SSP with the appropriate conversion factors to predict the amount of product that could be made from each reactant.

$$g\ N_2O_5 \longrightarrow mol\ N_2O_5 \longrightarrow mol\ HNO_3 \longrightarrow g\ HNO_3$$

$$g\ H_2O \longrightarrow mol\ H_2O \longrightarrow mol\ HNO_3 \longrightarrow g\ HNO_3$$

Step 1. Use the formulas to write the balanced chemical equation.

Work:

Step 1. The skeleton equation for the reaction is:

$N_2O_5 + H_2O \longrightarrow HNO_3$

Both N_2O_5 and H_2O are equally complicated species, so we could begin with either. N is balanced by using a coefficient of 2 for HNO_3. That also balances H and O.

$N_2O_5 + H_2O \longrightarrow 2\ HNO_3$ 1(2) N left = 2(1) N right

1(5) + 1(1) O left = 2(3) O right

1(2) H left = 2(1) H right

Step 2. The SSP for each reactant can be treated as a single 3 step conversion process. The molar masses required are:

N_2O_5 = 2(14.0067 N) + 5(15.9994 O) = 108.0104 g mol^{-1}

H_2O = 2(1.0079 H) + 1(15.9994 O) = 18.0152 g mol^{-1}

HNO_3 = 1(1.0079 H) + 1(14.0067 N) + 3(15.9994 O) = 63.0128 g mol^{-1}

The equivalence statements for moles are:

1 mol N_2O_5 ↔ 2 mol HNO_3 and 1 mol H_2O ↔ 2 mol HNO_3.

$$?\ g\ HNO_3 = 100.0\ g\ N_2O_5 \times \frac{1\ mol\ N_2O_5}{108.0104\ g\ N_2O_5} \times \frac{2\ mol\ HNO_3}{1\ mol\ N_2O_5} \times \frac{63.0128\ g\ HNO_3}{1\ mol\ HNO_3} =$$

116.7 g HNO_3

$$?\ g\ HNO_3 = 20.0\ g\ H_2O \times \frac{1\ mol\ H_2O}{18.0152\ g\ H} \times \frac{2\ mol\ HNO_3}{1\ mol\ H_2O} \times \frac{63.0128\ g\ HNO_3}{1\ mol\ HNO_3} =$$

140 g HNO_3

Step 3. Compare the amounts of product.

a. 116.7 g is less than 140 g. Only **116.7 g HNO_3** can be made.

b. **N_2O_5** is the limiting reactant because it limits the amount of product that can be made.

3.61 Answer the following questions.

a. How much Na_2O_2 is formed when 2.63 g Na reacts with 4.00 g O_2?

b. How much of the reactant in excess remains unreacted?

Analysis: For part a.

Target: The mass of Na_2O_2. (? g Na_2O_2) =

Knowns: The mass of Na and the mass of O_2. From the formulas of reactants and products, the molar masses of reactants and products can be determined and the balanced chemical equation can be written.

Relationships: The balanced chemical equation gives mole ratios of all species. The molar masses provide the equivalence statements g ↔ mol.

Plan: We recognize that this is a limiting reactant problem because starting amounts of both reactants are given. Consult the discussion in the plan for Problem 3.55.

Step 3. Compare the amount of Na_2O_2 made if each reactant could be totally consumed.

Step 2. Use the SSP with the appropriate conversion factors to predict the amount of product that could be made from each reactant.

g Na → mol Na → mol Na_2O_2 → g Na_2O_2

g O_2 → mol O_2 → mol Na_2O_2 → g Na_2O_2

Step 1. Use the formulas to write the balanced chemical equation.

Work:

Step 1. The skeleton equation for the reaction is:

$Na + O_2 \rightarrow Na_2O_2$

O is already balanced with 2 atoms on each side. The Na can be balanced by using a coefficient of 2 for Na.

$2\,Na + O_2 \rightarrow Na_2O_2$ — 2(1) Na left = 1(2) Na right; 1(2) O left = 2(1) O right

Step 2. The SSP for each reactant can be treated as a single 3 step conversion process. The molar masses required are:

Na = 22.98977 g mol^{-1}

O_2 = 2(15.9994 O) = 31.9988 g mol^{-1}

Na_2O_2 = 2(22.98977 g mol^{-1}) + 2(15.9994 O) = 77.9783 g mol^{-1}

The equivalence statements for moles are:

2 mol Na ↔ 1 mol Na_2O_2 and 1 mol O_2 ↔ 1 mol Na_2O_2

$$? \text{ g } Na_2O_2 = 2.63 \text{ g Na} \times \frac{1\text{mol Na}}{22.98977 \text{ g Na}} \times \frac{1 \text{ mol } Na_2O_2}{2 \text{ mol Na}} \times \frac{77.9783 \text{ g } Na_2O_2}{1 \text{ mol } Na_2O_2} =$$

4.46 g Na_2O_2

$$? \text{ g } Na_2O_2 = 4.00 \text{ g } O_2 \times \frac{1 \text{ mol } O_2}{31.9988 \text{ g } O_2} \times \frac{1 \text{ mol } Na_2O_2}{1 \text{ mol } O_2} \times \frac{77.9783 \text{ g } Na_2O_2}{1 \text{ mol } Na_2O_2} =$$

9.75 g Na_2O_2

Step 3. Compare the two amounts of product.
4.46 g is less than 9.75 g. Only **4.46 g Na_2O_2** can be made.

Analysis: For part b.

Target: The mass of excess reactant. (? g excess) =

Knowns: All the information from part a, including that Na is the limiting reactant. Therefore, O_2 is the excess reactant.

Relationships: All of the information from part a. The excess reactant can react until all the limiting reactant has been consumed.

Plan: If we can determine the amount of the excess reactant, O_2, that reacts with the limiting reactant, Na, we can subtract that amount from the initial amount of O_2 to determine the amount of excess reactant remaining.

Step 2. Subtract the mass of O_2 used from the initial mass of O_2.

Step 1. Use the SSP with the appropriate conversion factors to determine the amount of O_2 that reacted.

The equivalence statement for moles is: 1 mol O_2 ↔ 2 mol Na

Work:

Step 1. Determine the number of grams of O_2 that reacts.

$$? \text{ g } O_2 = 2.63 \text{ g Na} \times \frac{1 \text{ mol Na}}{22.98977 \text{ g Na}} \times \frac{1 \text{ mol } O_2}{2 \text{ mol Na}} \times \frac{31.9988 \text{ g } O_2}{1 \text{ mol } O_2} = 1.83 \text{ g } O_2 \text{ used}$$

Step 2. Subtract the mass of O_2 used from the original mass of O_2.

? g O_2 = 4.00 g O_2 initial - 1.83 g O_2 used = **2.17 g O_2 remains unreacted**

3.63 Answer the following questions.

a. If 195.8 g KNO_3, 60.1 g S and 41.9 g C react according to

$2\ KNO_3 + S + 3\ C \longrightarrow K_2S + N_2 + 3\ CO_2$, how much K_2S is formed?

b. Which is the limiting reactant?

Analysis:

Target: The mass of K_2S. (? g K_2S) =

Knowns: The mass of KNO_3, the mass of S, the mass of C, and the balanced chemical equation. The molar masses of the substances are implicit knowns.

Relationships: The balanced chemical equation gives mole ratios of all species. The molar masses provide the equivalence statements g ↔ mol.

Plan: We recognize that this is a limiting reactant problem because starting amounts of all three reactants are given. Consult the discussion in the plan for problem 3.55.

Step 2. Compare the amounts of K_2S made if each reactant could be totally consumed.

Step 1. Use the SSP with the appropriate conversion factors to predict the amount of product that could be made from each reactant.

g KNO_3 ⟶ mol KNO_3 ⟶ mol K_2S ⟶ g K_2S

g S ⟶ mol S ⟶ mol K_2S ⟶ g K_2S

g C ⟶ mol C ⟶ mol K_2S ⟶ g K_2S

Work:

Step 1. The SSP for each reactant can be treated as a single 3 step conversion process. The molar masses required are:

KNO_3 = 1(39.0983 K) + 1(14.0067 N) + 3(15.9994 O) = 101.1032 g mol^{-1}

S = 32.06 g mol^{-1}

C = 12.011 g mol^{-1}

K_2S = 2(39.0983 K) + 1(32.06 S) = 110.26 g mol^{-1}

The equivalence statements for moles are:

2 mol KNO_3 ↔ 1 mol K_2S, 1 mol S ↔ 1 mol K_2S and 3 mol C ↔ 1 mol K_2S.

$$?\ g\ K_2S = 195.8\ g\ KNO_3 \times \frac{1\ mol\ KNO_3}{101.1032\ g\ KNO_3} \times \frac{1\ mol\ K_2S}{2\ mol\ KNO_3} \times \frac{110.26\ g\ K_2S}{1\ mol\ K_2S} =$$

106.8 g K_2S

$$?\ g\ K_2S = 60.1\ g\ S \times \frac{1\ mol\ S}{32.06\ g\ S} \times \frac{1\ mol\ K_2S}{1\ mol\ S} \times \frac{110.26\ g\ K_2S}{1\ mol\ K_2S} = 207\ g\ K_2S$$

$$?\ g\ K_2S = 41.9\ g\ C \times \frac{1\ mol\ C}{12.011\ g\ C} \times \frac{1\ mol\ K_2S}{3\ mol\ C} \times \frac{110.26\ g\ K_2S}{1\ mol\ K_2S} = 128\ g\ K_2S$$

Step 2. Compare the amounts of product.

a. 106.8 g is less than 207 g or 128 g. Only **106.8 g K_2S** can be made.

b. **KNO_3** is the limiting reactant because it limits the amount of product that can be made.

3.65 If 15 g sodium carbonate is obtained from the thermal decomposition of 50 g sodium hydrogen carbonate, $2\ NaHCO_3 \longrightarrow Na_2CO_3 + H_2O + CO_2$, what is the percent yield?

Analysis:

Target: The percent yield of Na_2CO_3. (? % Na_2CO_3) =

Knowns: The mass of $NaHCO_3$ used, the actual yield of Na_2CO_3, and the balanced chemical equation. The molar masses of the substances are implicit knowns.

Relationships: The balanced chemical equation gives mole ratios of all species. The molar masses provide the equivalence statements g ↔ mol. The percent yield is defined as:

$$\frac{\text{actual yield}}{\text{theoretical yield}} \times 100.$$

Plan: If we knew the theoretical yield, we could use the actual yield and the definition of percent yield to answer the question. We can find the theoretical yield using the SSP.
Step 2. Use the definition of percent yield directly.
Step 1. Use the SSP with the appropriate conversion factors to find the theoretical yield.

g $NaHCO_3$ ⟶ mol $NaHCO_3$ ⟶ mol Na_2CO_3 ⟶ g Na_2CO_3

Work:
Step 1. The SSP can be treated as a single 3 step conversion process to find the theoretical yield. The molar masses required are:
$NaHCO_3$ = 1(22.98977 Na) + 1(1.0079 H) + 1(12.011 C) + 3(15.9994 O) = 84.007 g mol^{-1}
Na_2CO_3 = 2(22.98977 Na) + 1(12.011 C) + 3(15.9994 O) = 105.989 g mol^{-1}
The equivalence statement for moles is 2 mol $NaHCO_3$ ↔ 1 mol Na_2CO_3.

$$? \text{ g } Na_2CO_3 = 50 \text{ g } NaHCO_3 \times \frac{1 \text{ mol } NaHCO_3}{84.007 \text{ g } NaHCO_3} \times \frac{1 \text{ mol } Na_2CO_3}{2 \text{ mol } NaHCO_3} \times$$

$$\frac{105.989 \text{ g } Na_2CO_3}{1 \text{ mol } Na_2CO_3} = 32 \text{ g } Na_2CO_3 \text{ can theoretically be made}$$

Step 2. Use the definition of percent yield and the unrounded theoretical yield from Step 1 to determine the percent yield.

$$? \text{ percent yield} = \frac{15 \text{ g } Na_2CO_3 \text{ actual}}{32 \text{ g } Na_2CO_3 \text{ theoretical}} \times 100 = 48\%$$

3.67 Disulfur dichloride and carbon tetrachloride are the products of the reaction of carbon disulfide and chlorine. What is the percent yield if 75 kg S_2Cl_2 is obtained from the reaction of 145 kg Cl_2?

Analysis:
Target: The percent yield of S_2Cl_2. (? % S_2Cl_2) =
Knowns: The mass of Cl_2 used, the mass of S_2Cl_2 obtained, and the names of the reactants and products. The formulas of the compounds are implicitly known from the names. From the formulas of reactants and products, the molar masses of reactants and products can be determined and the balanced chemical equation can be written.
Relationships: The balanced chemical equation gives mole ratios of all species. The molar masses provide the equivalence statements g ↔ mol. The percent yield is defined as: $\frac{\text{actual yield}}{\text{theoretical yield}} \times 100$.

Plan: If we knew the theoretical yield of S_2Cl_2, we could use the actual yield and the definition of percent yield to answer the question. If we knew the balanced chemical equation, we could use the SSP to determine the theoretical yield of S_2Cl_2. From the names we can determine the formulas, and from the formulas we can write the balanced equation.
Step 4. Use the definition of percent yield directly.
Step 3. Use the SSP with the appropriate conversion factors to find the theoretical yield (in kg because the actual yield is given in kg).

kg Cl_2 ⟶ kmol Cl_2 ⟶ kmol S_2Cl_2 ⟶ kg S_2Cl_2

Step 2. Use the formulas to write the balanced chemical equation.
Step 1. Use the names to determine the formulas of the compounds.

Work:
Step 1. Use the names to determine the formulas that were not provided.
Carbon tetrachloride is CCl_4 and carbon disulfide is CS_2.

Step 2. Balance the equation.
The skeleton equation for the reaction is:

$Cl_2 + CS_2 \longrightarrow CCl_4 + S_2Cl_2$

C is balanced with 1 atom on each side and S is balanced with 2 atoms on each side. Cl can be balanced by a coefficient of 3 in front of the Cl_2, completing the process.

$3\ Cl_2 + CS_2 \longrightarrow CCl_4 + S_2Cl_2$ 3(2) Cl left = 1(4) + 1(2) Cl right

Step 3. The SSP can be treated as a single 3 step conversion process to find the theoretical yield. The molar masses required are:

Cl_2 = 2(35.453) = 70.906 g mol^{-1}

S_2Cl_2 = 2(32.06 S) + 2(35.453 Cl) = 135.03 g mol^{-1}

The equivalence statement for moles is 3 mol Cl_2 ↔ 1 mol S_2Cl_2.

$$? \text{ kg } S_2Cl_2 = 145 \text{ kg } Cl_2 \times \frac{1 \text{ kmol } Cl_2}{70.906 \text{ kg } Cl_2} \times \frac{1 \text{ kmol } S_2Cl_2}{3 \text{ kmol } Cl_2} \times \frac{135.03 \text{ kg } S_2Cl_2}{1 \text{ kmol } S_2Cl_2} =$$

92.0 kg S_2Cl_2

Step 4. Use the definition of percent yield and the unrounded theoretical yield from Step 1 to determine the percent yield.

$$? \%\ S_2Cl_2 = \frac{75 \text{ kg } S_2Cl_2 \text{ actual}}{92.0 \text{ kg } S_2Cl_2 \text{ theoretical}} \times 100 = \mathbf{82\%\ S_2Cl_2}$$

3.69 If, from a 50.0 g sample of an iron ore containing Fe_3O_4, 2.09 g of Fe is obtained by the reaction $Fe_3O_4 + 2\ C \longrightarrow 3\ Fe + 2\ CO_2$, what is the percent of Fe_3O_4 in the ore?

Analysis:

Target: The percent Fe_3O_4 in an ore. (? % Fe_3O_4) =

Knowns: The mass of iron ore used, the mass of Fe obtained and the balanced chemical equation. The molar masses are implicitly known from the formulas of the substances.

Relationships: The balanced chemical equation gives mole ratios of all species. The molar masses provide the equivalence statements g ↔ mol. The percent Fe_3O_4 in the ore is defined as: $\frac{\text{reactant mass}}{\text{mixture mass}} \times 100$ or $\frac{\text{mass } Fe_3O_4}{\text{mass sample}} \times 100$.

Plan: If we knew the mass of Fe_3O_4 in the ore, we could use the mass of ore and the definition of percent composition to answer the question. We can use the SSP to determine the mass of Fe_3O_4 in the sample.

Step 2. Use the definition of percent composition directly.

Step 1. Use the SSP with the appropriate conversion factors to find the mass of Fe_3O_4 in the ore.

g Fe ⟶ mol Fe ⟶ mol Fe_3O_4 ⟶ g Fe_3O_4

Work:

Step 1. The SSP can be treated as a single 3 step conversion process to find the mass of Fe_3O_4. The molar masses required are:

Fe = 55.847 g mol^{-1}

Fe_3O_4 = 3(55.847 Fe) + 4(15.9994 O) = 231.539 g mol^{-1}

The equivalence statement for moles is 3 mol Fe ↔ 1 mol Fe_3O_4.

$$? \text{ g } Fe_3O_4 = 2.09 \text{ g Fe} \times \frac{1 \text{ mol Fe}}{55.847 \text{ g Fe}} \times \frac{1 \text{ mol } Fe_3O_4}{3 \text{ mol Fe}} \times \frac{231.539 \text{ g } Fe_3O_4}{1 \text{ mol } Fe_3O_4} =$$

2.89 g Fe_3O_4

Step 2. Use the definition of percent composition and the unrounded mass of Fe_3O_4 from Step 1 to determine the percent Fe_3O_4 in the ore.

$$? \%\ Fe_3O_4 = \frac{2.89 \text{ g } Fe_3O_4}{50.0 \text{ g ore}} \times 100 = \mathbf{5.78\%\ Fe_3O_4}$$

3.71 Aqueous silver nitrate forms solid silver chloride when mixed with aqueous potassium chloride (or any other dissolved chloride salt), according to
$AgNO_3(aq) + KCl(aq) \longrightarrow AgCl(s) + KNO_3(aq)$. Suppose 0.546 g of silver nitrate, known to contain some sodium nitrate as an impurity, is dissolved in water. It is then treated with excess KCl(aq), and 0.444 g AgCl is obtained. What is the percent purity of the silver nitrate?

Analysis:
Target: The percent $AgNO_3$ in an impure sample. (? % $AgNO_3$) =
Knowns: The mass of $AgNO_3$ sample used, the mass of AgCl obtained and the balanced chemical equation. The molar masses are implicitly known from the formulas of the substances.
Relationships: The balanced chemical equation gives mole ratios of all species. The molar masses provide the equivalence statements g ↔ mol. The percent $AgNO_3$ in the sample is defined as: $\frac{\text{mass } AgNO_3}{\text{mass sample}} \times 100$.
Plan: If we knew the mass of $AgNO_3$ in the impure sample, we could use the mass of sample and the definition of percent composition to answer the question. We can use the SSP to determine the mass of $AgNO_3$ in the sample.
Step 2. Use the definition of percent composition directly.
Step 1. Use the SSP with the appropriate conversion factors to find the mass of $AgNO_3$ in the impure sample.

$$\text{g AgCl} \longrightarrow \text{mol AgCl} \longrightarrow \text{mol } AgNO_3 \longrightarrow \text{g } AgNO_3$$

Work:
Step 1. The SSP can be treated as a single 3 step conversion process to find the mass of $AgNO_3$. The molar masses required are:
AgCl = 1(107.8682 Ag) + 1(35.453 Cl) = 143.321 g mol^{-1}
$AgNO_3$ = 1(107.8682 Ag) + 1(14.0067 N) + 3(15.9994 O) = 169.8731 g mol^{-1}
The equivalence statement for moles is 1 mol $AgNO_3$ ↔ 1 mol AgCl

$$? \text{ g } AgNO_3 = 0.444 \text{ g AgCl} \times \frac{1 \text{ mol AgCl}}{143.321 \text{ g AgCl}} \times \frac{1 \text{ mol } AgNO_3}{1 \text{ mol AgCl}} \times \frac{169.8731 \text{ g } AgNO_3}{1 \text{ mol } AgNO_3} =$$

0.526 g $AgNO_3$

Step 2. Use the definition of percent composition and the unrounded mass of $AgNO_3$ from Step 1 to determine the percent $AgNO_3$ in the sample.

$$? \% \; AgNO_3 = \frac{0.526 \text{ g } AgNO_3}{0.546 \text{ g sample}} \times 100 = \mathbf{96.4 \; \% \; AgNO_3}$$

3.73 Suppose your company wishes to prepare 150 kg sodium peroxide by the direct reaction of sodium with oxygen. You know from past experience that the reaction proceeds in 92% yield. How much sodium should you use?

Analysis:
Target: A mass of Na. (? g Na) =
Knowns: The actual yield of sodium peroxide needed, the typical percent yield of the reaction and the names of the reactants and products. The formulas of the compounds are implicitly known from the names. From the formulas of reactants and products, the molar masses of reactants and products can be determined and the balanced chemical equation can be written.
Relationships: The balanced chemical equation gives mole ratios of all species. The molar masses provide the equivalence statements g ↔ mol. The percent yield is defined as: $\frac{\text{actual yield}}{\text{theoretical yield}} \times 100$.
Plan: If we knew the theoretical yield of sodium peroxide required to give the actual yield, we could determine the amount of sodium needed to produce that theoretical yield. If we knew the balanced chemical equation, we could use the SSP to determine the mass of Na required.

From the names, we can determine the formulas, and, from the formulas, we can write the balanced equation.

Step 4. Use the SSP with the appropriate conversion factors to find the mass of sodium (in kg because the actual yield is given in kg) needed to produce the theoretical yield.

kg sodium peroxide ⟶ kmol sodium peroxide ⟶ kmol sodium ⟶ kg sodium

Step 3. Use the actual yield and the percent yield to determine the theoretical yield needed to provide the requested amount of sodium peroxide.

Step 2. Use the formulas to write the balanced chemical equation.

Step 1. Use the names to determine the formulas of the compounds.

Work:

Step 1. Use the names to determine the formulas of the compounds.

Sodium is Na. The sodium ion is Na^+ and the peroxide ion is O_2^{2-}, so sodium peroxide is Na_2O_2. Oxygen is a diatomic element, O_2.

Step 2. The skeleton equation for the reaction is:

$Na + O_2 \longrightarrow Na_2O_2$

The O is already balanced with 2 atoms on each side. The Na can be balanced by placing a 2 in front of Na.

$2\ Na + O_2 \longrightarrow Na_2O_2$ 2(1) Na left = 1(2) Na right
1(2) O left = 2(1) O right

Step 3. Use the factor-label method to determine the theoretical yield required.

$$?\ g\ Na_2O_2 = 150\ kg\ Na_2O_2\ required \times \frac{100\ kg\ Na_2O_2\ theoretical}{92\ kg\ Na_2O_2\ actual} = 1.6 \times 10^2\ kg\ Na_2O_2$$

Step 4. Use the SSP to determine the mass of Na required to provide this (unrounded) theoretical yield. The SSP can be treated as a single 3 step conversion process.

The molar masses required are:

Na = 22.98977 g mol^{-1}

Na_2O_2 = 2(22.98977 g mol^{-1}) + 2(15.9994 O) = 77.9783 g mol^{-1}

The equivalence statement for moles is: 2 mol Na ↔ 1 mol Na_2O_2

$$?\ g\ Na = 1.6 \times 10^2\ kg\ Na_2O_2 \times \frac{1\ kmol\ Na_2O_2}{77.9783\ kg\ Na_2O_2} \times \frac{2\ kmol\ Na}{1\ kmol\ Na_2O_2}$$

$$\times \frac{22.98977\ kg\ Na}{1\ kmol\ Na} = \mathbf{96\ kg\ Na}$$

3.75 How much solute is contained in each of the following solutions? Express each answer in both moles and grams.

a. 75 mL of 0.036 M $RbBrO_3$

b. 500 g of 2.00% (m/m) $K_2Cr_2O_7$

c. 1.625 L of a 0.998 M solution of CsCl

Analysis:

Target: A number of moles and a mass of each solute. (? mol solute) = and (? g solute) =

Knowns: The nature of the solute, the volume of solution and the molarity of solute in the solution or the mass of solution and the mass percent of solute in the solution. The molar mass of the solute is implicit from its formula.

Relationships: M = molarity = $\frac{mol\ solute}{1\ L\ solution}$. The molarity provides the equivalence statement mol ↔ L. 1 mL ↔ 10^{-3} L. (The relationship n = M•V, in which n is the number of moles, M is the molarity and V is the volume in liters may be employed instead of the factor-label method.) The molar mass of the solute provides the equivalence statement g ↔ mol. The definition of mass percent is: $\frac{\text{mass solute}}{\text{mass solution}} \times 100$. This provides the equivalence statement mass solute ↔ 100 g solution.

Plan: Use the formula to determine the molar mass of the solute. Use the factor-label method to determine the number of moles and the mass of each solute.

Work:

a. **Molar Mass $RbBrO_3$** = 1(85.4878 Rb) + 1(79.904 Br) + 3(15.9994 O) = 213.370 g mol^{-1}

$$? \text{ mol } RbBrO_3 = \frac{0.036 \text{ mol } RbBrO_3}{1 \text{ L}} \times 75 \text{ mL} \times \frac{10^{-3}\text{L}}{1 \text{ mL}} = \mathbf{2.7 \times 10^{-3} \text{ mol } RbBrO_3}$$

$$? \text{ g } RbBrO_3 = 2.7 \times 10^{-3} \text{ mol } RbBrO_3 \times \frac{213.370 \text{ g } RbBrO_3}{1 \text{ mol } RbBrO_3} = \mathbf{0.58 \text{ g } RbBrO_3}$$

b. **Molar Mass $K_2Cr_2O_7$** = 2(39.0983 K) + 2(51.996 Cr) + 7(15.9994 O) = 294.184 g mol^{-1}

$$? \text{ g } K_2Cr_2O_7 = 500 \text{ g solution} \times \frac{2.00 \text{ g } K_2Cr_2O_7}{100 \text{ g solution}} = \mathbf{10.0 \text{ g } K_2Cr_2O_7}$$

$$? \text{ mol } K_2Cr_2O_7 = 10.0 \text{ g } K_2Cr_2O_7 \times \frac{1 \text{ mol } K_2Cr_2O_7}{294.184 \text{ g } K_2Cr_2O_7} = \mathbf{0.0340 \text{ mol } K_2Cr_2O_7}$$

c. **Molar Mass CsCl** = 1(132.9054 Cs) + 1(35.453 Cl) = 168.358 g mol^{-1}

$$? \text{ mol CsCl} = 1.625 \text{ L} \times \frac{0.998 \text{ mol CsCl}}{1 \text{ L}} = \mathbf{1.62 \text{ mol CsCl}}$$

$$? \text{ g CsCl} = 1.62 \text{ mol CsCl} \times \frac{168.358 \text{ g CsCl}}{1 \text{ mol CsCl}} = \mathbf{273 \text{ g CsCl}}$$

3.77 Solve the following problems.
a. What volume of 0.563 M solution contains 3.25 mol solute?
b. What volume of 0.0979 M KOH contains exactly 1 mol KOH?
c. What mass of 1.93% (m/m) solution contains 10.0 g solute?
d. What mass of 5.25% (m/m) NaClO contains exactly 1 mol NaClO?

Analysis:

Target: A volume of solution or a mass of solution. (? L solution) = or (? g solution) =

Knowns: The nature of the solute, the number of moles of the solute and the molarity of solute in the solution or the mass of solution and the mass percent of solute in the solution. The molar mass of the solute is implicit from its formula.

Relationships: M = molarity = $\frac{\text{mol solute}}{1 \text{ L solution}}$. The molarity provides the equivalence statement mol ↔ L. The definition of mass percent is: $\frac{\text{mass solute}}{\text{mass solution}} \times 100$. This provides the equivalence statement mass solute ↔ 100 g solution. The molar mass provides the equivalence statement g ↔ mol.

Plan: Use the factor-label method to determine the volume or the mass of each solution.

Work:

a. $$? \text{ L solution} = 3.25 \text{ mol solute} \times \frac{1 \text{ L solution}}{0.563 \text{ mol solute}} = \mathbf{5.77 \text{ L solution}}$$

b. $$? \text{ L solution} = 1 \text{ mol KOH} \times \frac{1 \text{ L solution}}{0.0979 \text{ mol KOH}} = \mathbf{10.2 \text{ L solution}}$$

c. $$? \text{ g solution} = 10.0 \text{ g solute} \times \frac{100 \text{ g solution}}{1.93 \text{ g solute}} = \mathbf{518 \text{ g solution}}$$

d. **Molar Mass NaClO** = 1(22.98977 Na) + 1(35.453 Cl) +1(15.9994 O) = 74.442 g mol^{-1}

$$? \text{ g solution} = 1 \text{ mol NaClO} \times \frac{74.442 \text{ g NaClO}}{1 \text{ mol NaClO}} \times \frac{100 \text{ g solution}}{5.25 \text{ g NaClO}} = \mathbf{1.42 \times 10^3 \text{ g solution}}$$

3.79 Give directions for the preparation of each of the following solutions.
a. 500 g of 2.5% aqueous $KClO_4$
b. 1 L of 5% (v/v) alcohol in water
c. 2 L of 0.200 M KOH

Analysis: Part a.
Target: Mass of solute and water needed to prepare a given amount of solution.
(? g solute) = and (? g water) =
Knowns: The mass of solution needed and the mass percent solution requested.
Relationships: Mass percent solution = $\frac{\text{mass solute}}{\text{mass solution}} \times 100$. This provides the equivalence statement g solute ↔ 100 g solution. Masses are additive.
Plan: Use the factor-label method. Determine the mass of solute needed to provide the requested solution. Subtract that mass of solute from the total mass of solution to find the mass of solvent (water) needed.
Work:

$$? \text{ g } KClO_4 = 500 \text{ g solution} \times \frac{2.5 \text{ g } KClO_4}{100 \text{ g solution}} = 12 \text{ g } KClO_4$$

$$? \text{ g water} = 500 \text{ g solution} - 12 \text{ g } KClO_4 = 488 \text{ g water}$$

The solution is made by dissolving 12 g $KClO_4$ in 488 g H_2O.

Analysis: Part b.
Target: The volume of alcohol in the given solution. (? mL alcohol) =
Knowns: The total volume of solution needed and the volume percent solution needed.
1 mL ↔ 10^{-3} L.
Relationships: Volume percent solution = $\frac{\text{volume solute}}{\text{volume solution}} \times 100$. This provides the equivalence statement mL solute ↔ 100 mL solution.
Plan: Determine the volume of alcohol needed using the factor-label method.
Work:

$$? \text{ mL alcohol} = 1 \text{ L solution} \times \frac{1 \text{ mL solution}}{10^{-3} \text{ L solution}} \times \frac{5.0 \text{ mL alcohol}}{100 \text{ mL solution}} = 50 \text{ mL alcohol}$$

The solution would be made by measuring 50 mL of alcohol and adding water until the total volume is 1000 mL.

Analysis: Part c.
Target: The mass of KOH needed to make the requested solution. (? g KOH) =
Knowns: The volume of solution required, the nature of the solute and the molarity of solution required. The molar mass of KOH is an implicit known.
Relationships: M = molarity = $\frac{\text{mol solute}}{1 \text{ L solution}}$. The molarity provides the equivalence statement mol ↔ L. The molar mass provides the equivalence statement g ↔ mol.
Plan: Use the factor-label method to determine the mass of KOH needed to make the requested solution.
Work:
Molar Mass KOH = 1(39.0983 K) + 1(15.9994 O) + 1(1.0079 H) = 56.1056 g mol^{-1}

$$? \text{ g KOH} = 2 \text{ L} \times \frac{0.0200 \text{ mol KOH}}{1 \text{ L}} \times \frac{56.1056 \text{ g KOH}}{1 \text{ mol KOH}} = 2.24 \text{ g KOH}$$

The solution would be prepared by dissolving 2.24 g KOH in enough water to give a total volume of 2 L.

3.81 Give laboratory instructions for the preparation of each of the following solutions.
a. 250 mL 0.100 M NaOH from 3.04 M stock solution
b. 750 mL 0.25 M KCl from 6.0 M stock solution
c. 300 mL 0.20 M H_2SO_4 from 4.00 M stock solution

Analysis:
Target: The volume of solution of known molarity required to make a certain volume with a different molarity. (? mL) =
Known: When a diluted solution is made from a stock solution, the number of moles of solute is unchanged.
Relationship: The relationship $M_{stock}V_{stock} = M_{dilute}V_{dilute}$ applies.
Plan: Solve the dilution relationship for V_{stock} and recognize that solvent must be added to make the required volume.

Work: In each case, $V_{stock} = \frac{M_{dilute}V_{dilute}}{M_{stock}}$.

a. $M_{dilute} = 0.100$ M, $V_{dilute} = 250$ mL and $M_{stock} = 3.04$ M

$$? \text{ mL} = V_{stock} = \frac{M_{dilute}V_{dilute}}{M_{stock}} = \frac{(0.100 \text{ M})(250 \text{ mL})}{3.04 \text{ M}} = 8.22 \text{ mL}$$

The solution would be made by taking 8.22 mL of the stock solution and adding enough water to make a total volume of 250 mL.

b. $M_{dilute} = 0.25$ M, $V_{dilute} = 750$ mL and $M_{stock} = 6.0$ M

$$? \text{ mL} = V_{stock} = \frac{M_{dilute}V_{dilute}}{M_{stock}} = \frac{(0.25 \text{ M})(750 \text{ mL})}{6.0 \text{ M}} = 31 \text{ mL}$$

The solution would be made by taking 31 mL of the stock solution and adding enough water to make a total volume of 750 mL.

c. $M_{dilute} = 0.20$ M, $V_{dilute} = 300$ mL and $M_{stock} = 4.00$ M

$$? \text{ mL} = V_{stock} = \frac{M_{dilute}V_{dilute}}{M_{stock}} = \frac{(0.20 \text{ M})(300 \text{ mL})}{4.00 \text{ M}} = 15 \text{ mL}$$

The solution would be made by taking 15 mL of the stock solution and adding enough water to make a total volume of 300 mL.

3.83 What volume of 0.0496 M $HClO_4$ is required to react with 25.0 mL 0.505 M KOH according to $KOH + HClO_4 \rightarrow KClO_4 + H_2O$?

Analysis:
Target: The volume of solution. (? mL $HClO_4$) =
Knowns: The molarity and volume of one reactant solution (KOH), the molarity of the $HClO_4$ solution and the balanced chemical equation.
Relationships: The balanced chemical equation provides the mole equivalence statements. The number of moles can be found from the molarity and the volume. The molarity gives the equivalence statements mol ↔ 1 L or mmol ↔ 1 mL.
Plan: Use a modified form of the SSP, relating the number of moles to the molarity and volume instead of to the mass and molar mass. Because mL are given, work in millimoles.

KOH (M,V) → mmol KOH → mmol $HClO_4$ → $HClO_4$ (M,V)

Work: The equivalency statement is 1 mmol KOH ↔ 1 mmol $HClO_4$.

$$? \text{ mL } HClO_4 = 25.0 \text{ mL} \times \frac{0.505 \text{ mmol KOH}}{1 \text{ mL}} \times \frac{1 \text{ mmol } HClO_4}{1 \text{ mmol KOH}} \times \frac{1 \text{ mL}}{0.0496 \text{ mmol } HClO_4} =$$

255 mL $HClO_4$ solution

3.85 If 36.2 mL KOH(aq) are required to react with 25.0 mL 0.0513 M HNO_3 according to $KOH + HNO_3 \longrightarrow KNO_3 + H_2O$, what is the molarity of the KOH?

Analysis:

Target: The molarity of a solution of KOH. (? M) =

Knowns: The molarity and volume of one reactant solution (HNO_3), the volume of the KOH solution and the balanced chemical equation.

Relationships: The balanced chemical equation provides the mole equivalence statements. The number of moles can be found from the molarity and the volume. The molarity gives the equivalence statement mol ↔ 1 L or mmol ↔ 1 mL

Plan: Use a modified form of the SSP, relating the number of moles to the molarity and volume instead of to the mass and molar mass. Because mL are given, work in millimoles.

HNO_3 (M,V) ⟶ mmol HNO_3 ⟶ mmol KOH ⟶ KOH (M,V)

Work: The equivalency statement is 1 mmol KOH ↔ 1 mmol HNO_3.

$$? \text{ M KOH} = 25.0 \text{ mL} \times \frac{0.0513 \text{ mmol } HNO_3}{1 \text{ mL } HNO_3} \times \frac{1 \text{ mmol KOH}}{1 \text{ mmol} HNO_3} \times \frac{1}{36.2 \text{ mL KOH}} =$$

0.0354 M KOH

3.87 What volume of 0.0974 M HCl is required to react with 0.250 g Zn according to $Zn + 2\,HCl \longrightarrow ZnCl_2 + H_2$?

Analysis:

Target: The volume of a solution of HCl. (? mL) =

Knowns: The mass of Zn, the molarity of the HCl solution and the balanced chemical equation. The molar mass of Zn is an implicit known.

Relationships: The balanced chemical equation provides the mole equivalence statements. The molar mass gives the equivalence statement g ↔ mol. The molarity gives the equivalence statement mol ↔ 1 L or mmol ↔ 1 mL. 10^{-3} L ↔ 1 mL.

Plan: Use a modified form of the SSP, relating the number of moles to the molarity and volume for the HCl and to the mass and molar mass for Zn.

g Zn ⟶ mol Zn ⟶ mol HCl ⟶ HCl (M,V)

Work: The molar mass of Zn is 65.38 g mol^{-1}. The equivalence statement needed is: 1 mol Zn ⟶ 2 mol HCl.

$$? \text{ mL HCl} = 0.250 \text{ g Zn} \times \frac{1 \text{ mol Zn}}{65.38 \text{ g Zn}} \times \frac{2 \text{ mol HCl}}{1 \text{ mol Zn}} \times \frac{1 \text{ L}}{0.0974 \text{ mol HCl}} \times \frac{1 \text{ mL}}{10^{-3} \text{ L}} =$$

78.5 mL HCl solution

3.89 If 36.1 mL of KOH is required to react with 0.247 g oxalic acid ($H_2C_2O_4$) according to $2\,KOH + H_2C_2O_4 \longrightarrow K_2C_2O_4 + 2\,H_2O$, what is the molarity of the KOH?

Analysis:

Target: The molarity of a solution of KOH. (? M) =

Knowns: The mass of $H_2C_2O_4$, the volume of the KOH solution and the balanced chemical equation. The molar mass of $H_2C_2O_4$ is an implicit known.

Relationships: The balanced chemical equation provides the mole equivalence statements. The molar mass gives the equivalence statement g ↔ mol. The molarity gives the equivalence statement mol ↔ 1 L or mmol ↔ 1 mL. 10^{-3} L ↔ 1 mL.

Plan: Use a modified form of the SSP, relating the number of moles to the molarity and volume for the KOH and to the mass and molar mass for $H_2C_2O_4$.

g $H_2C_2O_4$ ⟶ mol $H_2C_2O_4$ ⟶ mol KOH ⟶ KOH (M,V)

Work: The equivalence statement is 1 mol $H_2C_2O_4$ ↔ 2 mol KOH.

Molar Mass $H_2C_2O_4$ = 2(1.0079 H) + 2(12.011 C) + 4(15.994 O) = 90.035 g mol^{-1}

$$? \text{ M KOH} = 0.247 \text{ g } H_2C_2O_4 \times \frac{1 \text{ mol } H_2C_2O_4}{90.035 \text{ g } H_2C_2O_4} \times \frac{2 \text{ mol KOH}}{1 \text{ mol } H_2C_2O_4} \times \frac{1}{36.1 \text{ mL}} \times \frac{1 \text{ mL}}{10^{-3} \text{ L}} =$$

0.152 M KOH

3.91 Describe the reactions of the alkali metals with oxygen, and write balanced equations for each.

All the alkali metals make M+ cations. Lithium reacts to form the oxide, O^{2-}.

$4 \text{ Li} + O_2 \longrightarrow 2 \text{ Li}_2O$

Sodium reacts to form the peroxide, O_2^{2-}.

$2 \text{ Na} + O_2 \longrightarrow Na_2O_2$.

The other alkali metals (M) form the superoxide, O_2^{-}.

$M + O_2 \longrightarrow MO_2$.

3.93 Which of the Group 1A elements is most reactive toward water? Which is the least reactive? Write the general equation.

Cs is most reactive. Li is least reactive. The general equation is:

$2 \text{ M} + 2 \text{ H}_2O \longrightarrow 2 \text{ MOH} + H_2$.

3.95 What are the trends in melting point, boiling point, and density in Group 1A?

See Table 3.1 in the text. Melting point and boiling point decrease with increasing molar mass. The density increases with increasing molar mass.

3.97 Give two methods by which sodium carbonate is commercially produced.

Sodium carbonate is produced by the Solvay process. Consult the equations in the Summary Problem. It is also produced by thermal decomposition of *trona*, $NaHCO_3 \cdot Na_2CO_3 \cdot 2H_2O$. The reaction is shown below.

$2 \text{ NaHCO}_3 \cdot Na_2CO_3 \cdot 2H_2O(s) \longrightarrow 3 \text{ Na}_2CO_3(s) + CO_2(g) + 5 \text{ H}_2O(g)$

3.99 What polyatomic ion is similar in its chemical behavior to the alkali ions?

The ammonium ion, NH_4^+, has properties similar to those of the alkali metals.

3.101 How is sodium hydroxide produced, and what is an important byproduct of its manufacture?

NaOH is produced by electrolysis of an aqueous solution of NaCl. $Cl_2(g)$ is an important commercial byproduct. The reaction is shown below.

$$2\ NaCl(aq) + 2\ H_2O(\ell) \longrightarrow 2\ NaOH(aq) + Cl_2(g) + H_2(g)$$

3.103 Name two sodium compounds used in the curing and preservation of meat.

Sodium chloride, NaCl, and sodium nitrite, $NaNO_2$, are used in the curing and preservation of meat.

4 Energy, Heat and Chemical Change

4.1 What is meant by heat of reaction?

The heat of reaction is the amount of heat absorbed from or released to the surroundings by a chemical reaction or a physical change.

4.3 What is Hess's Law?

Hess's law states that the heat of a reaction is identical whether it occurs in a single step or in a series of steps.

4.5 What is meant by standard state?

The standard state of a substance is its most stable form at 25 °C and 1 atm.

4.7 How may heats of reaction be determined from heats of formation?

Heats of reaction can be determined from heats of formation by using the relationship, derivable from Hess's law, that:

$$\Delta H^\circ_{rxn} = \sum \left(n\Delta H^\circ_f \text{ products}\right) - \sum \left(n\Delta H^\circ_f \text{ reactants}\right)$$

4.9 Is the freezing of water an exothermic or endothermic process?

It is an **exothermic** process because heat must be removed for it to occur.

4.11 Define and give units for the following.
a. Heat capacity b. Specific heat c. Molar heat capacity at constant pressure

a. The heat capacity of a substance is the amount of heat required to change the temperature of that substance by 1 °C. It has units of **J °C^{-1}**.

b. The specific heat of a substance is the amount of heat required to change the temperature of 1 g of the substance by 1 °C. It has units of **J g^{-1} °C^{-1}**.

c. The molar heat capacity of a substance at constant pressure is the amount of heat required to change the temperature of 1 mole of the substance by 1 °C, with no change in pressure. It has units of **J mol^{-1} °C^{-1}**.

4.13 Describe how a coffee-cup calorimeter would be used to determine the specific heat of a 10 g piece of metal.

Place a known amount of water into the calorimeter cup and determine the initial temperature of the water. Determine the mass of the metal and then heat it to some temperature above that of the water. Determine the initial temperature of the metal. Quickly, add the metal to the water and stir carefully until thermal equilibrium (a constant temperature) is reached. Record that final temperature. Determine the heat change of the water, using the relationship heat = $q = mc_s\Delta T$. The heat change of the metal is equal in magnitude, but opposite in sign, to that of the water, so $q_{metal} = -q_{water}$. To determine the specific heat of the metal, divide the heat change of the metal by the mass of the metal and by its change in temperature. The result is that: $c_{s,metal} = -\frac{(m_{water}c_{s,water}\Delta T_{water})}{(m_{metal}\Delta T_{metal})}$. (In this relationship, m is mass, c_s is specific heat, and ΔT is the change in temperature, $T_{final} - T_{initial}$).

4.15 How much heat is required to decompose 33.9 g HBr by the following reaction?

$2\ HBr(g) \longrightarrow H_2(g) + Br_2(\ell);$ $\Delta H_{rxn} = +72.8\ kJ$

Analysis:

Target: The heat required for the decomposition process. (? kJ) =

Knowns: The mass of HBr, the heat of reaction and the balanced equation for the reaction. The molar mass is an implicit known.

Relationships: The molar mass gives the equivalence statement g ↔ mol and the balanced equation gives the equivalence statement 2 mol HBr ↔ 72.8 kJ.

Plan: Use the factor-label method for a 2 step conversion.

Work:

Molar Mass HBr = 1(1.0079 H) + 1(79.904 Br) = 80.912 g mol^{-1}.

$$? \text{ kJ} = 33.9 \text{ g HBr} \times \frac{1 \text{ mol HBr}}{80.912 \text{ g HBr}} \times \frac{72.8 \text{ kJ}}{2 \text{ mol HBr}} = \mathbf{15.3\ kJ}$$

4.17 How much heat is liberated when 0.0426 moles of sodium reacts with excess water according to the following equation?

$2\ Na(s) + 2\ H_2O(\ell) \longrightarrow H_2(g) + 2\ NaOH(aq);$ $\Delta H_{rxn} = -368\ kJ$

Analysis:

Target: The heat liberated in the reaction. (? kJ) =

Knowns: The number of moles of sodium, the heat of reaction and the balanced equation for the reaction.

Relationships: The balanced equation gives the equivalence statement 2 mol Na ↔ -368 kJ.

Plan: Use the factor-label method for this conversion.

Work:

$$? \text{ kJ} = 0.0426 \text{ mol Na} \times \frac{-368 \text{ kJ}}{2 \text{ mol Na}} = \mathbf{-7.84\ kJ}$$

(The negative sign indicates that heat is released in this process.)

4.19 How much heat is required for the production of one metric ton (1000 kg) of iron by the following reaction?

$FeO(s) + CO(g) \longrightarrow Fe(s) + CO_2(g);$ $\Delta H_{rxn} = 9.0\ kJ$

Analysis:

Target: The heat required for the iron production. (? kJ) =

Knowns: The mass of Fe, the heat of reaction and the balanced equation for the reaction. The molar mass is an implicit known.

Relationships: The molar mass gives the equivalence statement g ↔ mol and the balanced equation gives the equivalence statement 2 mol HBr ↔ 72.8 kJ. 1 kg ↔ 1000 g.

Plan: Use the factor label method for a 3 step conversion.

Work:

Molar Mass Fe = 55.847 g mol^{-1}.

$$? \text{ kJ} = 1000 \text{ kg Fe} \times \frac{1000 \text{ g}}{1 \text{kg}} \times \frac{1 \text{ mol Fe}}{55.847 \text{ g Fe}} \times \frac{9.0 \text{ kJ}}{1 \text{ mol Fe}} = \mathbf{1.6 \times 10^5\ kJ}$$

4.21 What is the heat of the reaction $2\ H_2(g) + O_2(g) \longrightarrow 2\ H_2O(g)$ if 672 kJ is released when 50.0 g water is produced?

Analysis:

Target: The heat of the reaction. (? kJ) =

Knowns: The mass of water, the heat released when that mass of water is formed, and the balanced equation for the reaction. The molar mass is an implicit known.
Relationships: The molar mass gives the equivalence statement g ↔ mol. The heat of the reaction is the amount released when 2 mol H_2O are produced. The equivalence statement 2 mol H_2O = x kJ is what we are looking for. When heat is released, the sign of the heat of reaction is negative.
Plan: Use the factor-label method for a 3 step conversion.
Work:
Molar Mass H_2O = 2(1.0079 H) + 1(15.9994 O) = 18.0152 g mol^{-1}.

$$? \text{ kJ} = \frac{-672 \text{ kJ}}{50.0 \text{ g } H_2O} \times \frac{18.0152 \text{ g } H_2O}{1 \text{ mol } H_2O} \times 2 \text{ mol } H_2O = \mathbf{-484 \text{ kJ}}$$

4.23 Given that

$N_2(g) + 2\ O_2(g) \longrightarrow 2\ NO_2(g)$; ΔH = 66.4 kJ

$2\ NO(g) + O_2(g) \longrightarrow 2\ NO_2(g)$; ΔH = -114.1 kJ

calculate the heat of the reaction $N_2(g) + O_2(g) \longrightarrow 2\ NO(g)$. Check your answer by looking up the heat of formation of NO in Table 4.1.

Analysis:
Target: The heat of the reaction. (? kJ) =
Knowns: Two different thermochemical equations.
Relationships: According to Hess's law, the heat of reaction is the same if the process occurs in one step or a series of steps.
Plan: Find some combination of the two steps given that adds to the reaction of interest by reversing and/or multiplying the equations by coefficients. If a reaction must be reversed, change the sign of its ΔH. If the reaction must be multiplied, multiply the value of its ΔH by the same number. Add the ΔH values thus obtained to determine the overall ΔH of the process. Begin with the species that occurs in the fewest number of equations.
Work:
N_2 occurs in only one equation and has the proper coefficient. Use the first equation as written with its ΔH value.

$N_2(g) + 2\ O_2(g) \longrightarrow 2\ NO_2(g)$ ΔH = 66.4 kJ

NO also occurs in just one equation. It has the correct coefficient, but appears as a reactant instead of as a product. Reverse the equation and change the sign of ΔH.

$2\ NO_2(g) \longrightarrow 2\ NO(g) + O_2(g)$ ΔH = -(-114.1 kJ) = +114.1 kJ

When the equations are added, the desired equation is obtained.

$N_2(g) + 2\ O_2(g) \longrightarrow 2\ NO_2(g)$

$2\ NO_2(g) \longrightarrow 2\ NO(g) + O_2(g)$

$N_2(g) + O_2(g) \longrightarrow 2\ NO(g)$ ΔH = 66.4 kJ + 114.1 kJ = **+180.5 kJ**

Table 4.1 gives a value for $\Delta H_f°$ = **+90.25 kJ/mol.** The difference is large because in this thermochemical equation **2 moles** of NO are formed. Multiplying by 2 gives the same value: 2 mol(90.25 kJ mol^{-1}) = +180.5 kJ.

4.25 Given that the heat of combustion of graphite is -393.5 kJ mol^{-1} and the heat of the phase transition C(graph) $\longrightarrow$ C(diamond) is 1.88 kJ mol^{-1}, calculate the heat of combustion of diamond. The mass of a gem-quality diamond is measured in the (otherwise obsolete) unit of carats: 1 carat ↔ 3.17 grains, and 1 lb ↔ 7000 grains. How much heat is released when a 1/2 carat diamond burns?

Analysis:

Target: The heat of combustion of diamond and the heat of combustion of a diamond with a specific mass. ΔH_c = (? kJ) and ΔH_{rxn} for diamond = (? kJ)

Knowns: The mass of the diamond, the heat of combustion of graphite, the nature of a combustion reaction and the heat of transformation of graphite to diamond. The molar mass of carbon is an implicit known.

Relationships: According to Hess's law, the heat of reaction is the same if the process occurs in one step or a series of steps. The relationships 1 carat ↔ 3.17 grains, and 1 lb ↔ 7000 grains are given in the problem. The equivalence statements 1 lb ↔ 0.45359 kg and 1 kg ↔ 1000 g also apply.

Plan: To determine the heat of combustion of a specific diamond, its mass must be converted to a number of moles and the heat of combustion of diamond must be determined. Once the equations for the combustion of graphite and of diamond are written, Hess's law can be used to determine the heat of combustion of diamond.

Step 4. Use the factor-label method with the heat of combustion, the mass of the diamond and its molar mass to determine the heat of combustion of the specific diamond.

Step 3. Use the factor-label method to convert the mass of the diamond in carats to a mass in grams.

Step 2. Use Hess's law to determine the heat of combustion of diamond.

Step 1. Write the thermochemical equations for the combustion of diamond and of graphite.

Work:

Step 1. In the combustion reaction, carbon reacts with oxygen to form CO_2.

$$C(graph) + O_2(g) \longrightarrow CO_2(g) \qquad \Delta H_{rxn} = -393.5 \text{ kJ mol}^{-1}$$

$$C(diamond) + O_2(g) \longrightarrow CO_2(g) \qquad \Delta H_{rxn} = ?$$

The equations are already balanced with 1 C atom and 2 O atoms on each side.

Step 2. Use Hess's law to determine the heat of combustion of diamond.

If the reaction, C(graph) ⟶ C(diamond) is reversed and added to the reaction for the combustion of graphite, the reaction for the combustion of diamond is obtained. When the reaction is reversed, the sign of ΔH_{rxn} is also changed.

$$C(graph) + O_2(g) \longrightarrow CO_2(g) \qquad \Delta H_{rxn} = -393.5 \text{ kJ mol}^{-1}$$

$$C(diamond) \longrightarrow C(graph) \qquad \Delta H_{rxn} = -(1.88 \text{ kJ mol}^{-1}) = -1.88 \text{ kJ mol}^{-1}$$

$$C(diamond) + O_2(g) \longrightarrow CO_2(g) \qquad \Delta H_{rxn} = -393.5 \text{ kJ mol}^{-1} + (-1.88 \text{ kJ mol})$$

$$\Delta H_{rxn} = \mathbf{-395.4 \text{ kJ mol}^{-1}}$$

Step 3. Determine the mass of the diamond in grams, using the factor-label method. (Assume three significant digits for the "1/2" carat.)

$$? \text{ g diamond} = 0.500 \text{ carat} \times \frac{3.17 \text{ grain}}{1 \text{ carat}} \times \frac{1 \text{ lb}}{7000 \text{ grain}} \times \frac{0.45359 \text{ kg}}{1 \text{ lb}} \times \frac{1000 \text{ g}}{1 \text{ kg}} =$$

0.103 g

Step 4. Determine the heat of combustion of the diamond using the factor-label method. Molar Mass C = 12.011 g mol^{-1}.

$$? \text{ kJ} = 0.103 \text{ g C} \times \frac{1 \text{ mol C}}{12.011 \text{ g C}} \times \frac{395.4 \text{ kJ}}{\text{mol}} = \mathbf{3.38 \text{ kJ}}$$

4.27 Use the data in Table 4.1 to calculate the heat of vaporization of water. Compare your answer with the value given in the text (Equation 4-4).

Analysis:

Target: The heat of vaporization of water. ΔH_{vap} = (? kJ mol^{-1})

Knowns: The nature of vaporization. From Table 4.1, we know that ΔH_f° for $H_2O(g)$ = (-241.818 kJ mol^{-1}) and for $H_2O(\ell)$ = (-285.830 kJ mol^{-1}).

Relationship: $\Delta H^\circ_{rxn} = \sum(n\Delta H^\circ_f \text{ products}) - \sum(n\Delta H^\circ_f \text{ reactants})$.

Plan: Write the vaporization reaction and then use the relationship.

Work:

Vaporization is for the process $(\ell) \longrightarrow (g)$. The vaporization reaction is: $H_2O(\ell) \longrightarrow H_2O(g)$.

$\Delta H_{vap} = \Delta H^\circ_f(H_2O(g)) - \Delta H^\circ_f(H_2O(\ell)) = -241.818 - (-285.830) =$ **+44.012 kJ mol^{-1}**

Equation 4.4 gives ΔH_{vap} as +44.0 kJ mol^{-1}. This agrees within the number of significant digits shown.

4.29 Given that ΔH of combustion of acetaldehyde, $CH_3CHO(\ell)$, is -1167 kJ mol^{-1}, calculate the heat of formation of acetaldehyde and check your answer against the value in Table 4.1.

Analysis:

Target: The heat of formation, ΔH°_f, for CH_3CHO. (? kJ mol^{-1}) =

Knowns: The nature of a combustion reaction and $\Delta H_c = \Delta H^\circ_{rxn}$ The heats of formation of the other substances can be determined using Table 4.1 or Appendix G.

Relationship: $\Delta H^\circ_{rxn} = \sum(n\Delta H^\circ_f \text{ products}) - \sum(n\Delta H^\circ_f \text{ reactants})$.

Plan: Write the balanced equation for the combustion process. Look up the necessary ΔH°_f values and then apply the relationship, solving for ΔH°_f of CH_3CHO.

Work:

The skeleton equation for the combustion reaction is:

$CH_3CHO(\ell) + O_2(g) \longrightarrow CO_2(g) + H_2O(\ell)$

$CH_3CHO(\ell)$ is the most complicated species. The C can be balanced by using a coefficient of 2 for CO_2.

$CH_3CHO(\ell) + O_2(g) \longrightarrow 2\,CO_2(g) + H_2O(\ell)$ 1(2) C left = 2(1) C right

H can be balanced by using a coefficient of 2 for H_2O.

$CH_3CHO(\ell) + O_2(g) \longrightarrow 2\,CO_2(g) + 2\,H_2O(\ell)$ 1(4) H left = 2(2) H right

The balancing is completed by using a coefficient of 5/2 for O_2.

$CH_3CHO(\ell) + 5/2\,O_2(g) \longrightarrow 2\,CO_2(g) + 2\,H_2O(\ell)$

1(1) + 5/2(2) O left = 2(2) + 2(1) O right

The necessary ΔH°_f values are:

Reactants: $CH_3CHO(\ell)$ = ? $O_2(g)$ = 0 kJ mol^{-1}

Products: $CO_2(g)$ = -393.509 kJ mol^{-1} $H_2O(\ell)$ = -285.830 kJ mol^{-1}

$-1167 \text{ kJ mol}^{-1} = 2\Delta H^\circ_f(CO_2(g)) + 2\Delta H^\circ_f(H_2O(\ell)) - [\Delta H^\circ_f(CH_3CHO(\ell)) + 5/2\Delta H^\circ_f(O_2(g))]$

$-1167 = 2(-393.509) + 2(-285.830) - [\Delta H_f^\circ(CH_3CHO(\ell)) + 5/2(0)]$

$-1167 = -787.018 - 571.660 - \Delta H_f^\circ(CH_3CHO(\ell)) - 0$

$-\Delta H_f^\circ(CH_3CHO(\ell)) = -1167 + 787.018 + 571.660 = +192$

$\Delta H_f^\circ(CH_3CHO(\ell))$ = **-192 kJ mol^{-1}**

Table 4.1 gives a value of **-192.3 kJ mol^{-1}** for the heat of formation of this compound. The slight difference occurs because the given heat of combustion limits the number of significant digits in the answer.

4.31 Aluminum reacts vigorously with many oxidizing agents, as for example

$4\ Al(s) + 3\ O_2(g) \longrightarrow 2\ Al_2O_3(s)$; $\Delta H = -3352$ kJ

$4\ Al(s) + 3\ MnO_2(s) \longrightarrow 3\ Mn\ (s) + 2\ Al_2O_3(s)$; $\Delta H = -1792$ kJ.

Use this information to determine the heat of formation of $MnO_2(s)$.

Analysis:

Target: The heat of formation, $\Delta H_f°$, for $MnO_2(s)$. (? kJ mol^{-1}) =

Knowns: The nature of a formation reaction and the thermochemical equations for two other reactions.

Relationship: According to Hess's law, the heat of reaction is the same if the process occurs in one step or a series of steps.

Plan: Write the formation reaction for $MnO_2(s)$. Find a combination of the two given thermochemical equations that adds to the formation reaction. Make the appropriate changes to the ΔH values if a reaction is multiplied or reversed.

Work:

The formation reaction for $MnO_2(s)$ is for making 1 mol of $MnO_2(s)$ from its element in their standard states: $Mn(s) + O_2(g) \longrightarrow MnO_2(s)$.

Inspection of the given equations indicates that Mn(s) and $MnO_2(s)$ will be in the proper places if the second reaction is reversed and all coefficients are divided by 3. The sign of ΔH must also be changed and its value divided by 3.

$Mn(s) + 2/3\ Al_2O_3(s) \longrightarrow 4/3\ Al(s) + MnO_2(s)$ $\Delta H = -1/3(-1792\ kJ) = +597.3$ kJ

To eliminate Al and to put O_2 in the correct place, the first equation can be used in the same order, but all coefficients must be divided by 3. The value of ΔH must also be divided by 3.

$4/3\ Al(s) + O_2(g) \longrightarrow 2/3\ Al_2O_3(s)$ $\Delta H = 1/3(-3352\ kJ) = -1117$ kJ

$Mn(s) + 2/3\ Al_2O_3(s) \longrightarrow 4/3\ Al(s) + MnO_2(s)$

$4/3\ Al(s) + O_2(g) \longrightarrow 2/3\ Al_2O_3(s)$

$Mn(s) + O_2(g) \longrightarrow MnO_2(s)$ $\Delta H = +597.3 + (-1117) =$ **-520 kJ mol^{-1}**

4.33 Use the data in Table 4.1 to determine the heat released per mole when glucose burns according to the reaction $C_6H_{12}O_6(s) + 6\ O_2(g) \longrightarrow 6\ CO_2(g) + 6\ H_2O(\ell)$.

Analysis:

Target: The heat of combustion of glucose. (? kJ mol^{-1}) =

Knowns: The balanced equation and the heats of formation of the reactants and products.

Relationship: $\Delta H^\circ_{rxn} = \sum \left(n\Delta H^\circ_f \text{ products}\right) - \sum \left(n\Delta H^\circ_f \text{ reactants}\right)$.

Plan: Look up the heats of formation and apply the relationship.

Work:

The necessary ΔH°_f values are:

Reactants: $C_6H_{12}O_6(s) = -1273.3$ kJ mol^{-1} $O_2(g) = 0$ kJ mol^{-1}

Products: $CO_2(g) = -393.509$ kJ mol^{-1} $H_2O(\ell) = -285.830$ kJ mol^{-1}

$\Delta H^\circ_{rxn} = 6\Delta H^\circ_f(CO_2(g)) + 6\Delta H^\circ_f(H_2O(\ell)) - [1\Delta H^\circ_f(C_6H_{12}O_6(s)) + 6\Delta H^\circ_f(O_2(g))]$

$\Delta H^\circ_{rxn} = 6(-393.509) + 6(-285.830) - [1(-1273.3) + 6(0)]$

$\Delta H^\circ_{rxn} =$ **-2802.7 kJ mol^{-1}**

(The negative sign of ΔH°_{rxn} indicates that heat is released.)

4.35 Use the data in Table 4.1 to calculate the amount of heat required to decompose 10.0 g of HgO(s) according to 2 HgO(s) $\longrightarrow$ 2 Hg(ℓ) + O_2(g).

Analysis:

Target: The heat required to decompose a certain mass of HgO(s). (? kJ) =

Knowns: The balanced equation and the heats of formation of all species. The molar mass of HgO is an implicit known.

Relationship: $\Delta H^\circ_{rxn} = \sum\left(n\Delta H^\circ_f \text{ products}\right) - \sum\left(n\Delta H^\circ_f \text{ reactants}\right)$

Plan: Use known heats of formation to determine the ΔH°_{rxn} for the decomposition process. Use the factor-label method to determine the amount of heat required for the given mass.

Work:

The necessary ΔH°_f values are:

Reactant: HgO(s) = -90.83 kJ mol^{-1}
Products: Hg(ℓ) = 0 kJ mol^{-1} $\qquad$ O_2(g) = 0 kJ mol^{-1}

$\Delta H^\circ_{rxn} = 2\Delta H^\circ_f(Hg(\ell)) + 1\Delta H^\circ_f(O_2(g)) - [2\Delta H^\circ_f(HgO(s))]$

$\Delta H^\circ_{rxn} = 2(0) + 1(0) - 2(-90.83) = +\ 181.7$ kJ

Now the factor-label method can be applied.

Molar Mass HgO = 1(200.59 Hg) +1(15.9994 O) = 216.59 g mol^{-1}

$$? \text{ kJ} = 10.0 \text{ g HgO} \times \frac{1 \text{ mol HgO}}{216.59 \text{ g HgO}} \times \frac{+181.7 \text{ kJ}}{2 \text{ mol HgO}} = \mathbf{+4.19 \text{ kJ}}$$

4.37 Suppose that the combustion of acetylene supplies the heat needed to produce mercury by the decomposition of mercury(II) oxide: 2 HgO(s) $\longrightarrow$ 2 Hg(ℓ) + O_2(g). How much acetylene must be burned in order to produce one kilogram of mercury? Use the data in Table 4.1.

Analysis:

Target: A mass of acetylene. (? g acetylene) =

Knowns: The balanced equation for the decomposition, the nature of a combustion reaction and ΔH°_f values for reactants and products. The formula of acetylene is in Table 4.1.

The molar masses of acetylene and HgO are implicit knowns. The ΔH°_{rxn} for the decomposition of HgO was determined in Problem 4.35. 1 kg $\leftrightarrow$ 1000 g.

Relationships: $\Delta H^\circ_{rxn} = \sum\left(n\Delta H^\circ_f \text{ products}\right) - \sum\left(n\Delta H^\circ_f \text{ reactants}\right)$.

Plan: The balanced equation for the combustion of acetylene must be written. ΔH°_f values can be used to determine the heat of combustion of acetylene. The factor-label method can be used to determine the amount of heat required and the mass of acetylene needed to provide that much heat.

Step 4. Use the factor-label method to determine the mass of acetylene needed.

Step 3. Use the factor-label method to determine the amount of heat required for the decomposition.
Step 2. Use the heats of formation to determine $\Delta H°_{rxn}$ for the combustion process.
Step 1. Write the balanced equation for the combustion of acetylene.

Work:

Step 1. From Table 4.1, acetylene is C_2H_2. The skeleton equation for the combustion is:

$$C_2H_2(g) + O_2(g) \longrightarrow CO_2(g) + H_2O(\ell).$$

C can be balanced by using a coefficient of 2 for CO_2.

$$C_2H_2(g) + O_2(g) \longrightarrow 2\ CO_2(g) + H_2O(\ell) \qquad 1(2)\text{ C left} = 2(1)\text{ C right}$$

H is already balanced with 2 atoms on each side. O can be balanced by using a coefficient of 5/2 for O_2.

$$C_2H_2(g) + 5/2\ O_2(g) \longrightarrow 2\ CO_2(g) + H_2O(\ell) \qquad 5/2(2)\text{ O left} = 2(2) + 1(1)\text{ O right}$$

Step 2. Look up ΔH_f^o values and use the relationship above to find $\Delta H°_{rxn}$ for the combustion. The ΔH_f^o values needed are:

Reactants: $C_2H_2(g) = +226.73\ kJ\ mol^{-1}$ $\quad O_2(g) = 0\ kJ\ mol^{-1}$
Products: $CO_2(g) = -393.509\ kJ\ mol^{-1}$ $\quad H_2O(\ell) = -285.830\ kJ\ mol^{-1}$

$$\Delta H^o_{rxn} = 2\Delta H_f^o(CO_2(g)) + 1\Delta H_f^o(H_2O(\ell)) - [1\Delta H_f^o(C_2H_2(g)) + 5/2\Delta H_f^o(O_2(g))]$$

$$\Delta H^o_{rxn} = 2(-393.509) + 1(-285.830) - [1(+226.73) + 5/2(0)]$$

$$\Delta H^o_{rxn} = -1299.58\ kJ\ mol^{-1}$$

Step 3. Use the ΔH^o_{rxn} from Problem 4.35 with the factor-label method to determine the heat required for the decomposition. (Assume 4 significant digits for the 1 kilogram of Hg because the heat of the reaction is known to 4 significant digits.) The molar mass of Hg is 200.59 g mol^{-1}.

$$?\ kJ = 1.000\ kg \times \frac{1000\ g}{1\ kg} \times \frac{1\ mol\ Hg}{200.59\ g\ Hg} \times \frac{181.7\ kJ}{2\ mol\ Hg} = 452.9\ kJ$$

Step 4. Use the factor-label method to determine the mass of acetylene needed.
Molar Mass C_2H_2 = 2(12.011 C) + 2(1.0079 H) = 26.038 g mol^{-1}

$$?\ g\ C_2H_2 = 452.9\ kJ \times \frac{1\ mol\ C_2H_2}{1299.58\ kJ} \times \frac{26.038\ g\ C_2H_2}{1\ mol\ C_2H_2} = \mathbf{9.074\ g\ C_2H_2}$$

4.39 The heat capacity of a pair of pliers is 205 J K^{-1}. How much heat is required to heat it from 25 °C to 150 °C?

Analysis:

Target: The heat required for a heating process. (? kJ) =
Knowns: The heat capacity of the pliers and the temperature change requested.
Relationships: Heat = q = $C\Delta T$. The magnitude of °C is the same as the magnitude of K, so the heat capacity may also be expressed as 205 J °C^{-1}. $\Delta T = T_{final} - T_{initial}$. 1 kJ ↔ 1000 J.

Plan: Insert the known values into the equation and solve. Use the factor-label method to express the answer in kJ.

Work:

$$?\ kJ = 205\ J\ °C^{-1} \times (150°C - 25°C) \times \frac{1\ kJ}{1000\ J} = \mathbf{25.6\ kJ}$$

4.41 If 2.50 kJ heat is added to the pair of pliers in Problem 4.39, by how many degrees will its temperature rise?

Analysis:
Target: The temperature change of a process. (? °C) =
Knowns: The heat capacity of the pliers and the heat applied.
Relationships: Heat = q = CΔT. The magnitude of °C is the same as the magnitude of K, so the heat capacity may also be expressed as 205 J °C^{-1}. 1 kJ ↔ 1000 J.
Plan: Rearrange the equation to solve for ΔT. Use the factor-label method to convert the heat to J and then insert the known values into the equation and solve.
Work:

$$? \text{ °C} = \Delta T = \frac{q}{C} = \frac{2.50 \text{ kJ} \times \frac{1000 \text{ J}}{1 \text{ kJ}}}{205 \text{ J °C}^{-1}} = \mathbf{12.2 \text{ °C}}$$

4.43 The heat capacity of a pair of pliers is 205 J K^{-1}. If it weighs 350 g, what is the specific heat of the metal?

Analysis/Plan: The heat capacity and the sample mass are given. The heat capacity, C, is equal to the specific heat, c_s, times the mass or C = c_sm. Algebraically rearrange this relationship and solve for c_s.
Work:

$$? \text{ J °C}^{-1} \text{ g}^{-1} = c_s = \frac{C}{m} = \frac{205 \text{ J °C}^{-1}}{350 \text{ g}} = \mathbf{0.586 \text{ J °C}^{-1} \text{ g}^{-1}}$$

4.45 What is the specific heat of a material, 15.0 g of which experiences a 3.52 °C temperature rise when 50.0 J heat is added?

Analysis/Plan: Heat (q), mass (m), and temperature change (ΔT) are given in the problem. These are related by the equation: q = m$c_s\Delta$T. Algebraically rearrange the relationship to solve for the specific heat, c_s.
Work:

$$? \text{ J °C}^{-1} \text{ g}^{-1} = c_s = \frac{q}{m\Delta T} = \frac{50.0 \text{ J}}{(15.0 \text{ g})(3.52 \text{ °C})} = \mathbf{0.947 \text{ J °C}^{-1} \text{ g}^{-1}}$$

4.47 A 1.50 kg sample of a substance having a specific heat of 1.75 J g^{-1} K^{-1} is heated from 0 °C to 100 °C; how much heat is required?

Analysis/Plan: Specific heat (c_s), mass (m), and initial and final temperatures are given in the problem. These are related by the equation: q = m$c_s\Delta$T. Use the equation in its current form to determine q. Because the mass is given in kilograms, use the fact that 1.75 J g^{-1} is equivalent to 1.75 kJ kg^{-1} to express the answer in kilojoules.
Work:

$$? \text{ kJ} = q = mc_s\Delta T = (1.50 \text{ kg})(1.75 \text{ kJ kg}^{-1} \text{ °C}^{-1})(100 \text{ °C} - 0 \text{ °C}) = \mathbf{263 \text{ kJ}}$$

4.49 What is the final temperature if 750 J heat is added to 25 g of a substance, originally at -20.0 °C? The specific heat is 2.04 J g^{-1} K^{-1}.

Analysis/Plan: Specific heat (c_s), mass (m), heat, and initial temperature are given in the problem. These are related by the equation: q = m$c_s\Delta$T. Use this equation to determine ΔT, then use the relationship $\Delta T = T_{final} - T_{initial}$ to determine the final temperature. °C^{-1} has the same magnitude as K^{-1} and may be substituted for it.

Work:

$$\Delta T = \frac{q}{mc_s} = \frac{750 \text{ J}}{(25 \text{ g})(2.04 \text{ J g}^{-1} \text{ °C}^{-1})} = 15 \text{ °C}$$

$$T_{final} = T_{initial} + \Delta T = -20.0 \text{ °C} + 15 \text{ °C} = \mathbf{-5 \text{ °C}}$$

4.51 What will be the final temperature if 200 g H_2O at 30 °C is mixed with 100 g H_2O at 50 °C?

Analysis:

Target: The final temperature of a mixture of two water samples. (? °C) =

Knowns: The initial temperature and the mass of two water samples. The specific heat of water can be looked up in Table 4.2 and is 4.184 J g^{-1} $°C^{-1}$.

Relationships: The heat lost by the warmer water, $-q_{warm\ water}$, is equal to the heat gained by the cooler water, $q_{cool\ water}$. They come to the same final temperature, T_f. The relationship $q = mc_s\Delta T$ applies to both the warmer and the cooler water.

Plan: Use the knowns and the relationship in the following manner.

Step 3. Solve the equation for T_{final}.

Step 2. Represent each of the heats by $q = mc_s\Delta T$, using T_f for the final temperature.

Step 1. Write the equality between the heats as $-q_{warm\ water} = q_{cool\ water}$.

Work:

Step 1. $-q_{warm\ water} = q_{cool\ water}$

Step 2. Substitute the relationship $mc_s\Delta T$ for q on each side, using known values of m, c_s and $T_{initial}$.

$$-[(100 \text{ g})(4.184 \text{ J g}^{-1} \text{ °C}^{-1})(T_f - 50 \text{ °C})] = (200 \text{ g})(4.184 \text{ J g}^{-1} \text{ °C}^{-1})(T_f - 30 \text{ °C})$$

Step 3. Recognize that the equation can be simplified by dividing both sides by 4.184 J g^{-1} $°C^{-1}$; then algebraically solve for T_f.

$$(-100 \text{ J °C}^{-1})T_f + 5.0 \times 10^3 \text{ J} = (200 \text{ J °C}^{-1})T_f - 6.0 \times 10^3 \text{ J}$$

$$5.0 \times 10^3 \text{ J} + 6.0 \times 10^3 \text{ J} = (200 \text{ J °C}^{-1})T_f + (-100 \text{ J °C}^{-1})T_f$$

$$1.1 \times 10^4 \text{ J} = (300 \text{ J °C}^{-1})T_f \qquad \text{therefore, } T_f = \frac{1.1 \times 10^4 \text{ J}}{300 \text{ J °C}^{-1}} = \mathbf{37 \text{ °C}}$$

4.53 Suppose a piece of limestone weighing 250 g is heated to 95 °C and dropped into a beaker containing 1 liter of water at 25.0 °C. What will be the final temperature of the system? The specific heat of water is 4.184 J g^{-1} K^{-1}, while that of limestone is 0.90 J g^{-1} K^{-1}.

Analysis:

Target: The final temperature of a mixture of water and limestone. (? °C) =

Knowns: The initial temperature of each sample, the mass of the limestone, the volume of water, and the specific heat of each substance.

Relationships: The heat lost by the limestone, $-q_{limestone}$, is equal to the heat gained by the cooler water, q_{water}. They come to the same final temperature, T_f. The relationship $q = mc_s\Delta T$ applies to both the water and the limestone. The density of water can be looked up and is 1.00 g mL^{-1} at 25 °C. 1 L ↔ 1000 mL. $°C^{-1}$ has the same magnitude as K^{-1} and may be substituted for it.

Plan: Use the knowns and the relationship in the following manner.

Step 3. Solve the equation for T_{final}.

Step 2. Represent each of the heats by $q = mc_s\Delta T$, using T_f for the final temperature of each. Write the equality between the heats as $-q_{limestone} = q_{water}$.

Step 1. Determine the mass of water by the factor-label method.

Work:

Step 1. Determine the mass of water. Assume the volume to be 3 significant digits.

$$? \text{ g } H_2O = 1.00 \text{ L} \times \frac{1000 \text{ mL}}{1 \text{ L}} \times \frac{1.00 \text{ g}}{1 \text{ mL}} = 1.00 \times 10^3 \text{ g}$$

Step 2. Substitute the relationship $mc_s\Delta T$ for q on each side, using known values of m, c_s and $T_{initial}$ into the $-q_{limestone} = q_{water}$ equation.

$$-[(250 \text{ g})(0.90 \text{ J g}^{-1} \text{ °C}^{-1})(T_f - 95\text{°C})] = (1.00 \times 10^3 \text{ g})(4.184 \text{ J g}^{-1} \text{ °C}^{-1})(T_f - 25.0 \text{ °C})$$

Step 3. Algebraically solve for T_f.

$$(-2.2 \times 10^2 \text{ J °C}^{-1})T_f + 2.1 \times 10^4 \text{ J} = (4.18 \times 10^3 \text{ J °C}^{-1})T_f - 1.05 \times 10^5 \text{ J}$$

$$2.1 \times 10^4 \text{ J} + 1.05 \times 10^5 \text{ J} = (4.18 \times 10^3 \text{ J °C}^{-1})T_f + (2.2 \times 10^2 \text{ J °C}^{-1})T_f$$

$$1.26 \times 10^5 \text{ J} = (4.40 \times 10^3 \text{ J °C}^{-1})T_f \quad \text{therefore, } T_f = \frac{1.26 \times 10^5 \text{ J}}{4.40 \times 10^3 \text{J °C}^{-1}} = \mathbf{28.6 \text{ °C}}$$

4.55 The heat of fusion of lead is 22.9 J g^{-1}. How much heat is required to melt one mole of lead?

Analysis:

Target: The heat required to melt one mole of lead. (? kJ) =

Knowns: The heat of fusion and the amount of lead to be melted. The molar mass of lead is an implicit known.

Relationships: For a heat of fusion given in g, $q = m\Delta H_{fus}$. 1 kJ ↔ 1000 J.

Plan: The problem can be solved by the factor-label method.

Work:

Molar Mass Pb = 207.2 g mol^{-1}. Assume one mole has 3 significant digits.

$$? \text{ kJ} = 1.00 \text{ mol Pb} \times \frac{207.2 \text{ g Pb}}{1 \text{ mol Pb}} \times \frac{22.9 \text{ J}}{\text{g}} \times \frac{1 \text{ kJ}}{1000 \text{ J}} = \mathbf{4.74 \text{ kJ}}$$

4.57 How much water can be vaporized by the application of 2.53 kJ of heat? The heat of vaporization is 540 cal g^{-1}.

Analysis:

Target: The mass or number of moles of water that can be vaporized. (? g) = or (? mol) =

Knowns: The amount of heat applied and the heat of vaporization. The molar mass of water is an implicit known.

Relationships: For a heat of vaporization given in g, $q = m\Delta H_{vap}$. 1 kJ ↔ 1000 J. The equivalence statement 1 cal ↔ 4.184 J is needed and can be looked up.

Plan: The problem can be solved by the factor-label method.

Work:

Molar Mass H_2O = 2(1.0079 H) + 1(15.9994 O) = 18.0152 g mol^{-1}

$$? \text{ g } H_2O = 2.53 \text{ kJ} \times \frac{1000 \text{ J}}{1 \text{ kJ}} \times \frac{1 \text{ cal}}{4.184 \text{ J}} \times \frac{1 \text{ g}}{540 \text{ cal}} = \mathbf{1.12 \text{ g } H_2O}$$

$$? \text{ mol } H_2O = 1.12 \text{ g } H_2O \times \frac{1 \text{ mol } H_2O}{18.0152 \text{ g } H_2O} = \mathbf{0.0622 \text{ mol } H_2O}$$

4.59 Calculate the energy input required to heat 10 g H_2O from -10 °C to +110 °C. The specific heats of ice, water, and steam are 0.49, 1.0, and 0.48 cal g^{-1} °C^{-1}, respectively, while the heats of fusion and vaporization are 79.7 and 540 cal g^{-1}, respectively.

Analysis:

Target: The heat required for conversion of water from ice to steam. (? kcal) =

Knowns: The mass of water, the initial temperature of the ice, the final temperature of the system, the specific heats of all three states of water and the heats of vaporization and fusion. The normal melting point of water (0 °C) and the normal boiling point of water (100 °C) are the basis of the Celsius temperature scale.

Relationships: In the temperature ranges in which a phase change is not occurring, the relationship $q = mc_s\Delta T$ applies. At the temperature at which a state change occurs, the following relationships apply. At the melting point, $q = m\Delta H_{fus}$ and, at the boiling point, $q = m\Delta H_{vap}$. 1 kcal ↔ 1000 cal.

Plan: This is a five step process. The overall heat required is the sum of the heat needed for each of the five steps. In each step, use the factor-label method to convert the heat to kilocalories.

Step 6. Sum the results of the five steps to obtain the total amount of heat required.
Step 5. Heat the steam to the final temperature, using the relationship: $q = mc_{s,steam}\Delta T$.
Step 4. Boil the water, using the relationship: $q = m\Delta H_{vap}$.
Step 3. Heat the water to the boiling point, using the relationship: $q = mc_{s,water}\Delta T$.
Step 2. Melt the ice, using the relationship: $q = m\Delta H_{fus}$.
Step 1. Heat the ice to the melting point, using the relationship: $q = mc_{s,ice}\Delta T$.

Work:

Step 1. Heat the ice to the melting point.

$$? \text{ kcal} = 10 \text{ g} \times \frac{0.49 \text{ cal}}{\text{g °C}} \times \frac{1 \text{ kcal}}{1000 \text{ cal}} \times (0 \text{ °C} - (-10 \text{ °C})) = 0.049 \text{ kcal}$$

Step 2. Melt the ice.

$$? \text{ kcal} = 10 \text{ g} \times \frac{79.7 \text{ cal}}{\text{g}} \times \frac{1 \text{ kcal}}{1000 \text{ cal}} = 0.80 \text{ kcal}$$

Step 3. Heat the water to the boiling point.

$$? \text{ kcal} = 10 \text{ g} \times \frac{1.00 \text{ cal}}{\text{g °C}} \times \frac{1 \text{ kcal}}{1000 \text{ cal}} \times (100 \text{ °C} - 0 \text{ °C}) = 1.0 \text{ kcal}$$

Step 4. Boil the water.

$$? \text{ kcal} = 10 \text{ g} \times \frac{540 \text{ cal}}{\text{g}} \times \frac{1 \text{ kcal}}{1000 \text{ cal}} = 5.4 \text{ kcal}$$

Step 5. Heat the steam to the final temperature.

$$? \text{ kcal} = 10 \text{ g} \times \frac{0.48 \text{ cal}}{\text{g °C}} \times \frac{1 \text{ kcal}}{1000 \text{ cal}} \times (110 \text{ °C} - 100 \text{ °C}) = 0.048 \text{ kcal}$$

Step 6. Sum the heats of the five individual steps.

$$? \text{ kcal} = 0.049 \text{ kcal} + 0.80 \text{ kcal} + 1.0 \text{ kcal} + 5.4 \text{ kcal} + 0.048 \text{ kcal} = \mathbf{7.3 \text{ kcal}}$$

4.61 If 1.00 g ice at 0 °C is dropped into a beaker containing 100 mL water at 35 °C, what will be the final temperature after the system has reached equilibrium? The heat of fusion of water is 6.02 kJ mol^{-1}.

Analysis:

Target: The final temperature of the ice/water mixture. (? °C) =

Knowns: The initial temperature and the mass of the ice, the initial temperature and the volume of the water and the heat of fusion. The molar mass of water is an implicit known. The specific heat of water, 4.184 J g^{-1} $°C^{-1}$, and the density of water, 1.00 g mL^{-1}, can be looked up.

Relationships: The magnitude of the heat gained by the ice will be equal to the heat lost by the liquid water. The ice must undergo a phase change and then be heated to the final temperature which will be the same as that of the liquid water. For melting, $q = n\Delta H_{fus}$ and for temperature changes without phase changes, $q = mc_s\Delta T$. Because the mass of ice is much less than the mass of water, assume that all of the ice will be converted to liquid water. The melting point of ice, 0 °C, is the basis of the Celsius temperature scale.

Plan: Write equations for q gained by the ice and q lost by the liquid water, making appropriate conversions for the given units. Set $q_{gained} = -q_{lost}$ and solve for the final temperature.

Step 6. Solve the equation for T_{final}.
Step 5. Set $-q_{lost} = q_{gained}$.
Step 4. Represent heat lost by the water by $q_{lost} = mc_s\Delta T$, using T_f for the final temperature.
Step 3. Determine the mass of water by the factor-label method.
Step 2. Represent the heat gained by the ice by $q_{gained} = n\Delta H_{fus} + mc_s\Delta T$, using T_f for the final temperature.
Step 1. Determine the number of moles of ice using the factor-label method.

Work:

Step 1. Determine the number of moles of ice.

Molar Mass H_2O = 2(1.0079 H) + 1(15.9994 O) = 18.0152 g mol^{-1}

$$? \text{ mol} = 1.00 \text{ g} \times \frac{1 \text{ mol}}{18.0152 \text{ g}} = 0.0555 \text{ mol}$$

Step 2. Determine the heat gained by the ice.

$$q_{gained} = 0.0555 \text{ mol} \times \frac{6.02 \text{ kJ}}{\text{mol}} \times \frac{1000 \text{ J}}{1 \text{ kJ}} + (1.00 \text{ g})(4.184 \text{ J g}^{-1} \text{ °C}^{-1})(T_f - 0\text{°C})$$

$$q_{gained} = 334 \text{ J} + (4.18 \text{ J °C}^{-1})T_f$$

Step 3. Determine the mass of water.

$$? \text{ g } H_2O = 100 \text{ mL} \times \frac{1.00 \text{ g}}{1 \text{ mL}} = 100 \text{ g}$$

Step 4. Determine the heat lost by the water.

$$q_{lost} = (100 \text{ g})(4.184 \text{ J g}^{-1} \text{ °C}^{-1})(T_f - 35 \text{ °C}) = (418 \text{ J °C}^{-1})T_f - 1.5 \times 10^4 \text{ J}$$

Step 5. Set heat lost by water equal to heat gained by ice. ($-q_{lost} = q_{gained}$)

$$-[(418 \text{ J °C}^{-1})T_f - 1.5 \times 10^4 \text{ J} = 334 \text{ J} + (4.18 \text{ J °C}^{-1})T_f$$

Step 6. Solve for the final temperature of the mixture.

$$1.5 \times 10^4 \text{ J} - 334 \text{ J} = (418 \text{ J °C}^{-1})T_f + (4.18 \text{ J °C}^{-1})T_f$$

$$1.4 \times 10^4 \text{ J} = (422 \text{ J °C}^{-1})T_f \qquad \text{therefore, } T_f = \frac{1.4 \times 10^4 \text{ J}}{(422 \text{ J °C}^{-1})} = 34 \text{ °C}$$

(Intermediate numbers are shown rounded to the correct number of significant digits, however, the answer is obtained by not rounding until the end of the process (14310/422.584).)

4.63 A typical Olympic-size swimming pool is 50 m long, 25 yards wide, and has an average depth of 8 feet. How much heat energy is required to increase the temperature of the water in such a pool by 1.00 °F?

Analysis:

Target: A heat energy. (? kJ) =

Knowns: The length, width and depth of the pool and the temperature change requested. The specific heat and the density of water can be looked up and are 4.184 J g^{-1} $°C^{-1}$ and 1.00 g mL^{-1}, respectively.

Relationships: For a system not undergoing a phase change, $q = mc_s\Delta T$. The following equivalence statements can be looked up on the end cover of the text: $°C = \frac{°F - 32}{1.8}$, 1 mL ↔ 1 cm^3, 1 yd ↔ 36 in, 1 in ↔ 2.54 cm, 1 m ↔ 100 cm and 1 ft ↔ 12 in. Volume is equal to length times width times depth.

Plan: Convert lengths to appropriate units to determine the volume of the pool. Use the volume and the density to determine the mass of water in the pool. Convert 1°F to °C and use the $q = mc_s\Delta T$ relationship.

Step 7. Determine the heat required for the process using $q = mc_s\Delta T$.
Step 6. Convert the temperature change to °C.
Step 5. Determine the mass of water in the pool, using the density and the volume.
Step 4. Determine the volume of the pool in mL, using the volume formula and the factor-label method.
Step 3. Determine the length of the pool in cm, using the factor-label method.
Step 2. Determine the depth of the pool in cm, using the factor-label method.
Step 1. Determine the width of the pool in cm, using the factor-label method..

Work:

Step 1. Determine the width of the pool in cm, using the factor-label method.

$$? \text{ cm} = 25 \text{ yd} \times \frac{36 \text{ in}}{1 \text{ yd}} \times \frac{2.54 \text{ cm}}{1 \text{ in}} = 2.3 \times 10^3 \text{ cm}$$

Step 2. Determine the depth of the pool in cm, using the factor-label method.

$$? \text{ cm} = 8 \text{ ft} \times \frac{12 \text{ in}}{1 \text{ ft}} \times \frac{2.54 \text{ cm}}{1 \text{ in}} = 2 \times 10^2 \text{ cm}$$

Step 3. Determine the length of the pool in cm, using the factor-label method.

$$? \text{ cm} = 50 \text{ m} \times \frac{100 \text{ cm}}{1 \text{ m}} = 5.0 \times 10^3 \text{ cm}$$

Step 4. Determine the volume of the pool in mL, using the volume formula and the factor-label method.

$$? \text{ mL} = (2.3 \times 10^8 \text{ cm})(2 \times 10^2 \text{ cm})(5.0 \times 10^3 \text{ cm}) \times \frac{1 \text{ mL}}{1 \text{ cm}^3} = 3 \times 10^9 \text{ mL}$$

Step 5. Determine the mass of water in the pool, using the density and the volume.

$$? \text{ g} = 3 \times 10^9 \text{ mL} \times \frac{1.00 \text{ g}}{1 \text{ mL}} = 3 \times 10^9 \text{ g}$$

Step 6. Convert the temperature change to °C. Since this is a change in temperature, the 32 would appear in both the initial and final temperatures and can be omitted.

$$? \Delta \text{ °C} = \frac{\Delta° \text{ F}}{1.8} = \frac{1.00 \text{ °F}}{1.8} = 0.556 \text{ °C}$$

Step 7. Determine the heat required for the process using $q = mc_s\Delta T$.

$$? \text{ kJ} = (3 \times 10^9 \text{ g})(4.184 \text{ J g}^{-1} \text{ °C}^{-1})(0.556 \text{ °C}) \times \frac{1 \text{ kJ}}{1000 \text{ J}} = \mathbf{6 \times 10^6 \text{ kJ}}$$

(Unrounded lengths, etc., were used in successive calculations.)

4.65 For gaseous argon, c_v is 12.5 J mol^{-1} K^{-1}. How much heat is required to raise the temperature of 100 g Ar from 15 °C to 50 °C, provided there is no change in volume?

Analysis:

Target: The heat required for a process at constant volume. (? kJ) =

Knowns: The molar heat capacity and the mass of argon, the initial and final temperatures and the heat capacity at constant volume. The molar mass of argon is an implicit known.

Relationships: Heat = q = $mc_v\Delta T$. The magnitude of K is the same as that of °C.
1 kJ ↔ 1000 J.

Plan: Use the factor-label method to determine the number of moles of argon and then use the relationship to find the heat.

Work:

Molar Mass Ar = 39.948 g mol^{-1}.

$$? \text{ mol Ar} = 100 \text{ g Ar} \times \frac{1 \text{ mol Ar}}{39.948 \text{ g Ar}} = 2.50 \text{ mol Ar}$$

$$? \text{ kJ} = 2.50 \text{ mol} \times \frac{12.5 \text{ J}}{\text{mol °C}} \times (50 \text{ °C} - 15 \text{ °C}) \times \frac{1 \text{ kJ}}{1000 \text{ J}} = \mathbf{1.1 \text{ kJ}}$$

4.67 A 436 g chunk of lead was removed from a beaker of boiling water, quickly dried, and dropped into a styrofoam cup containing 50.0 g water at 25.0 °C. As the system reached equilibrium, the water temperature rose to 40.8 °C. Calculate the heat capacity and the specific heat of the lead.

Analysis:

Target: The heat capacity and the specific heat of lead. (? J °C^{-1}) = and (? J g^{-1} °C^{-1}) =

Knowns: The mass of the lead and the mass and initial temperature of the water. The specific heat of water, 4.184 J g^{-1} °C^{-1}, is a constant that can be looked up. Water boils normally at 100.0°C, so that is the initial temperature of lead.

Relationships: $-q_{\text{lost lead}} = q_{\text{gained water}}$. Because there is no phase change, q = $mc_s\Delta T$.
C = mc_s.

Plan: Use the known relationships as follows.

Step 2. Use the relationship C = mc_s to determine the heat capacity.

Step 1. Put the appropriate data into the q = $mc_s\Delta T$ relationships, then rearrange the relationship $-q_{\text{lost lead}} = q_{\text{gained water}}$ to solve for $c_{s,\text{lead}}$.

Work:

Step 1. Use the relationships q = $mc_s\Delta T$ and $-q_{\text{lost lead}} = q_{\text{gained water}}$.

$$-[(436 \text{ g})(c_{s,\text{lead}})(100 \text{ °C} - 40.8 \text{ °C})] = (50.0 \text{ g})(4.184 \text{ J g}^{-1} \text{ °C}^{-1})(40.8 \text{ °C} - 25.0 \text{ °C})$$

$$? \text{ J g}^{-1} \text{ °C}^{-1} = c_{s,\text{lead}} = \frac{(50.0 \text{ g})(4.184 \text{ J g}^{-1} \text{ °C}^{-1})(40.8 \text{ °C} - 25.0 \text{ °C})}{(436 \text{ g})(100.0 \text{ °C} - 40.8 \text{ °C})} = \mathbf{0.128 \text{ J g}^{-1} \text{ °C}^{-1}}$$

Step 2. Use the relationship C = mc_s to determine the heat capacity of lead.

$$? \text{ J °C}^{-1} = C_{\text{lead}} = (436 \text{ g})(0.128 \text{ J g}^{-1} \text{ °C}^{-1}) = \mathbf{55.8 \text{ J °C}^{-1}}$$

4.69 A 5.00 g sample of KCl at 25.0 °C was dissolved in 110 g H_2O contained in a styrofoam cup. The water was originally at 25.0 °C, and after the salt dissolved, the temperature of the solution was 22.5 °C. Is the dissolution of KCl an exo- or endothermic process? Calculate the heat of solution of KCl; express the answer in (a) J g^{-1} and (b) kJ mol^{-1}.

Analysis:

Target: Whether the process is endo- or exothermic and the heat of solution of KCl.
(a) (? J g^{-1}) = and (b) (? kJ mol^{-1}) =

Knowns: The original mass and temperature of the KCl and the original temperature and mass of the water. The molar mass of KCl is an implicit known.

Relationships: Since there is no phase change, the heat change is given by $q_{solution} = mc_s\Delta T$, where the mass is the total mass of the solution. $q_{rxn} = -q_{solution}$. If the temperature goes down, the dissolution requires heat, therefore, the process is **endothermic** and the heat of solution should have a positive sign. 1 kJ ↔ 1000 J.

Plan: Assume that the solution's specific heat is the same as that of water, 4.184 J g^{-1} °C^{-1}. Use the known relationships and the factor-label method to convert units.

Step 4. Use the factor-label method to convert J g^{-1} to kJ mol^{-1}.

Step 3. Use the factor-label method to determine the heat of solution in J g^{-1} from $\frac{q}{m}$.

Step 2. Determine q_{rxn} from the relationship $q_{rxn} = -q_{solution}$.

Step 1. Determine $q_{solution}$ using the relationship $q_{solution} = mc_s\Delta T$.

Work:

Step 1. Determine $q_{solution}$ using the relationship $q_{solution} = mc_s\Delta T$.

$$q_{solution} = (5.00\text{ g KCl} + 110\text{ g }H_2O)(4.184\text{ J g}^{-1}\text{ °C}^{-1})(22.5\text{ °C} - 25.0\text{ °C}) = -1.2 \times 10^3\text{ J}$$

Step 2. Determine q_{rxn} from the relationship $q_{rxn} = -q_{solution}$.

$$q_{rxn} = -(-1.2 \times 10^3\text{ J}) = +1.2 \times 10^3\text{ J}$$

Step 3. Use the factor-label method to determine the heat of solution in J g^{-1} from $\frac{q}{m}$.

$$?\text{ J g}^{-1} = \frac{+1.2 \times 10^3\text{ J}}{5.00\text{ g KCl}} = \mathbf{+2.4 \times 10^2\text{ J g}^{-1}}$$

Step 4. Use the factor-label method to convert J g^{-1} to kJ mol^{-1}.
Molar Mass KCl = 1(39.0983 K) + 1(35.453 Cl) = 74.551 g mol^{-1}

$$?\text{ kJ mol}^{-1} = \frac{+2.4 \times 10^2\text{ J}}{\text{g KCl}} \times \frac{1\text{ kJ}}{1000\text{ J}} \times \frac{74.551\text{ g KCl}}{1\text{ mol KCl}} = \mathbf{+18\text{ kJ mol}^{-1}}$$

4.71 A small quantity, 0.100 g, of CaO was added to 125 g H_2O at 23.6 °C in a styrofoam cup. The reaction $CaO + H_2O \longrightarrow Ca(OH)_2$ has $\Delta H_{rxn} = -350$ kJ mol^{-1}. What was the final temperature of the solution?

Analysis:

Target: The final temperature of the solution. (? °C) =

Knowns: The original mass of the KCl, the original temperature and mass of the water and the ΔH_{rxn}. The molar mass of CaO is an implicit known.

Relationships: Since there is no phase change, the heat change is given by $q_{solution} = mc_s\Delta T$, where the mass is the total mass of the solution. $\Delta T = T_{final} - T_{initial}$. $q_{rxn} = -q_{solution}$. The heat of the chemical reaction is equal to $n\Delta H_{rxn}$. 1 kJ ↔ 1000 J.

Plan: Assume that the solution's specific heat is the same as that of water, 4.184 J g^{-1} °C^{-1}. Use the known relationships and the factor-label method to convert units.

Step 4. Rearrange $\Delta T = T_{final} - T_{initial}$ to find T_{final}.

Step 3. Rearrange the relationship $q_{solution} = mc_s\Delta T$ to solve for ΔT.

Step 2. Determine $q_{solution}$ from the relationship $q_{rxn} = -q_{solution}$.

Step 1. Determine q_{rxn} using the factor-label method.

Work:

Step 1. Determine q_{rxn} using the factor-label method.

Molar Mass CaO = 1(40.078 Ca) + 1(15.9994 O) = 56.077 g mol^{-1}

$$? \text{ J} = 0.100 \text{ g CaO} \times \frac{1 \text{ mol CaO}}{56.077 \text{ g CaO}} \times \frac{-350 \text{ kJ}}{1 \text{ mol CaO}} \times \frac{1000 \text{ J}}{1 \text{ kJ}} = -624 \text{ J}$$

Step 2. Determine $q_{solution}$ from the relationship $q_{rxn} = -q_{solution}$.

? J = -(-624 J) = +624 J

Step 3. Rearrange the relationship $q_{solution} = mc_s\Delta T$ to solve for ΔT.
The total solution mass is 125 g H_2O + 0.100 g CaO = 125 g solution.

$$\Delta T = \frac{q_{solution}}{mc_s} = \frac{+624 \text{ J}}{(125 \text{ g solution})(4.184 \text{ J g}^{-1} \text{ °C}^{-1})} = +1.19 \text{ °C}$$

Step 4. Rearrange $\Delta T = T_{final} - T_{initial}$ to find T_{final}.

? °C = $T_{final} = T_{initial} + \Delta T$ = 23.6 °C + 1.19 °C = **24.8 °C**

4.73 A bomb calorimeter was calibrated by burning 0.509 g benzoic acid, ($C_7H_6O_2$), whose heat of combustion is -3227 kJ mol^{-1}. If the calorimeter temperature rose by 2.81 °C as a result, what was its heat capacity?

Analysis:

Target: A heat capacity. (? J °C^{-1}) =

Knowns: The mass of acid burned, the heat of combustion of the acid and the temperature change of the calorimeter. The molar mass of $C_7H_6O_2$ is an implicit known.

Relationships: The heat gained by the calorimeter is $q_{calorimeter} = C_{calorimeter}\Delta T$. The heat gained by the calorimeter is equal in magnitude but opposite in sign to the heat lost by the combustion reaction, so $q_{calorimeter} = -q_{rxn}$. The heat of the chemical reaction is equal to $n\Delta H_{rxn}$.

Plan: If we can find the heat released in the combustion process, we can use the relationship $q_{calorimeter} = -q_{rxn}$ to find the heat gained by the calorimeter. Rearranging the relationship $q_{calorimeter} = C_{calorimeter}\Delta T$ yields the heat capacity of the calorimeter.

Step 3. Rearrange the relationship $q_{calorimeter} = C_{calorimeter}\Delta T$ to find the heat capacity of the calorimeter.

Step 2. Use the relationship $q_{calorimeter} = -q_{rxn}$ to determine the heat absorbed by the calorimeter.

Step 1. Use the factor-label method to determine q_{rxn}.

Work:

Step 1. Use the factor-label method to determine q_{rxn}.
Molar Mass $C_7H_6O_2$ = 7(12.011 C) + 6(1.0079 H) + 2(15.9994 O) = 122.123 g mol^{-1}

$$q_{rxn} = 0.509 \text{ g } C_7H_6O_2 \times \frac{1 \text{ mol } C_7H_6O_2}{122.123 \text{ g } C_7H_6O_2} \times \frac{-3227 \text{ kJ}}{\text{mol } C_7H_6O_2} = -13.4 \text{ kJ}$$

Step 2. Use the relationship $q_{calorimeter} = -q_{rxn}$ to determine the heat absorbed by the calorimeter.

$q_{calorimeter}$ = -(-13.4 kJ) = +13.4 kJ

Step 3. Rearrange the relationship $q_{calorimeter} = C_{calorimeter}\Delta T$ to find the heat capacity of the calorimeter.

$$? \text{ kJ °C}^{-1} = C_{calorimeter} = \frac{q_{calorimeter}}{\Delta T} = \frac{+13.4 \text{ kJ}}{2.81 \text{ °C}} = \mathbf{4.79 \text{ kJ °C}^{-1}}$$

4.75 Naphthalene ($C_{10}H_8$) has a heat of combustion of 5.157×10^3 kJ mol^{-1}. What would be the temperature rise of a bomb calorimeter having a heat capacity of 4532 J K^{-1} if 0.25 g naphthalene were burned in it?

Analysis:

Target: The temperature rise of a calorimeter. (? °C) =

Knowns: The mass of naphthalene burned, the heat of combustion of naphthalene and the heat capacity of the calorimeter. The molar mass of $C_{10}H_8$ is an implicit known.

Relationships: The heat gained by the calorimeter is $q_{calorimeter} = C_{calorimeter}\Delta T$. The heat gained by the calorimeter is equal in magnitude but opposite in sign to the heat lost by the combustion reaction, so $q_{calorimeter} = -q_{rxn}$. The heat of the chemical reaction is equal to $n\Delta H_{rxn}$. 1 kJ ↔ 1000 J. The magnitude of K is the same as for °C.

Plan: If we can find the heat released in the combustion process, we can use the relationship $q_{calorimeter} = -q_{rxn}$ to find the heat gained by the calorimeter. Rearranging the relationship $q_{calorimeter} = C_{calorimeter}\Delta T$ yields the temperature change of the calorimeter.

Step 3. Rearrange the relationship $q_{calorimeter} = C_{calorimeter}\Delta T$ to find the change in temperature of the calorimeter.

Step 2. Use the relationship $q_{calorimeter} = -q_{rxn}$ to determine the heat absorbed by the calorimeter.

Step 1. Use the factor-label method to determine q_{rxn}.

Work:

Step 1. Use the factor-label method to determine q_{rxn}.

Molar Mass $C_{10}H_8$ = 10(12.011 C) + 8(1.0079 H) = 128.173 g mol^{-1}

$$q_{rxn} = 0.25 \text{ g } C_{10}H_8 \times \frac{1 \text{ mol } C_{10}H_8}{128.173 \text{ g } C_{10}H_8} \times \frac{-5.157 \times 10^3 \text{ kJ}}{\text{mol } C_{10}H_8} \times \frac{1000 \text{ J}}{1 \text{ kJ}} = -1.0 \times 10^4 \text{ J}$$

Step 2. Use the relationship $q_{calorimeter} = -q_{rxn}$ to determine the heat absorbed by the calorimeter.

$$q_{calorimeter} = -(-1.0 \times 10^4 \text{ J}) = +1.0 \times 10^4 \text{ J}$$

Step 3. Rearrange the relationship $q_{calorimeter} = C_{calorimeter}\Delta T$ to find the change in temperature of the calorimeter.

$$\Delta T = \frac{q_{calorimeter}}{C_{calorimeter}} = \frac{+1.0 \times 10^4 \text{ J}}{4532 \text{ J °C}^{-1}} = 2.2 \text{ °C}$$

4.77 Which, if any, of the Group 2A metals occur(s) naturally in elemental form?

All of the Group 2A elements are too reactive to exist in nature in elemental form.

4.79 What is the chief source of (a) calcium and (b) magnesium for commercial use?

a. The chief source of calcium is limestone,
b. The chief source of magnesium is seawater.

4.81 In commercial operations, how are the pure Group 2A metals prepared from their compounds? Write a general equation, using M to represent the element, that describes the process.

The Group 2A metals are prepared by electrolysis of the molten chloride.

$$MCl_2(\ell) \longrightarrow M(\ell) + Cl_2(g)$$

4.83 Name a Group 2A compound that is used in sugar refining.

Strontium hydroxide, $Sr(OH)_2$, is used in sugar refining.

4.85 What are the chemical name and formula of "lime"?

Two different compounds compose lime. "Quicklime" is calcium oxide, CaO, and "slaked lime" is calcium hydroxide, $Ca(OH)_2$.

4.87 Write general equations, using M to represent the metal, for the reaction of the alkaline earth elements with (a) oxygen and (b) nitrogen.

a. $2\ M + O_2 \longrightarrow 2\ MO$

b. $3\ M + N_2 \longrightarrow M_3N_2$

4.89 Write a general equation, using M to represent the metal, for the reaction of the Group 2A oxides with water. Are the reactions of the Group 2A oxides with water endo- or exothermic?

$MO + H_2O \longrightarrow M(OH)_2$ These reactions are very exothermic.

4.91 What is the mass of $Ca(OH)_2$ formed when 750 g CaO reacts with excess water?

Analysis:
Target: A mass of $Ca(OH)_2$. (? g $Ca(OH)_2$) =
Knowns: The mass of CaO and the balanced equation for the process (see Problem 4.89). The molar masses of CaO and $Ca(OH)_2$ are implicit knowns.
Relationships: The molar masses provide the equivalence statements g ↔ mol and the balanced equation gives the mole ratios.
Plan: Apply the standard stoichiometric procedure (SSP). See Chapter 3 if necessary.

g CaO ⟶ mol CaO ⟶ mol $Ca(OH)_2$ ⟶ g $Ca(OH)_2$

Work:
The molar masses needed are:
CaO = 1(40.078 Ca) + 1(15.9994 O) = 56.077 g mol^{-1}
$Ca(OH)_2$ = 1(40.078 Ca) +2(15.9994 O) +2(1.0079 H) = 74.093 g mol^{-1}

$$? \text{ g Ca(OH)}_2 = 750 \text{ g CaO} \times \frac{1 \text{ mol CaO}}{56.077 \text{ g CaO}} \times \frac{1 \text{ mol Ca(OH)}_2}{1 \text{ mol CaO}} \times \frac{74.093 \text{ g Ca(OH)}_2}{1 \text{ mol Ca(OH)}_2} =$$

991 g $Ca(OH)_2$

4.93 How much $Ca(OH)_2$ is required to react with one ton (2000 pounds) of SO_2?

Analysis:
Target: A mass of $Ca(OH)_2$. (? ton $Ca(OH)_2$) =
Knowns: The mass of SO_2. The balanced equation for the process, given in Section 4.5, is

$Ca(OH)_2 + SO_2 \longrightarrow CaSO_2 + H_2O$. The molar masses of SO_2 and $Ca(OH)_2$ are implicit knowns.

Relationships: The molar masses provide the equivalence statements g ↔ mol and the balanced equation gives the mole ratios. Because data is given in the English system and it makes sense to report the final answer in the English system, we can use the lb-mol. (1 lb-mol of H has a mass of 1.0079 lb, 1 lb-mol of C has a mass of 12.011 lb, etc.) The molar mass, therefore, gives the mass of a lb-mol of any substance. Coefficients in the balanced equation give molar ratios in terms of lb-mol as well as the "g-mol" we have characteristically used.

Plan: Apply the standard stoichiometric procedure (SSP). Also, use the factor-label method to convert pounds back to tons.

$$\text{lb } SO_2 \longrightarrow \text{lb-mol } SO_2 \longrightarrow \text{lb-mol } Ca(OH)_2 \longrightarrow \text{lb } Ca(OH)_2 \longrightarrow \text{ton } Ca(OH)_2$$

Work:

Molar Mass SO_2 = 1(32.06 S) + 2(15.9994 O) = 64.06 lb lb-mol^{-1}

Molar Mass $Ca(OH)_2$ = 1(40.078 Ca) +2(15.9994 O) +2(1.0079 H) = 74.093 lb lb-mol^{-1}

$$? \text{ ton } Ca(OH)_2 = 2000 \text{ lb } SO_2 \times \frac{1 \text{ lb-mol } SO_2}{64.06 \text{ lb } SO_2} \times \frac{1 \text{ lb-mol } Ca(OH)_2}{1 \text{ lb-mol } SO_2} \times \frac{74.093 \text{ lb } Ca(OH)_2}{1 \text{ lb-mol } Ca(OH)_2} \times$$

$$\frac{1 \text{ ton } Ca(OH)_2}{2000 \text{ lb } Ca(OH)_2} = \mathbf{1.16 \text{ ton } Ca(OH)_2}$$

5 The Gaseous State

5.1 What are the three states of matter and their distinguishing characteristics?

A **gas** has an indefinite volume; it assumes the shape and volume of the container into which it is placed. A **liquid** has a definite volume, but a shape which depends upon the container into which it is placed. A **solid** has a definite shape and a definite volume.

5.3 Define pressure. Give a precise scientific definition, and also one that could be understood by someone without scientific training.

Pressure is defined as the force per unit area. It can be considered to be how "hard" something pushes against something else.

5.5 Describe the preparation and operational principles of a mercury barometer.

A mercury barometer is prepared by filling a tube with mercury and inverting it into a container of mercury that is open to the air. The mercury tries to run out of the tube, but the force of the atmosphere pushing on the mercury in the container keeps it raised. The difference between the top of the mercury in the container and that in the tube is a measure of the atmospheric pressure--how "hard" the atmosphere is pushing on the surface.

5.7 One statement of Boyle's Law is that for a given sample of gas at a given temperature, the product of pressure and volume is a constant. What is the meaning of the word "constant" in this context?

This "constant" has a value which does not change when pressure and volume change.

5.9 What is the ideal gas law? To what gases does it apply? Are there any limitations to the validity of the law?

The ideal gas law is **PV = nRT**. It applies fairly well to all gases. The ideal gas law becomes less accurate at high pressures and low temperatures.

5.11 What is meant by the "absolute zero" of temperature? How can it be measured?

"Absolute zero" is 0 K and is the lowest temperature that it is theoretically possible to obtain. Its value is found by plotting the volume of a gas at constant pressure versus temperature and extrapolating to the temperature at which the volume of the gas would be zero.

5.13 List the postulates of the kinetic molecular theory. Are these postulates "true"? Are they reasonable? Are they useful?

The postulates are:

1. A gas consists of tiny, discrete particles called molecules (or atoms).
2. On the average, the particles in a gas are separated by distances that are large compared to the size of the particles.
3. There are no attractive or repulsive forces between particles.
4. The particles are in constant, rapid, random motion.
5. Particles travel in straight lines between elastic collisions, which means that no energy of motion is lost during the collision.
6. The average kinetic energy of the particles in a gas is directly proportional to the absolute temperature and is the same for any gas at the same temperature.

It is **not** true that the particles have no attractions. The volume of a real gas will **not** be zero at absolute zero. While not absolutely true, the postulates are both reasonable and useful because they explain many known properties of gases and they allow us to make predictions about gas behavior.

5.15 Is there a relationship between temperature and molecular velocity? What is it?

The relationship is that the average velocity, ⟨v⟩, is equal to $(8RT/\pi M)^{1/2}$.

5.17 What experimental evidence supports the validity of the KMT?

According to the text (Section 5.4), the best evidence comes from molecular beam measurements of the velocity distribution in gas samples.

5.19 Describe the procedure for collecting gases over water.

The reaction is initiated in one container. That container is connected via tubing to another container that is originally filled with water and inverted in a container of water. As the gas is produced by the reaction in the first container, it displaces water from the second container. Calibration marks on the container allow one to determine the amount of gas produced.

5.21 How do real gases differ from ideal gases? What is the relationship between these differences and the validity of the postulates of the KMT?

Real gases do not follow the ideal gas law exactly because they have non-zero volumes and because they do have intermolecular attractions. Rather than having their volumes continue to decrease as the temperature is lowered, they condense to liquids or solids. The KMT becomes less valid at high pressure and low temperature. Under these conditions, intermolecular attractions hold the particles together and they are not widely spaced.

5.23 What aspects of gas behavior are associated with the following names? Boyle, Charles, Avogadro, Dalton, Torricelli, van der Waals, and Graham.

Boyle: PV = constant (at constant temperature).
Charles: V/T = constant (at constant pressure).
Avogadro: V/n = constant (at constant pressure and temperature).
Dalton: The total pressure of a system of gases is the sum of the partial pressures of the gases.
Torricelli: Developed the barometer and determined the pressure of the atmosphere.
van der Waals: Corrected the ideal gas law for nonideality.
Graham: Studied effusion and diffusion, determining that the velocity of a gas is inversely proportional to its density (or molar mass).

5.25 Express the pressure 5.00 atm in torr, Pa, kPa, and bars.

Analysis/Plan: This is an exercise in unit conversion using the factor-label method. The appropriate equivalence statements can be looked up and are: 1 atm ↔ 760 torr, 1 atm ↔ 1.01325×10^5 Pa, 1 kPa ↔ 1000 Pa, and 1 bar ↔ 10^5Pa.

Work:

$$? \text{ torr} = 5.00 \text{ atm} \times \frac{760 \text{ torr}}{1 \text{ atm}} = \mathbf{3.80 \times 10^3 torr}$$

$$? \text{ Pa} = 5.00 \text{ atm} \times \frac{1.01325 \times 10^5 \text{ Pa}}{1 \text{ atm}} = \mathbf{5.07 \times 10^5 \text{ Pa}}$$

$$? \text{ kPa} = 5.07 \times 10^5 \text{ Pa} \times \frac{1 \text{ kPa}}{1000 \text{ Pa}} = \mathbf{507 \text{ kPa}}$$

$$? \text{ bar} = 5.07 \times 10^5 \text{ Pa} \times \frac{1 \text{ bar}}{10^5 \text{ Pa}} = \mathbf{5.07 \text{ bar}}$$

5.27 The average pressure of the atmosphere at an altitude of 20 km is only 5.5% of the sea-level value. Express the pressure at 20 km in torr.

Analysis/Plan: The value of pressure at sea level is 760 torr. Use the percent directly, remembering that a percent can be expressed as parts per 100.
Work:

$$? \text{ torr} = 760 \text{ torr} \times \frac{5.5}{100} = \mathbf{42\ torr}$$

5.29 The downward force exerted by a column of liquid is given by the formula $f = DghA$, where h and A are the height and cross-sectional area of the column, D is the density of the liquid, and g is the acceleration of gravity, 9.819 m s^{-2}. The *pressure* (force per unit area) is $P = f/A = Dgh$. If SI units (kg, m, s) are used, the calculated pressure is in pascals. Calculate the pressure exerted by a column of mercury 0.7600 m high, given that the density of mercury is 13.58 g mL^{-1}.

Analysis:
Target: The pressure of a column of mercury. (? Pa) =
Knowns: The density and height of a mercury column and the formula for pressure.
Relationships: The given units must be converted to SI units to be used in the formula $P = Dgh$. Appropriate equivalence statements are: 1000 g ↔ 1 kg, 1 mL ↔ 1 cm^3, and 10^{-2} m ↔ 1 cm.
Plan: Use the formula, converting the given units to the appropriate SI units with the factor-label method.
Work:

$$? \text{ Pa} = \frac{13.58 \text{ g}}{\text{mL}} \times \frac{1 \text{ mL}}{1 \text{ cm}^3} \times \left(\frac{1 \text{ cm}}{10^{-2} \text{ m}}\right)^3 \times \frac{1 \text{ kg}}{1000 \text{ g}} \times \frac{9.819 \text{ m}}{\text{s}^2} \times 0.7600 \text{ m} \times \frac{1 \text{ Pa}}{\frac{\text{kg}}{\text{m} \cdot \text{s}^2}} =$$

$$\mathbf{1.013 \times 10^5\ Pa}$$

5.31 Steel tanks for storage of gases are capable of withstanding pressures greater than 150 atm. Express this pressure in psi.

Analysis/Plan: This is an exercise in unit conversion, using the factor-label method. The appropriate equivalence statement (Table 5.1) is 1 atm ↔ 14.70 psi.
Work:

$$? \text{ psi} = 150 \text{ atm} \times \frac{14.70 \text{ psi}}{1 \text{ atm}} = \mathbf{2.21 \times 10^3\ psi}$$

5.33 A gas occupies 5.00 L at 25 °C and 1.00 atm. What is the volume if the gas is compressed to 5.00 atm with no change in temperature?

Analysis:
Target: The final volume of a gas. (? L) =
Knowns: The initial volume and pressure of the gas and the final pressure.
Relationship: $P_iV_i = P_fV_f$.
Plan: Solve for the final volume, insert the known values and calculate.
Work:

$$? \text{ L} = V_f = \frac{P_iV_i}{P_f} = \frac{(1.00 \text{ atm})(5.00 \text{ L})}{5.00 \text{ atm}} = \mathbf{1.00\ L}$$

5.35 A gas fills a 7.5 L container at 100 °C and 25 atm. What is the new volume if the gas undergoes a constant temperature expansion until its pressure is 1.00 atm?

Analysis:
Target: The final volume of a gas. (? L) =

Knowns: The initial volume and pressure of the gas and the final pressure.
Relationship: $P_iV_i = P_fV_f$.
Plan: Solve for the final volume, insert the known values and calculate.
Work:

$$? \text{ L} = V_f = \frac{P_iV_i}{P_f} = \frac{(25 \text{ atm})(7.5 \text{ L})}{1.00 \text{ atm}} = \mathbf{1.9 \times 10^2 \text{ L}}$$

5.37 By what factor does the volume of a gas increase if the pressure is halved?

Analysis:
Target: The final volume of a gas. (? L)=
Knowns: The initial volume and pressure of the gas and the final pressure.
Relationships: $P_iV_i = P_fV_f$. If the pressure is halved, $P_f = 1/2\ P_i$.
Plan: Solve for the final volume, insert the known values and calculate.
Work:

$$? \text{ L} = V_f = \frac{P_iV_i}{P_f} = \frac{(P_i)(V_i)}{1/2\ P_i} = \mathbf{2\ V_i}$$ (Volume increases by a factor of 2.)

5.39 A gas is initially at 5.3 L and 75 kPa. What pressure must be exerted in order to compress the gas to 4.0 L? Assume that the temperature is unchanged during the compression.

Analysis:
Target: The final pressure of a gas. (? atm) =
Knowns: The initial volume and pressure of the gas and the final volume.
Relationship: $P_iV_i = P_fV_f$.
Plan: Solve for the final pressure, insert the known values and calculate.
Work:

$$? \text{ atm} = P_f = \frac{P_iV_i}{V_f} = \frac{(75 \text{ kPa})(5.3 \text{ L})}{4.00 \text{ L}} = \mathbf{99 \text{ kPa}}$$

5.41 If 0.500 L of a gas at 2.77 atm pressure is expanded into a vessel whose volume is 5.00 L, what will the pressure be? Assume no change in temperature.

Analysis:
Target: The final pressure of a gas. (? atm) =
Knowns: The initial volume and pressure of the gas and the final volume.
Relationship: $P_iV_i = P_fV_f$.
Plan: Solve for the final pressure, insert the known values and calculate.
Work:

$$? \text{ atm} = P_f = \frac{P_iV_i}{V_f} = \frac{(2.77 \text{ atm})(0.500 \text{ L})}{5.00 \text{ L}} = \mathbf{0.277 \text{ atm}}$$

5.43 If all the gas in a container is transferred to a new container having 1/3 the volume of the original, by what factor does the pressure change?

Analysis:
Target: The final pressure of a gas. (? atm) =
Knowns: The initial volume and pressure of the gas and the final volume.
Relationships: $P_iV_i = P_fV_f$. If the volume is changed by 1/3, then $V_f = 1/3\ V_i$.
Plan: Solve for the final pressure, insert the known values and calculate.
Work:

$$? \text{ atm} = P_f = \frac{P_iV_i}{V_f} = \frac{(P_i)(V_i)}{(1/3\ V_i)} = \mathbf{3\ P_i}$$ (Pressure increases by a factor of 3.)

5.45 If 500 mL of gas is heated from 300 to 400 K in a constant pressure process, what is the new volume?

Analysis:

Target: The final volume of a gas. (? mL) =

Knowns: The initial volume and temperature and the final temperature of a gas.

Relationship: $\frac{V_i}{T_i} = \frac{V_f}{T_f}$. Temperature **must** be in K.

Plan: Solve for the final volume, insert the known values and calculate.

Work:

$$? \text{ L} = V_f = \frac{V_i T_f}{T_i} = \frac{(500 \text{ mL})(400 \text{ K})}{(300 \text{ K})} = \mathbf{667\ mL}$$

5.47 A 375 mL sample of gas is cooled from STP to 150 K in a constant pressure process. What is the new volume?

Analysis:

Target: The final volume of a gas. (? mL) =

Knowns: The initial volume and temperature and the final temperature of a gas. STP is defined as a temperature of 0 °C, which is equal to 273.15 K, and a pressure of 1 atm.

Relationship: $\frac{V_i}{T_i} = \frac{V_f}{T_f}$. Temperature **must** be in K.

Plan: Solve for the final volume, insert the known values and calculate.

Work:

$$? \text{ mL} = V_f = \frac{V_i T_f}{T_i} = \frac{(375 \text{ mL})(150 \text{ K})}{(273.15 \text{ K})} = \mathbf{206\ mL}$$

5.49 A sample of gas occupying 2.58 L at 300 K is allowed to expand into a 4.00 L vessel. If the pressure is unchanged during the process, what is the final temperature?

Analysis:

Target: The final temperature of a gas. (? K) =

Knowns: The initial volume and temperature and the final volume of a gas.

Relationship: $\frac{V_i}{T_i} = \frac{V_f}{T_f}$. Temperature **must** be in K.

Plan: Solve for the final temperature, insert the known values and calculate.

Work:

$$? \text{ K} = T_f = \frac{V_f T_i}{V_i} = \frac{(4.00 \text{ L})(300 \text{ K})}{(2.58 \text{L})} = \mathbf{465\ K}$$

5.51 A sample of oxygen occupying 876 mL at 20 atm and 50 °C undergoes a constant-pressure expansion to a final volume of 1.60 L. What is the final temperature?

Analysis:

Target: The final temperature of a gas. (? K) =

Knowns: The initial volume and temperature and the final volume of a gas.

Relationships: $\frac{V_i}{T_i} = \frac{V_f}{T_f}$. Temperature **must** be in K. The equivalence statements needed are: 10^{-3} L ↔ 1 mL and K ↔ 273.15 + °C.

Plan: Convert the initial temperature to K and the initial volume to liters using the factor-label method. Solve for the final temperature, insert the known values and calculate.

Work:

$$T_i = 50 \text{ °C} + 273.15 = 323 \text{ K} \qquad V_i = 876 \text{ mL} \times \frac{10^{-3} \text{ L}}{1 \text{ mL}} = 0.876 \text{ L}$$

$$? \text{ K} = T_f = \frac{V_f T_i}{V_i} = \frac{(1.60 \text{ L})(323 \text{ K})}{(0.876 \text{ L})} = \mathbf{590\ K}$$

5.53 A sample of gas occupies 0.893 L at 500 K and 1 atm. If the pressure remains unchanged, to what volume must the gas be reduced in order to decrease the temperature by 20%?

Analysis:

Target: The final volume of a gas. (? L) =

Knowns: The initial volume and temperature and the percent by which the temperature is to be decreased.

Relationships: $\frac{V_i}{T_i} = \frac{V_f}{T_f}$. Temperature **must** be in K. Percent is parts per 100.

Plan: Find the final temperature, using the percent. Solve for the final volume, insert the known values and calculate.

Work:

The final temperature is $500\text{ K} - \frac{20}{100}(500\text{ K}) = 400\text{ K}$.

$$? \text{ L} = V_f = \frac{V_i T_f}{T_i} = \frac{(0.893\text{ L})(400\text{ K})}{(500\text{ K})} = \mathbf{0.714\text{ L}}$$

5.55 At what temperature will a gas have exactly half the volume it occupies at room temperature (25 °C)? The pressure is the same at both temperatures.

Analysis:

Target: The final temperature of a gas. (? K) =

Knowns: The initial volume and temperature and the final volume of a gas.

Relationships: $\frac{V_i}{T_i} = \frac{V_f}{T_f}$. Temperature **must** be in K. The equivalence statement needed is: K ↔ 273.15 + °C. If the volume is to be halved, then $V_f = 1/2V_i$.

Plan: Convert the initial temperature to K. Solve for the final temperature, insert the known values and calculate.

Work:

$T_i = 25°\text{C} + 273.15 = 298\text{ K}$

$$? \text{ K} = T_f = \frac{V_f T_i}{V_i} = \frac{(1/2\ V_i)(298\text{ K})}{(V_i)} = \mathbf{149\text{ K}}$$

5.57 A 5.00 L sample of gas at 3.35 atm and 307 K is compressed and cooled to 6.73 atm and 290 K. What is the new volume of the gas?

Analysis:

Target: The final volume of the gas. (? L) =

Knowns: The initial volume, pressure and temperature and the final pressure and temperature of the gas.

Relationships: $\frac{P_i V_i}{T_i} = \frac{P_f V_f}{T_f}$. Temperature **must** be in K.

Plan: Solve for the final volume, insert the known values and calculate.

Work:

$$? \text{ L} = V_f = \frac{P_i V_i T_f}{P_f T_i} = \frac{(3.35\text{ atm})(5.00\text{ L})(290\text{ K})}{(6.73\text{ atm})(307\text{ K})} = \mathbf{2.35\text{ L}}$$

5.59 A gas sample occupying 15.0 L at 2.00 atm and 298 K is compressed to 10.0 L. During the process the temperature rises to 333 K. What is the new pressure?

Analysis:

Target: The final pressure of the gas. (? atm) =

Knowns: The initial volume, pressure and temperature and the final volume and temperature of the gas.

Relationships: $\frac{P_i V_i}{T_i} = \frac{P_f V_f}{T_f}$. Temperature **must** be in K.

Plan: Solve for the final pressure, insert the known values and calculate.
Work:

$$? \text{ atm} = P_f = \frac{P_iV_iT_f}{V_fT_i} = \frac{(2.00 \text{ atm})(15.0 \text{ L})(333 \text{ K})}{(10.0 \text{ L})(298 \text{ K})} = \mathbf{3.35 \text{ atm}}$$

5.61 If 0.958 L of gas at 500 torr and 35 °C is expanded into a 1.00 L vessel while the pressure drops to 400 torr, what is the new temperature?

Analysis:
Target: The final temperature of the gas. (? K) =
Knowns: The initial volume, pressure and temperature and the final volume and pressure of the gas.
Relationships: $\frac{P_iV_i}{T_i} = \frac{P_fV_f}{T_f}$. Temperature **must** be in K. The equivalence statement needed is K ↔ 273.15 + °C.
Plan: Convert the initial temperature to K. Solve for the final temperature, insert the known values and calculate.
Work:

T_i = 35 °C + 273.15 = 308 K

$$? \text{ K} = T_f = \frac{P_fV_fT_i}{P_iV_i} = \frac{(400 \text{ torr})(1.00 \text{ L})(308 \text{ K})}{(500 \text{ torr})(0.958 \text{ L})} = \mathbf{257 \text{ K}} \quad \mathbf{(-16\ °C)}$$

5.63 15 L of gas at STP is expanded to a pressure of 0.75 atm and a temperature of 100 °C. What is its volume under these conditions?

Analysis:
Target: The final volume of a gas. (? L) =
Knowns: The gas is initially at STP. The final temperature and pressure are given.
Relationships: $\frac{P_iV_i}{T_i} = \frac{P_fV_f}{T_f}$. Temperature **must** be in K. The equivalence statement needed is K ↔ 273.15 + °C. STP conditions are 273.15 K and 1.00 atm.
Plan: Convert the final temperature to K. Solve for the final volume, insert the known values and calculate.
Work:

T_f = 100 °C + 273.15 = 373 K

$$? \text{ L} = V_f = \frac{P_iV_iT_f}{P_fT_i} = \frac{(1.00 \text{ atm})(15 \text{ L})(373 \text{ K})}{(0.75 \text{ atm})(273.15 \text{ K})} = \mathbf{27 \text{ L}}$$

5.65 Suppose the absolute temperature of a gas and its volume are both doubled. By what factor does the pressure increase?

Analysis:
Target: The factor by which the pressure of the gas increases. (? atm) =
Knowns: The final volume is twice the initial volume ($V_f = 2\ V_i$) and the final absolute temperature is twice the initial absolute temperature ($T_f = 2\ T_i$).
Relationship: $\frac{P_iV_i}{T_i} = \frac{P_fV_f}{T_f}$.
Plan: Solve for the final pressure, insert the known values and calculate.
Work:

$$? \text{ atm} = P_f = \frac{P_iV_iT_f}{V_fT_i} = \frac{(P_i)(V_i)(2\ T_i)}{(2\ V_i)(T_i)} = P_i$$

This combination causes **no change** in the pressure.

5.67 A gas at 1500 torr and 30 °C is compressed to 75% of its initial volume and heated to 100 °C. What is the new pressure?

Analysis:

Target: The final pressure of the gas. (? torr) =

Knowns: The initial volume, pressure and temperature of the gas. The final volume is 75% of the initial volume (V_f = 0.75 V_i) and the final temperature of the gas is given.

Relationships: $\frac{P_iV_i}{T_i} = \frac{P_fV_f}{T_f}$. Temperature **must** be in K. The equivalence statement is K ↔ 273.15 + °C.

Plan: Convert the initial and final temperatures to K. Solve for the final pressure, insert the known values and calculate.

Work:

T_i = 273.15 + 30°C = 303 K; T_f = 273.15 + 100 °C = 373 K

$$? \text{ torr} = P_f = \frac{P_iV_iT_f}{V_fT_i} = \frac{(1500 \text{ torr})(V_i)(373 \text{ K})}{(0.75\ V_i)(303 \text{ K})} = \mathbf{2.5 \times 10^3 \text{ torr}}$$

5.69 If a gas is initially at 2.00 atm and its temperature increases from 300 K to 400 K, what must the new pressure be in order to keep the volume the same?

Analysis:

Target: The final pressure of the gas. (? atm) =

Knowns: The initial pressure and temperature of the gas. The final volume is the same as the initial volume (V_f = V_i) and the final temperature of the gas is given.

Relationships: $\frac{P_iV_i}{T_i} = \frac{P_fV_f}{T_f}$. Temperature **must** be in K.

Plan: Solve for the final pressure, insert the known values and calculate.

Work:

$$? \text{ atm} = P_f = \frac{P_iV_iT_f}{V_fT_i} = \frac{(2.00 \text{ atm})(V_i)(400 \text{ K})}{(V_i)(300 \text{ K})} = \mathbf{2.67 \text{ atm}}$$

5.71 What is the pressure if 0.115 mol gas is confined to a 2.20 L volume at 150 °C?

Analysis:

Target: The pressure of a gas. (? atm) =

Knowns: The number of moles of the gas, its volume and the temperature.

Relationships: The ideal gas law, PV = nRT. R = 0.0820578 L atm mol^{-1} K^{-1}. The equivalence statement K ↔ 273.15 + °C is needed.

Plan: Convert the temperature to K. Solve the ideal gas law for the pressure, P, insert the known values and calculate.

Work:

? K = 273.15 + 150 °C = 423 K

$$? \text{ atm} = P = \frac{nRT}{V} = \frac{(0.115 \text{ mol})(0.0820578 \text{ L atm mol}^{-1} \text{ K}^{-1})(423 \text{ K})}{(2.20 \text{ L})} = \mathbf{1.81 \text{ atm}}$$

5.73 What volume is occupied by 2.00 mol gas at 200 K and 130 kPa?

Analysis:

Target: The volume of a gas. (? L) =

Knowns: The number of moles of the gas, its pressure and the temperature.

Relationships: The ideal gas law, PV = nRT. R = 0.0820578 L atm mol^{-1} K^{-1}. The equivalence statements 1 atm ↔ 1.01325 x 10^5 Pa and 1 kPa ↔ 1000 Pa are needed.

Plan: Convert the pressure to atm, using the factor-label method. Solve the ideal gas law for the volume, V, insert the known values and calculate.

Work:

$$? \text{ atm} = 130 \text{ kPa} \times \frac{1000 \text{ Pa}}{1 \text{ kPa}} \times \frac{1 \text{ atm}}{1.01325 \times 10^5 \text{ Pa}} = 1.28 \text{ atm}$$

$$? \text{ L} = V = \frac{nRT}{P} = \frac{(2.00 \text{ mol})(0.0820578 \text{ L atm mol}^{-1} \text{ K}^{-1})(200 \text{ K})}{(1.28 \text{ atm})} = \mathbf{25.6 \text{ L}}$$

5.75 A gas sample occupies 30.0 L at 0 °C and 2.75 atm. How many moles of gas are there?

Analysis:

Target: The number of moles in a gas sample. (? mol) =

Knowns: The volume of the gas, its pressure and its temperature.

Relationships: The ideal gas law, PV = nRT. R = 0.0820578 L atm mol^{-1} K^{-1}. The equivalence statement K ↔ 273.15 + °C is needed.

Plan: Convert the temperature to K. Solve the ideal gas law for the number of moles of gas, n, insert the known values and calculate.

Work:

? K = 273.15 + 0 °C = 273 K

$$? \text{ mol} = n = \frac{PV}{RT} = \frac{(2.75 \text{ atm})(30.0 \text{ L})}{(0.0820578 \text{ L atm mol}^{-1} \text{ K}^{-1})(273 \text{ K})} = \mathbf{3.68 \text{ mol}}$$

5.77 What must the temperature be to accommodate 0.0750 mol of gas in a volume of 1.50 L at 700 torr pressure?

Analysis:

Target: The temperature of a gas sample. (? K) =

Knowns: The number of moles in a gas sample, its volume and its pressure.

Relationships: The ideal gas law, PV = nRT. R = 0.0820578 L atm mol^{-1} K^{-1}. The equivalence statement 1 atm ↔ 760 torr is needed.

Plan: Convert the pressure to atm, using the factor-label method. Solve the ideal gas law for the temperature, T, insert the known values and calculate.

Work:

$$? \text{ atm} = 700 \text{ torr} \times \frac{1 \text{ atm}}{760 \text{ torr}} = 0.921 \text{ atm}$$

$$? \text{ K} = T = \frac{PV}{nR} = \frac{(0.921 \text{ atm})(1.50 \text{ L})}{(0.0750 \text{ mol})(0.0820578 \text{ L atm mol}^{-1} \text{ K}^{-1})} = \mathbf{224 \text{ K}}$$

5.79 It takes 2.61 g NO to fill a 5.00 L vessel to a pressure of 913 torr. What is the temperature?

Analysis:

Target: The temperature of a gas sample. (? K) =

Knowns: The mass of a NO sample, its volume and its pressure. The molar mass of NO is an implicit known.

Relationships: The ideal gas law, PV = nRT. R = 0.0820578 L atm mol^{-1} K^{-1}. The equivalence statement 1 atm ↔ 760 torr is needed. The molar mass gives the equivalence statement g ↔ mol.

Plan: **Step 2.** Solve the ideal gas law for T, insert the known values and calculate.
Step 1. Convert the pressure to atm and the mass of NO to moles, using the factor-label method.

Work:

Step 1. Make the necessary conversions.

$$? \text{ atm} = 913 \text{ torr} \times \frac{1 \text{ atm}}{760 \text{ torr}} = 1.20 \text{ atm}$$

Molar Mass NO = 1(14.0067 N) + 1(15.9994 O) = 30.0061 g mol^{-1}

$$? \text{ mol NO} = 2.61 \text{ g NO} \times \frac{1 \text{ mol NO}}{30.0061 \text{ g NO}} = 0.0870 \text{ mol NO}$$

Step 2. Use the ideal gas law to determine the temperature.

$$? \text{ K} = T = \frac{PV}{nR} = \frac{(1.20 \text{ atm})(5.00 \text{ L})}{(0.0870 \text{ mol})(0.0820578 \text{ L atm mol}^{-1} \text{ K}^{-1})} = \mathbf{841 \text{ K}}$$

5.81 What is the mass of 945 mL Xe at 650 torr and 230 °C?

Analysis:
Target: The mass of a Xe sample. (? g) =
Knowns: The volume of a Xe sample, its pressure and its temperature. The molar mass of Xe is an implicit known.
Relationships: The ideal gas law, PV = nRT. R = 0.0820578 L atm mol^{-1} K^{-1}. The equivalence statements 1 atm ↔ 760 torr, K ↔ 273.15 + °C and 10^{-3} L ↔ 1 mL are needed. The molar mass gives the equivalence statement g ↔ mol.
Plan:
Step 3. Use the molar mass and the number of moles in the gas sample to determine the mass by the factor-label method.
Step 2. Solve the ideal gas law for n, insert the known values and calculate.
Step 1. Convert the pressure, volume and temperature to appropriate units to use with R.
Work:
Step 1. Make the necessary conversions.

$$? \text{ atm} = 650 \text{ torr} \times \frac{1 \text{ atm}}{760 \text{ torr}} = 0.855 \text{ atm}$$

$$? \text{ L} = 945 \text{ mL} \times \frac{10^{-3} \text{ L}}{1 \text{ mL}} = 0.945 \text{ L}$$

$$? \text{ K} = 273.15 + 230 \text{ °C} = 503 \text{ K}$$

Step 2. Solve the ideal gas equation for the number of moles.

$$? \text{ mol} = n = \frac{PV}{RT} = \frac{(0.855 \text{ atm})(0.945 \text{ L})}{(0.0820578 \text{ L atm mol}^{-1} \text{ K}^{-1})(503 \text{ K})} = 0.0196 \text{ mol}$$

Step 3. Use the molar mass and the number of moles to determine the mass.
Molar Mass Xe = 131.29 g mol^{-1}

$$? \text{ g Xe} = 0.0196 \text{ mol Xe} \times \frac{131.29 \text{ g Xe}}{1 \text{ mol Xe}} = \mathbf{2.57 \text{ g Xe}}$$

5.83 What is the STP density of gaseous SF_6?

Analysis:
Target: The density of SF_6 at STP. (? g L^{-1}) =
Knowns: The molar mass of SF_6 is an implicit known.
Relationships: At STP, 1 mol of gas ↔ 22.4 L. The molar mass gives the equivalence statement g ↔ mol.
Plan: Determine the molar mass of SF_6 and then use the factor-label method with the molar mass and standard molar volume.
Work:
Molar Mass SF_6 = 1(32.06 S) + 6(18.998403 F) = 146.05 g mol^{-1}

$$? \text{ g L}^{-1} = \frac{146.05 \text{ g SF}_6}{1 \text{ mol SF}_6} \times \frac{1 \text{ mol SF}_6}{22.4 \text{ L}} = \mathbf{6.52 \text{ g SF}_6 \text{ L}^{-1}}$$

5.85 What must the pressure be in order for CO_2 to have a room temperature (25 °C) density of exactly 1.00 g L^{-1}?

Analysis:

Target: The pressure of a CO_2 sample. (? atm) =

Knowns: The density of the gas and the temperature. The molar mass of CO_2 is an implicit known.

Relationships: The ideal gas law, PV = nRT. R = 0.0820578 L atm mol^{-1} K^{-1}. The equivalence statement K ↔ 273.15 + °C is needed. The molar mass gives the equivalence statement g ↔ mol.

Plan: Assume there is exactly 1 L of gas. Then the mass of the gas would be 1.00 g.

Step 3. Solve the ideal gas law for the pressure, P, insert the known values and calculate.

Step 2. Use the factor-label method to determine the number of moles in the sample.

Step 1. Convert the temperature to K.

Work:

Step 1. Convert the temperature to K.

$$? \text{ K} = 273.15 + 25°\text{C} = 298 \text{ K}$$

Step 2. Determine the number of moles in the sample.

Molar Mass CO_2 = 1(12.011 C) + 2(15.9994 O) = 44.010 g mol^{-1}

$$? \text{ mol } CO_2 = 1.00 \text{ g } CO_2 \text{ x } \frac{1 \text{ mol } CO_2}{44.010 \text{ g } CO_2} = 0.0227 \text{ mol } CO_2$$

Step 3. Use the ideal gas law to determine the pressure.

$$? \text{ atm} = P = \frac{nRT}{V} = \frac{(0.0227 \text{ mol})(0.0820578 \text{ L atm mol}^{-1} \text{ K}^{-1})(298 \text{ K})}{(1.00 \text{ L})} = \mathbf{0.556 \text{ atm}}$$

5.87 If 1.563 g of an unknown gas is placed in a 2.509 L vessel at 25.00 °C and the measured pressure is 677.8 torr, what is the molar mass?

Analysis:

Target: The molar mass of a gas sample. (? g mol^{-1}) =

Knowns: The mass of a gas sample, its volume, its pressure and its temperature.

Relationships: The ideal gas law, PV = nRT. R = 0.0820578 L atm mol^{-1} K^{-1}. The equivalence statements 1 atm ↔ 760 torr and K ↔ 273.15 + °C are needed.

Plan: The units for molar mass dictates the procedure to be used.

Step 3. Divide the mass of the sample by the number of moles.

Step 2. Solve the ideal gas law for n, insert the known values and calculate.

Step 1. Convert the pressure and temperature to appropriate units to use with R.

Work:

Step 1. Make the necessary conversions.

$$? \text{ atm} = 677.8 \text{ torr x } \frac{1 \text{ atm}}{760 \text{ torr}} = 0.8918 \text{ atm}$$

$$? \text{ K} = 273.15 + 25.00°\text{C} = 298.15 \text{ K}$$

Step 2. Solve the ideal gas equation for the number of moles.

$$? \text{ mol} = n = \frac{PV}{RT} = \frac{(0.8918 \text{ atm})(2.509 \text{ L})}{(0.0820578 \text{ L atm mol}^{-1} \text{ K}^{-1})(298.15 \text{ K})} = 0.09146 \text{ mol}$$

Step 3. Divide the mass by the number of moles to find the molar mass.

$$? \text{ g mol}^{-1} = \frac{1.563 \text{ g}}{0.09146 \text{ mol}} = \mathbf{17.09 \text{ g mol}^{-1}}$$

5.89 What is the molar mass of a gas if a 789 mg sample has a volume of 751 mL at 280 torr and 30.0° C?

Analysis:

Target: The molar mass of a gas sample. (? g mol^{-1}) =

Knowns: The mass of a gas sample, its volume, its pressure and its temperature.

Relationships: The ideal gas law, PV = nRT. R = 0.0820578 L atm mol^{-1} K^{-1}. The equivalence statements 1 atm ↔ 760 torr, K ↔ 273.15 + °C, 10^{-3} L ↔ 1 mL, and 10^{-3} g ↔ 1 mg are needed.

Plan: The units for molar mass dictate the procedure to be used.

Step 3. Convert the mass of the sample to grams and divide it by the number of moles.

Step 2. Solve the ideal gas law for the number of moles, insert the known values and calculate.

Step 1. Convert the pressure, volume and temperature to appropriate units to use with R.

Work:

Step 1. Make the necessary conversions to use with R.

$$? \text{ atm} = 280 \text{ torr} \times \frac{1 \text{ atm}}{760 \text{ torr}} = 0.368 \text{ atm}$$

$$? \text{ L} = 751 \text{ mL} \times \frac{10^{-3} \text{ L}}{1 \text{ mL}} = 0.751 \text{ L}$$

$$? \text{ K} = 273.15 + 30.0 \text{ C} = 303.2 \text{ K}$$

Step 2. Solve the ideal gas equation for the number of moles.

$$? \text{ mol} = n = \frac{PV}{RT} = \frac{(0.368\text{atm})(0.751 \text{ L})}{(0.0820578 \text{ L atm mol}^{-1} \text{ K}^{-1})(303.2 \text{ K})} = 0.0111 \text{ mol}$$

Step 3. Convert the mass to grams and divide by the number of moles to find the molar mass.

$$? \text{ g mol}^{-1} = \frac{789 \text{ mg} \times \frac{10^{-3} \text{ g}}{1 \text{ mg}}}{0.0111 \text{ mol}} = \mathbf{70.9 \text{ g mol}^{-1}}$$

5.91 What is the molar mass of a gas whose STP density is 2.86 g L^{-1}?

Analysis:

Target: The molar mass of a gas. (? g mol^{-1}) =

Knowns: The density of the gas at STP.

Relationship: At STP, 1 mol of gas ↔ 22.4 L.

Plan: Use the factor-label method with the density and the standard molar volume.

Work:

$$? \text{ g mol}^{-1} = \frac{22.4 \text{ L}}{1 \text{ mol}} \times \frac{2.86 \text{ g}}{\text{L}} = \mathbf{64.1 \text{ g mol}^{-1}}$$

5.93 What is the density of acetylene (C_2H_2) at 5.00 atm and 25 °C?

Analysis:

Target: The density of a C_2H_2 sample. (? g L^{-1}) =

Knowns: The pressure and the temperature of the sample. The molar mass of C_2H_2 is an implicit known.

Relationships: The ideal gas law, PV = nRT. R = 0.0820578 L atm mol^{-1} K^{-1}. The equivalence statement K ↔ 273.15 + °C is needed. The molar mass gives the equivalence statement g ↔ mol.

Plan: Assume exactly 1 L of the gas and determine the mass of C_2H_2 in that volume.

Step 3. Use the molar mass and the number of moles in the gas sample to determine the mass by the factor-label method.
Step 2. Solve the ideal gas law for n, insert known values and calculate.
Step 1. Convert the temperature K.

Work:

Step 1. Convert the temperature to K.

$$? \text{ K} = 273.15 + 25\ °\text{C} = 298 \text{ K}$$

Step 2. Solve the ideal gas equation for the number of moles.

$$? \text{ mol} = n = \frac{PV}{RT} = \frac{(5.00 \text{ atm})(1.00 \text{ L})}{(0.0820578 \text{ L atm mol}^{-1} \text{ K}^{-1})(298 \text{ K})} = 0.204 \text{ mol}$$

Step 3. Use the molar mass and the number of moles to determine the mass.
Molar Mass C_2H_2 = 2(12.011 C) + 2(1.0079 H) = 26.038 g mol^{-1}

$$? \text{ g } C_2H_2 = 0.204 \text{ mol } C_2H_2 \times \frac{26.038 \text{ g } C_2H_2}{1 \text{ mol } C_2H_2} = 5.32 \text{ g } C_2H_2.$$

Since we assumed 1.00 L, the density of C_2H_2 under these conditions is **5.32 g L^{-1}**.

5.95 Calculate the average velocity of nitrogen molecules at STP. (*Hint:* Recall that, in SI, the unit of mass is the kg; the molar mass of N_2 is 0.0280134 kg mol^{-1}.)

Analysis:

Target: The average velocity of a N_2 molecule. (? m s^{-1}) =
Knowns: The temperature of the gas. The molar mass of N_2 in SI units is given.
Relationships: $\langle v \rangle = (8RT/\pi M)^{1/2}$. The value of R needed to give the appropriate units for velocity is 8.31451 J mol^{-1} K^{-1}. At STP, T = 273.15 K. 1 J ↔ 1 kg m^2 s^{-2}.

Plan: Replace J by its equivalent units, then use the formula directly to determine $\langle v \rangle$.

Work:

$$? \text{ m s}^{-1} = \sqrt{\frac{(8)(8.31451 \text{ kg m}^2 \text{ s}^{-2} \text{ mol}^{-1} \text{ K}^{-1})(273.15 \text{ K})}{(3.14159)(0.0280134 \text{ kg mol}^{-1})}} = \mathbf{454 \text{ m s}^{-1}}$$

5.97 Suppose conditions are such that the molecules of a certain gas have an average velocity of 500 m s^{-1}. What will the average velocity be if the pressure is doubled with no change in temperature?

The velocity is independent of the pressure and only depends on the molar mass of the gas and its absolute temperature. Therefore, the average velocity would be the same-- **500 m s^{-1}**.

5.99 What is the molar mass of a gas whose average molecular velocity at STP is 514 m s^{-1}?

Analysis:

Target: The molar mass of a molecule. (? g mol^{-1}) =
Knowns: The temperature of the gas. The average velocity of the gas.
Relationships: $\langle v \rangle = (8RT/\pi M)^{1/2}$. The value of R needed to give the appropriate units for velocity is 8.31451 J mol^{-1} K^{-1}. At STP, T = 273.15 K. The equivalence statements 1 kg ↔ 1000 g and 1 J ↔ 1 kg m^2 s^{-2} are needed.

Plan: Rearrange the formula to determine the molar mass. Replace J by its equivalent units, then use the factor-label method to convert to the more typical units of g mol^{-1}.

Work:

$$\langle v \rangle = (8RT/\pi M)^{1/2} \quad \text{therefore, } \langle v \rangle^2 = \frac{8RT}{\pi M} \quad \text{and } \pi M \langle v \rangle^2 = 8RT$$

$$M = \frac{8RT}{\pi \langle v \rangle^2} = \frac{(8)(8.31451\ \text{kg m}^2\ \text{s}^{-2}\ \text{mol}^{-1}\ \text{K}^{-1})(273.15\ \text{K})}{(3.14159)(514\ \text{m s}^{-1})^2} = 0.0219\ \text{kg mol}^{-1}$$

$$?\ \text{g mol}^{-1} = 0.0219\ \text{kg mol}^{-1} \times \frac{1000\ \text{g}}{1\ \text{kg}} = \mathbf{21.9\ g\ mol^{-1}}$$

5.101 The velocity an object (or a molecule) must have in order to overcome the earth's gravitational field and escape into space is called the "escape velocity"; it is 11.2 km s^{-1}. What must the temperature be for an average H_2 molecule to escape from the earth?

Analysis:

Target: The temperature of escape velocity for H_2. (? K) =

Knowns: The velocity needed for escape. The molar mass of H_2 is an implicit known.

Relationships: $\langle v \rangle = (8RT/\pi M)^{1/2}$. The value of R needed to give the appropriate units for velocity is 8.31451 J mol^{-1} K^{-1}. Equivalence statements 1 J ↔ 1 kg m^2 s^{-2} and 1000 m ↔ 1 km are needed. The molar mass gives the equivalence statement g ↔ mol.

Plan: Rearrange the formula to determine the temperature, using the factor-label method to convert the velocity to m s^{-1}. Replace J by its equivalent units.

Work:

Molar Mass H_2 = 2(0.0010079) = 0.0020158 kg mol^{-1}

$$\langle v \rangle = (8RT/\pi M)^{1/2} \quad \text{therefore, } \langle v \rangle^2 = \frac{8RT}{\pi M} \quad \text{and } T = \frac{\langle v \rangle^2 \pi M}{8R}$$

$$?\ K = T = \frac{\left(11.2\ \text{km s}^{-1} \times \frac{1000\ \text{m}}{1\ \text{km}}\right)^2 (3.14159)(0.0020158\ \text{kg mol}^{-1})}{(8)(8.31451\ \text{kg m}^2\ \text{s}^{-2}\ \text{mol}^{-1}\ \text{K}^{-1})} = \mathbf{1.19 \times 10^4\ K}$$

5.103 What is the ratio of the average speeds of O_2 and N_2 molecules at 298 K? What is the ratio of their average energies at this temperature?

Analysis:

Target: The ratio of the average speeds of O_2 and N_2. $\left(?\ \frac{\langle v \rangle_{N_2}}{\langle v \rangle_{O_2}}\right)$ =

Knowns: The molar masses of O_2 and N_2 are implicit knowns.

Relationships: The rates of effusion are directly related to the velocity. The Graham's law relationship yields $\frac{\langle v \rangle_A}{\langle v \rangle_B} = \left(\frac{M_B}{M_A}\right)^{1/2}$.

Plan: Compute the molar masses of N_2 and O_2 and use this relationship.

Work:

Molar Mass O_2 = 2(15.9994) = 31.9988 g mol^{-1}
Molar Mass N_2 = 2(14.0067) = 28.0134 g mol^{-1}

$$\frac{\langle v \rangle_{N_2}}{\langle v \rangle_{O_2}} = \left(\frac{M_{O_2}}{M_{N_2}}\right)^{1/2} = \left(\frac{31.9988\ \text{g mol}^{-1}}{28.0134\ \text{g mol}^{-1}}\right)^{1/2} = \mathbf{1.06877}$$

The N_2 has a velocity 1.06877 times as fast as the O_2. Both have the **same average kinetic energy** because they are at the same temperature.

5.105 An unknown gas effuses from a container at a rate of 6.46 x 10^{-5} mol s^{-1}. Under the same conditions argon effuses at 1.38 x 10^{-4} mol s^{-1}. What is the molar mass of the unknown gas?

Analysis:

Target: The molar mass of an unknown gas. (? g mol^{-1}) =

Length

Metric

$1\,km = 10^3\,m$

$1\,cm = 10^{-2}\,m$

$1\,mm = 10^{-3}\,m$

$1\,nm = 10^{-9}\,m = 10\,Å$

English

$1\,ft = 12\,in$

$1\,yd = 3\,ft$

$1\,mile = 5280\,ft$

Metric - English

$1\,in = 2.54\,cm$

$1\,m = 39.37\,in$

$1\,mile = 1.609\,km$

Volume

Metric

$1\,m^3 = 10^6\,cm^3 = 10^3\,L$

$1\,cm^3 = 1\,mL = 10^{-3}\,L$

English

$1\,gallon = 4\,qt = 8\,pt$

$1\,qt\,(Can.) = 69.35\,in^3$

$1\,qt\,(US\ liq) = 57.75\,in^3$

Metric - English

$1\,ft^3 = 28.32\,L$

$1\,L = 0.8799\,qt\,(Can)$

$1\,L = 1.057\,qt\,(US\ liq)$

Knowns: The rate of effusion of the unknown gas and the rate of effusion of Ar at the same temperature. The molar mass of Ar is an implicit known.

Relationship: Graham's law $\frac{rate_{eff\ A}}{rate_{eff\ B}} = \left(\frac{M_B}{M_A}\right)^{1/2}$.

Plan: Rearrange this relationship to solve for the molar mass of the unknown. Let Ar be **Ar** and the unknown gas be **B**.

Work:

Molar Mass Ar = 39.948 g mol^{-1}

$$\frac{rate_{eff\ Ar}}{rate_{eff\ B}} = \left(\frac{M_B}{M_{Ar}}\right)^{1/2}, \text{ therefore } \left(\frac{rate_{eff\ Ar}}{rate_{eff\ B}}\right)^2 = \left(\frac{M_B}{M_{Ar}}\right)$$

$$M_B = (M_{Ar})\left(\frac{rate_{eff\ Ar}}{rate_{eff\ B}}\right)^2 = 39.948 \text{ g mol}^{-1}\left(\frac{1.38 \times 10^{-4} \text{ mol s}^{-1}}{6.46 \times 10^{-5} \text{ mol s}^{-1}}\right)^2 = \mathbf{182\ g\ mol^{-1}}$$

5.107 A gas is collected over water when the atmospheric pressure is 745 torr and the temperature is such that the vapor pressure of water is 25 torr. What is the partial pressure of the unknown gas?

Analysis/Plan: Dalton's law states that the total pressure of a gas mixture is equal to the sum of the partial pressures. Subtract the pressure due to water vapor from the total pressure to determine the pressure of the gas.

Work:

? torr = 745 torr total - 25 torr H_2O = **720 torr**

5.109 The mole fraction of helium in a certain gas mixture is 0.974. What is the partial pressure of helium when the total pressure of the mixture is 2.28 atm?

Analysis/Plan: The relationship $P_A = X_A P_T$ relates the mole fraction, X_A, to the partial pressure, P_A, and the total pressure, P_T. Use this relationship directly.

Work:

? atm = P_{He} = 0.974(2.28 atm) = **2.22 atm**

5.111 A mixture of 1.00 mol He and 2.00 mol N_2 is put into a 10.0 L container at 298 K. Calculate the total pressure, and the partial pressure and mole fraction of each gas. Calculate the mass percent of He in the mixture.

Analysis:

Targets: The total pressure, the partial pressure of each gas, the mole fraction of each gas and the mass percent He.

Knowns: The number of moles of each gas, the container volume and the temperature. The molar mass of each gas is an implicit known.

Relationships: The total number of moles of gas is the sum of the number of moles of the individual gases. The total pressure can be found from the ideal gas law, $P_{Total}V = n_{Total}RT$. The mole fraction, X_A, is defined as $\frac{n_A}{n_{Total}}$. The partial pressure of each gas is found from the relationship, $P_A = X_A P_T$. The mass percent of A is defined as $\frac{mass_A}{mass_{Total}} \times 100$. The molar mass gives the equivalence statement g ↔ mol.

Plan: The following series of steps is used to find all of the information requested.

Step 6. Find the mass percent He, using the relationship that % He = $\frac{mass_{He}}{mass_{Total}} \times 100$.

Step 5. Find the mass of each gas, using its molar mass, its number of moles and the factor-label method.

Step 4. Find the partial pressure of each gas, using the relationship $P_A = X_A P_T$.

Step 3. Find the mole fraction of each gas, using the relationship $X_A = \frac{n_A}{n_{Total}}$.
Step 2. Find the total pressure, by rearranging the ideal gas law to solve for P, inserting the known values and calculating.
Step 1. Find the total number of moles.

Work:

Step 1. Find the total number of moles.

$$? \text{ mol total} = 1.00 \text{ mol He} + 2.00 \text{ mol } N_2 = 3.00 \text{ mol total}$$

Step 2. Find the total pressure.

$$? \text{ atm} = P = \frac{nRT}{V} = \frac{(3.00 \text{ mol})(0.0820578 \text{ L atm mol}^{-1} \text{ K}^{-1})(298 \text{ K})}{(10.0 \text{ L})} = \mathbf{7.34 \text{ atm}}$$

Step 3. Find the mole fraction of each gas.

$$X_{He} = \frac{1.00 \text{ mol He}}{3.00 \text{ mol total}} = \mathbf{0.333}$$

$$X_{N_2} = \frac{2.00 \text{ mol } N_2}{3.00 \text{ mol total}} = \mathbf{0.667}$$

Step 4. Find the partial pressure of each gas.

$$? \text{ atm He} = X_{He}P_T = 0.333(7.34 \text{ atm}) = \mathbf{2.45 \text{ atm}}$$

$$? \text{ atm } N_2 = X_{N_2}P_T = 0.667(7.34 \text{ atm}) = \mathbf{4.89 \text{ atm}}$$

Step 5. Find the mass of each gas.
Molar Mass He = 4.00260 g mol^{-1}

$$? \text{ g He} = 1.00 \text{ mol He} \times \frac{4.00260 \text{ g He}}{1 \text{ mol He}} = 4.00 \text{ g He}$$

Molar Mass N_2 = 2(14.0067) = 28.0134 g mol^{-1}

$$? \text{ g } N_2 = 2.00 \text{ mol } N_2 \times \frac{28.0134 \text{ g } N_2}{1 \text{ mol } N_2} = 56.0 \text{ g } N_2$$

Step 6. Find the mass percent He, using the relationship that % He = $\frac{\text{mass He}}{\text{mass total}} \times 100$.

$$? \text{ \% He} = \frac{4.00 \text{ g He}}{4.00 \text{ g He} + 56.0 \text{ g } N_2} \times 100 = \mathbf{6.67\% \text{ He}}$$

5.113 The STP density of a mixture of hydrogen and oxygen is 1.00 g L^{-1}. Calculate the mole fraction and partial pressure of each gas.

Analysis:

Target: The mole fraction and the partial pressure of H_2 and O_2 in a mixture.
Knowns: The STP density of the mixture and the pressure at STP. The molar masses of H_2 and O_2 are implicit knowns.
Relationships: At STP, 1 mol of gas ↔ 22.4 L and the pressure is defined as 1 atm. The mole fraction, X_A is defined as $X_A = \frac{n_A}{n_{Total}}$. The partial pressure of a gas is $P_A = X_AP_T$. The molar mass gives the equivalence statement g ↔ mol.

Plan: Assume that we have exactly 1 L of the mixture. It would have a total mass of 1.00 g. We can find the total number of moles in the sample by using the standard molar volume. We can let the number of moles of one gas = x and the other = n_T -x. The knowledge that **Molar Mass x n = mass** and algebra can be used to determine **x**, and then the above relationships can be used.

Step 4. Find the partial pressure of each gas, using the relationship $P_A = X_A P_T$.

Step 3. Find the mole fraction of each gas, using the relationship $X_A = \frac{n_A}{n_{Total}}$.

Step 2. Use the molar masses and algebra to determine the number of moles of each gas.

Step 1. Use the standard molar volume to determine the total number of moles in the mixture.

Work:

Step 1. Use the standard molar volume to determine the total number of moles in the mixture.

$$? \text{ mol} = 1.00 \text{ L} \times \frac{1.00 \text{ mol}}{22.4 \text{ L}} = 0.0446 \text{ mol total}$$

Step 2. Use the molar masses and algebra to determine the number of moles of each gas.

Molar Mass O_2 = 2(15.9994) = 31.9988 g mol^{-1}; let x = number of moles of O_2

Molar Mass H_2 = 2(1.0079) = 2.0158 g mol^{-1}; let 0.0446 - x = number of moles of H_2

$$\frac{31.9988 \text{ g } O_2}{1 \text{ mol } O_2}(x) + \frac{2.0158 \text{ g } H_2}{1 \text{ mol } H_2}(0.0446 - x) = 1.00 \text{ g}$$

$$31.9988x + 0.0899 - 2.0158x = 1.00$$

$$29.9830x = 1.00 - 0.0899 = 0.91$$

$$x = \frac{0.91}{29.9830} = 0.030 \text{ mol} = \text{number of moles of } O_2$$

$$0.0446 - x = 0.0446 \text{ mol} - 0.030 \text{ mol} = 0.015 \text{ mol} = \text{number of moles of } H_2$$

Step 3. Find the mole fraction of each gas, using the relationship $X_A = \frac{n_A}{n_{Total}}$.

$$X_{H_2} = \frac{0.015 \text{ mol } H_2}{0.015 \text{ mol } H_2 + 0.030 \text{ mol } O_2} = \mathbf{0.32}$$

$$X_{O_2} = \frac{0.030 \text{ mol } O_2}{0.015 \text{ mol } H_2 + 0.030 \text{ mol } O_2} = \mathbf{0.68}$$

Step 4. Find the partial pressure of each gas, using the relationship $P_A = X_A P_T$.

$$P_{H_2} = (0.32)(1.00 \text{ atm}) = \mathbf{0.32 \text{ atm}}$$

$$P_{O_2} = (0.68)(1.00 \text{ atm}) = \mathbf{0.68 \text{ atm}}$$

5.115 What volume of CO_2, measured at 30 °C and 745 torr, is produced by the action of acid on 1.00 g $CaCO_3$? The reaction is $CaCO_3 + 2\,HCl \longrightarrow CaCl_2 + H_2O + CO_2$.

Analysis:

Target: The volume of CO_2. (? L) =

Knowns: The mass of $CaCO_3$, the balanced equation for the process, and the pressure and the temperature of the product gas. The molar mass of $CaCO_3$ is an implicit known.

Relationships: The ideal gas law, PV = nRT. R = 0.0820578 L atm mol^{-1} K^{-1}. The equivalence statements 1 atm ↔ 760 torr and K ↔ 273.15 + °C are needed. The mole ratios are given by coefficients in the balanced equation. The standard stoichiometric procedure (SSP) can be used to find the number of moles of CO_2 produced. The molar mass gives the equivalence statement g ↔ mol.

Plan: The following steps will give the desired information.

Step 3. Solve the ideal gas law for V, insert the known values and calculate.
Step 2. Convert temperature and pressure to appropriate units to be used with R.
Step 1. Use the SSP to find the number of moles of CO_2.

$$g\ CaCO_3 \longrightarrow mol\ CaCO_3 \longrightarrow mol\ CO_2 \longrightarrow L\ CO_2$$

Work:

Step 1. Use the SSP to find the number of moles of CO_2.
Molar Mass $CaCO_3$ = 1(40.078 Ca) + 1(12.011 C) + 3(15.9994 O) = 100.087 g mol^{-1}

$$?\ mol\ CO_2 = 1.00\ g\ CaCO_3 \times \frac{1\ mol\ CaCO_3}{100.087\ g\ CaCO_3} \times \frac{1\ mol\ CO_2}{1\ mol\ CaCO_3} = 9.99 \times 10^{-3}\ mol\ CO_2$$

Step 2. Convert temperature and pressure to appropriate units to be used with R.

$$?\ atm = 745\ torr \times \frac{1\ atm}{760\ torr} = 0.980\ atm$$

$$?\ K = 273.15 + 30\ °C = 303\ K$$

Step 3. Solve the ideal gas law for V, insert the known values and calculate.

$$?\ L = V = \frac{nRT}{P} = \frac{(9.99 \times 10^{-3}\ mol)(0.0820578\ L\ atm\ mol^{-1}\ K^{-1})(303\ K)}{(0.980\ atm)} = \mathbf{0.253\ L}$$

5.117 If 15.0 L (STP) of SO_2 is formed when S reacts with O_2, how much S was consumed in the reaction?

Analysis:

Target: The mass of S which reacted. (? g S)=
Knowns: The volume of SO_2 formed and the nature of the reactants and the product. The molar mass of S is an implicit known.
Relationships: The balanced equation can be written from knowledge of the reactants and product. The mole ratios are given by coefficients in the balanced equation. The standard molar volume (1 mol of gas ↔ 22.4 L) can be used with the SSP. The molar mass gives the equivalence statement g ↔ mol.

Plan: The following steps will give the desired information.

Step 2. Use standard molar volume with the SSP to find the mass of S.

$$L\ SO_2 \longrightarrow mol\ SO_2 \longrightarrow mol\ S \longrightarrow g\ S$$

Step 1. Write the balanced chemical equation.

Work:

Step 1. Write the balanced chemical equation.
The skeleton equation is $S + O_2 \longrightarrow SO_2$. (Balanced: 1 S atom; 2 O atoms each side)

Step 2. Use the standard molar volume and the SSP to find the mass of S.
Molar Mass S = 32.06 g mol^{-1}

$$?\ g\ S = 15.0\ L\ SO_2 \times \frac{1\ mol\ SO_2}{22.4\ L\ SO_2} \times \frac{1\ mol\ S}{1\ mol\ SO_2} \times \frac{32.06\ g\ S}{1\ mol\ S} = \mathbf{21.5\ g\ S}$$

5.119 How many liters of H_2, measured at 600 K and 7.5 atm, are required to react with 100 L N_2 (also measured at 600 K and 7.5 atm) to form NH_3? What volume of NH_3 is produced?

Analysis:

Target: The volumes of H_2 and of NH_3. (? L H_2) = and (? L NH_3) =

Knowns: The nature of the reactants and the products and the volume of N_2, measured at the same temperature and pressure as the other gases.
Relationships: If gases are at the same pressure and temperature, coefficients give the combining volumes as well as the mole ratios. From the nature of the reactants and product, the balanced equation can be written.
Plan: Write the balanced equation, then use the coefficients in the balanced equation as volume ratios.
Work:

The skeleton equation is $N_2 + H_2 \longrightarrow NH_3$. The N can be balanced by using a coefficient of 2 for the NH_3.

$N_2 + H_2 \longrightarrow 2\ NH_3$ 1(2) N left = 2(1) N right

The balancing is completed by using a coefficient of 3 for H_2.

$N_2 + 3\ H_2 \longrightarrow 2\ NH_3$ 3(2) H left = 2(3) H right.

$$?\ L\ H_2 = 100\ L\ N_2 \times \frac{3\ L\ H_2}{1\ L\ N_2} = \mathbf{300\ L\ H_2}$$

$$?\ L\ NH_3 = 100\ L\ N_2 \times \frac{2\ L\ NH_3}{1\ L\ N_2} = \mathbf{200\ L\ NH_3}$$

5.121 A certain type of coal contains 1.89% (w/w) of sulfur. When the coal burns, all the sulfur is converted to gaseous SO_2. What STP volume of SO_2 is produced when a 500 mg sample of coal is burned?

Analysis:
Target: A volume of SO_2. (? mL) =
Knowns: The mass percent S in the coal, the nature of the reactant and product and the nature of STP conditions.
Relationships: The balanced equation can be written from the nature of the reactants and products. The mass percent provides the factor g S/100 g coal. The standard molar volume (1 mol of gas ↔ 22.4 L) can be used with the SSP. The molar mass gives the equivalence statement g ↔ mol.
Plan:
Step 2. Use the SSP, with both the mass percent S and the standard molar volume as factors.

mg coal ⟶ mg S ⟶ mmol S ⟶ mmol SO_2 ⟶ mL SO_2

Step 1. Write the balanced equation for the reaction.
Work:
Step 1. The balanced equation is $S + O_2 \longrightarrow SO_2$. (See Problem 5.117)

Step 2. Use the modified standard stoichiometric procedure. Note that if we leave the mass in mg, units of mL result. Molar Mass S = 32.06 g mol^{-1} or 32.06 mg $mmoL^{-1}$.

$$?\ mL\ SO_2 = 500\ mg\ coal \times \frac{1.89\ mg\ S}{100\ mg\ coal} \times \frac{1\ mmol\ S}{32.06\ mg\ S} \times \frac{1\ mmol\ SO_2}{1\ mmol\ S} \times \frac{22.4\ mL\ SO_2}{1\ mmol\ SO_2} =$$

6.60 mL SO_2

5.123 A gaseous compound containing carbon and hydrogen is analyzed and found to contain 85.7% C. The mass of 100 mL of this gas, measured at 100 °C and 100 torr, is 0.04 g. What is the molecular formula?

Analysis:
Target: The molecular formula of the compound.

Knowns: The mass of a gas sample, its volume, its pressure and its temperature. The percent composition of the gas is given.
Relationships: The mass percent composition in a compound must add to 100%. From mass percent composition and molar masses, the empirical formula can be determined. The molecular formula is a whole number multiple of the empirical formula. Molar mass has units of g mol^{-1}. R = 0.0820578 L atm mol^{-1} K^{-1}. The equivalence statements 1 atm ↔ 760 torr, K ↔ 273.15 + °C and 10^{-3} L ↔ 1 mL are needed.

Plan:
Step 8. Multiply each subscript by the number of empirical units in one molecule.
Step 7. Divide the molar mass of the compound by the molar mass of the empirical unit to determine the number of empirical units in one molecule.
Step 6. Determine the mass of the empirical unit.
Step 5. Divide each number of moles by the smaller number to find the empirical formula.
Step 4. Assume 100 g of compound, then use the molar masses and the percent composition data to determine the number of moles of each element.
Step 3. Convert the mass of the sample to grams and divide it by the number of moles to determine the molar mass of the compound.
Step 2. Solve the ideal gas law for n, insert the known values and calculate.
Step 1. Convert the pressure, volume and temperature to appropriate units to use with R.

Work:
Step 1. Make the necessary conversions to use with R.

$$? \text{ atm} = 100 \text{ torr} \times \frac{1 \text{ atm}}{760 \text{ torr}} = 0.132 \text{ atm}$$

$$? \text{ L} = 100 \text{ mL} \times \frac{10^{-3} \text{ L}}{1 \text{ mL}} = 0.100 \text{ L}$$

$$? \text{ K} = 273.15 + 100 \text{ °C} = 373 \text{ K}$$

Step 2. Solve the ideal gas equation for the number of moles.

$$? \text{ mol} = n = \frac{PV}{RT} = \frac{(0.132 \text{ atm})(0.100 \text{ L})}{(0.0820578 \text{ L atm mol}^{-1} \text{ K}^{-1})(373 \text{ K})} = 4.31 \times 10^{-4} \text{ mol}$$

Step 3. Divide the mass by the number of moles to find the molar mass.

$$? \text{ g mol}^{-1} = \frac{0.04 \text{ g}}{4.31 \times 10^{-4} \text{ mol}} = 9 \times 10^{1} \text{ g mol}^{-1}$$

Step 4. Assume 100 g of compound. Then use the molar masses and the percent composition data to determine the number of moles of each element.
Molar Mass C = 12.011 g mol^{-1}

$$? \text{ mol C} = 85.7 \text{ g C} \times \frac{1 \text{ mol C}}{12.011 \text{ g C}} = 7.14 \text{ mol C}$$

Molar Mass H = 1.0079 g mol^{-1}

$$? \text{ mol H} = 100.0 \text{ g compound} - 85.7 \text{ g C} = 14.3 \text{ g H} \times \frac{1 \text{ mol H}}{1.0079 \text{ g H}} = 14.2 \text{ mol H}$$

Step 5. Divide each number of moles by the smaller number to find the empirical formula.

$$\frac{7.14 \text{ mol C}}{7.14 \text{ mol C}} = 1.00 \qquad \frac{14.2 \text{ mol H}}{7.14 \text{ mol C}} = 1.99$$

The empirical formula is CH_2.

Step 6. Determine the mass of the empirical unit.
Molar Mass CH_2 = 1(12.011 C) + 2(1.0079 H) = 14.027 g mol^{-1}

Step 7. Divide the molar mass of the compound by the molar mass of the empirical unit to determine the number of empirical units in one molecule.

$$\frac{9 \times 10^1 \text{ g mol}^{-1}}{14.027 \text{ g mol}^{-1}} \approx 7$$

Step 8. Multiply each subscript by the number of empirical units in one molecule.

The molecular formula is $\mathbf{C_7H_{14}}$.

5.125 Given that constants in the van der Waals equation have values of a = 3.59 L^2 atm mol^{-2} and b = 0.043 L mol^{-1} for carbon dioxide, calculate the pressure exerted by 5.0 mol CO_2 when confined to a 1.00 L container at 300 K. Compare this to the pressure that would be exerted if the gas behaved ideally.

Analysis:

Target: The pressure of a gas. (? atm) =

Knowns: The number of moles, the volume and the temperature and the van der Waals constants for CO_2.

Relationships: The ideal gas law, PV = nRT, can be used for ideal conditions and the van der Waals equation, $(P + \frac{an^2}{V^2})(V - nb) = nRT$ can be used for the nonideal conditions.

Plan: Solve the appropriate equation for P and insert the known values.

Work:

If the ideal gas law is used, the pressure is:

$$? \text{ atm} = P = \frac{nRT}{V} = \frac{(5.0 \text{ mol})(0.0820578 \text{ L atm mol}^{-1} \text{ K}^{-1})(300 \text{ K})}{1.00 \text{ L}} = \mathbf{1.2 \times 10^2 \text{ atm}}$$

If the van der Waals equation is used, the pressure is:

$$\left(P + \frac{(3.59 \text{ L}^2 \text{ atm mol}^{-2})(5.0 \text{ mol})^2}{(1.00 \text{ L})^2}\right)[1.00 \text{ L} - (0.043 \text{ L mol}^{-1})(5.00 \text{ L})] =$$

$$(5.0 \text{ mol})(0.0820578 \text{ L atm mol}^{-1} \text{ K}^{-1})(300 \text{ K})$$

$$(P + 89.8)(0.78) = 1.2 \times 10^2$$

$$P + 89.8 = \frac{1.2 \times 10^2}{0.78} = 1.6 \times 10^2$$

$P = 1.6 \times 10^2 - 89.8 = \mathbf{7 \times 10^1 \text{ atm}}$ (This is a **much** smaller pressure.)

5.127 One of the Group 3A elements is the third most abundant element on Earth. Which one?

Aluminum is the third most abundant element on Earth.

5.129 Which of the Group 3A elements is very toxic?

Thallium is very toxic.

5.131 What properties of aluminum account for its wide-spread use in our society?

Aluminum has a low density, high electrical conductivity and is resistant to corrosion.

5.133 Each of the Group 3A metals is produced by electrolysis of a solution of one of its compounds. For each element, identify the solvent.

For aluminum, the solvent is cryolite (Na_3AlF_6). For gallium, indium and thallium, the solvent is water.

5.135 All the metals of Group 3A react with chlorine to form chlorides of empirical formula MCl_3. Which two elements form *molecular* compounds having the formula M_2Cl_6?

Aluminum forms Al_2Cl_6 and gallium forms Ga_2Cl_6.

5.137 Suppose that the electrolysis of 1.00 kg of an impure sample of bauxite yields 416 g of aluminum metal. What is the percentage of AlO(OH) in the bauxite sample?

Analysis:
Target: A percent. (? %) =
Knowns: The mass of ore and the mass of aluminum produced from it. The molar masses are implicit knowns.
Relationships: The mass percent is $\frac{\text{mass AlO(OH)}}{\text{mass ore}}$ x 100. In the absence of other aluminum compounds, we can assume that 1 mole AlO(OH) provides 1 mole of Al. The molar mass gives the equivalency statement g ↔ mol. The equivalence statement 1 kg ↔ 1000 g is needed.
Plan: Assume that bauxite is the only source of aluminum. Determine the mass of AlO(OH) in the original sample using the SSP, and then use the definition of mass percent, converting the mass of ore to grams with the factor-label method.

g Al ⟶ mol Al ⟶ mol AlO(OH) ⟶ g AlO(OH)

Work:
Molar Mass Al = 26.98154 g mol^{-1}
Molar Mass AlO(OH) = 1(26.98154 Al) + 2(15.9994 O) + 1(1.0079 H) = 59.9882 g mol^{-1}

$$? \text{ g AlO(OH} = 416 \text{ g Al} \times \frac{1 \text{ mol Al}}{26.98154 \text{ g Al}} \times \frac{1 \text{ mol AlO(OH)}}{1 \text{ mol Al}} \times \frac{59.9882 \text{ g AlO(OH)}}{1 \text{ mol AlO(OH)}} =$$

925 g AlO(OH)

$$? \text{ \% AlO(OH)} = \frac{925 \text{ g AlO(OH)}}{1.00 \text{ kg ore} \times \frac{1000 \text{ g}}{1 \text{ kg}}} \times 100 = \mathbf{92.5 \text{ \% AlO(OH)}}$$

5.139 What volume (STP) of hydrogen is produced when 15.4 g Al dissolves in hot NaOH(aq)?

Analysis:
Target: The volume of H_2 produced at STP. (? L) =
Knowns: The mass of Al reacted. The molar mass of Al is an implicit known. Section 5.8 gives the appropriate reaction: 2 Al(s) + 6 NaOH(aq) ⟶ 2 Na_3AlO_3(aq) + 3 H_2(g).
Relationships: The balanced equation provides the mole ratios. The molar mass gives the equivalency statement g ↔ mol. At STP, 1 mol of gas ↔ 22.4 L.
Plan: Use the SSP with the standard molar volume.

g Al ⟶ mol Al ⟶ mol H_2 ⟶ L H_2

Work:
Molar Mass Al = 26.98154 g mol^{-1}

$$? \text{ L } H_2 = 15.4 \text{ g Al} \times \frac{1 \text{ mol Al}}{26.98154 \text{ g Al}} \times \frac{3 \text{ mol } H_2}{2 \text{ mol Al}} \times \frac{22.4 \text{ L } H_2}{1 \text{ mol } H_2} = \mathbf{19.2 \text{ L } H_2}$$

5.141 Aluminum reacts with the oxides of other metals, for example iron, to produce aluminum oxide and the other metal. This reaction, called the *thermite* reaction, is sufficiently exothermic that it can be used for welding of steel rails and pipes.

a. Write the balanced equation for the reaction of aluminum with iron(III) oxide.
b. Calculate the standard enthalpy change of the reaction in (a).
c. How much heat is released per gram of aluminum that reacts?
d. Given that the heat of vaporization of water at 100 °C is 540 cal g^{-1}, how much water could be boiled by the heat released when one pound (454 g) of aluminum reacts with iron oxide?

a. **Analysis/Plan:** The formulas of the reactants and products can be determined from their names. The equation can then be balanced in the usual manner.

Work:

Aluminum makes a 3+ cation, and oxygen makes a 2- anion. Aluminum oxide is Al_2O_3. Iron(III) is a 3+ cation, so iron(III) oxide is Fe_2O_3. The skeleton equation is:

$$Al + Fe_2O_3 \longrightarrow Al_2O_3 + Fe.$$

A coefficient of 2 for both Al and Fe balances the equation:

$$\mathbf{2\ Al + Fe_2O_3 \longrightarrow Al_2O_3 + 2\ Fe.} \qquad \text{(2 Al, 2 Fe and 3 O on each side)}$$

b. **Analysis:**

Target: The standard enthalpy change for the process. $\Delta H°_{rxn}$ = (? kJ) =

Knowns: The balanced equation and the heats of formation of the reactants and products.

Relationship: $\Delta H°_{rxn} = \sum(n\Delta H°_{f\ products}) - \sum(n\Delta H°_{f\ reactants})$

Plan: Look up the heats of formation and apply the relationship.

Work:

The necessary $\Delta H_f°$ values are:

Reactants: $Al(s) = 0\ kJ\ mol^{-1}$ $\qquad$ $Fe_2O_3(s) = -824.2\ kJ\ mol^{-1}$

Products: $Al_2O_3(s) = -1675.7\ kJ\ mol^{-1}$ $\qquad$ $Fe(s) = 0\ kJ\ mol^{-1}$

$$\Delta H°_{rxn} = 1\Delta H_f°(Al_2O_3(s)) + 2\Delta H_f°(Fe(s)) - [1\Delta H_f°(Fe_2O_3(s)) + 2\Delta H_f°(Al(s))]$$

$$?\ kJ = \Delta H°_{rxn} = 1(-1675.7) + 2(0) - [1(-824.2) + 2(0)] = \mathbf{-851.5\ kJ}$$

c. **Analysis/Plan:** In part (b), we calculated the heat released when, 2 moles of Al react. We can use the factor-label method with the known molar mass of aluminum to solve this.

Work:

Molar Mass Al = 26.98154 g mol^{-1}

$$?\ kJ\ g^{-1} = \frac{-851.5\ kJ}{2\ mol\ Al} \times \frac{1\ mol\ Al}{26.98154\ g\ Al} = \mathbf{-15.78\ kJ\ g^{-1}}$$

d. **Analysis:**

Target: A mass of H_2O. (? kg H_2O) =

Knowns: The mass of Al, the heat released per gram of Al (from part (c)) and the heat of vaporization of water .

Relationships: The heat lost by the aluminum is equal in magnitude to the heat gained by the water ($-q_{Al} = q\ H_2O$). The equivalence statements 1 cal ↔ 4.184 J is needed.

Plan: We can use the factor-label method, converting units as required, to first find the heat released by the aluminum and then the mass of water that can be vaporized.

Step 2. Use the factor-label method to determine the mass of water.

Step 1. Use the factor-label method and the relationship $-q_{Al} = q_{water}$ to determine the heat absorbed by the water.

Work:

Step 1. Determine the heat absorbed by the water.

$$?\ kJ\ H_2O = -q_{Al} = -(454\ g\ Al \times \frac{-15.78\ kJ}{g\ Al}) = -(-7.16 \times 10^3\ kJ) = +\ 7.16 \times 10^3\ kJ$$

Step 2. Use the factor-label method to determine the mass of water that can be boiled. Note that, because the heat is given in terms of kilojoules, it is appropriate to report the mass of H_2O in kg. The factors can be changed appropriately.

$$? \text{ kg } H_2O = 7.16 \times 10^3 \text{ kJ} \times \frac{1 \text{ kcal}}{4.184 \text{ kJ}} \times \frac{1 \text{ kg } H_2O}{540 \text{ kcal}} = \mathbf{3.17 \text{ kg } H_2O}$$

6 Particles, Waves and the Structure of Atoms

6.1 Distinguish between reflection and refraction of light.

Reflection is the change of direction of light at an interface between two substances, such as air and a mirror. The light does not pass through the interface, but stays in the same medium. **Refraction** is the change in direction of light at an interface between two transparent substances such as air and water. In refraction, the light moves into the second medium.

6.3 Other than its effect on the eye, what are two differences between blue light and green light?

See Figure 6.8 in the text. Blue light has a shorter wavelength than green light. That means that it also has a higher frequency and more energy per photon.

6.5 Describe the photoelectric effect.

The photoelectric effect is the absorption of an photon by a metal surface, resulting in the ejection of an electron. The kinetic energy of the ejected electron depends upon the energy of the absorbed light.

6.7 What is meant by diffraction?

Diffraction is the spreading of light waves into a shadowed area.

6.9 Distinguish between two forms of interference.

In **constructive interference**, two light rays of the same wavelength combine to produce one ray with an intensity that is the **sum** of the intensities of the constituent rays. In **destructive interference**, two light rays of the same wavelength combine to produce one ray with an intensity that is the **difference** between the intensities of the constituent rays.

6.11 Criticize the statement, "The speed of light is 3×10^{8} m s^{-1}".

This speed is true only in a vacuum. In other media, the speed is less, causing refraction as a ray passes from one medium into another.

6.13 What is a spectrum? Distinguish between emission and absorption spectra.

A spectrum is the display of the wavelengths or frequencies in a sample of light and usually also includes the relative intensities or "brightness". An absorption spectra is a display of the specific wavelengths and their intensities that are absorbed by a sample. An emission spectra is a display of the specific wavelengths and their intensities that are emitted by a sample.

6.15 What type of light source emits a continuous spectrum? A discrete spectrum?

Almost every heated object emits a continuous spectrum. Discrete spectra are characteristic of heated gases or gases that have had electricity passed through them.

6.17 Name two behavioral characteristics of light that cannot be fully explained by the particle theory, and two that cannot be fully explained by the wave theory.

The particle theory of light cannot explain how light can be diffracted or how interference occurs. The wave theory of light cannot explain the photoelectric effect or blackbody radiation.

6.19 Define the term, "wave-particle duality".

"Wave-particle duality" means that a substance (such as light or electrons) exhibits both wave and particle properties Neither wave theory nor particle theory alone can account for all observed properties.

6.21 List the feature(s) of the Bohr theory that survived to become a part of quantum theory, and the feature(s) that did not survive.

The Bohr features that are used in quantum theory are the idea of quantized energy, stationary states and the fact that discrete amounts of energy are absorbed or emitted in moving between states. The Bohr feature that is no longer used is the idea that electrons orbit around the nucleus as the earth orbits the sun.

6.23 What is a matter wave?

A matter wave consists of moving particles which have momentum such that wave properties become important. The uncertainty in the position of the particle is on the same order or larger than the size of the particle itself.

6.25 What is a wave function? What is its physical significance?

A wave function is a solution to the Schrödinger equation. It has no real physical significance, but the square of the function does give a "picture" of the negative charge density in the atom.

6.27 What are quantum numbers?

Quantum numbers are interdependent integer solutions to the Schrödinger equation. These integers describe the energy, size, shape and orientation of orbitals.

6.29 What is meant by shell and subshell?

A shell is the set of all orbitals having the same principal quantum number (same energy level). A subshell is the set of all orbitals have both the same principal quantum number and the same angular momentum quantum number ("shape").

6.31 Give the formulas for three one-electron species with atomic number greater than 10.

Remember that the atomic number gives the number of protons (positive charges) in the nucleus. If a species has only 1 electron then the charge on the ion is equal to the atomic number minus 1. Possible species are ${}_{11}Na^{10+}$, ${}_{12}Mg^{11+}$, ${}_{13}Al^{12+}$, etc.

6.33 The visible region of the spectrum extends over the approximate wavelength range 400 - 700 nm. What is this range in frequency units?

Analysis:

Target: Two frequencies. (? Hz) =

Knowns: The wavelength of the light and the speed of light (2.99792458×10^{8} m s^{-1}).

Relationships: $c = \lambda\nu$, where c is the speed of light, λ is the wavelength and ν is the frequency. Appropriate equivalence statements are: 10^{-9} m ↔ 1 nm and 1 Hz ↔ 1 s^{-1}.

Plan: Solve the formula for ν, insert the known values for wavelength, using the factor-label method to convert the wavelength to meters and s^{-1} to Hz and calculate.

Work:

For 400 nm, $? \text{ Hz} = \nu = \frac{c}{\lambda} = \dfrac{2.99792458 \times 10^8 \text{ m s}^{-1} \times \frac{1 \text{ Hz}}{1 \text{ s}^{-1}}}{400 \text{ nm} \times \frac{10^{-9} \text{ m}}{1 \text{ nm}}} = \mathbf{7.50 \times 10^{14} \text{ Hz}}$

For 700 nm, $? \text{ Hz} = \nu = \frac{c}{\lambda} = \dfrac{2.99792458 \times 10^8 \text{ m s}^{-1} \times \frac{1 \text{ Hz}}{1 \text{ s}^{-1}}}{700 \text{ nm} \times \frac{10^{-9} \text{ m}}{1 \text{ nm}}} = \mathbf{4.28 \times 10^{14} \text{ Hz}}$

6.35 What is the frequency of x-rays having a wavelength of 1.224 Å (Angstroms)?

Analysis:
Target: A frequency. (? Hz) =
Knowns: The wavelength of the light and the speed of light (2.99792458×10^8 m s^{-1}).
Relationships: $c = \lambda\nu$, where c is the speed of light, λ is the wavelength and ν is the frequency. Appropriate equivalence statements are: 1 Å ↔ 10^{-10} m and 1 Hz ↔ 1 s^{-1}.
Plan: Solve the formula for ν, insert the known value of wavelength, using the factor-label method to convert the wavelength to meters and s^{-1} to Hz, and calculate.
Work:

$$? \text{ Hz} = \nu = \frac{c}{\lambda} = \frac{2.99792458 \times 10^8 \text{ m s}^{-1} \times \frac{1 \text{ Hz}}{1 \text{ s}^{-1}}}{1.224 \text{ Å} \times \frac{10^{-10} \text{ m}}{\text{Å}}} = \mathbf{2.449 \times 10^{18} \text{ Hz}}$$

6.37 What is the wavelength of a television signal that is broadcast at 600 MHz?

Analysis:
Target: A wavelength. (? m) =
Knowns: The frequency of the light and the speed of light (2.99792458×10^8 m s^{-1}).
Relationships: $c = \lambda\nu$, where c is the speed of light, λ is the wavelength and ν is the frequency. Equivalence statements needed are: 1 MHz ↔ 10^6 Hz and 1 Hz ↔ 1 s^{-1}.
Plan: Solve the formula for λ, insert the known value for the frequency, using the factor-label method to convert the frequency to Hz and Hz to s^{-1}, and calculate.
Work:

$$? \text{ m} = \lambda = \frac{c}{\nu} = \frac{2.99792458 \times 10^8 \text{ m s}^{-1}}{600 \text{ MHz} \times \frac{10^6 \text{ Hz}}{1 \text{ MHz}} \times \frac{1 \text{ s}^{-1}}{1 \text{ Hz}}} = \mathbf{0.500 \text{ m}}$$

6.39 Sound travels through air at about 330 m s^{-1}. What is the wavelength of the note "A" (440 Hz)?

Analysis:
Target: A wavelength. (? m) =
Knowns: The frequency, ν, of the sound and the speed, c, of sound.
Relationship: $c = \lambda\nu$. The appropriate equivalence statement is: 1 Hz ↔ 1 s^{-1}.
Plan: Solve the formula for λ, insert the known value of the frequency and calculate.
Work:

$$? \text{ m} = \lambda = \frac{c}{\nu} = \frac{330 \text{ m s}^{-1}}{440 \text{ Hz} \times \frac{1 \text{ s}^{-1}}{1 \text{ Hz}}} = \mathbf{0.750 \text{ m}} \text{ or } \mathbf{75.0 \text{ cm}}$$

6.41 How long did it take for TV signals to return to earth from the Explorer landing craft on Mars, 100 million km away?

Analysis:

Target: A time. (? s) =

Knowns: The distance to Mars and the speed of light (2.99792458×10^8 m s^{-1}).

Relationships: Speed = $\frac{\text{distance}}{\text{time}}$. Appropriate equivalence statements are: 1 km ↔ 10^3 m, and 1 million ↔ 10^6.

Plan: Rearrange the formula to solve for time. Insert known values, use the factor-label method to convert the distance to meters, and calculate.

Work:

$$? \text{ s} = \text{time} = \frac{\text{distance}}{\text{speed}} = \frac{100 \times 10^6 \text{ km} \times \frac{10^3 \text{ m}}{1 \text{ km}}}{(2.99792458 \times 10^8 \text{ m s}^{-1})} = \mathbf{334 \text{ s } (5.56 \text{ min})}$$

6.43 It takes radio waves about 1.5 seconds to travel to the moon (which introduces a three-second delay in spoken communication with astronauts on the moon). Use this information to estimate the distance to the moon.

Analysis:

Target: A distance. (? km) =

Knowns: The time to the moon and the speed of radio waves (2.99792458×10^8 m s^{-1}).

Relationships: Speed = $\frac{\text{distance}}{\text{time}}$. The equivalence statement needed is 1 km ↔ 10^3 m.

Plan: Rearrange the formula to solve for distance, use the factor-label method to convert the distance to km, and calculate.

Work:

$$? \text{ km} = \text{distance} = \text{speed} \times \text{time} = (2.99792458 \times 10^8 \text{ m s}^{-1}) \times 1.5 \text{ s} \times \frac{1 \text{ km}}{10^3 \text{m}} = \mathbf{4.5 \times 10^5 \text{ km}}$$

6.45 Calculate the energy carried by one photon of the following electromagnetic radiations.

a. Blue light (450 nm)
b. FM broadcast (92.7 MHz)
c. Lyman α-line (121 nm)

Analysis:

Target: The energy of a photon. (? J) =

Knowns: Either the wavelength or the frequency of the radiation. Planck's constant, $h = 6.626076 \times 10^{-34}$ J s, and $c = 2.99792458 \times 10^8$ m s^{-1}.

Relationships: $E = h\nu$ or $E = \frac{hc}{\lambda}$. The following equivalence statements are needed: 1 Hz ↔ 1 s^{-1}, 1 MHz ↔ 10^6 Hz, and 1 nm ↔ 10^{-9} m.

Plan: Use the appropriate relationship and the factor-label method to convert units as necessary.

Work:

a. $$? \text{ J} = \frac{hc}{\lambda} = \frac{(6.626076 \times 10^{-34} \text{ J s})(2.99792458 \times 10^8 \text{ m s}^{-1})}{450 \text{ nm} \times \frac{10^{-9} \text{ m}}{1 \text{ nm}}} = \mathbf{4.41 \times 10^{-19} \text{ J}}$$

b. $$? \text{ J} = h\nu = (6.626076 \times 10^{-34} \text{ J s})(92.7 \text{ MHz}) \times \frac{10^6 \text{ Hz}}{1 \text{ MHz}} \times \frac{1 \text{ s}^{-1}}{1 \text{ Hz}} = \mathbf{6.14 \times 10^{-26} \text{ J}}$$

c. $$? \text{ J} = \frac{hc}{\lambda} = \frac{(6.626076 \times 10^{-34} \text{ J s})(2.99792458 \times 10^8 \text{ m s}^{-1})}{121 \text{ nm} \times \frac{10^{-9} \text{ m}}{1 \text{ nm}}} = \mathbf{1.64 \times 10^{-18} \text{ J}}$$

6.47 Calculate the energy carried by one photon of the following electromagnetic radiations.
a. Red light (692 nm)
b. AM broadcast (880 kHz)
c. Radar (1 cm)

Analysis:
Target: The energy of a photon. (? J) =
Knowns: Either the wavelength or the frequency of the radiation. Planck's constant, $h = 6.626076 \times 10^{-34}$ J s, and $c = 2.99792458 \times 10^{8}$ m s^{-1}.
Relationships: $E = h\nu$ or $E = \frac{hc}{\lambda}$. The following equivalence statements are needed:
1 Hz ↔ 1 s^{-1}, 1 kHz ↔ 10^3 Hz, 1 cm ↔ 10^{-2} m, and 1 nm ↔ 10^{-9} m.
Plan: Use the appropriate relationship and the factor-label method to convert units as necessary.
Work:

$$\text{a.}\quad ?\ \text{J} = \frac{hc}{\lambda} = \frac{(6.626076 \times 10^{-34}\ \text{J s})(2.99792458 \times 10^{8}\ \text{m s}^{-1})}{692\ \text{nm} \times \frac{10^{-9}\ \text{m}}{1\ \text{nm}}} = \mathbf{2.87 \times 10^{-19}\ J}$$

$$\text{b.}\quad ?\ \text{J} = h\nu = (6.626076 \times 10^{-34}\ \text{J s})(880\ \text{kHz}) \times \frac{10^{3}\ \text{Hz}}{1\ \text{kHz}} \times \frac{1\ \text{s}^{-1}}{1\ \text{Hz}} = \mathbf{5.83 \times 10^{-28}\ J}$$

$$\text{c.}\quad ?\ \text{J} = \frac{hc}{\lambda} = \frac{(6.626076 \times 10^{-34}\ \text{J s})(2.99792458 \times 10^{8}\ \text{m s}^{-1})}{1\ \text{cm} \times \frac{10^{-2}\ \text{m}}{\text{cm}}} = \mathbf{2 \times 10^{-23}\ J}$$

6.49 When compounds of barium are heated in a flame, green light ($\lambda = 554$ nm) is emitted. How much energy does one mole of these photons carry?

Analysis:
Target: The energy of a mole of photons. (? kJ mol^{-1}) =
Knowns: The wavelength of the radiation. Planck's constant, $h = 6.626076 \times 10^{-34}$ J s, and $c = 2.99792458 \times 10^{8}$ m s^{-1}.
Relationships: $E = \frac{hc}{\lambda}$ gives the energy per photon. Necessary equivalence statements are:
1 mol ↔ 6.022137×10^{23} photons, 1 kJ ↔ 10^3 J and 1 nm ↔ 10^{-9} m.
Plan: Use the relationship and the factor-label method to convert units as necessary.
Work:

$$?\ \text{kJ mol}^{-1} = \frac{hc}{\lambda} = \frac{6.022137 \times 10^{23}\ \text{photons}}{\text{mol photons}} \times$$

$$\frac{(6.626076 \times 10^{-34}\ \text{J s})(2.99792458 \times 10^{8}\ \text{m s}^{-1})}{554\ \text{nm} \times \frac{10^{-9}\ \text{m}}{\text{nm}}} \times \frac{1\ \text{kJ}}{10^{3}\ \text{J}} = \mathbf{216\ kJ\ mol^{-1}}$$

6.51 Chemical changes stimulated by radiation are called *photochemical reactions*. How much energy per mole, approximately, is available from light in the following spectral regions? (In each case choose a wavelength that is somewhere in the middle of the region, for example 550 nm for visible light.)
a. X-ray   b. Visible

Analysis:
Target: The energy of a mole of photons. (? kJ mol^{-1}) = or (? MJ mol^{-1}) =
Knowns: The wavelength of the radiation. Planck's constant, $h = 6.626076 \times 10^{-34}$ J s, and $c = 2.99792458 \times 10^{8}$ m s^{-1}.
Relationships: $E = \frac{hc}{\lambda}$ gives the energy per photon. Necessary equivalence statements are:

1 mol ↔ 6.022137 x 10^{23} photons, 1 kJ ↔ 10^3 J, 1 MJ ↔ 10^6 J, and 1 nm ↔ 10^{-9} m.

Plan: Use the relationship and the factor-label method to convert units as necessary.

Work:

a. From Figure 6.9, the middle of the x-ray region is about 10^{-10} m.

$$? \text{ MJ mol}^{-1} = \frac{hc}{\lambda} = \frac{6.022137 \times 10^{23} \text{ photons}}{\text{mol photons}} \times$$

$$\frac{(6.626076 \times 10^{-34} \text{ J s})(2.99792458 \times 10^{8} \text{ m s}^{-1})}{10^{-10} \text{ m}} \times \frac{1 \text{ MJ}}{10^{6} \text{ J}} = \mathbf{10^3 \text{ MJ mol}^{-1}}$$

b.

$$? \text{ kJ mol}^{-1} = \frac{hc}{\lambda} = \frac{6.022137 \times 10^{23} \text{ photons}}{\text{mol photons}} \times$$

$$\frac{(6.626076 \times 10^{-34} \text{ J s})(2.99792458 \times 10^{8} \text{ m s}^{-1})}{550 \text{ nm} \times \frac{10^{-9} \text{ m}}{\text{nm}}} \times \frac{1 \text{ kJ}}{10^{3} \text{ J}} = \mathbf{2.20 \times 10^2 \text{ kJ mol}^{-1}}$$

6.53 Suppose a particular photochemical reaction requires 350 kJ mol^{-1} to proceed. What is the maximum wavelength of light that will be effective? In what region of the spectrum is this?

Analysis:

Target: The maximum wavelength of light and the region of the spectrum it is in. (? m) =

Knowns: The energy of the radiation. Planck's constant, h = 6.626076 x 10^{-34} J s, and c = 2.99792458 x 10^8 m s^{-1}.

Relationships: $E = \frac{hc}{\lambda}$ gives the energy per photon. Necessary equivalence statements are: 1 mol ↔ 6.022137 x 10^{23} photons and 1 kJ ↔ 10^3 J.

Plan: Rearrange the relationship to solve for λ. Use the factor-label method to convert units as necessary.

Work:

$$? \text{ m} = \lambda = \frac{hc}{E} = \frac{(6.626076 \times 10^{-34} \text{ J s})(2.99792458 \times 10^{8} \text{ m s}^{-1})}{350 \frac{\text{kJ}}{\text{mol}} \times \frac{1 \text{ mol}}{6.022 \times 10^{23} \text{ photons}} \times \frac{10^3 \text{ J}}{1 \text{ kJ}}} = \mathbf{3.42 \times 10^{-7} \text{ m}}$$

Figure 6.8 indicates that this is in the **ultraviolet region**.

6.55 If a microwave oven emits 10 W of 60 GHz radiation, how many photons are emitted per second? Express this number also in moles s^{-1}. (1 W ↔ 1 J s^{-1}).

Analysis:

Target: The number of photons per second. (? photon s^{-1}) = and (? mol s^{-1}) =

Knowns: The frequency and the power of the radiation. Planck's constant, h = 6.626076 x 10^{-34} J s, and c = 2.99792458 x 10^8 m s^{-1}.

Relationships: E = hν gives the energy per photon and ν is photons per second. The following equivalence statements are needed: 1 mol ↔ 6.022137 x 10^{23} photons, 1 Hz ↔ 1 s^{-1} and 1 GHz ↔ 10^9 Hz.

Plan: Inspection of the units shows that energy is equal to the wattage divided by the frequency. Determine the energy of the radiation, then rearrange the relationship E = hν to solve for ν. Use the factor-label method to convert units as necessary.

Work:

$$E = \frac{10 \text{ W}}{60 \text{ GHz}} \times \frac{1 \text{ J s}^{-1}}{1 \text{ W}} \times \frac{1 \text{ GHz}}{10^9 \text{ Hz}} \times \frac{1 \text{ Hz}}{\text{s}^{-1}} = 1.7 \times 10^{-10} \text{ J}$$

$$? \text{ photons s}^{-1} = \nu = \frac{E}{h} = \frac{(1.7 \times 10^{-10} \text{ J})}{(6.626076 \times 10^{-34} \text{ J s})} = \mathbf{2.5 \times 10^{23} \text{ photons s}^{-1}}$$

$$? \text{ mol s}^{-1} = 2.5 \times 10^{23} \text{ photons s}^{-1} \times \frac{1 \text{ mol}}{6.022137 \times 10^{23} \text{ photons}} = \mathbf{0.42 \text{ mol s}^{-1}}$$

6.57 The "light" in an electron microscope consists of a beam of electrons traveling with a velocity around 20% of the speed of light. What is the wavelength of this "matter wave"? How does this compare with the diameter of a hydrogen atom, which is about 75 pm?

Analysis:

Target: A wavelength. (? pm) =

Knowns: The velocity of the electron. Planck's constant, $h = 6.626076 \times 10^{-34}$ J s, and $c = 2.99792458 \times 10^{8}$ m s^{-1}. The electron mass (9.109390×10^{-31} kg) is an implicit known.

Relationships: The de Broglie equation for matter waves is $\lambda = \frac{h}{mv}$. The percent gives parts per hundred and can be used as a conversion factor. The necessary equivalence statements are: 1 pm ↔ 10^{-12} m and 1 J ↔ 1 kg m^2 s^{-2}.

Plan: Determine the velocity of the electron using the percent of the speed of light and then use the de Broglie equation and the factor-label method to make necessary conversions.

Work:

$$v = 2.99792458 \times 10^{8} \text{ m s}^{-1} \times \frac{20}{100} = 6.0 \times 10^{7} \text{ m s}^{-1}$$

$$? \text{ pm} = \lambda = \frac{h}{mv} = \frac{(6.626076 \times 10^{-34} \text{ J s}) \times \left(\frac{\frac{1 \text{ kg m}^2}{\text{s}^2}}{\text{J}}\right) \times \frac{1 \text{ pm}}{10^{-12} \text{ m}}}{(9.109390 \times 10^{-31} \text{ kg})(6.0 \times 10^{7} \text{ m s}^{-1})} = \mathbf{12 \text{ pm}}$$

This is **about 15%** of the diameter of a hydrogen atom.

6.59 The wave properties of a beam of neutrons are used in the technique of *neutron diffraction* to investigate the geometrical structure of molecules. What velocity must a neutron have in order to have a de Broglie wavelength of 12 pm?

Analysis:

Target: A velocity. (? m s^{-1}) =

Knowns: The de Broglie wavelength. Planck's constant, $h = 6.626076 \times 10^{-34}$ J s. The mass of the neutron (1.674929×10^{-27} kg) is an implicit known.

Relationships: The de Broglie equation for matter waves is $\lambda = \frac{h}{mv}$. The necessary equivalence statements are: 1 pm ↔ 10^{-12} m and 1 J ↔ 1 kg m^2 s^{-2}.

Plan: Rearrange the de Broglie equation to solve for v. Insert known values, using the factor-label method to make necessary conversions.

Work:

$$? \text{ m s}^{-1} = v = \frac{h}{m\lambda} = \frac{(6.626076 \times 10^{-34} \text{ J s}) \times \left(\frac{\frac{1 \text{ kg m}^2}{\text{s}^2}}{\text{J}}\right)}{(1.674929 \times 10^{-27} \text{ kg})(12 \text{ pm}) \times \frac{10^{-12} \text{ m}}{1 \text{ pm}}} = \mathbf{3.3 \times 10^{4} \text{ m s}^{-1}}$$

6.61 Suppose a 150 g baseball is travelling with a speed between 89 and 91 km per hour; what would be the minimum possible uncertainty in knowing its location? Comment on the magnitude of the answer.

Analysis:

Target: The uncertainty in the position of a ball. (? m) =

Knowns: The mass of the baseball and the velocity range of the baseball. Planck's constant, $h = 6.626076 \times 10^{-34}$ J s.

Relationships: The uncertainty principal, with mathematical form: $\Delta x \cdot m\Delta v > h$.

Necessary equivalence statements are: $1\ J \leftrightarrow 1\ kg\ m^2\ s^{-2}$, $1\ kg \leftrightarrow 10^3\ g$, $1\ km \leftrightarrow 10^3\ m$, $60\ s \leftrightarrow 1\ min$ and $60\ min \leftrightarrow 1\ hr$.

Plan: Find the difference in speed and convert it to $m\ s^{-1}$. Solve for Δx, then use the known values, using the factor-label method to convert units as necessary.

Work:

$\Delta v = 91\ km\ hr^{-1} - 89\ km\ hr^{-1} = 2\ km\ hr^{-1}$

$$?\ m\ s^{-1} = \frac{2\ km}{hr} \times \frac{1\ hr}{60\ min} \times \frac{1\ min}{60\ s} \times \frac{10^3\ m}{1\ km} = 0.6\ m\ s^{-1}$$

$$?\ m = \Delta x > \frac{h}{m\Delta v} = \frac{(6.626076 \times 10^{-34}\ J\ s) \times \left(\frac{\frac{1\ kg\ m^2}{s^2}}{J}\right)}{(150\ g \times \frac{1\ kg}{10^3\ g})(0.6\ m\ s^{-1})} = \mathbf{8 \times 10^{-33}\ m}$$

This is a very small size compared to the diameter of the baseball. You should readily be able to tell where the baseball is.

6.63 The threshold energy for the photoelectric effect in platinum is about 545 kJ mol^{-1}. What is the threshold wavelength (the maximum wavelength that causes the effect)? Will a platinum phototube respond to visible light?

Analysis:

Target: The maximum wavelength of light. (? nm) =

Knowns: The energy of the radiation. Planck's constant, $h = 6.626076 \times 10^{-34}$ J s, and $c = 2.99792458 \times 10^8\ m\ s^{-1}$.

Relationships: $E = \frac{hc}{\lambda}$ gives the energy per photon. The following equivalence statements are needed: $1\ mol \leftrightarrow 6.022137 \times 10^{23}$ photons, $1\ nm \leftrightarrow 10^{-9}\ m$, and $1\ kJ \leftrightarrow 10^3\ J$.

Plan: Rearrange the relationship to solve for λ. Use the factor-label method to convert units as necessary.

Work:

$$?\ m = \lambda = \frac{hc}{E} = \frac{(6.626076 \times 10^{-34}\ J\ s)(2.99792458 \times 10^8\ m\ s^{-1}) \times \frac{1\ nm}{10^{-9}\ m}}{545\ \frac{kJ}{mol} \times \frac{1\ mol}{6.022137 \times 10^{23}\ photons} \times \frac{10^3\ J}{1\ kJ}} = \mathbf{219\ nm}$$

Because visible light is longer in wavelength (lower in energy), it will **not** cause the effect.

6.65 Calculate the difference in energy (kJ mol^{-1}) between two atomic states, if light of wavelength 850 nm is emitted in the transition between the levels. In what spectral region is the light?

Analysis:

Target: The energy between two atomic states. (? kJ mol^{-1}) =

Knowns: The wavelength of the radiation. Planck's constant, $h = 6.626076 \times 10^{-34}$ J s, and $c = 2.99792458 \times 10^8\ m\ s^{-1}$.

Relationships: $E = \frac{hc}{\lambda}$ gives the energy per photon. The following equivalence statements are needed: $1\ mol \leftrightarrow 6.022137 \times 10^{23}$ photons, $1\ kJ \leftrightarrow 10^3\ J$, and $1\ nm \leftrightarrow 10^{-9}\ m$.

Plan: Use the relationship, with the factor-label method to convert units as necessary.

Work:

$$? \text{ kJ mol}^{-1} = E = \frac{hc}{\lambda} = \frac{6.022137 \times 10^{23} \text{ photons}}{\text{mol photons}} \times$$

$$\frac{(6.626076 \times 10^{-34} \text{ J s})(2.99792458 \times 10^{8} \text{ m s}^{-1})}{850 \text{ nm} \times \frac{10^{-9} \text{ m}}{1 \text{ nm}}} \times \frac{1 \text{ kJ}}{10^{3} \text{ J}} = \mathbf{141 \text{ kJ mol}^{-1}}$$

According to Figure 6.8, 800 nm is in the **infrared** range.

6.67 What wavelength of light is emitted when a hydrogen atom drops from the $n = 6$ to the $n = 4$ state? In what region of the spectrum does this occur?

Analysis:

Target: The wavelength of light and the region of the spectrum it is in. (? m)=

Known: The energy levels between which the electron moves.

Relationship: The Rydberg equation, $\frac{1}{\lambda} = -\Re \cdot \left(\frac{1}{n^2_U} - \frac{1}{n^2_L}\right)$ with $\Re = 1.097 \times 10^7 \text{ m}^{-1}$.

Plan: Use the Rydberg equation to find λ^{-1} and then solve for λ.

Work:

$$\frac{1}{\lambda} = -1.097 \times 10^7 \text{ m}^{-1} \cdot \left(\frac{1}{6^2} - \frac{1}{4^2}\right) = 3.809 \times 10^5 \text{ m}^{-1}$$

$$\lambda = \frac{1}{3.809 \times 10^5 \text{ m}^{-1}} = \mathbf{2.625 \times 10^{-6} \text{ m} = 2.625 \ \mu\text{m}}$$

According to Figure 6.8, this wavelength is in the **infrared** region.

6.69 What wavelength is emitted when a He^+ ion undergoes a transition from the $n = 2$ to the $n = 1$ level?

Analysis:

Target: The wavelength of light (? m) =

Known: The energy levels between which the electron moves. He^+ is a single electron species, with a nuclear charge of 2+.

Relationship: The Rydberg equation can be modified by adding a Z^2 term for the nuclear charge, so: $\frac{1}{\lambda} = -\Re \cdot Z^2 \cdot \left(\frac{1}{n^2_U} - \frac{1}{n^2_L}\right)$ with $\Re = 1.097 \times 10^7 \text{ m}^{-1}$

Plan: Use this version of the Rydberg equation to find λ^{-1} and then solve for λ.

Work:

$$\frac{1}{\lambda} = -1.097 \times 10^7 \text{ m}^{-1} \cdot 2^2 \cdot \left(\frac{1}{2^2} - \frac{1}{1^2}\right) = 3.291 \times 10^7 \text{ m}^{-1}$$

$$\lambda = \frac{1}{3.291 \times 10^7 \text{ m}^{-1}} = \mathbf{3.039 \times 10^{-8} \text{ m} \ (30.39 \text{ nm})}$$

6.71 What wavelength of light is required to excite a hydrogen atom from the $n = 2$ to the $n = 6$ level? In what region of the spectrum is this light?

Analysis:

Target: The wavelength of light and the region of the spectrum it is in. (? m) =

Known: The energy levels between which the electron moves.

Relationship: The Rydberg equation, $\frac{1}{\lambda} = -\Re \cdot \left(\frac{1}{n^2_U} - \frac{1}{n^2_L}\right)$ with $\Re = 1.097 \times 10^7 \text{ m}^{-1}$.

Plan: Use the Rydberg equation to find λ^{-1} and then solve for λ.

Work:

$\frac{1}{\lambda} = -1.097 \times 10^{7}\ m^{-1} \bullet \left(\frac{1}{6^2} - \frac{1}{2^2}\right) = 2.438 \times 10^{6}\ m^{-1}$

$\lambda = \frac{1}{2.438 \times 10^{6}\ m^{-1}} =$ **4.102×10^{-7} m (410.2 nm)**

This is on the borderline **between the ultraviolet and visible regions.**

6.73 One of the lines in the Lyman series of the hydrogen spectrum has a wavelength of 93.8 nm; what is the upper level of this transition? [Note: Since λ is given to only 3 significant digits, the calculated value of n_u will not be an exact whole number.]

Analysis:

Target: The upper n level in the hydrogen atom. (? n) =

Known: The wavelength of a line in the Lyman series.

Relationships: The Rydberg equation, $\frac{1}{\lambda} = -\Re \bullet \left(\frac{1}{n^2_U} - \frac{1}{n^2_L}\right)$ with $\Re = 1.097 \times 10^{7}\ m^{-1}$.

In the Lyman series, transitions are between the $n = 1$ level and other levels, so $n_L = 1$. The equivalence statement needed is 1 nm ↔ 10^{-9} m.

Plan: Insert known values and solve the Rydberg equation for n_U.

Work:

$$\frac{1}{\lambda} = \frac{1}{93.8\ nm \times \frac{10^{-9}\ m}{1\ nm}} = 1.07 \times 10^{7} = -1.097 \times 10^{7}\ m^{-1} \bullet \left(\frac{1}{n_U^2} - \frac{1}{1^2}\right) =$$

$$\frac{1}{n_U^2} - 1 = \frac{1.07 \times 10^{7}}{-1.097 \times 10^{7}} = -0.972, \text{ therefore } \frac{1}{n_U^2} = -0.972 + 1 = 0.028$$

$$n^2_U = \frac{1}{0.028} = 36 \quad \text{and} \quad n_U = \mathbf{6}$$

6.75 Infrared radiation of 1876 nm is observed in emission from an electrical discharge in hydrogen. What are the quantum numbers of the upper and lower levels? *Hint*: The most practical way to solve this is by trial and error, using pairs of small integers.

Analysis:

Target: The quantum numbers of the level between which the electron moves.

Known: The wavelength of a line in the hydrogen spectrum.

Relationships: The Rydberg equation, $\frac{1}{\lambda} = -\Re \bullet \left(\frac{1}{n^2_U} - \frac{1}{n^2_L}\right)$ with $\Re = 1.097 \times 10^{7}\ m^{-1}$.

The equivalence statement needed is 1 nm ↔ 10^{-9} m.

Plan: Insert known values and solve the Rydberg equation for $\left(\frac{1}{n^2_U} - \frac{1}{n^2_L}\right)$. Use the trial and error method suggested in the hint.

Work:

$$\frac{1}{\lambda} = \frac{1}{1876\ nm \times \frac{10^{-9}\ m}{1\ nm}} = 5.330 \times 10^{5} = -1.097 \times 10^{7}\ m^{-1} \bullet \left(\frac{1}{n_U^2} - \frac{1}{n^2_L{}^2}\right)$$

$$\left(\frac{1}{n^2_U} - \frac{1}{n^2_L}\right) = \frac{5.330 \times 10^{6}}{-1.097 \times 10^{7}} = -0.04859$$

The combination that works is $n_U = $ **4** and $n_L = $ **3.**

Check: $\frac{1}{4^2} - \frac{1}{3^2} = 0.06250 - 0.1111 = -0.04861$

6.77 Which of the following sets of quantum numbers (n, ℓ, m_ℓ) are not allowed for hydrogen and other one-electron species? In each case, state why not.
a. (1,1,1) b. (3,1,0) c. (1,0,1) d. (0,-1,0) e. (3,2,1) f. (3,2,-2)

Analysis/Plan: The quantum number convention is (n, ℓ, m_ℓ). Use stepwise application of the quantum number rules in the book.
Work:
a. The maximum value of ℓ is n - 1. Since n = 1, ℓ can only be 0.
b. This is an acceptable set of quantum numbers.
c. The maximum value of m_ℓ is +ℓ. Since ℓ = 0, m_ℓ cannot be anything but 0.
d. The value of ℓ can never be less than 0, because ℓ ranges from 0 to n - 1.
e. This is an acceptable set of quantum numbers.
f. This is an acceptable set of quantum numbers.

6.79 How many different orbitals can have the designation 4*d*?

Analysis/Plan: Use the quantum number rules as detailed in the book.
Work:
For *d*, ℓ = 2. Each m_ℓ defines a specific orbital. The values of m_ℓ are integers from -ℓ...0...+ℓ. That is **five** different 4 *d* orbitals: (4,2,-2), (4,2,-1), (4,2,0), (4,2,+1), and (4,2,+2).

6.81 How many different orbitals can have m_ℓ = -2 and n less than 5?

Analysis/Plan: Use the quantum number rules as detailed in the book.
Work: To have m_ℓ = -2, since the minimum value of m_ℓ is -ℓ, ℓ must be 2 or greater. Because the maximum value of ℓ is n - 1, this means that n must be either 3 or 4. Possible orbitals are: (3,2,-2), (4,2,-2) and (4,3,-2) for a total of **three** orbitals.

6.83 Sketch each of the following orbitals.
a. 2*s* b. 2*p*

See Figure 6.23 in the text for the 2*s* orbital and Figure 6.25 for the 2*p* orbital for an artist's conception of these shapes.

2*s*

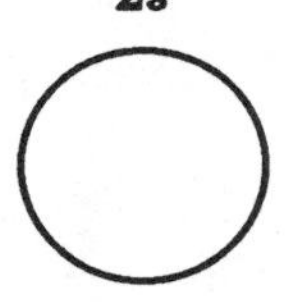

2*p*

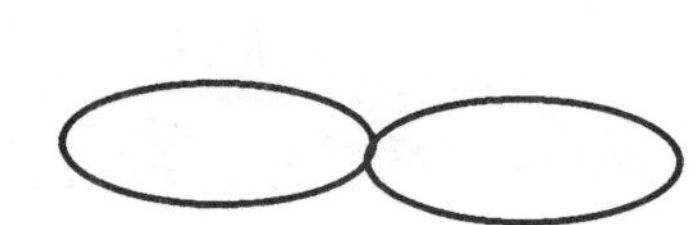

6.85 Identify the values of the principal and the angular momentum quantum number in each of the following: (a) 3*d*, (b) 3*f*, (c) 2*p*. Identify each as allowed or forbidden.

Analysis/Plan: Use the quantum number rules as detailed in the book.
Work:
a. n = 3 and ℓ = 2. This is an allowed set of quantum numbers.
b. n = 3 and ℓ = 3. This is not allowed because the maximum value of ℓ is n - 1.
c. n = 2 and ℓ = 1. This is an allowed set of quantum numbers.

6.87 What is the maximum value of m_ℓ in an *f* orbital? In a *p* orbital?

Analysis/Plan: Use the quantum number rules as detailed in the book.

Work:
The values of m_ℓ are integers from $-\ell...0...+\ell$.
For *f*, $\ell = 3$. The maximum value of $m_\ell = +3$.
For *p*, $\ell = 1$. The maximum value of $m_\ell = +1$.

6.89 What is the principal quantum number of an *s* orbital with 3 spherical nodes?

Analysis/Plan: Use the quantum number rules as detailed in the book.
Work:
An *s* orbital with quantum number *n* has $n - 1$ nodes so, to have three nodes, $n = 4$.

6.91 A fellow student, having some conceptual troubles, asks you "If the probability of an electron being on a nodal surface is truly zero, how can an electron get from one 2*p* orbital lobe to the other?" Your job: Answer this question in your own words.

Remember that an electron has **wave properties** as well as particle properties. It exists throughout the atom instead of being localized at a particular point. It doesn't have to move from *p* lobe to *p* lobe; it is in both simultaneously. Additionally, if we take quantum mechanics into account, the probability is not really zero because of the uncertainty involved.

7 Electron Configurations and Periodic Properties

7.1 What is meant by the term "many-electron species"? Give two examples each of one-electron and many-electron species.

A many-electron species is any species which has more than one electron. Mg and U are many-electron species. One-electron species include H and Ca^{19+}.

7.3 Describe any differences between the set of orbital energies in hydrogen and those in fluorine.

Because fluorine has a higher nuclear charge than hydrogen, the orbital energies in fluorine are less than those in the corresponding ones hydrogen. For example, a $1s$ orbital in F is lower in energy than a $1s$ orbital in H.

7.5 Other than its magnitude, is there any difference between the magnetic moment of a bar magnet and that of an electron?

There is **no** difference.

7.7 What is the numerical value of the combined magnetic moments of two paired electrons?

If the electrons are paired, their spins are opposite of each other, but of equal magnitude. This makes the magnetic moments opposite and equal, giving rise to a net **0** magnetic moment.

7.9 Criticize the statement, "The condition of paired/unpaired electrons is called the 'electron configuration' of an atom".

The configuration of an atom refers to the orbitals that the electrons are occupying, **not** just whether they are paired or unpaired.

7.11 An electron configuration in which the total energy of all electrons is the least possible is called the **ground** state of an atom.

7.13 Describe the interaction with a magnetic field of substances which are (a) diamagnetic, (b) paramagnetic, and (c) ferromagnetic.

a. Diamagnetic substances are **slightly repelled** by a magnetic field.
b. Paramagnetic substances are **moderately attracted** to a magnetic field.
c. Ferromagnetic substances are **strongly attracted** to a magnetic field.

7.15 What is meant by the term "aufbau"?

"Aufbau" is the systematic process of "building up" electrons in an atom. It provides a means of predicting electron configurations.

7.17 Elements are arranged in the periodic table according to their **electron configurations** and their **chemical properties**.

7.19 What is shielding? What is its effect on the strength of attraction between the nucleus and the outer electrons?

Shielding is the result of inner (core) electrons partially blocking the outer electrons from the positive charge of the nucleus. This decreases the attraction between the nucleus and the outer electrons and allows them to be removed more easily.

7.21 How is it possible for the effective nuclear charge to be different for different electrons in the outer shell?

Effective nuclear charge depends both on the energy level and on the sublevel. Outer electrons with different ℓ values penetrate the core to different degrees. This changes the extent to which they are shielded by inner electrons and alters the effective nuclear charge.

7.23 Use a chemical equation to illustrate the term, "third ionization energy."

This is the energy required to remove a third electron from an atom. An example is below.

$Al^{2+} + IE_3 \longrightarrow Al^{3+} + e^-$

7.25 How is covalent radius defined? How is it measured?

Atoms are approximated with a hard sphere model. The covalent radius is the distance from the center of the nucleus to the outside of the hard sphere. Covalent radii are estimated by using x-ray diffraction techniques to measure the internuclear distances in different compounds.

7.27 What is the relationship between the covalent radius of an atom and the ionic radii of the anion and cation formed from that atom?

Anions are always **larger** in radius than the atom because each additional electron is subjected to an increasingly smaller effective nuclear charge. There are more electrons attracted by the constant nuclear charge. Cations are always **smaller** in radius than the atom because, each time an electron is removed, the remaining electrons are subjected to a higher effective nuclear charge. There are fewer electrons attracted by the constant nuclear charge.

7.29 What is an isoelectronic series? Give an example.

An isoelectronic series is any series in which all the species have the **same number of electrons**. Members of an isoelectronic series often have the **same electron configuration**. Rb^+, Sr^{2+}, Kr, Br^- and Se^{2-} constitute an isoelectronic series with each species having 36 electrons. In each case, the electron configuration is [Kr].

7.31 Under what circumstances does an orbital diagram give more information than an electron configuration?

The orbital diagram shows the number of paired and unpaired electrons and, therefore, indicates which species are paramagnetic and which are diamagnetic. Electron configurations only give the total number of electrons in each subshell.

7.33 Write the set of four quantum numbers {n, ℓ, m_ℓ, m_s} for all the electrons in the ground state of boron.

Analysis:

Target: A set of four quantum numbers for all the electrons in the ground state of boron.
Knowns: The name of the element and, implicitly, its atomic number, 5, are given.
Relationships: The quantum number rules, discussed in Chapter 6, apply. The aufbau principle and Hund's rule must be followed. The general rules for writing electron configurations are:

1) Electrons occupy the lowest available orbitals. The relative orbital energies are determined by learning the "filling diagram", Figure 7.6, or the positions on the periodic table, Figure 7.7 in the text and in the figure below.
2) The maximum occupation of an orbital is: *s*(2), *p*(6), *d*(10), and *f*(14).
3) The total spin is maximized in partly-filled subshells.

Plan: Beginning with the first energy level, assign electrons until that level is full. Proceed through succeeding levels and sublevels, following the rules, until all electrons have been assigned. This same procedure will be used for writing quantum numbers, electron configurations and orbital occupation diagrams.

	s^1	s^2
1	H	He
2	Li	Be
3	Na	Mg
4	K	Ca
5	Rb	Sr
6	Cs	Ba
7	Fr	Ra

	d^1	d^2	d^3	§	d^5	d^6	d^7	d^8	‡	d^{10}
3	Sc	Ti	V	Cr	Mn	Fe	Co	Ni	Cu	Zn
4	Y	Zr	Nb	Mo	Tc	Ru	Rh	Pd	Ag	Cd
5	La*	Hf	Ta	W	Re	Os	Ir	Pt	Au	Hg
6	Ac^									

	p^1	p^2	p^3	p^4	p^5	p^6
2	B	C	N	O	F	Ne
3	Al	Si	P	S	Cl	Ar
4	Ga	Ge	As	Se	Br	Kr
5	In	Sn	Sb	Te	I	Xe
6	Tl	Pb	Bi	Po	At	Rn

	f^1	f^2	f^3	f^4	f^5	f^6	f^7	f^8	f^9	f^{10}	f^{11}	f^{12}	f^{13}	f^{14}
4 *	Ce	Pr	Nd	Pm	Sm	Eu	Gd	Tb	Dy	Ho	Er	Tm	Yb	Lu
5 ^	Th	Pa	U	Np	Pu	Am	Cm	Bk	Cf	Es	Fm	Md	No	Lr

§ These elements have an s^1d^5 configuration.
‡These elements have an s^1d^{10} configuration.

Note that helium is placed by its configuration to stress the trends, rather than with the rest of the noble gases as is more typically shown.

Work:
The configuration is: $1s^22s^22p^1$. The quantum numbers are: {1, 0, 0, +1/2}, {1, 0, 0, -1/2}, {2, 0, 0,+ 1/2}, {2, 0, 0, -1/2} and {2, 1, x, y}, where x = -1, 0 or +1 and y = +1/2 or -1/2.

7.35 What are the quantum numbers {n, ℓ, m_ℓ, m_s} of the lowest-lying electron in the ground state of calcium. Of the highest-lying?

Analysis/Plan: Calcium has atomic number 20. Follow the general plan outlined in the answer to Problem 7.33. Write a set of quantum numbers for the highest (last filled) and lowest (first filled) lying electrons.

Work:
The configuration is $1s^22s^22p^63s^23p^64s^2$. The lowest-lying electron has quantum numbers {1, 0, 0, x}, where x = +1/2 or -1/2. The highest-lying electron has quantum numbers {4, 0, 0, x}, where x = +1/2 or -1/2.

7.37 How many core electrons are there in the ground state of phosphorus?

Analysis/Plan: Phosphorus has atomic number 15. The core electrons are those electrons which occupy completely filled shells. Follow the general plan outlined in the answer to Problem 7.33. Write the electron configuration for P and count the number of electrons in filled shells.

Work:

The configuration is $1s^22s^22p^63s^23p^3$. The electrons in the first and second shells (n = 1 and n = 2) are the core electrons. This is a total of **10 core electrons.**

7.39 Identify each electron in the element with atomic number 9 as "core" or "outer", and give its quantum numbers {n, ℓ, m_ℓ, m_s}.

Analysis/Plan: The element is fluorine which has atomic number 9. Core electrons are those in completely filled shells, while outer electrons are those in incompletely filled shells. Follow the general plan outlined in the answer to Problem 7.33. Write a set of quantum numbers for each electron.

Work:

The configuration is $1s^22s^22p^5$. Quantum numbers are:

{1, 0, 0, +1/2}, core **{1, 0, 0, -1/2}, core**
{2, 0, 0, +1/2}, outer **{2, 0, 0, -1/2}, outer**
{2, 1, -1, y} outer **{2, 1, 0, y}, outer**
{2, 1, +1, y}, outer; y = either +1/2 or -1/2 for all these
2 electrons with **{2, 1, z, a}, outer**
z = two choices out of -1, 0 and +1 and **a = 1/2 with the opposite sign of y.**

7.41 One of the following sets of quantum numbers cannot occur. Identify it and explain why it is impossible.

a. {5, 3, -3, +1/2} b. {3, -2, 0, +1/2} c. {1, 0, 0, -1/2} d. {5, 0, 0, -1/2}

Analysis/Plan: Follow the quantum number rules introduced in Chapter 6 and those for spin. Determine if any combination does not agree with the rules.

Work:

a. This is possible and represents a 5*f* electron.
b. **This is the impossible quantum number because ℓ cannot be negative.**
c. This is possible and represents a 1*s* electron.
d. This is possible and represents a 5*s* electron.

7.43 The highest-lying electron in a certain atom has the quantum numbers {4, 1, 0, -1/2}. What element(s) could this be?

The quantum numbers tell us that the highest-lying electron is in the 4*p* (n = 4, ℓ = 1) subshell. Consultation with the periodic table indicates that the six elements in the *p*-block in the fourth period are: Ga, Ge, As, Se, Br, and Kr.

7.45 How many unpaired electrons are there in manganese? What are their quantum numbers?

Analysis/Plan: Manganese has atomic number 25. Follow the general plan outlined in the answer to Problem 7.33. Write the electron configuration for Mn and determine how many of the outer electrons are unpaired. Use the quantum number rules to write an acceptable set of quantum numbers for each unpaired electron.

Work:

The configuration is $1s^22s^22p^63s^23p^64s^23d^5$. The *d* sublevel can hold 10 electrons. By Hund's rule, each *d* orbital must be singly occupied, with spins parallel, before any pairing occurs. Each of the five electrons must singly occupy an orbital; **five electrons are unpaired**. Quantum numbers are: **{3, 2, -2, x}**, **{3, 2, -1, x}**, **{3, 2, 0, x}**, **{3, 2, +1, x}**, and **{3, 2, +2, x}**, where **x = either +1/2 or -1/2.**

7.47 Which of the following elements is/are paramagnetic: Cl, Rn, Ra?

Analysis:

Target: Determine if any of the elements are paramagnetic.
Knowns: The atomic number of each element is an implicit known.

Relationships: Paramagnetic species must have at least one unpaired electron.

Plan: If the number of electrons is odd, the species **must** be paramagnetic. For other elements, follow the general plan outlined in the answer to Problem 7.33. Write the electron configuration for the element and determine if any of the outer electrons are unpaired.

Work:

Cl has atomic number 17. With an odd number of electrons, it must be **paramagnetic**.

Rn is a noble gas with atomic number 86. Its configuration is: $[Xe]6s^24f^{14}5d^{10}6p^6$ or simply [Rn]. The *p* subshell only holds 6 electrons, so they must all be paired; therefore, Rn is **diamagnetic.**

Ra has atomic number 88. Its configuration is $[Rn]7s^2$. The *s* subshell only holds 2 electrons, so they must be paired; therefore, Ra is **diamagnetic.**

7.49 Consider the hypothetical element, the next higher noble gas beyond radon. What would be its outer-electron configuration? What would be its atomic number?

To be a noble gas, it would have to be in the *p* block with a $\mathbf{7s^27p^6}$ outer shell configuration. The $6d$ and $5f$ sublevels would also have to be full, which means that the complete configuration would be: $[Rn]7s^25f^{14}6d^{10}7p^6$ for a total of 118 electrons and atomic number **118**.

7.51 Give the electron configurations of the ground state in F, Cl, I, and draw their orbital diagrams.

Analysis/Plan: Look up the atomic number of each element. For each element, follow the general plan outlined in the answer to Problem 7.33. When drawing orbital diagrams, for *filled* subshells, place an up-down arrow pair in the orbital boxes for each filled subshell. For partly-filled subshells, place "spin-up" arrows one-by-one into the orbital boxes. Only pair electrons if there are more electrons than there are orbitals of that type.

Work:

$_9$F: $1s^22s^22p^5$ F: [↑↓] ($1s$) [↑↓] ($2s$) [↑↓][↑↓][↑] ($2p$)

$_{17}$Cl: $[Ne]3s^23p^5$ Cl: [Ne] [↑↓] ($3s$) [↑↓][↑↓][↑] ($3p$)

$_{53}$I: $[Kr]5s^24d^{10}5p^5$ I: [Ne] [↑↓] ($5s$) [↑↓][↑↓][↑↓][↑↓][↑↓] ($4d$) [↑↓][↑↓][↑] ($5p$)

7.53 Draw orbital diagrams for the following elements: calcium, chromium, cesium.

Analysis/Plan: Look up the atomic number of each element. Follow the general rules given in Problem 7.33 and the orbital filling guidelines presented in Problem 7.51. Remember that Cr is an exception to the normal filling order (Section 7.2 text) with a configuration that gives it two half-filled sets of subshells, the *s* and the *d*.

Work:

$_{20}$Ca: [Ar] [↑↓] ($4s$) $_{24}$Cr: [Ar] [↑] ($4s$) [↑][↑][↑][↑][↑] ($3d$) $_{55}$Cs: [Xe] [↑] ($6s$)

7.55 Identify each of the following electron configurations as possible or impossible. For those that are possible, further identify them as ground-state or excited-state. For those that are impossible, explain why.

a. $1s^22s^22p^63s^1$
b. $1s^22s^22p^63s^23p^63d^14s^24p^3$
c. $1s^22s^22p^63s^33p^6$
d. $[Ne]2p^63s^2$

Analysis:

Target: Determine the acceptability of an electron configuration.
Knowns: Potential electron configurations are given.
Relationships: The guidelines in the text and discussed in the answer to Problem 7.33 give the necessary relationships.

Plan: See if each configuration meets the guidelines.

Work:

a. This is a **possible ground state** configuration, because each sublevel is filled in proper order. (The element could be Na if this is not an ion configuration.)
b. This is a **possible** configuration, but would be an **excited state** because the $3d$ subshell should fill before the $4p$ subshell. (The element could be Cr if this is not an ion configuration.)
c. This is an **impossible** configuration because the maximum occupancy of the s subshell is 2.
d. This is an **impossible** configuration because the [Ne] core already contains the $2p$ subshell.

7.57 One or more of the following orbital diagrams represents an impossible electron configuration. Identify it and tell why it is impossible.

a. [Ar] [↑↓] ($4s$) [↑↓][↑↓][↑↓][↑][↑] ($3d$) [ ][ ][ ] ($4p$)

b. [Ar] [↑↓] ($4s$) [↑][↑][↑][↑][↑] ($3d$) [ ][ ][ ] ($4p$)

c. [↑↓] ($1s$) [↑↑] ($2s$) [↑][↑][↑] ($2p$)

d. [↑↓] ($1s$) [↑] ($2s$) [↑][↑][↑] ($2p$)

Analysis:

Target: Determine the acceptability of an electron configuration.
Knowns: Some potential electron configurations are given.
Relationships: The guidelines in the text and discussed in the answer to Problem 7.33 give the necessary relationships.

Plan: See if each configuration meets the guidelines.

Work:

a. This is a **possible ground state** configuration, because each sublevel is filled in proper order. (This could be the element Ni if it is not an ion configuration.)
b. This is a **possible ground state** configuration, because each sublevel is filled in proper order. (This could be the element Mn if it is not an ion configuration.)
c. This is an **impossible** configuration because the $2s$ subshell contains two electrons with parallel spins. If electrons are in the same orbital, their spins must be opposite (paired).
d. This is a **possible** configuration, but it would be an excited state since one electron has been promoted from the $2s$ level to the $2p$ level. (This could be the element C if it is not an ion configuration.)

7.59 Name the element having each of the following configurations.

a. $1s^22s^22p^63s^1$
b. $[Rn]7s^25f^{13}$

Analysis/Plan: The electron configuration gives orbital occupancy for each electron in an atom. Add the superscripts to determine the total number of electrons and consult the periodic table to determine the element.

Work:

a. There are **11** electrons. The element must be **neon** (**Ne**) with atomic number 11.
b. There are **101** electrons. The element must be **mendelevium** (**Md**) with atomic number 101.

7.61 Answer this question without consulting a periodic table or the text: How many electrons are there in each of the noble-gas cores symbolized by [He], [Ar], [Xe]

Analysis/Plan: This information is learned by repeated use of the periodic table.
Work:
[He] = **2 electrons** [Ar] = **18 electrons** [Xe] = **54 electrons**

7.63 Which of the following configurations of the carbon atom has the higher energy?
a. [↑↓] [↑↓] [↑↓][ ][ ]
1*s* 2*s* 2*p*
b. [↑↓] [↑↓] [↑][↑][ ]
1*s* 2*s* 2*p*

Analysis:
Target: Determine which configuration has the higher energy.
Knowns: Two different possible electron configurations are given.
Relationships: The guidelines in the text and discussed in the answer to Problem 7.33 give the necessary relationships.
Plan: Apply the guidelines to determine which has the higher energy.
Work:
Configuration **a** has higher energy because two 2*p* electrons occupy the same orbital. This gives rise to extra repulsions which raise the energy of **a** relative to b.

7.65 Identify the group and period of the elements having each of the following configurations:
a. [Ar][↑↓] [↑↓][↑↓][↑↓][↑↓][↑↓] [↑↓][↑][↑]
4*s* 3*d* 4*p*

b. [↑↓] [↑↓] [↑↓][↑↓][↑↓] [↑↓]
1*s* 2*s* 2*p* 3*s*

Analysis/Plan: The electron configuration gives orbital occupancy for each electron in an atom. Add the superscripts to determine the total number of electrons and consult the periodic table to determine the element. The alternative procedure is to learn the positions on the periodic table and count directly on the table.
Work:
a. There are **34** electrons. The element is **selenium** (Se) with atomic number 34 in the **fourth period** and **Group 6A.**
b. There are **12** electrons. The element is **magnesium** (Mg) with atomic number 12 in the **third period** and **Group 2A.**

7.67 Match one or more of the terms "main-group", "transition", "inner transition", "alkali", "alkaline earth", "halogen", or "noble gas" to each of the following outer-shell configurations.
a. $ns^2(n - 1)d^3$ b. ns^2np^5 c. ns^1 d. ns^2np^4

Analysis/Plan: The electron configuration gives orbital occupancy for each electron in an atom. The locations and names of specific groups of elements must be memorized. The best approach is to learn the positions on the periodic table and count directly on the table to determine to which classification the element belongs.
Work:
a. These are elements in the *d* block. They are **transition** elements.
b. These are elements in the fifth column of the *p* block. They are **main group elements** and also **halogens.**
c. These are elements in the first column of the *s* block. They are **main group elements** and also **alkali** metals.
d. These are elements in the fourth column of the *p* block. They are **main group elements.**

7.69 Which class of elements is characterized by an outer shell having a partly filled *p* subshell? Name several elements in this class.

Analysis/Plan: Learn the position of subshells on the periodic table and the names of specific groups. Determine which elements fit within this category.

Work:
These are the **main-group elements** (except for those in Group 1A and Group 2A, whose electron configurations end in *s*). They include **nitrogen, silicon, arsenic, tin** and all the rest of the elements in **Groups 3A - 7A.**

7.71 Elements having a partly-filled 5*d* subshell are known as **the third series of transition elements**.

7.73 What outer-shell configuration(s) is/are characteristic of the main-group elements?

Analysis/Plan: Learn the names and characteristics of the specific groups.
Work: The main-group elements are in Groups 1A - 7A. Their outer-shell configurations are: ns^1, ns^2, ns^2np^1, ns^2np^2, ns^2np^3, ns^2np^4, ns^2np^5.

7.75 What groups in the periodic table are called the "*p*-block" elements? The "*f*-block" elements?

Analysis/Plan: Learn the position of subshells on the periodic table and the names of specific groups. Determine which elements fit within each category.
Work:
The *p*-block elements are in **Groups 3A - 8A.**
The *f*-block elements are the **inner transition elements**, which include the **lanthanides** and the **actinides**.

7.77 In which group of the periodic table would a hypothetical element of atomic number Z = 117 belong? What is the name of this group?

Analysis/Plan: The locations and names of specific groups of elements must be memorized. The best approach is to use the rules discussed in Problem 7.33 to write out the complete electron configuration. Then consult the chart to see into which group the elements fits.
Work:
The entire configuration of the element would be $[Rn]7s^2 4f^{14} 6d^{10} 7p^5$. This would put the element in **Group 7A**, the **halogens**.

7.79 Arrange the following elements in order of increasing covalent radius: N, Si, P.

Analysis/Plan: General trends in atomic radius are summarized in Figure 7.20 and must be learned. Find the elements on the periodic table and apply the trends.
Work:
P is larger than N because it is further down in the same group. Si is larger than P because it is more to the left in the same period. The radii are: **N < P < Si**.

7.81 Identify the larger of each of the following pairs of species.
a. (Mg, Na) b. (F, Cl)

Analysis/Plan: General trends in atomic radius are summarized in Figure 7.20 and must be learned. Find the elements on the periodic table and apply the trends.
Work:
a. **Na is larger than Mg** because it is more to the left in the same period.
b. **Cl is larger than F** because it is further down in the same group.

7.83 Identify the larger species in each of the following pairs.
a. (S^{2-}, Cl^-) b. (Se^{2-}, Se^+)

Analysis/Plan: Anions are always larger than the atom from which they were made and cations are smaller than the atom from which they were made. For isoelectronic species in the same period, the most negative has the largest radius and the most positive has the smallest radius. Identify the relation between the given ions and apply these guidelines.

Work:

a. S^{2-} and Cl^- both have 18 electrons, however, Cl has more protons attracting those electrons, and, therefore, the electrons in Cl^- are held closer to the nucleus. **S^{2-} is larger than Cl^-.**

b. These are ions made from the same atom, so each has 34 protons in the nucleus. There are 36 electrons in Se^{2-} and only 33 electrons in Se^+. The greater effective nuclear charge per electron in Se^+ pulls the electrons in more tightly. **Se^{2-} is larger than Se^+.**

7.85 In each of the following pairs, identify the species with the larger ionization energy.
a. (S, O) b. (Rb, Sr)

Analysis/Plan: General trends in ionization energy are summarized in Figure 7.22 and must be learned. Find the elements on the periodic table and apply the trends.

Work:

a. O is above S in the same group. **O** has the larger ionization energy.

b. Sr is to the right of Rb in the same period. **Sr** has the larger ionization energy.

7.87 Rank the elements Ne, Cl, Se in order of increasing ionization energy.

Analysis/Plan: General trends in ionization energy are summarized in Figure 7.22 and must be learned. Find the elements on the periodic table and apply the trends.

Work:

Cl is above and to the right of Se, while Ne is above and to the right of Cl. Both trends indicate a higher ionization energy. **Se < Cl < Ne.**

7.89 Which of the following processes requires the larger energy input? Explain your answer.
$Co^{2+}(g) \longrightarrow Co^{3+}(g) + e^-$; $Cu^+(g) \longrightarrow Cu^{2+}(g) + e^-$.

Analysis/Plan: The general trends for ionization energy must be learned. Each successive ionization requires more energy than the previous one.

Work:

Cu is to the right of Co. Even though the ionization energy of Cu is greater than that of Co, **Co^{2+}** probably has a higher ionization energy than Cu^+. This is because the electron removed in Co^{2+} is the third ionization, while in Cu^+ it is only the second.

7.91 Are there any elements for which the second ionization energy is less that the first? If so, name them. If not, explain.

There are **no** elements that fit this description. When an electron is removed from a neutral atom, remaining electrons are subjected to a higher effective nuclear charge. They are held more tightly and are more difficult to remove, and, therefore, require more energy for ionization.

7.93 How much energy is required to ionize 5.00 mg neon?

Analysis:

Target: An ionization energy. (? J) =

Knowns: The mass of neon is given. The first ionization energy of Ne is 2.081 MJ mol^{-1} (Table 7.1). The molar mass of neon is an implicit known.

Relationships: The necessary equivalence statements are 1 mg ↔ 10^{-3} g, 1 MJ ↔ 10^6J, and 1 mol Ne ↔ 20.180 g Ne.

Plan: Use the factor-label method, converting units as necessary.

Work:

$$? \text{ J} = 5.00 \text{ mg Ne} \times \frac{10^{-3} \text{ g}}{1 \text{ mg}} \times \frac{1 \text{ mol Ne}}{20.180 \text{ g Ne}} \times \frac{2.081 \text{ MJ}}{1 \text{ mol Ne}} \times \frac{10^{6} \text{ J}}{1 \text{ MJ}} = \mathbf{516 \text{ J}}$$

7.95 Compared to others in the same period, which elements have the highest ionization energy?

Analysis/Plan: The general trends for ionization energy must be learned.
Work:
The **noble gases** always have the highest ionization energies in a period because they have the highest nuclear charge (highest atomic number) of any element in the period.

7.97 Estimate the internuclear distances in HBr and HI.

Analysis/Plan: The internuclear distance is roughly the sum of the covalent radii. Look up the radii in Figure 7.18 and then add them.
Work:
Radius H = 37 pm and radius Br = 114 pm. Internuclear distance ≈ 37 + 114 = **151 pm**.
Radius H = 37 pm and radius I = 133 pm. Internuclear distance ≈ 37 + 133 = **170 pm**.

7.99 Predict the value of the ionization energy of the (hypothetical) next higher halogen after astatine.

Analysis/Plan: Study the values and the trends in Figure 7.22. Extrapolate the data from the given values.
Work:
The ionization energy drops by about 100 kJ mol^{-1} between Cl and Br. Another drop of about 150 kJ mol^{-1} occurs between Br and I. At would be expected to be 100 or 150 kJ mol^{-1} less than I. The hypothetical element would be about 100 or 150 kJ mol^{-1} less than that. This would put it in the range of 200 - 300 kJ mol^{-1} less than I. This would be in the range of about 700 - 800 kJ mol^{-1}.

7.101 Consider the elements B, C, Al, and Si. Which is largest? Which has the greatest ionization energy?

Analysis/Plan: The general trends for radius and ionization energy must be learned. Find the elements on the periodic table and apply the trends. Note that the radius and ionization trends are essentially opposite from each other.
Work:
B is to the left of C, so it is larger than C. Al is to the left of Si, so it is larger than Si. Al is below B in the same group, so it is larger than B. **Al has the largest radius.** C is above Si in the same period, so it is smaller than Si. C is the smallest of the elements and, therefore, **C has the highest ionization energy**.

7.103 Use Figure 7.16 to estimate the molar volume (which has never been measured) of francium. Estimate its covalent radius.

Analysis/Plan: Study the values and the trends in Figure 7.16. The covalent radius can be estimated using Figure 7.18. In both cases, extrapolate from the given data.
Work:
The molar volume of K (~45 mL mol^{-1}) is 10 less than that of Rb (~55 mL mol^{-1}) while that of Rb is about 15 mL mol^{-1} less than Cs (~70 mL mol^{-1}). Fr would be expected to be about 10 - 15 mL mol^{-1} larger than Cs or about **80 - 85 mL mol^{-1}**.

Cs (262 pm) is 18 pm larger than Rb (244 pm), which is 12 pm larger than K (231 pm). Fr might be 10 - 20 pm larger than Cs for a covalent radius of about **270 - 280 pm**.

7.105 Write the symbols for 5 species having 25 electrons.

Analysis/Plan: Consult the periodic table to determine the atomic number and, therefore, the number of electrons in the neutral atom. Remember that metals tend to make positive ions. For each electron removed, increase the charge by 1.

Work:

Isoelectronic species with 25 electrons are: ${}_{25}\mathrm{Mn}$, ${}_{26}\mathrm{Fe}^{+}$, ${}_{27}\mathrm{Co}^{2+}$, ${}_{28}\mathrm{Ni}^{3+}$, and ${}_{29}\mathrm{Cu}^{4+}$.

7.107 While periodic trends offer a relatively easy way to predict that the ionization energy of Cl is greater than that of Se, it is not so easy to predict whether S or Br has the greater ionization energy. Why is the latter comparison more difficult? Which actually has the larger IE?

Analysis/Plan: Learn the trends, then consult the periodic table to determine the positions of the atoms. Use Figure 7.21 to determine the actual ionization energies.

Work:

Cl is both **above** and to the **right** of Se. Both trends indicate a **higher** ionization energy. S is **above** Br and would be expected to have a higher ionization energy from that trend; however, it is also to the **left** of Br, which indicates a **lower** ionization energy. With competing trends, it is difficult to predict which will have the dominant effect. Figure 7.21 indicates that **Br has a higher ionization energy than S**.

7.109 Name the halogen elements and describe each as solid, liquid, or gaseous at room temperature.

Fluorine (gas), chlorine (gas), bromine (liquid), iodine (solid) and astatine (probably solid).

7.111 What is the major source of the halogens?

The major source is the oceans or deposits left by the evaporation of previous seas.

7.113 Rank the hydrogen halides in order of increasing internuclear distance. Which, if any, are soluble in water? Which, if any, are gaseous at room temperature?

HF < HCl < HBr < HI. All are soluble in water and all are gases at room temperature.

7.115 How is bromine produced? Give any relevant chemical equations.

Bromine is typically produced by bubbling chlorine through sea water.

$$2\ \mathrm{NaBr(aq)} + \mathrm{Cl_2(g)} \longrightarrow \mathrm{Br_2}(\ell) + 2\ \mathrm{NaCl(aq)}$$

7.117 Which, if any, of the hypohalous acids reacts with water?

Only HOF reacts with water and, therefore, cannot exist in aqueous solution.

7.119 Define disproportionation.

Disproportionation is a process in which a single species reacts with itself to give two different products. For example, sodium hypochlorite, NaClO, disproportionates according to the equation below.

$$3\ \mathrm{NaClO(aq)} \longrightarrow \mathrm{NaClO_3(aq)} + 2\ \mathrm{NaCl(aq)}$$

8 The Chemical Bond

8.1 Define valence.

Valence, when used as an adjective, refers to chemical bonding. When used as a noun, it refers to the "combining power" of an element.

8.3 Discuss the statement, "Ionic compounds do not consist of molecules."

In the solid state, an ionic compound is composed of a combination of anions and cations such that the overall assemblage is neutral. There are no specific neutral species. In the gas phase, which is not encountered at room temperature under normal conditions, the ions do form clusters or pairs that could be considered to have a stable existence and, therefore, might be called molecules.

8.5 Distinguish between formula unit and empirical formula.

A formula unit is the discrete molecule of a molecular compound, but it is the lowest whole number ratio of ions in an ionic compound. An empirical formula is the lowest whole number ratio of atoms in either type of compound. For most ionic compounds, the empirical formula is the same as the formula unit. Exceptions include compounds with peroxide, O_2^{2-}; mercury(I) ion, Hg_2^{2+}; and oxalate ion, $C_2O_4^{2-}$. Na_2O_2, Hg_2Cl_2 and $K_2C_2O_4$ would be the formula units, but NaO, HgCl and KCO_2 would be the empirical formulas for these compounds.

8.7 Define each of the following characteristics of the covalent bond: order, length, and energy.

Bond order is the number of electron pairs which comprise the bond.
Bond length is the distance between the two nuclei joined by the bond.
Bond strength is the energy required to break the bond, thereby, separate the bonded atoms.

8.9 What is the relationship between bond order and bond energy?

A higher bond order requires that a greater amount of energy be used to break the bond.

8.11 What is formal charge? Is the actual charge on an atom different from the formal charge?

Formal charge is a means of determining the residual charge on an atom in a molecule or polyatomic ion. It is found by considering that all electrons in a bond are equally shared by the two bonded atoms and that all unshared electrons belong solely to that atom. The actual charge on the atom may be quite different from the formal charge except in homonuclear species (O_3, N_2, S_8, etc.) When electronegativities of the bonded atoms are different, the electrons in the bond are shared unequally, giving rise to partial charges rather than the whole numbers predicted by the formal charge definition.

8.13 What is resonance? How can you tell whether or not a species is resonant?

Resonance is the delocalization of an odd electron or a shared pair of electrons into more than one bond. If more than one "correct" Lewis structure can be written for the species and the different structures have the same geometrical arrangements of atoms, but vary in location of electrons, the species is resonant.

8.15 List some observable characteristics that could be used to distinguish resonant from non-resonant species.

Resonant species tend to be particularly stable. They also have bonds which are the same in length and strength in spite of being "drawn" with varying order and, consequently, varying bond energy.

8.17 What is meant by electron delocalization?

Electron delocalization is the spreading out of electron density. In resonant species, it means that an electron or electron pair is spread over two or more bonds rather than being shared between only two atoms.

8.19 What is bond polarity and how does it arise?

Bond polarity is the unequal sharing of electrons by two atoms. This occurs because the bonded atoms have different electronegativities and, therefore, different attractions for the electron pair(s). This gives rise to a partial positive charge on the less electronegative atom and a partial negative charge on the more electronegative atom.

8.21 Describe the general trend of electronegativity within the periodic table.

Electronegativity increases as the atomic number increases across a period. (The noble gases are not included in this trend because they tend not to make bonds). Electronegativity decreases as atomic number increases within a group.

8.23 "Bond moments cannot be measured." True or false? Explain.

This is generally true, because the dipole moment of a molecule is the sum of all of the individual bond moments in the molecule. Individual bond moments can be determined in diatomic species which just have one bond contributing to the dipole moment of the molecule (HCl, IF, NO) and in species with identical bonds and known geometry (SCl_2, NF_3).

8.25 Comment on the statement, "Most covalent bonds are polar".

A polar bond results when two atoms with different electronegativities are bonded. Because almost all elements have different electronegativities (see Figure 8.5 in the text), bonds between different elements are almost always polar.

8.27 What is electronegativity, and how is it measured?

Electronegativity is a measure of an atom's relative attraction for shared electrons in a covalent bond. Electronegativity is not measured. Instead, electronegativity is assigned on the basis of the atom's ionization energy and other properties.

8.29 Alkali and alkaline earth halides are all ionic compounds. Use this fact together with the electronegativity values in Figure 8.5 to suggest a minimum value of ΔEN, below which ionic bonding is unlikely.

Be has the highest electronegativity (1.5) of the metals, while I (2.5) has the lowest electronegativity of the common halogens. This suggests that the lower limit for ΔEN for ionic bonds is about 1.0. More commonly, the guideline for ionic bonding is considered to be a ΔEN of about 1.7 to 1.9.

8.31 Explain, in general terms, the relationship between bond energy and heat of reaction.

In a chemical reaction, usually some bonds are broken, while other bonds are made. Energy is required to break bonds and energy is released when bonds are made. In an endothermic ($\Delta H > 0$) process, generally more or more energetic bonds are broken and fewer or less energetic bonds are made. In an exothermic ($\Delta H < 0$) process, fewer or less energetic bonds are broken and more or more energetic bonds are made.

8.33 For which molecular shape(s) is it appropriate to use the terms "axial" and "equatorial"? Illustrate your answer with sketches.

The trigonal bipyramidal structure is the only nonregular geometric shape. In this shape, there is a distinction between the axial bonds which are 180° apart and the equatorial bonds which are 120° apart and 90° from the axial bonds.

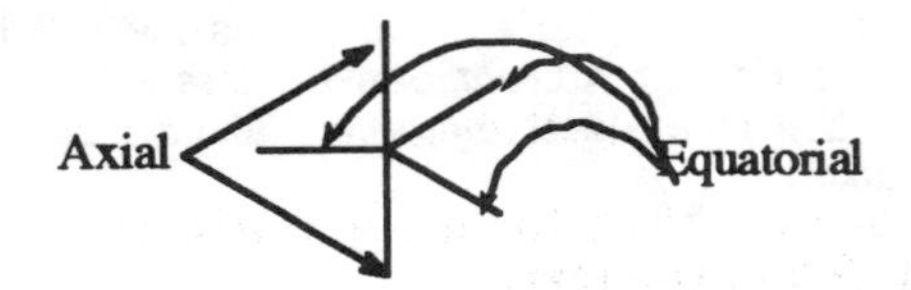

8.35 Which of the following are measurable quantities: bond moment, dipole moment, and distance between the centers of positive and negative charge in a polar molecule?

Dipole moments of molecules can be measured. Only in diatomic molecules can the bond moment can be measured because it is the same as the dipole moment.

8.37 What monatomic ions would you expect to form from each of the following atoms?
a. Li b. Sr c. S d. Al

Analysis/Plan: The charge of monatomic cations of the main group metals is equal to the group number. The charge of the monatomic anions of the main group nonmetals is equal to the group number minus eight. Find the element on the periodic table and apply these rules to write the charge.

Work:

a. Li^+ (Group 1A) b. Sr^{2+} (Group 2A) c. S^{2-} (Group 6A) d. Al^{3+} (Group 3A)

8.39 Write the electron configurations of
a. Mg and Mg^{2+} b. Li and Li^+ c. I and I^- d. Al and Al^{3+}

Analysis/Plan: Determine the number of electrons from the atomic number and the ion charge. Remember that a negative charge means that electrons have been added and a positive charge means that electrons have been removed. Write the electron configuration following the guidelines discussed in Chapter 7.

Work:

a. Mg (12 electrons) = $[Ne]3s^2$ — Mg^{2+} (12 - 2 = 10 electrons) = [Ne]
b. Li (3 electrons) = $[He]2s^1$ — Li^+ (3 - 1 = 2 electrons) = [He]
c. I (53 electrons) = $[Kr]5s^24d^{10}5p^5$ — I^- (53 + 1 = 54 electrons) = [Xe]
d. Al (13 electrons) = $[Ne]3s^23p^1$ — Al^{3+} (13 - 3 = 10 electrons) = [Ne]

8.41 Give the empirical formula of the ionic compound consisting of
a. Ba and O b. Al and O c. Cs and S d. Sodium and $Cr_2O_7^{2-}$

Analysis/Plan: The ionic compound must be electrically neutral. Charges of the main group ions can be determined from the periodic table as detailed in Problem 8.37. Find the element on the periodic table, determine its charge and then choose subscripts to make the total positive charge equal the total negative charge.

Work:

a. Ba is a Group 2A metal and has a 2+ charge. O is a Group 6A nonmetal and has a 2- charge. Since they have equal and opposite charges, they must be present in equal number in the compound. The compound is **BaO**.

b. Al is a Group 3A metal and has a 3+ charge. O (part a) has a 2- charge. The lowest common multiple of 2 and 3 is 6, so the empirical formula has $2Al^{3+}$ ion and $3O^{2-}$ ions. The compound is $\mathbf{Al_2O_3}$.

c. Cs is a Group 1A metal and has a + charge. S is a Group 6A nonmetal and has a 2- charge. Two Cs^+ cations are required for one S^{2-} anion. The formula is $\mathbf{Cs_2S}$.

d. Sodium, Na, is a Group 1A metal and has a + charge. The $Cr_2O_7^{2-}$ ion has a 2- charge. Two Na^+ cations are required for one $Cr_2O_7^{2-}$ anion. The formula is $\mathbf{Na_2Cr_2O_7}$.

8.43 Write the Lewis structures of the following molecules.

a. CH_2F_2 b. COClBr   c. CO   d. NH_2Cl

Analysis/Plan: Determine the number of valence electrons from the group numbers and any charge. Follow the guidelines presented in Table 8.2 of the text.

Work:

a. C is the central atom because it is further down and to the left on the periodic table. The total number of valence electrons is

1(4 C) + 2(1 H) + 2(7 F) = 20.

Using pairs to hold the atoms together gives the skeletal structure on the right.

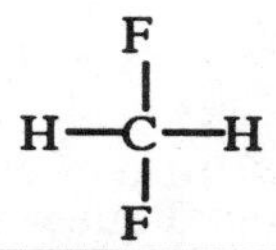

C and H have complete valence shells. Filling in the remaining 12 electrons completes the valence shells of the F atoms.

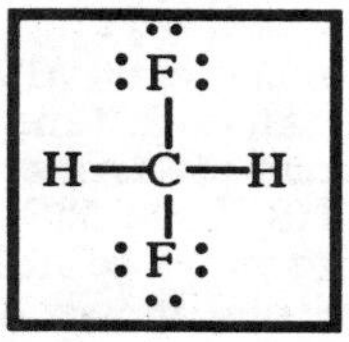

b. C is the central atom because it is to the left and because the halogens tend to make single bonds. The total number of valence electrons is

1(4 C) + 1(6 O) + 1(7 Br) + 1(7 Cl) = 24.

Using pairs to hold the atoms together gives the skeletal structure shown at the right.

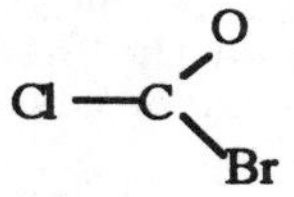

No atom has a complete valence shell. Filling in the remaining 16 electrons, terminal atoms first, gives the structure:

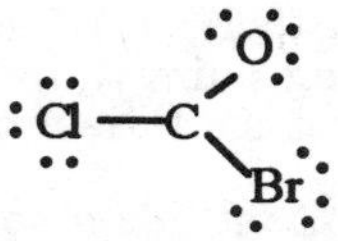

Br, O and Cl have complete valence shells, but C does not. Moving one pair of electrons from the O to between the C and O completes the valence shell of all atoms. (The double bond is made between O and C rather than between C and a halogen because a halogen tends to make 1 bond and oxygen tends to make two bonds to give the atom a formal charge of zero.)

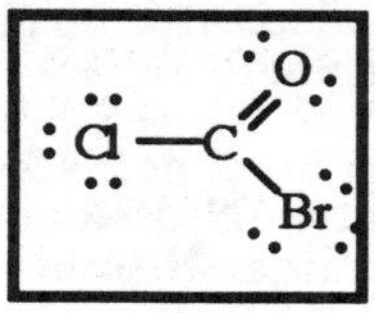

c. The total number of valence electrons is

1(4 C) + 1(6 O) = 10.

Using pairs to hold the atoms together gives the skeletal structure shown on the right.

No atom has a complete valence shell. Filling in the remaining 8 electrons gives:

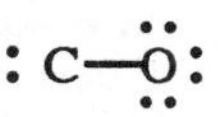

O has a complete valence shell, but C does not. Moving two pairs of electrons from the O to between the C and the O completes the valence shells of both atoms.

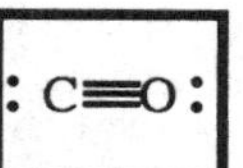

d. N is the central atom because it is to the left and because the halogens tend to make single bonds. The total number of valence electrons is

$1(5\ N) + 1(7\ Cl) + 2(1\ H) = 14.$

Using pairs to hold the atoms together gives the skeletal structure shown on the right.

H—N(—H)—Cl

The H atom has a complete valence shell. Filling in the remaining 8 electrons, terminal atoms first, completes the valence shells of N and Cl.

8.45 Write the Lewis structures of the following species. If there is resonance, show all contributing structures.

a. XeF_2 b. N_3^- c. ClO_3^- d. HN_3

Analysis/Plan: Determine the number of valence electrons from the group numbers and any charge. Follow the guidelines presented in Table 8.2 of the text.

Work:

a. Xe is the central atom because it is further down and because the halogens tend to make single bonds. The total number of valence electrons is

$1(8\ Xe) + 2(7\ F) = 22.$

Using pairs to hold the atoms together gives the skeletal structure shown on the right.

F—Xe—F

Using the remaining 18 valence electrons completes the valence shell for the fluorine and requires that three unshared pairs of electrons be placed around the central Xe atom.

b. One N is the central atom and the others are bonded to it. The total number of valence electrons is

$3(5\ N) + 1(1\ charge) = 16.$

Using pairs to hold the atoms together gives the skeletal structure shown on the right.

N—N—N $^-$

No atom has a complete valence shell. Filling in the remaining 12 electrons, terminal atoms first, gives the structure shown on the right.

The terminal N atoms have complete valence shells, but the central N does not. Moving one pair of electrons from each terminal N to between the N atoms completes the valence shells of all atoms.

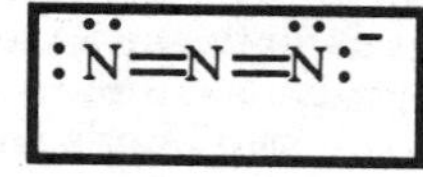

c. The total number of valence electrons is

$1(7\ Cl) + 3(6\ O) + 1(1\ charge) = 26.$

Using pairs to hold the atoms together gives the skeletal structure shown on the right.

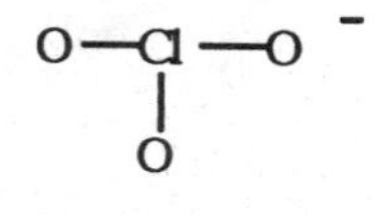

No atom has a complete valence shell. Filling in the remaining 20 electrons completes the valence shells of all atoms

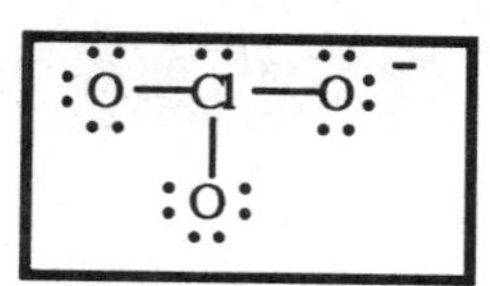

d. One N is the central atom and the H is bonded to one of the terminal atoms. The total number of valence electrons is
$3(5\ N) + 1(1\ H) = 16.$
Using pairs to hold the atoms together gives the skeletal structure shown on the right.

H—N—N—N

The H atom has a complete valence shell. Filling in the remaining 10 electrons, terminal N atoms first, gives the structure shown on the right.

H—N—N—N:

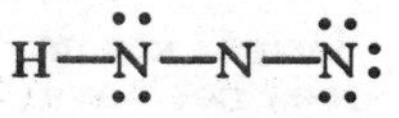

The terminal N atoms have complete valence shells, but the central N does not. Moving one pair of electrons from each terminal N to between the N atoms completes the valence shells of all atoms.

H—N=N=N:

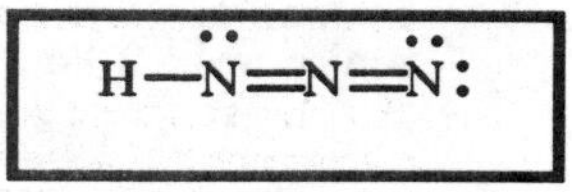

8.47 Write the Lewis structures of the following species. If there is resonance, show all contributing structures.

a. HO_2^- b. formate ion, HCO_2^-
c. acetic acid, CH_3CO_2H (this molecule has 3 C-H bonds and one O-H bond)
d. H_2SO_4 (H is bonded to O, and there are no O-O bonds)

Analysis/Plan: Determine the number of valence electrons from the group numbers and any charge. Follow the guidelines presented in Table 8.2 of the text.

Work:

a. The H cannot be the central atom because it can only make one covalent bond. The total number of valence electrons is
$2(6\ O) + 1(1\ H) + 1(1\ charge) = 14.$
Using pairs to hold the atoms together gives the skeletal structure shown on the right.

H—O—O $^-$

Using the remaining 10 valence electrons completes the valence shells of both O atoms.

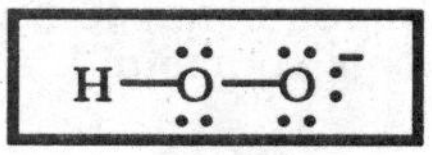

H—Ö—Ö: $^-$

b. C is the central atom and the others are bonded to it. The total number of valence electrons is
$1(1\ H) + 1(4\ C) + 2(6\ O) + 1(1\ charge) = 18.$
Using pairs to hold the atoms together gives the skeletal structure shown on the right.

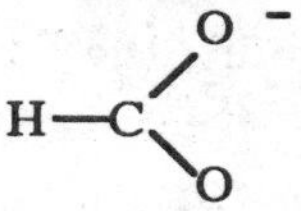

H has a complete valence shell. Filling in the remaining 12 electrons, terminal atoms first, gives the structure shown on the right.

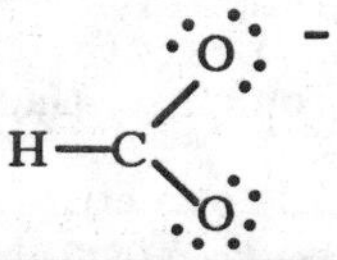

The terminal O atoms have complete valence shells, but the central C does not. Moving one pair of electrons from a terminal O to between the O and C completes the valence shells of all atoms. Because either O can be used to make the double bond, there are two resonance structures for this compound.

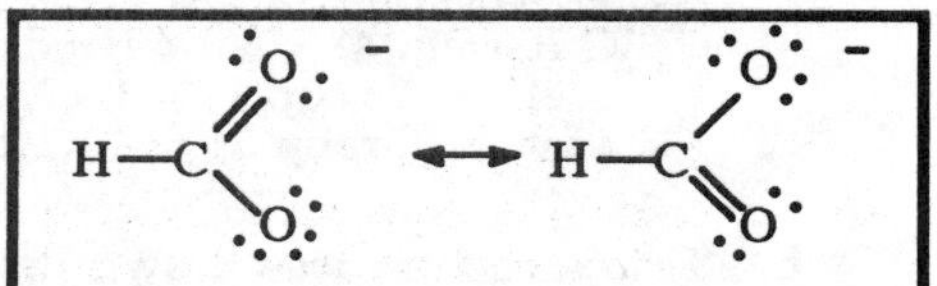

c. The way the formula is written indicates that 1 C atom has the 3 C-H bonds and the other has both O atoms bonded to it. The total number of valence electrons is
$4(1\ H) + 2(4\ C) + 2(6\ O) = 24.$
Using pairs to hold the atoms together gives the skeletal structure shown on the right.

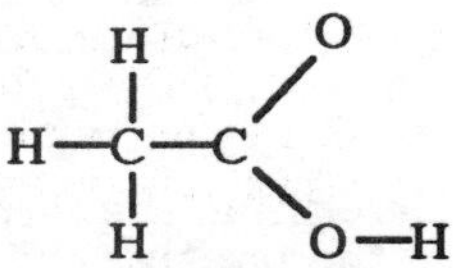

All H atoms have complete valence shells, as does the left C atom. Filling in the remaining 10 electrons, terminal atoms first, gives the structure shown on the right.

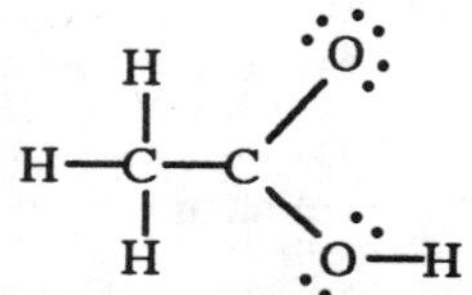

Both O atoms have complete valence shells, but the right hand C does not have a complete valence shell. One pair is moved from the O not bonded to H to complete the C valence shell. The O bonded to H is not used for the double bond because O tends to make only 2 bonds. This gives it a formal charge of zero.

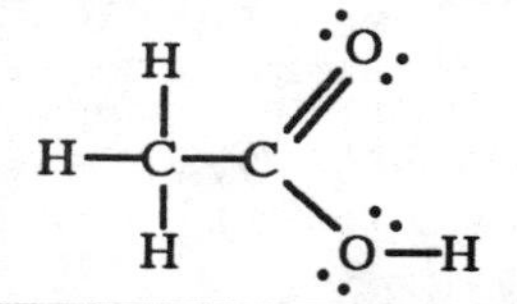

d. S is the central atom. The total number of valence electrons is
$2(1\ H) + 4(6\ O) + 1(6\ S) = 32.$
Using pairs to hold the atoms together gives the skeletal structure shown on the right.

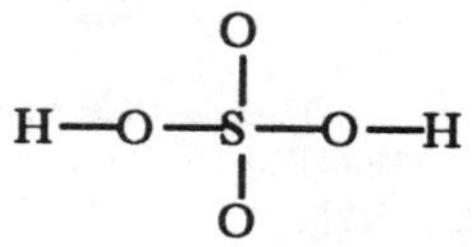

The H atoms have complete valence shells. Filling in the remaining 20 electrons, terminal atoms first, gives the following structure which completes the valence shells of all atoms.

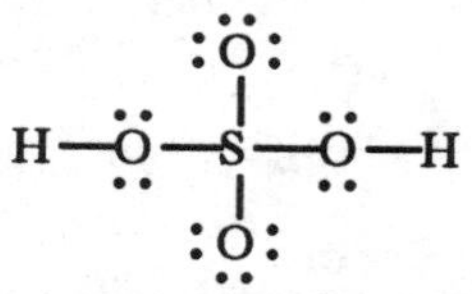

While the previous structure follows the octet rule for all atoms, the oxygens not bonded to hydrogen have only one bond. This gives them a formal charge of 6 - 1 - 6 = -1. The sulfur atom has a formal charge of 6 - 4 = +2. These formal charges can be made equal to zero by violating the octet rule. The structure for this compound is traditionally shown with two double bonds.

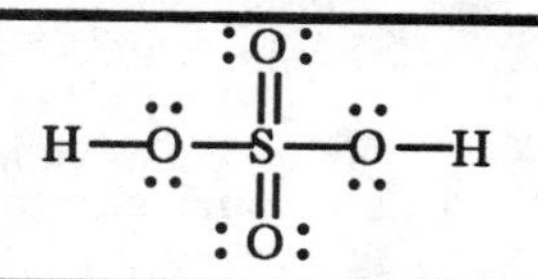

8.49 Determine the formal charges on all atoms in the following species.
a. CH_2F_2 b. CO

Analysis/Relationships: Formal charge is determined from the formula:
Formal charge = group number - number of bonds - number of unshared electrons. The group number gives the information required to draw the Lewis structure.
Plan: Draw the Lewis structure of the compound using the techniques discussed in previous problems. Use the formula to determine the formal charge of each atom.
Work:
a. This structure was drawn in Problem 8.43a and is shown on the right. Formal charges are:
C (Group 4) = 4 - 4 bonds = **0**
each H (Group 1) = 1 - 1 bond = **0**
each F (Group 7) = 7 - 1 bond - 6 unshared electrons = **0**

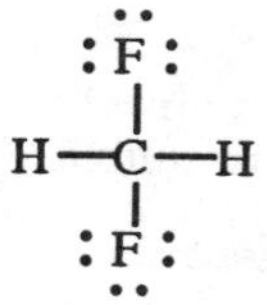

b. This structure was drawn in Problem 8.43c and is shown on the right. Formal charges are:
C (Group 4) = 4 - 3 bonds - 2 unshared electrons = **-1**
O (Group 6) = 6 - 3 bonds - 2 unshared electrons = **+1**

8.51 Determine the formal charges on the atoms in the ammonium ion. Which atom carries the bulk of the positive charge?

Analysis/Plan: The definition of formal charge and necessary considerations were presented in Problem 8.49. Draw the Lewis structure of the compound using the techniques discussed in previous problems. Use the formula to determine the formal charge of each atom.

Work:

The total number of valence electrons is

1(5 N) + 4(1 H) - 1(1 charge) = 8.

The Lewis structure is shown on the right. Formal charges are:

N (Group 5) = 5 - 4 bonds = **+1**

each H (Group 1) = 1 - 1 bond = **0**

N has the bulk of the positive charge.

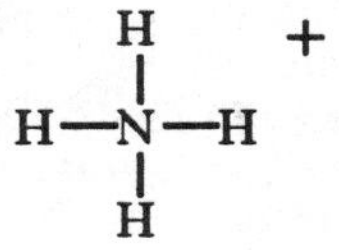

8.53 Determine the formal charges on all atoms in the following species. If there is resonance, show formal charges on all contributors.

a. SO_2 b. HCO_3^- c. $BeCl_2$ d. N_2H_2

Analysis/Relationships: The definition of formal charge and necessary considerations were presented in Problem 8.49. If there are resonance structures, the formal charge of an atom is the average of its formal charges in the different structures.

Plan: Draw the Lewis structure of the compound using the techniques discussed in previous problems. Use the formula to determine the formal charge of each atom.

Work:

a. S is the central atom. The total number of valence electrons is

1(6 S) + 2(6 O) = 18.

The two resonant Lewis structures are shown on the right. Formal charges are:

S (Group 6) = 6 - 3 bonds - 2 unshared electrons = **+1**

In each structure, one O (Group 6) has formal charge =

6 - 1 bond - 6 unshared electrons = -1

and the other O has formal charge =

6 - 2 bonds - 4 unshared electrons = 0

The average formal charge per O = $\frac{-1 + 0}{2}$ = **-1/2**

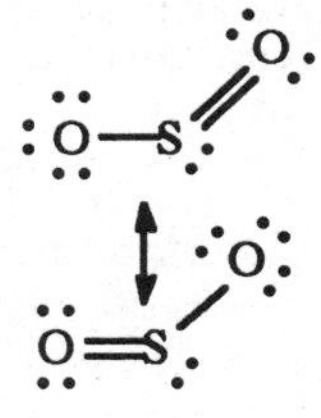

b. C is the central atom. The H is bonded to one of the O atoms. The total number of valence electrons is

1(1 H) + 1(4 C) + 3(6 O) + 1(1 charge) = 24.

The two resonance structures are shown on the right. Formal charges are:

C (Group 4) = 4 - 4 bonds = **0**

H (Group 1) = 1 - 1 bond = **0**

O (Group 6) bonded to H =

6 - 2 bonds - 4 unshared electrons = **0**

In each structure, 1 O not bonded to H has formal charge =

6 - 1 bond - 6 unshared electrons = -1

and the other O not bonded to H has formal charge =

6 - 2 bonds - 4 unshared electrons = 0

The average formal charge per O not bonded to H = $\frac{-1 + 0}{2}$

= **-1/2**

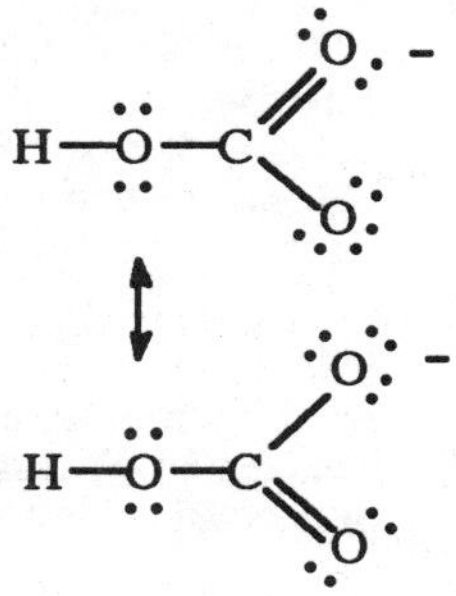

c. The total number of valence electrons is

1(2 Be) + 2(7 Cl) = 16.

The Lewis structure is shown on the right. Formal charges are:

Be (Group 2) = 2 - 2 bonds = **0**

each Cl (Group 7) = 7 - 1 bond - 6 unshared electrons = **0**

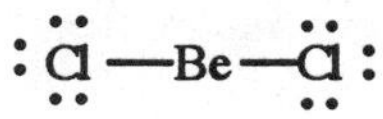

d. The total number of valence electrons is

2(5 N) + 2(1 H) = 12.

The Lewis structure is shown on the right. Formal charges are:

each N (Group 5) = 5 - 3 bonds - 2 unshared electrons = **0**

each H (Group 1) = 1 - 1 bond = **0**

8.55 Determine the bond order for all bonds in the following species. Watch for resonance and its effect on bond order.

a. CH_2F_2 b. COClBr c. CO d. NH_2Cl

Analysis/Plan: The bond order is readily determined from the Lewis structure. All these structures were drawn in answer to Problem 8.43. Look at the structure and determine the bond orders.

Work:

a. There are only single bonds, so all bond orders are **1**.
b. The CO bond is a double bond, so its bond order is **2**. All other bonds are single bonds with bond order **1**.
c. There is a triple bond, so the bond order is 3.
d. There are only single bonds, so all bond orders are **1**.

8.57 Determine the bond order for all bonds in the following species. Watch for resonance and its effect on bond order.

a. IF_5 b. NO_2^- c. acetic acid, CH_2CO_2H d. H_2SO_4

Analysis/Plan: The bond order is readily determined from the Lewis structure. Structures for compounds c and d were drawn in the answer to Problem 8.47. Draw the other Lewis structures following the previously discussed rules. Look at the structure and determine the bond orders.

Work:

a. I is the central atom. The total number of valence electrons is
$1(7\ I) + 5(7\ F) = 42.$
The structure is shown on the right. There are only single bonds, so each bond order is equal to **1**.

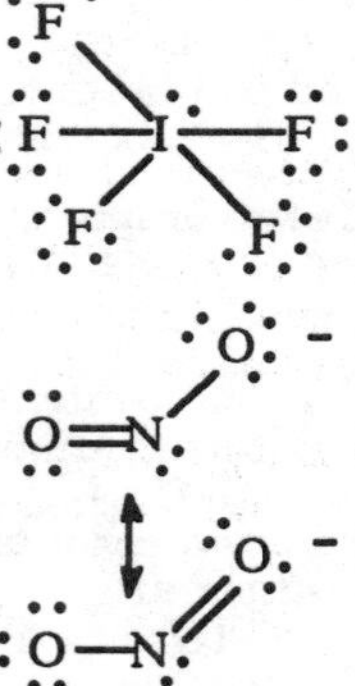

b. N is the central atom. The total number of valence electrons is
$1(5\ N) + 2(6\ O) + 1(1\ \text{charge}) = 18.$
The two resonant Lewis structures are shown on the right. The bond order of the NO bonds is an average of the double bond in one structure and the single bond in the other.
The average bond order of the NO bond is $\frac{2 + 1}{2} = \mathbf{3/2}$.

c. All CH bonds, the CC bond and the OH bond are single bonds with bond order = **1**. The O bonded to H has a single bond with C, so that bond has bond order = **1**. The CO double bond has bond order = **2**.

d. The H-O bonds and the S-O single bonds have bond order = **1**. The two S-O bonds shown as double bonds have bond order = **2**.

8.59 Why are some bond energies different from the values given in Table 8.4?

The bond energy depends not only on the bonded atoms, but also on their environment within the molecule. The tabulated bond energies are averages of the measured bond energies in many different molecules.

8.61 In which compound, SO_2 or SO_3, would you expect to find the shorter S-O bond?

Analysis/Plan: Bond lengths are roughly proportional to bond order. Bond orders can be determined from the Lewis structures. We need to draw the Lewis structures for the two substances, using the rules previously discussed. The structures can then be compared to see which has the shorter bonds.

Work:

The Lewis structure of SO_2 is a combination of two resonance structures (Problem 8.53a). In each structure, one SO bond is a single bond and the other SO bond is a double bond. The average bond order is $\frac{2 + 1}{2}$ = 3/2.

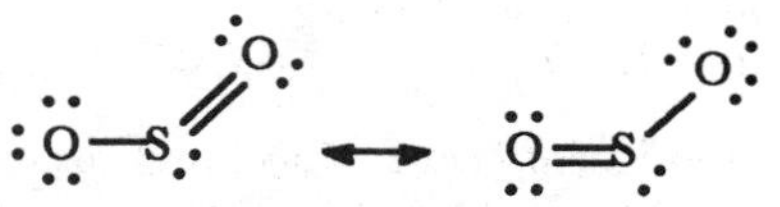

In SO_3, S is the central atom. The total number of valence electrons is

1(6 S) + 3(6 O) = 24.

The three resonant Lewis structures are shown on the right. In each structure, two SO bonds are single bonds and the other is a double bond. The average bond order is $\frac{2 + 1 + 1}{3}$ = 4/3.

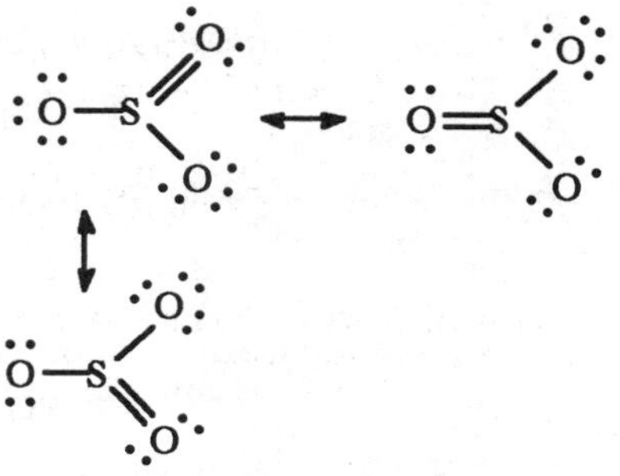

Because the bond order is greater in SO_2, it has the shorter bonds.

8.63 Estimate the heat of reaction for the following gas-phase reactions.

a. $H_2C{=}CH_2 + Br_2 \longrightarrow BrH_2C\text{-}CH_2Br$

b. $H_2O_2 \longrightarrow H_2O + 1/2\ O_2$

Analysis/Relationship: The heat of the reaction is the sum of the energies required to break all the bonds in the reactants minus the energy released when all the bonds in the products are formed. ΔH_{rxn} = Σ[energy of broken bonds] - Σ[energy of formed bonds]. We need the Lewis structures of all reactants and products so that we can determine whether bonds are single, double or triple and so that we can count the number of bonds of each type.

Plan:

Step 2. Using Table 8.4 in the text, find the sum of the bond energies of all broken bonds and subtract from it the sum of all energies of the bonds made.

Step 1. Use previously discussed rules to draw the Lewis structures of all reactants and products to determine the number and order of each bond.

Work:

a. **Step 1.** The number of valence electrons in each substance is found first. $H_2C{=}CH_2$ has 4(1 H) + 2(4 C) = 12, Br_2 has 2(7) = 14, and $BrH_2C\text{-}CH_2Br$ has 4(1 H) + 2(4 C) + 2(7 Br) = 26. The Lewis structures are:

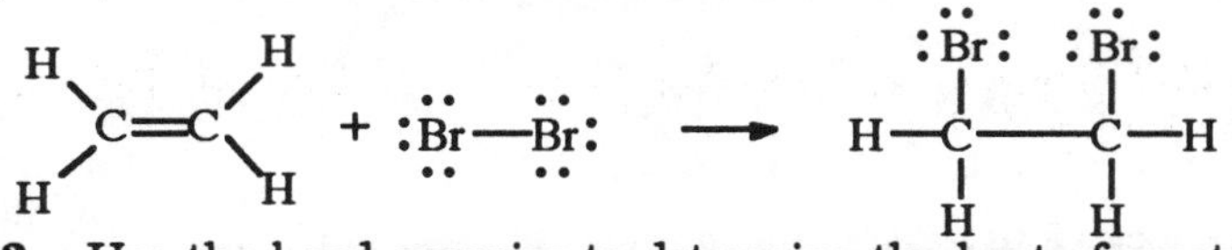

Step 2. Use the bond energies to determine the heat of reaction.

ΔH_{rxn} = Σ[energy of broken bonds] - Σ[energy of formed bonds].

$\Delta H°_{rxn}$ = Σ[4C-H + 1C=C + 1Br-Br] - Σ[4C-H + 2C-Br + 1C-C]

$\Delta H°_{rxn}$ = 4(436) + 1(615) + 1(193) - [4(436) + 2(276) + 1(347)] = **-91 kJ**

b. **Step 1.** The number of valence electrons in each substance is found first. H_2O_2 has 2(1 H) + 2(6 O) = 14 , H_2O has 2(1 H) + 1(6 O) = 8 and O_2 has 2(6 O) = 12. The Lewis structures are:

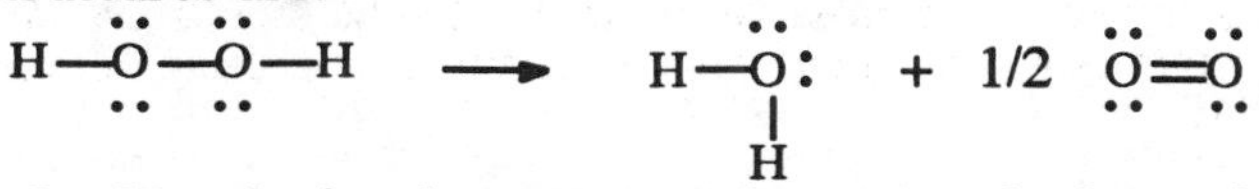

Step 2. Use the bond energies to determine the heat of reaction.

ΔH_{rxn} = Σ[energy of broken bonds] - Σ[energy of formed bonds].

$\Delta H°_{rxn}$ = Σ[2O-H + 1O-O] - Σ[2O-H + 1/2O=O]

$\Delta H°_{rxn}$ = 2(467) + 1(146) - [2(467) + 1/2(498)] = **-103 kJ**

8.65 Use the table of average bond energies to estimate the heats of formation of the following species.
a. $ClF_3(g)$ b. $N_2H_2(g)$

Analysis/Relationship: The heat of the reaction is the sum of the energies required to break all the bonds in the reactants minus the energy released when all the bonds in the products are formed. ΔH_{rxn} = Σ[energy of broken bonds] - Σ[energy of formed bonds]. In a formation reaction, 1 mole of a compound is made from its elements in their standard states. We need the Lewis structures of all reactants and products so that we can determine whether bonds are single, double or triple and so that we can count the number of bonds of each type.

Plan:

Step 3. Using Table 8.4 in the text, find the sum of the bond energies of all broken bonds and subtract from it the sum of all energies of the bonds made.

Step 2. Use previously discussed rules to draw the Lewis structures of all reactants and products to determine the number and order of each bond.

Step 1. Write the formation reaction for the compound.

Work:

a. **Step 1.** The formation reaction for ClF_3 is: $1/2Cl_2(g) + 3/2F_2(g) \longrightarrow ClF_3(g)$

Step 2. The number of valence electrons in each substance is found first. Cl_2 has 2(7)= 14, F_2 has 2(7) = 14, and ClF_3 has 1(7 Cl) + 3(7 F) = 28. The Lewis structures are:

1/2 :Cl—Cl: + 3/2:F—F: ⟶ :F—Cl—F: (with :F: bonded below Cl)

Step 3. Use the bond energies to determine the heat of reaction.
ΔH_{rxn} = Σ[energy of broken bonds] - Σ[energy of formed bonds].

$\Delta H°_{rxn}$ = Σ[1/2Cl-Cl + 3/2F-F - Σ[3Cl-F]

$\Delta H°_{rxn}$ = 1/2(239) + 3/2(154) - 3(253) = **-408 kJ mol^{-1}**

b. **Step 1.** The formation reaction is: $N_2 + H_2 \longrightarrow N_2H_2$

Step 2. The number of valence electrons in each substance is found first. N_2 has 2(5) = 10, H_2 has 2(1) = 2 and N_2H_2 has 2(5 N) + 2(1 H) = 12. The Lewis structures are:

:N≡N: + H—H ⟶ H—N=N—H

Step 2. Use the bond energies to determine the heat of reaction.
ΔH_{rxn} = Σ[energy of broken bonds] - Σ[energy of formed bonds].

$\Delta H°_{rxn}$ = Σ[1N≡N] + 1(H-H) - Σ[2N-H + 1N=N]

$\Delta H°_{rxn}$ = 1(945) + 1(436) - [2(391) + 1(418)] = **+181 kJ mol^{-1}**

8.67 Without looking at a list of electronegativities, identify the more electronegative element in the following pairs.
a. N, O b. K, Rb c. Mg, K d. O, S

Analysis/Relationship: Electronegativity is a measure of an element's tendency to attract electrons to itself in a covalent bond. In general, electronegativities increase as atomic number increases in a period, but decrease as atomic number increases in a group.

Plan: The elements need to be located on the periodic table and the trend applied.
Work:
a. **O is more electronegative**, because it has a higher atomic number than N (same period).
b. **K is more electronegative**, because it has a lower atomic number than Rb (same group).
c. **Mg is more electronegative**, because it has a higher atomic number than Na (same period). Na is more electronegative than K because it has a lower atomic number than K (same group).
d. **O is more electronegative**, because it has a lower atomic number than S (same group).

8.69 Which is the more polar bond in each of the following? Try to answer this question without reference to Figure 8.5.
a. C-F, H-F b. O-H, S-H c. S-Cl, S-Br d. C-H, O-H

Analysis/Relationship: A more polar bond has a greater electronegativity difference between the two bonded atoms. Generally, elements which are farther apart on the periodic table have a greater electronegativity difference. If they appear equally close, the trends or the electronegativity table (Figure 8.5) must be considered.
Plan: Locate the elements on the periodic table and analyze their position. Use the electronegativity table to confirm the answer.
Work:
a. **H-F is more polar**, because H (EN = 2.1) and F (EN 4.0) are further apart than C (EN 2.5) and F.
b. **O-H is more polar**, even though S and H are further apart. O (EN 3.5) is the second most electronegative element. The electronegativity of S (2.5) is lower than that of O(3.5) making it closer to that of H (2.1), causing a less polar bond.
c. **S-Cl is more polar**. Both S (2.5) and Br (2.8) are less electronegative than Cl (3.0), therefore, the electronegativity difference between them is less than that between S and Cl.
d. **O-H is more polar**, because H (EN 2.1) and O (EN. 3.5) are further apart than C (EN 2.5) and H.

8.71 In each of the following pairs, which bond has the greater ionic character?
a. C-Br, Si-O b. N-O, Cl-F c. C-B, C-H d. Li-Br, C-Cl

Analysis/Relationship: A more polar bond has more ionic character and results from a greater electronegativity difference between the two bonded atoms. A metal/nonmetal bond is generally more ionic than a nonmetal/nonmetal bond.
Plan: Analyze the atoms bonded and, if necessary, consult the electronegativity table, Figure 8.5, in the text.
Work:
a. **Si-O is more ionic.** The electronegativities are: C (2.5), Br (2.8), O (3.5) and Si(1.8). Therefore, ΔEN Si-O = 3.5 - 1.8 = 1.7, while ΔEN C-Br is only 2.5 - 2.3 = 0.2.
b. **Cl-F is more ionic.** The electronegativities are: Cl (3.0), F(4.0), O (3.5) and N(3.0). Therefore, ΔEN Cl-F = 4.0 - 3.0 = 1.0, while ΔEN N-O is only 3.5 - 3.0 = 0.5.
c. **C-B is more ionic.** The electronegativities are: C (2.5), B(2.0), and H(2.1). Therefore, ΔEN C-B= 2.5 - 2.0 = 0.5, while ΔEN C-H = 2.5 - 2.1 = 0.4.
d. **Li-Br is more ionic.** The electronegativities are: Li (1.0), Br(2.8), C (2.5) and Cl(3.0). Therefore, ΔEN Li-Br = 2.8 - 1.0 = 1.8, while ΔEN C-Cl is only 3.0 - 2.5 = 0.5. LiBr would also be predicted to be more ionic because it has a metal bonded to a nonmetal.

8.73 Give the electron-pair geometry and the molecular shape of each of the following.
a. IF_5 b. CCl_2F_2 c. NO_2^- d. CH_2O

Analysis/Relationship: Electron pair geometry is given solely by the number of electron regions around the central atom.
Plan: We need the Lewis structure to determine the number of electron regions.
Step 3. Read the electron geometry from Tables 8.6 and 8.7 in the text.
Step 2. Count the number of electron regions around the central atom.
Step 1. Draw the Lewis structure of the molecule, using the previously discussed rules.

Work:

a. **Step 1.** The Lewis structure of IF_5 was drawn in Problem 8.57a and is shown on the right.

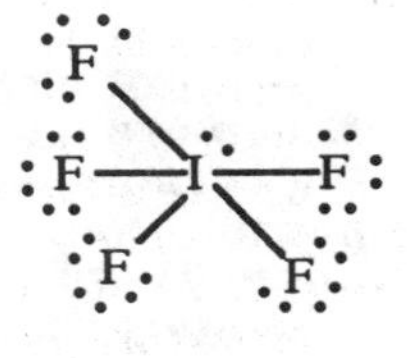

Step 2. There are 6 electron regions (5 bonding pairs and 1 lone pair).
Step 3. With 5 bonding pairs and 1 lone pair:
electron-pair geometry = octahedral
molecular shape = square pyramidal

b. **Step 1.** C is the central atom. The total number of valence electrons is 1(4 C) + 2(7 F) + 2(7 Cl) = 32. The structure is shown on the right.

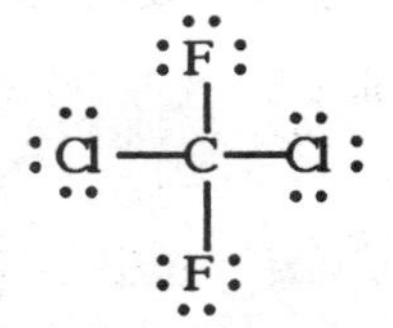

Step 2. There are 4 electron regions (4 bonding pairs).
Step 3. With 4 bonding pairs and no lone pairs:
electron-pair geometry = tetrahedral
molecular shape = tetrahedral

c. **Step 1.** The Lewis structure of NO_2^- consists of two resonance structures and was drawn in Problem 8.57b. It is shown on the right.

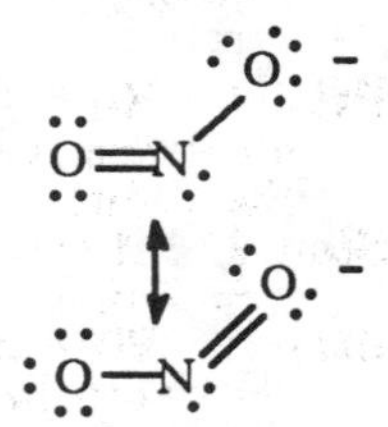

Step 2. In either structure, there are 3 electron regions (2 bonding regions and 1 lone pair).
Step 3. With 2 bonding regions and 1 lone pair:
electron-pair geometry = trigonal planar
molecular shape = bent

d. **Step 1.** C is the central atom. The total number of valence electrons is 1(4 C) + 2(1 H) + 1(6 O) = 12. The structure is shown on the right.

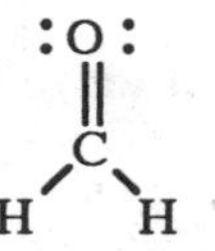

Step 2. There are 3 electron regions (3 bonding regions).
Step 3. With 3 bonding regions and no lone pairs:
electron-pair geometry = trigonal planar
molecular shape = trigonal planar

8.75 Give the electron-pair geometry and the molecular shape of each of the following.

a. SF_4 b. $POClBr_2$ c. OCN^- d. XeF_2

Analysis/Relationship: Electron pair geometry is given solely by the number of electron regions around the central atom.
Plan: We need the Lewis structure to determine the number of electron regions.
Step 3. Read the electron geometry from Tables 8.6 and 8.7 in the text.
Step 2. Count the number of electron regions around the central atom.
Step 1. Draw the Lewis structure of the molecule, using the previously discussed rules.
Work:

a. **Step 1.** S is the central atom. The total number of valence electrons is 1(6 S) + 4(7 F) = 34. The structure is shown on the right.

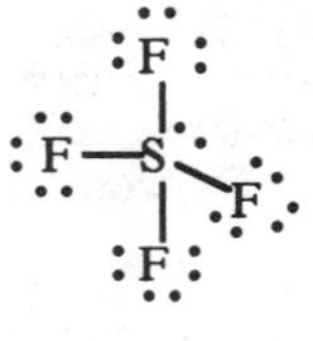

Step 2. There are 5 electron regions (4 bonding pairs and 1 lone pair).
Step 3. With 5 bonding pairs and 1 lone pair:
electron-pair geometry = trigonal bipyramidal
molecular shape = see-saw

b. **Step 1.** P is the central atom. The total number of valence electrons is 1(5 P) + 1(6 O) + 1(7 Cl) + 2(7Br) = 32. The structure is shown on the right. (The double bond is shown even though it violates the octet rule, because it gives both P and O a formal charge of 0.)
Step 2. There are 4 electron regions (4 bonding pairs).
Step 3. With 4 bonding pairs and no lone pairs:
electron-pair geometry = tetrahedral
molecular shape = tetrahedral

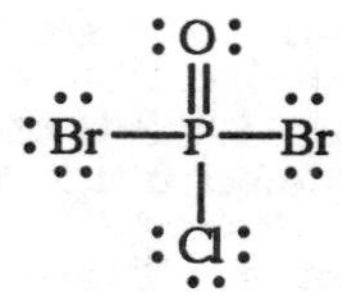

c. **Step 1.** C is the central atom. The total number of valence electrons is 1(4 C) + 1(6 O) + 1(5 N) + 1 = 16. The structure is shown on the right.
Step 2. There are 2 electron regions (2 bonding regions).
Step 3. With 2 bonding regions and no lone pairs:
electron-pair geometry = linear
molecular shape = linear

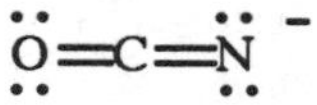

d. **Step 1.** Xe is the central atom. The total number of valence electrons is 1(8 Xe) + 2(7 F) = 22. The structure is shown on the right.
Step 2. There are 5 electron regions (2 bonding pairs and 3 lone pairs).
Step 3. With 2 bonding pairs and 3 lone pairs:
electron-pair geometry = trigonal bipyramidal
molecular shape = linear

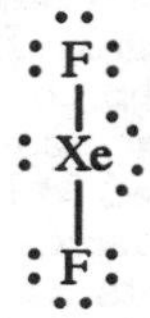

8.77 What value does VSEPR predict for the indicated bond angle in each of the following molecules?
a. H_2Te; <H-Te-H   b. H_2CO; <H-C-H

Analysis/Relationship: The angles depend on the electron pair geometry, which is given solely by the number of electron regions around the central atom.
Plan: We need the Lewis structure to determine the number of electron regions.
Step 3. Read the electron geometry and angles from Tables 8.6 and 8.7 in the text.
Step 2. Count the number of electron regions around the central atom.
Step 1. Draw the Lewis structure of the molecule, using the previously discussed rules.
Work:
a. **Step 1.** Te is the central atom. The total number of valence electrons is 1(6 Te) + 2(1 H) = 8. The structure is shown on the right.
Step 2. There are 4 electron regions.
Step 3. **With** 4 electron regions, the electron pair geometry is a tetrahedral with angles about **109°**. 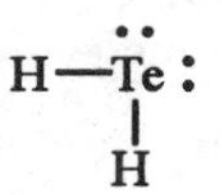
b. **Step 1.** This Lewis structure was drawn in Problem 8.73d and is shown on the right.
Step 2. There are 3 electron regions.
Step 3. With 3 electron regions, the electron pair geometry is trigonal planar with angles about **120°**.

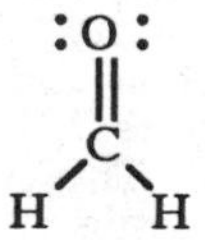

8.79 What are the bond angles in the PCl_5 molecule?
a. <Cl(axial)-P-Cl(equatorial)   b. <Cl(equatorial)-P-Cl(equatorial)

c. <Cl(axial)-P-Cl(axial)

Analysis/Relationship: The angles depend on the electron pair geometry which is given solely by the number of electron regions around the central atom. Axial bonds are the "top" and "bottom" bonds, while equatorial bonds are in the perpendicular plane.
Plan: We need the Lewis structure to determine the number of electron regions.
Step 3. Read the electron geometry and angles from Tables 8.6 and 8.7 in the text.
Step 2. Count the number of electron regions around the central atom.

Step 1. Draw the Lewis structure of the molecule, using the previously discussed rules.

Work:

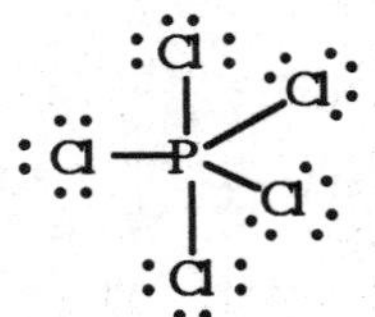

Step 1. P is the central atom. The total number of valence electrons is 1(5 P) + 5(7 Cl) = 40. The structure is shown on the right.

Step 2. There are 5 electron regions.

Step 3. With 5 electron regions, the electron pair geometry is a trigonal bipyramid.

a. The axial/equatorial angle is **90°**.

b. The equatorial/equatorial angle is **120°**.

c. The axial/axial angle is **180°**.

8.81 Identify each of the following as polar or non-polar. If polar, which end of the molecule is negative?

a. XeF_4 b. CS_2 c. H_2S d. H_2CO

Analysis/Relationship: Almost all covalent bonds between different elements have a non-zero bond moment. If the shape of the molecule is such that the bond moments do not cancel out to zero, then the molecule is polar and the end with the atom(s) of higher electronegativity is the negative end.

Plan: We need to know the Lewis structure to determine the molecular geometry.

Step 2. Decide if the molecule has the correct symmetry for cancellation of its bond moments and which end would be negative if it does not.

Step 1. Determine the molecular shape from the Lewis structure and VSEPR.

Work:

a. **Step 1.** Xe is the central atom. The total number of valence electrons is 1(8 Xe) + 4(7 F) = 36. The structure is shown on the right. There are 6 electron regions, with 4 bonding pairs and 2 lone pairs. The electron-pair geometry is octahedral and the molecular shape is square planar.

Step 2. The lone pairs occupy opposite positions and the 4 Xe-F bond moments cancel. The molecule is **nonpolar**.

b. **Step 1.** C is the central atom. The total number of valence electrons is 1(4 C) + 2(6 S) = 16. The structure is shown on the right. There are 2 electron regions, with 2 bonding regions and no lone pairs. Both electron-pair geometry and the molecular shape are linear.

Step 2. The two C-S bond moments cancel, so the molecule is **nonpolar**.

c. **Step 1.** S is the central atom. The total number of valence electrons is 1(6 S) + 2(1 H) = 8. The structure is shown on the right. There are 4 electron regions, with 2 bonding regions and 2 lone pairs. The electron-pair geometry is tetrahedral and the molecular shape is angular.

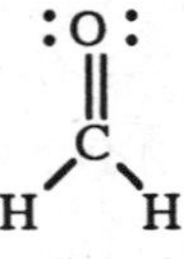

Step 2. The two S-H bond moments do not cancel, so the molecule is **polar**. S has a higher electronegativity (2.5) than H (2.1), so the **sulfur** end is **negative**.

d. **Step 1.** This molecule was drawn in Problem 8.73a and is shown on the right. The electron-pair geometry and the molecular shape are both trigonal planar.

Step 2. The two C-H bond moments do not cancel the C-O bond moment, so the molecule is **polar**. O has a higher electronegativity (3.5) than H (2.1), so the **oxygen** end is **negative**.

8.83 In which of the following is the dipole moment equal to zero?

a. HF b. SF_4

Analysis/Relationship: Almost all covalent bonds between different elements have a non-zero bond moment. If the shape of the molecule is such that the bond moments cancel out to zero, there is no net dipole moment.

Plan: We need to know the Lewis structure to determine the molecular geometry.

Step 2. Decide if the molecule has the correct symmetry for cancellation of its bond moments.

Step 1. Determine the molecular shape from the Lewis structure and VSEPR.

Work:

a. **Step 1.** The total number of valence electrons is 1(7 F) + 1(1 H) = 8. The structure is shown on the right. This molecule is linear.

H—F:

Step 2. Because the electronegativity of F (4.0) is greater than that of H (2.1), the molecule has a **nonzero dipole moment** with F being the negative end of the dipole.

b. **Step 1.** This structure was drawn and analyzed in Problem 8.75a. The structure is shown on the right.

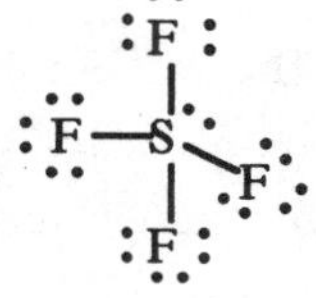

Step 2. Because the electronegativity of F (4.0) is greater than that of S (2.5), the molecule has a **nonzero dipole moment** with its negative end directed between the two equatorial S-F bonds in the molecule.

8.85 In which of the following is the dipole moment greater than zero?

a. SO_4^{2-} b. SF_2 c. HCN d. BeF_2

Plan: We need to know the Lewis structure to determine the molecular geometry.

Step 2. Decide if the molecule has the correct symmetry for cancellation of its bond moments.

Step 1. Determine the molecular shape from the Lewis structure and VSEPR.

Work:

a. **Step 1.** S is the central atom. The total number of valence electrons is 1(6 S) + 4(6 O) + 1(2 charge) = 32. The structure is shown on the right. The electron pair geometry and the ion shape are both tetrahedral.

Step 2. Because the S-O bond moments cancel, the dipole moment **is not greater than zero**.

b. **Step 1.** S is the central atom. The total number of valence electrons is 1(6 S) + 2(7 F) = 20. The structure is shown on the right. The electron pair geometry is tetrahedral, but the molecular shape is angular.

Step 2. Because the S-F bond moments do not cancel, the dipole moment **is greater than zero**, with its negative end directed between the two more electronegative F atoms.

c. **Step 1.** C is the central atom. The total number of valence electrons is 1(1 H) + 1(4 C) + 1(5 N) = 10. The structure is shown on the right. The electron pair geometry and the molecular shape are both linear.

H—C≡N:

Step 2. Because the C-N bond moment is greater than the C-H bond moment, the dipole moment **is greater than zero**, with its negative end directed toward the more electronegative N atom.

d. **Step 1.** Be is the central atom. The total number of valence electrons is 1(2 Be) + 2(7 F) = 16. The structure is shown on the right. The electron pair geometry and the molecular shape are both linear.

:F—Be—F:

Step 2. Because the Be-F bond moments cancel, the dipole moment **is not greater than zero**.

8.87 Which of the following contains the lowest percentage of oxygen: the atmosphere, the hydrosphere, or the lithosphere?

The atmosphere has the lowest percentage of oxygen (25% by mass). The hydrosphere is 89% oxygen by mass and the lithosphere is 50% oxygen by mass.

8.89 Describe the process by which oxygen is produced commercially.

Air is cooled by successive processes of compression and expansion until it liquefies. The oxygen has a higher boiling point than nitrogen and liquefies first. This makes a relatively pure product. Further purification of oxygen can be obtained by distilling liquid air.

8.91 Give the names and formulas of three ions that consist only of oxygen.

O^{2-} is the oxide ion. O_2^{2-} is the peroxide ion. O_2^- is the superoxide ion.

8.93 Give the names and formulas of six anions that contain oxygen and one other element.

Oxygen is found in numerous polyatomic ions such as: nitrate (NO_3^-), sulfate (SO_4^{2-}), carbonate (CO_3^{2-}), perchlorate (ClO_4^-), hydroxide (OH^-), and dichromate ($Cr_2O_7^{2-}$).

9.95 What type of binary oxide reacts with water to form acids?

Nonmetal oxides react with water to form acids.

9.97 How much oxygen is formed when 50.0 g sodium peroxide reacts completely with water?

Analysis:
Target: A mass of oxygen. (? g O_2) =
Knowns: The mass of sodium peroxide is given. Its formula is implicit from its name. The balanced equation is Equation 8-13 in the text. Molar masses are implicit knowns.
Relationships: The balanced equation gives the mole ratios and the molar masses provide the equivalence statements g ↔ mol.
Plan: The standard stoichiometric procedure can be used.

g Na_2O_2 ⟶ mol Na_2O_2 ⟶ mol O_2 ⟶ g O_2

Work:

The balanced equation is: $2\ Na_2O_2 + 2\ H_2O(\ell) \longrightarrow 4\ NaOH + O_2$.

Molar Mass Na_2O_2 = 2(22.98977 Na) + 2(15.9994 O) = 77.9783 g mol^{-1}

Molar Mass O_2 = 2(15.9994) = 31.9988 g mol^{-1}

$$? \text{ g } O_2 = 50.0 \text{ g } Na_2O_2 \times \frac{1 \text{ mol } Na_2O_2}{77.9783 \text{ g } Na_2O_2} \times \frac{1 \text{ mol } O_2}{2 \text{ mol } Na_2O_2} \times \frac{31.9988 \text{ g } O_2}{1 \text{ mol } O_2} = \mathbf{10.3 \text{ g } O_2}$$

8.99 Complete and balance the following equations.

a. $P_4O_{10}(s) + 6\ H_2O(\ell) \longrightarrow$

b. $3\ NO_2(g) + H_2O(\ell) \longrightarrow$

Analysis/Plan: These are nonmetal oxides and will make acids when reacted with water.
Work:

a. $\mathbf{P_4O_{10}(s) + 6\ H_2O(\ell) \longrightarrow 4\ H_3PO_4(aq)}$ (Equation 8-27 in the text)

b. $\mathbf{3\ NO_2(g) + H_2O(\ell) \longrightarrow 2\ HNO_3(aq) + NO(g)}$ (Equation 8-26 in the text)

9 Further Considerations of Bonding

9.1 Using as few words as possible, define what is meant by the term "covalent bond" in (a) Lewis theory, (b) valence bond theory, and (c) molecular orbital theory.

a. In Lewis theory, a covalent bond is a shared electron pair.
b. In valence bond theory, a covalent bond is the overlap of two atomic orbitals.
c. In molecular orbital theory, a covalent bond is the result of electrons occupying a molecular orbital.

9.3 Name some objects, other than those in Figure 9.4, that have cylindrical symmetry.

Objects with cylindrical symmetry would be such things as a soft drink can, a fluorescent light bulb, a toilet paper roll, an uncooked hot dog, and a lemon.

9.5 Criticize the statement, "Before it can form any bonds, a carbon atom must absorb sufficient energy to reach the valence state."

This is inaccurate because bonding occurs in stages, not all at once.

9.7 What are hydrogenlike orbitals? What is a parent orbital?

Hydrogenlike orbitals have characteristics similar to those with the same name in the hydrogen atom. Energy, shape, orientation and size are akin to those in the hydrogen atom. Parent orbitals are the hydrogenlike orbitals from which the hybrid orbitals are made.

9.9 What is the relationship between the number of hybrid orbitals in a set and the number of corresponding parent orbitals?

The number of hybrid orbitals is the same as the number of parent orbitals from which the hybrids were made.

9.11 Comment on the statement, "Resonance is an artificial concept in Lewis theory and an unnecessary one in molecular orbital theory."

It is artificial in Lewis theory because it indicates an average of several drawn structures. Molecular theory accounts for resonance by use of delocalized molecular orbitals, however, the concept is **not** unnecessary, because species which have such delocalized molecular orbitals are unusually stable.

9.13 What type of hybridization is associated with double bonds? With triple bonds?

Either *sp* or sp^2 hybridization may be associated with double bonds, while triple bonds are always *sp* hybridized.

9.15 Describe the shape, including the location of important nodal surfaces, of π and π^* orbitals.

π orbitals have two lobes, with the greatest electron density above and below the internuclear axis. There is a planar node on the internuclear axis. π^* orbitals have four lobes of high electron density. There is a planar node on the internuclear axis and a second planar node perpendicular to the internuclear axis.

9.17 Criticize the statement, "σ orbitals have a single lobe, while π orbitals have two."

The statement is only true for bonding orbitals. σ^* orbitals have two lobes and π^* orbitals have four lobes.

9.19 Why is sp^3d hybridization not used to describe the valence states of second-period elements?

There is no d orbital available in the second energy level. The d orbital closest in energy to that of the $2s$ and $2p$ is the $3d$. The energy required for promotion to $3d$ from $2s$ or $2p$ is too high to be useful for hybridization.

9.21 What aspects, if any, of the sea of electrons model are contradicted by the band theory of metals?

In the sea of electrons model, nuclei only weakly hold the electrons, so they are free to move at any time. In the band theory of metals, electrons cannot move unless there is a vacant orbital of comparable energy nearby in the sample into which they can move.

9.23 Compare the way in which each of the theories of metals accounts for high electrical conductivity.

In the sea of electrons model, electrons are so weakly held that they are free to move. Since electrical conduction depends on the movement of charged particles, the free mobility of electrons gives metals high conductivity. In the band theory, even though electrons are not completely free to move, there are numerous low-energy vacant orbitals in the conduction band and the valence band of metals. Electrons can readily move into these bands, providing the necessary current.

9.25 How do orbitals in the conduction band differ from those in the valence band?

Orbitals in the valence band of a metal are bonding molecular orbitals, while those in the conduction band are antibonding molecular orbitals. As a consequence, orbitals in the conduction band have higher energy than those in the valence band.

9.27 What is the meaning of the term, "saturated", when used to describe hydrocarbons?

A "saturated" hydrocarbon has no multiple bonds. All bonds are single bonds.

9.29 Distinguish between aromatic and aliphatic hydrocarbons in terms of their bonding.

Aromatic hydrocarbons have apparent alternating double and single bonds in their Lewis structures, therefore, they are resonating species. Aliphatic hydrocarbons, if they have multiple bonds, do not exhibit resonance.

9.31 Define, with examples, the term "isomer".

Isomers are compounds which have the same molecular formula but a different structural formula. Examples are the isomers of C_5H_{12} shown below.

pentane | 2-methylbutane | 2,2-dimethylpropane

9.33 Describe the hybridization of the atom designated by boldface type in each of the following species.

a. $\mathbf{N}H_3$ b. $\mathbf{N}H_4^+$

Analysis:

Target: The hybridization of an atom.

Knowns: The formula of the molecule or ion.

Relationships: The hybridization is related to the geometry of the species through the known orientation of orbitals in hybrid sets. This information is summarized in Table 9.1 in the text. The geometry of the species is related to the Lewis structure through VSEPR theory.

Plan: We need to know the Lewis structure and the appropriate relationships.

Step 3. Identify the hybrid orbital set having the proper orientation.
Step 2. Use VSEPR to determine the shape around the central atom.
Step 1. Draw the Lewis structure of the species.

Work:

a. **Step 1.** N is the central atom. The total number of valence electrons is 1(5 N) + 3(1 H) = 8. The structure is shown on the right.

H
|
H—N—H
 ··

Step 2. With 4 electron regions, the electronic geometry is tetrahedral.
Step 3. The hybridization is sp^3.

b. **Step 1.** N is the central atom. The total number of valence electrons is 1(5 N) + 4(1 H) - 1(1 charge) = 8. The structure is shown on the right.

H +
|
H—N—H
|
H

Step 2. With 4 electron regions, the electronic geometry is tetrahedral.
Step 3. The hybridization is sp^3.

9.35 Describe the hybridization of the atom designated by boldface type in each of the following species.

a. $\mathbf{S}F_4$ b. $\mathbf{S}F_6$

Analysis/Plan: Use the general procedure discussed in Problem 9.33.

Work:

a. **Step 1.** S is the central atom. The total number of valence electrons is 1(6 S) + 4(7 F) = 34. The structure is shown on the right.
Step 2. With 5 electron regions, the electronic geometry is trigonal bipyramidal.
Step 3. The hybridization is sp^3d.

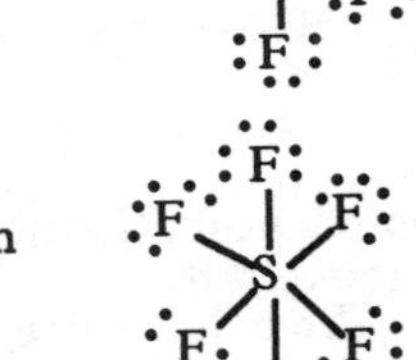

b. **Step 1.** S is the central atom. The total number of valence electrons is 1(6 S) + 6(7 F) = 48. The structure is shown on the right.
Step 2. With 6 electron regions, the electronic geometry is octahedral.
Step 3. The hybridization is sp^3d^2.

9.37 Describe the hybridization of the atom designated by boldface type in each of the following species.

a. $H_2\mathbf{N}NH_2$ b. N_3^- (central atom)

Analysis/Plan: Use the general procedure discussed in Problem 9.33.

Work:

a. **Step 1.** The N atoms are bonded together. The total number of valence electrons is 2(5 N) + 4(1 H) = 14. The structure is shown on the right.
Step 2. With 4 electron regions, the electronic geometry is tetrahedral.
Step 3. The hybridization is ***sp*³**

H H
H—N—N—H

b. **Step 1.** The total number of valence electrons is 3(5 N) + 1 = 16. The structure is shown on the right.
Step 2. With 2 electron regions, the electronic geometry is linear.
Step 3. The hybridization is ***sp***.

$:\ddot{N}=N=\ddot{N}:^-$

9.39 Describe the hybridization of the atom designated by boldface type in each of the following species.
a. **P**F_5 b. H**C**Cl_3

Analysis/Plan: Use the general procedure discussed in Problem 9.33.
Work:

a. **Step 1.** P is the central atom. The total number of valence electrons is 1(5 P) + 5(7 F) = 40. The structure is shown on the right.
Step 2. With 5 electron regions, the electronic geometry is trigonal bipyramidal.
Step 3. The hybridization is ***sp*³*d***.

b. **Step 1.** C is the central atom. The total number of valence electrons is 1(4 C) + 1(1 H) + 3(7 Cl) = 26. The structure is shown on the right.
Step 2. With 4 electron regions, the electronic geometry is tetrahedral.
Step 3. The hybridization is ***sp*³**.

9.41 Give the hybridization of each of the carbon atoms in the molecule whose structural formula is

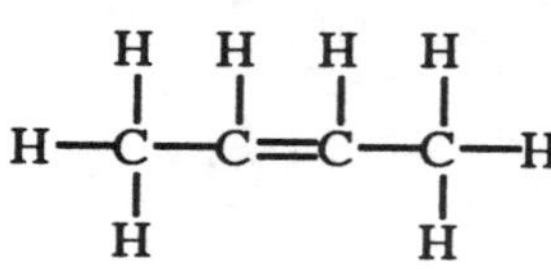

Analysis/Plan: Count the electron regions and determine the hybridization as discussed in Problem 9.33.
Work: From left to right, the carbons have the following hybridizations:
4 electron regions; tetrahedral geometry; ***sp*³** hybridization.
3 electron regions; trigonal planar geometry; ***sp*²** hybridization.
3 electron regions; trigonal planar geometry; ***sp*²** hybridization.
4 electron regions; tetrahedral geometry; ***sp*³** hybridization.

9.43 What is the hybridization of each of the carbon atoms in the molecule whose modified molecular formula is CH_2CH_2?

Analysis/Plan: We need to rewrite this formula into the structural formula so we can count the number of electron regions and determine the hybridization as in Problem 9.33. Remember that carbon makes 4 bonds in almost all of its compounds.

Work:
The expanded structural formula is shown on the right. Note that there must be a double bond between the two carbons to give each the necessary four bonds. Each carbon has three electron regions for a trigonal planar electron geometry and ***sp***^{2} hybridization.

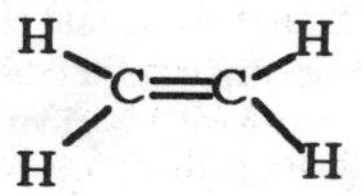

9.45 Describe the hybridization of each of the carbon atoms in the following molecules.
a. CH_3CHO b. $CCl_2CHCH_2CHCH_2$

Analysis/Plan: We need to rewrite each formula as a structural formula so that we can count the number of electron regions and determine the hybridization as in Problem 9.33. Remember that carbon makes 4 bonds in almost all of its compounds.
Work:
a. The expanded structural formula is shown on the right. There must be a double bond between the carbon and the oxygen to give carbon the necessary four bonds. The left-hand carbon has four electron regions for tetrahedral geometry and ***sp***^{3} hybridization. The right-hand carbon has three electron regions for trigonal planar electron geometry and ***sp***^{2} hybridization.

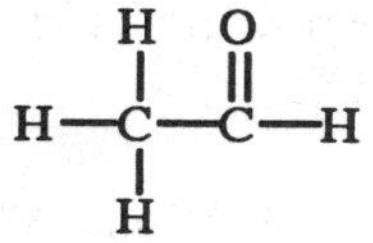

b. The expanded structural formula is shown on the right. The hybridizations from left to right are:
three electron regions for ***sp***^{2} hybridization,
three electron regions for ***sp***^{2} hybridization,
four electron regions for ***sp***^{3} hybridization,
three electron regions for ***sp***^{2} hybridization, and,
three electron regions for ***sp***^{2} hybridization.

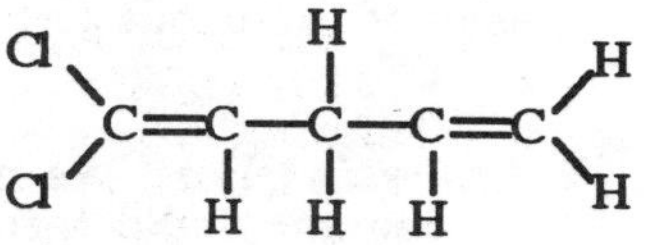

9.47 For each of the following molecules, give the hybridization of the atom in bold type, and describe each of the bonds formed by that atom as σ, π, or some combination of σ and π.
a. HCCH b. F_2CCF_2

Analysis/Plan: We need to rewrite each formula as a structural formula so that we can count the number of electron regions and determine the hybridization as in Problem 9.33. All single bonds are σ bonds. In a multiple bond, one of the bonds is a σ bond and the other bonds are π bonds. Carbon makes 4 bonds in almost all of its compounds.
Work:
a. The expanded structural formula is shown on the right. There must be a triple bond between the carbons to give each one four bonds. Both carbons have two electron regions for linear geometry and ***sp*** hybridization. Each C-H bond is a σ bond. The triple bond is composed of **1** σ and **2** π bonds.

b. The expanded structural formula is shown on the right. There must be a double bond between the carbons to give each one four bonds. Both carbons have three electron regions for trigonal planar geometry and ***sp***^{2} hybridization. Each C-F bond is a σ bond. The C-C double bond is composed of **1** σ and **1** π bond.

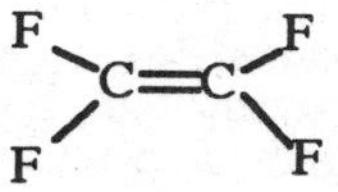

9.49 What is the hybridization of the sulfur atom in each of the following compounds? Identify each of the bonds as σ, π, or some combination of σ and π.
a. SO_2 b. SO_3

Analysis/Plan: We need each structural formula so that we can count the number of electron regions and determine the hybridization as in Problem 9.33. All single bonds are σ bonds. In a multiple bond, one of the bonds is a σ bond and the other bonds are π bonds.

Work:

a. The Lewis structure of SO_2 is a combination of two resonant structures (Problem 8.53a). With three electron regions, the hybridization is ***sp***^{2}. One bond between S and each O is a σ bond. There is also, on average, 1/2 π bond between S and each O.

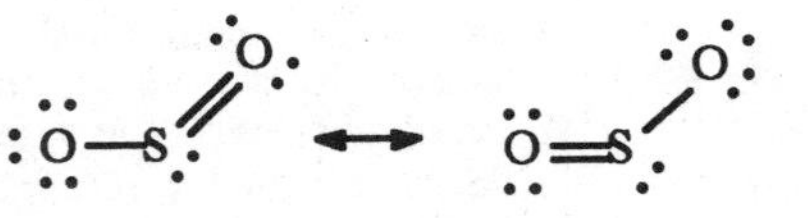

b. SO_3 is a combination of three resonant structures (Problem 8.61). With three electron regions, the hybridization is ***sp***^{2}. One bond between S and each O is a σ bond. There is also, on the average, 1/3 π bond between S and each O.

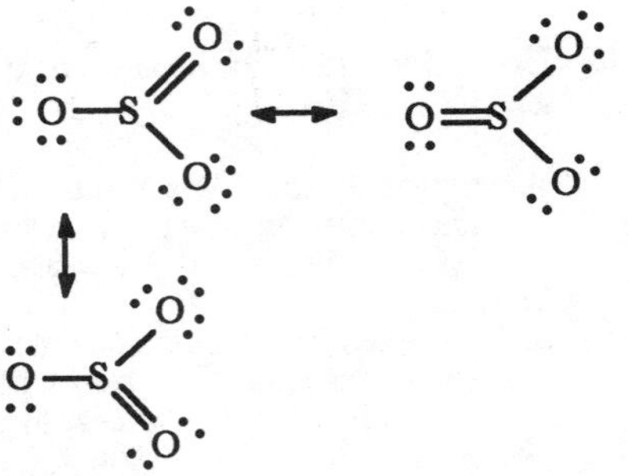

9.51 Give the hybridization of the carbon atom in each of the following molecules. Then describe each of the bonds formed by that atom as σ, π, or some combination of σ and π.

a. CS_2 b. $COCl_2$

Analysis/Plan: We need each structural formula so we can count the number of electron regions and determine the hybridization as in Problem 9.33. All single bonds are σ bonds. In a multiple bond, one of the bonds is a σ bond and the other bonds are π bonds.

Work:

a. This structure was drawn in Problem 8.81b and is shown on the right. There are two electron regions for a linear geometry and ***sp*** hybridization. Each C-S bond consists of 1 σ and 1 π bond.

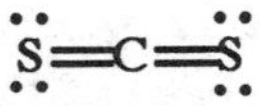

b. C is the central atom. The total number of valence electrons is 1(4 C) + 2(7 Cl) + 1(6 O) = 24 The structure is shown on the right. With three electron regions, the hybridization is ***sp***^{2}. The C-Cl bonds are σ bonds and the C-O bond consists of 1 σ bond and 1 π bond.

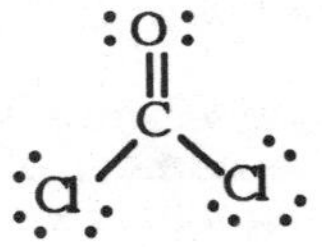

9.53 Give the hybridization of the carbon atoms, resonance structures (if any), and bond types (σ, π, or combination of σ and π), of the allene molecule, whose σ-framework is

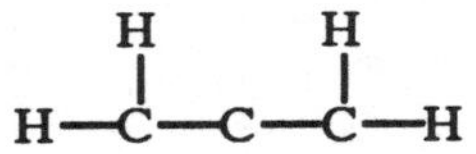

Analysis/Plan: We need to rewrite this into the complete structural formula, with each C having four bonds, so we can count the number of electron regions and determine the hybridization as in Problem 9.33. All single bonds are σ bonds. In a multiple bond, one of the bonds is a σ bond and the other bonds are π bonds.

Work:

The expanded structural formula is shown on the right. The outer carbon atoms have three electron regions for ***sp***^{2} hybridization. Each C-H bond is a σ bond. The central carbon has two electron regions for ***sp*** hybridization. The C-C bonds are each composed of 1 σ and 1 π bond.

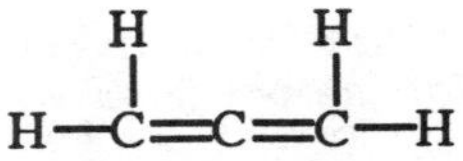

9.55 Describe the bond order and magnetism of each of the following diatomic ions.

a. H_2^+ b. H_2^- c. He_2^+

Analysis:

Target: The bond order and magnetism of each ion.

Knowns: The formula of each ion.

Relationships: The aufbau scheme for molecular orbitals leads to the electron configuration, from which the number of bonding and antibonding electrons and the number of unpaired electrons can be counted. Paramagnetism occurs when there is at least one unpaired electron. If all electrons are paired, the species is diamagnetic. The formula for determining bond order is:

$$\frac{\text{total number of electrons in bonding MO's - total number of electrons in antibonding MO's}}{2}$$

Plan:

Step 4. Count the number of electrons in bonding and antibonding MO's and apply the formula to determine the bond order.
Step 3. Determine the magnetism of the ion.
Step 2. Use the aufbau scheme to write the electron configuration for the ion.
Step 1. Count the number of valence electrons.

Work:

a. **Step 1.** H_2^+ has 2(1) - 1 = 1 electron.
Step 2. The electron configuration is $(\sigma_{1s})^1$.
Step 3. H_2^+ is **paramagnetic** because it has 1 unpaired electron.
Step 4. Bond order = $\frac{1 - 0}{2}$ = **1/2**

b. **Step 1.** H_2^- has 2(1) + 1 = 3 electrons.
Step 2. The electron configuration is $(\sigma_{1s})^2(\sigma_{1s}^*)^1$.
Step 3. H_2^+ is **paramagnetic** because it has 1 unpaired electron.
Step 4. Bond order = $\frac{2 - 1}{2}$ = **1/2**

c. **Step 1.** He_2^+ has 2(2) - 1 = 3 electrons.
Step 2. The electron configuration is $(\sigma_{1s})^2(\sigma_{1s}^*)^1$.
Step 3. He_2^+ is **paramagnetic** because it has 1 unpaired electron.
Step 4. Bond order = $\frac{2 - 1}{2}$ = **1/2**

9.57 Use MO theory to predict the relative bond energy in the following pairs of species.

a. F_2, F_2^+ b. C_2, C_2^-

Analysis/Plan: A species with a higher bond order also has a higher bond energy. The relationships and general plan are summarized in Problem 9.55. For this question, we need to carry out the following steps.

Step 4. Compare the bond order in the two species. The species with higher bond order also has higher bond energy.
Step 3. Count the number of electrons in bonding and antibonding MO's and apply the formula to calculate the bond order.
Step 2. Use the aufbau scheme to write the electron configuration for the ion.
Step 1. Count the number of valence electrons.

Work:

a. **Step 1.** F_2 has 2(7) = 14 valence electrons and F_2^+ has 2(7) - 1 = 13 valence electrons.
Step 2. Electron configurations are: $F_2 = (\sigma_{2s})^2(\sigma_{2s}^*)^2(\pi_{2p})^4(\sigma_{2p})^2(\pi_{2p}^*)^4$ and $F_2^+ = (\sigma_{2s})^2(\sigma_{2s}^*)^2(\pi_{2p})^4(\sigma_{2p})^2(\pi_{2p}^*)^3$.
Step 3. F_2 bond order = $\frac{8 - 6}{2}$ = 1 and F_2^+ bond order = $\frac{8 - 5}{2}$ = 1.5.
Step 4. **F_2^+ has the higher bond order and, therefore, the higher bond energy.**

b. **Step 1.** C_2 has 2(4) = 8 valence electrons and C_2^- has 2(4) + 1 = 9 valence electrons.
Step 2. Electron configurations are: $C_2 = (\sigma_{2s})^2(\sigma_{2s}^*)^2(\sigma_{2p})^2(\pi_{2p})^2$ and
$C_2^- = (\sigma_{2s})^2(\sigma_{2s}^*)^2(\sigma_{2p})^2(\pi_{2p})^3$.
Step 3. C_2 bond order = $\frac{6 - 2}{2}$ = 2 and C_2^- bond order = $\frac{7 - 2}{2}$ = 2.5.
Step 4. C_2^- has the higher bond order and, therefore, the higher bond energy.

9.59 Use MO theory to predict the relative bond energy in the following pairs of species.
a. CO, CO^+ b. NO, NO^+

Analysis/Plan: This question is the same as that in Problem 9.57 and we proceed in exactly the same manner.
Work:
a. **Step 1.** CO has 1(4 C) + 1(6 O) = 10 valence electrons and CO^+ has 1(4 C) + 1(6 O) - 1 = 9 valence electrons.
Step 2. Electron configurations are: CO = $(\sigma_{2s})^2(\sigma_{2s}^*)^2(\pi_{2p})^4(\sigma_{2p})^2$ and
$CO^+ = (\sigma_{2s})^2(\sigma_{2s}^*)^2(\pi_{2p})^4(\sigma_{2p})^1$.
Step 3. CO bond order = $\frac{8 - 2}{2}$ = 3 and CO^+ bond order = $\frac{7 - 2}{2}$ = 2.5.
Step 4. CO has the higher bond order and, therefore, the higher bond energy.

b. **Step 1.** NO has 1(5 N) + 1(6 O) = 11 valence electrons and NO^+ has 1(5 N) + 1(6 O) - 1 = 10 valence electrons.
Step 2. Electron configurations are: NO = $(\sigma_{2s})^2(\sigma_{2s}^*)^2(\pi_{2p})^4(\sigma_{2p})^2(\pi_{2p}^*)^1$ and
$NO^+ = (\sigma_{2s})^2(\sigma_{2s}^*)^2(\pi_{2p})^4(\sigma_{2p})^2$.
Step 3. NO bond order = $\frac{8 - 3}{2}$ = 2.5 and NO^+ bond order = $\frac{8 - 2}{2}$ = 3.
Step 4. NO^+ has the higher bond order and, therefore, the higher bond energy.

9.61 A number of diatomic oxygen species are known: Describe the magnetic behavior of each of the following, and rank them in order of increasing bond energy. O_2, O_2^+, O_2^-, O_2^{2-}

Analysis/Plan: The relationships and general plan are summarized in Problem 9.55. For this question, we need to carry out the following steps.
Step 5. Compare the bond order in the four species.
Step 4. Count the number of electrons in bonding and antibonding MO's and apply the formula to determine the bond order.
Step 3. Determine the magnetic behavior by finding the number of unpaired electrons.
Step 2. Use the aufbau scheme to write the electron configuration for the species.
Step 1. Count the number of valence electrons.
Work:
a. **Step 1.** O_2 has 2(6) = 12 valence electrons; O_2^+ has 2(6) - 1 = 11 valence electrons; O_2^- has 2(6) + 1 = 13 valence electrons; and, O_2^{2-} has 1(6) + 2 = 14 valence electrons.
Step 2. Electron configurations are: $O_2 = (\sigma_{2s})^2(\sigma_{2s}^*)^2(\pi_{2p})^4(\sigma_{2p})^2(\pi_{2p}^*)^2$,
$O_2^+ = (\sigma_{2s})^2(\sigma_{2s}^*)^2(\pi_{2p})^4(\sigma_{2p})^2(\pi_{2p}^*)^1$, $O_2^- = (\sigma_{2s})^2(\sigma_{2s}^*)^2(\pi_{2p})^4(\sigma_{2p})^2(\pi_{2p}^*)^3$ and
$O_2^{2-} = (\sigma_{2s})^2(\sigma_{2s}^*)^2(\pi_{2p})^4(\sigma_{2p})^2(\pi_{2p}^*)^4$.
Step 3. All are paramagnetic except for O_2^{2-}. (Remember that, by Hund's rule, one electron must be placed in each (π_{2p}^*) orbital before either is paired.)
Step 4. O_2 bond order = $\frac{8 - 4}{2}$ = 2, O_2^+ bond order = $\frac{8 - 3}{2}$ = 2.5; O_2^- bond order = $\frac{8 - 5}{2}$ = 1.5; and O_2^{2-} bond order = $\frac{8 - 6}{2}$ = 1.
Step 5. Bond orders increase: $O_2^{2-} < O_2^- < O_2 < O_2^+$.

9.63 Which of the following species would you expect to increase in stability on the addition of one electron? C_2, N_2, O_2, F_2, CO, NO.

Analysis/Plan: Stability increases as bond order increases. A species will be more stable if the electron is added to a bonding MO, while it will be less stable if the electron is added to an antibonding MO. We need to write the MO electron configuration for the species and determine which type of orbital the next electron will fill.

Work:

a. C_2 has 2(4) = 8 valence electrons. Its electron configuration is $(\sigma_{2s})^2(\sigma_{2s}^*)^2(\pi_{2p})^4$. The next electron added would be in the (σ_{2p}) bonding orbital so C_2^- would be **more stable.**

b. N_2 has 2(5) = 10 valence electrons. Its electron configuration is $(\sigma_{2s})^2(\sigma_{2s}^*)^2(\pi_{2p})^4(\sigma_{2p})^2$. The next electron added would be in the (π_{2p}^*) antibonding orbital, so N_2^- would be **less stable.**

c. O_2 has 2(6) = 12 valence electrons. Its electron configuration is $(\sigma_{2s})^2(\sigma_{2s}^*)^2(\pi_{2p})^4(\sigma_{2p})^2(\pi_{2p}^*)^2$. The next electron added would be in the (π_{2p}^*) antibonding orbital, so O_2^- would be **less stable.**

d. F_2 has 2(7) = 14 valence electrons. Its electron configuration is $(\sigma_{2s})^2(\sigma_{2s}^*)^2(\pi_{2p})^4(\sigma_{2p})^2(\pi_{2p}^*)^4$. The next electron added would be in the (σ_{2p}^*) antibonding orbital, so F_2^- would be **less stable.**

e. CO has 1(4 C) + 1(6 O) = 10 valence electrons. Its electron configuration is the same as N_2 (part b) and it will also be **less stable** if an electron is added.

f. NO has 1(5 N) + 1(6 O) = 11 valence electrons. Its electron configuration is $(\sigma_{2s})^2(\sigma_{2s}^*)^2(\pi_{2p})^4(\sigma_{2p})^2(\pi_{2p}^*)^1$. The next electron added would be in the (π_{2p}^*) antibonding orbital, so NO^- would be **less stable**

9.65 When carbon vaporizes, at extremely high temperatures, among the species present in the vapor is the diatomic molecule C_2. Is C_2 paramagnetic? Which would you expect to have the greatest bond energy: C_2, C_2^-, or C_2^{2-}?

Analysis/Plan: The species with the highest bond order also has the highest bond energy. The other relationships and general plan are summarized in Problem 9.55. For this question, we need to carry out the following steps.

Step 5. Compare the bond order and, therefore, the bond energy in the three species.
Step 4. Count the number of electrons in bonding and antibonding MO's and apply the formula.
Step 3. Determine the magnetic behavior by finding the number of unpaired electrons.
Step 2. Use the aufbau scheme to write the electron configuration for the species.
Step 1. Count the number of valence electrons.

Work:

a. **Step 1.** C_2 has 2(4) = 8 valence electrons; C_2^- has 2(4) + 1 = 9 valence electrons; and, C_2^{2-} has 2(4) + 2 = 10 valence electrons.

Step 2. Electron configurations are: $C_2 = (\sigma_{2s})^2(\sigma_{2s}^*)^2(\pi_{2p})^4$, $C_2^- = (\sigma_{2s})^2(\sigma_{2s}^*)^2(\pi_{2p})^4(\sigma_{2p})^1$, and $C_2^{2-} = (\sigma_{2s})^2(\sigma_{2s}^*)^2(\pi_{2p})^4(\sigma_{2p})^2$.

Step 3. C_2 has all electrons paired, so it is **diamagnetic.**

Step 4. C_2 bond order $= \frac{6-2}{2} = 2$, C_2^- bond order $= \frac{7-2}{2} = 2.5$, and C_2^{2-} bond order $= \frac{8-2}{2} = 3$.

Step 5. Bond orders and bond energies increase: $C_2 < C_2^- < C_2^{2-}$

9.67 Name each of the following hydrocarbons.

a. b. 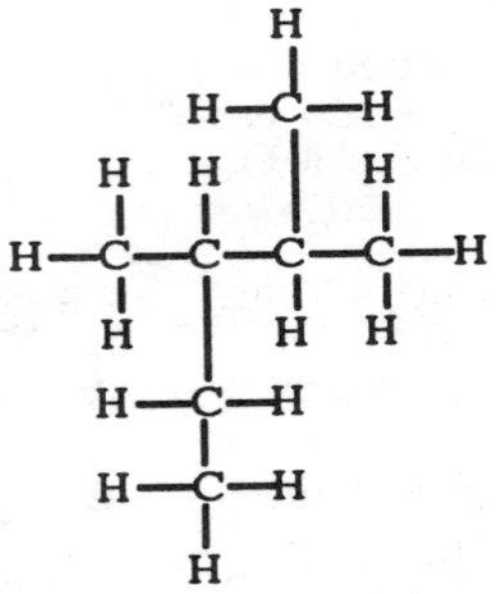 c.

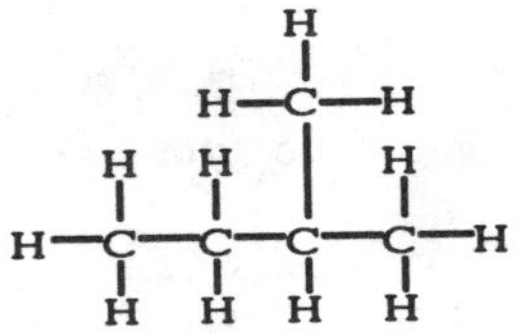

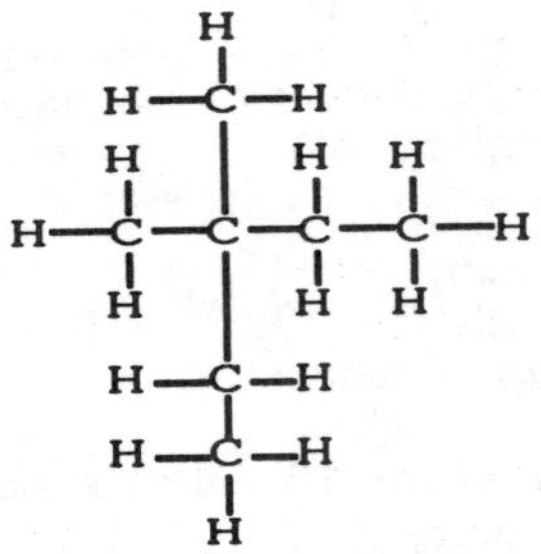

Analysis/Plan: The IUPAC rules for naming must be obeyed. We need to find the longest continuous chain of carbons first and then follow the rules as detailed in the book.

Work:

a. The longest continuous chain of carbons is 4 and all bonds are single bonds, so this is a butane. There is one single carbon (methyl) group. This compound is **2-methylbutane**

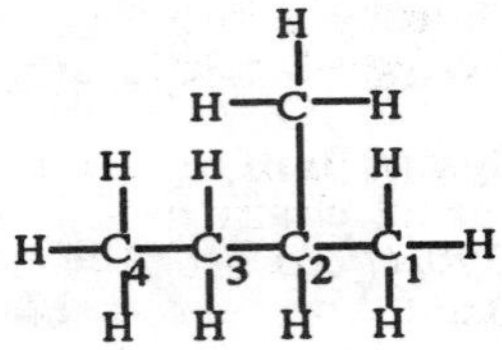

b. The longest continuous chain of carbons is 5 and all bonds are single bonds, so this is a pentane. This numbering gives the lowest total number for the positions of the substituents, each of which is a one carbon (methyl) group. The name of the compound is **2,3-dimethylpentane.**

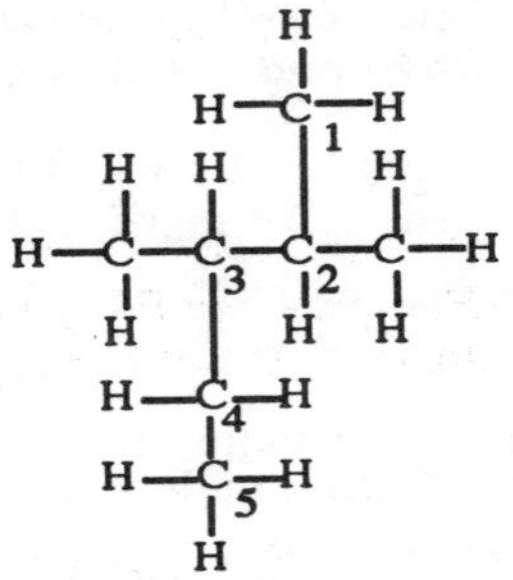

c. The longest continuous chain of carbons is 5 and all bonds are single bonds, so this is a pentane. Each of the substituents is a one carbon (methyl) group. The name of the compound is **3,3-dimethylpentane.**

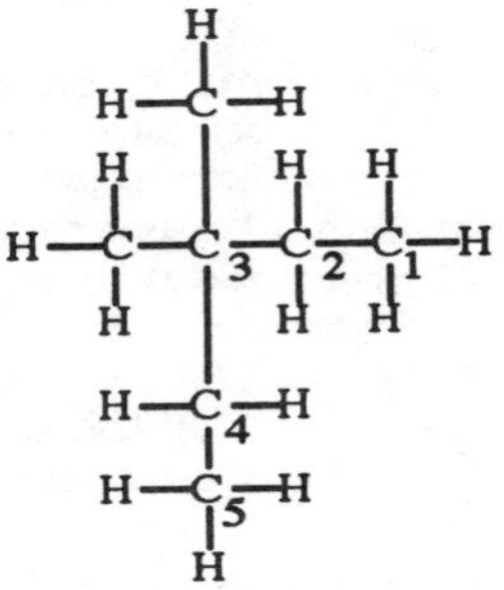

9.69 What is the total number of σ bonds in each of the molecules below? The number of π bonds?

a. b.

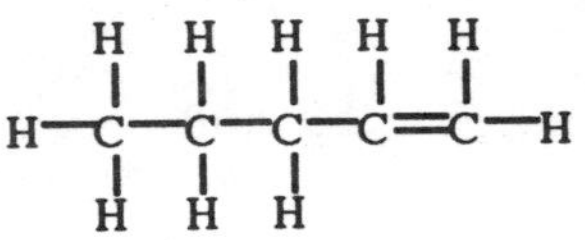

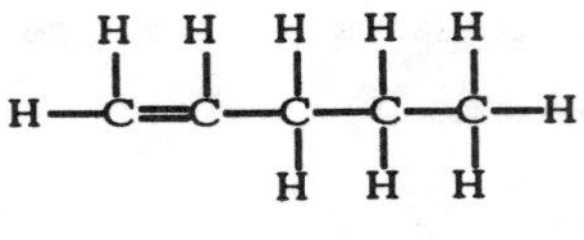

c.

```
    H   H   H   H   H
    |   |   |   |   |
H — C — C = C — C — C — H
    |           |   |
    H           H   H
```

Analysis/Plan: Remember that all single bonds are σ bonds. In a multiple bond, one bond is a σ bond and the other bonds are π bonds. Simply count the number of bonds in the structure.

Work:

a. This structure has **14** σ bonds and **1** π bond.
b. This structure has **14** σ bonds and **1** π bond.
c. This structure has **14** σ bonds and **1** π bond.

9.71 Draw the structural formulas of each of the following compounds.

a. 2-methylhexane
b. 2,3-dimethylpentane
c. 2-chloro-3-methylnonane
d. 2-bromo-4-chloro-4,5,7-trimethyl decane

Analysis/Plan: The IUPAC rules for naming must be obeyed. We need to write down the parent chain first, number it, and then add the substituents at the correct locations. Hydrogens are then added to give each carbon four bonds.

Work:

a. Hexane has 6 carbons and a methyl group has 1 carbon.

```
                         H
                         |
                     H — C — H
    H    H    H    H     |     H
    |    |    |    |     |     |
H — C6 — C5 — C4 — C3 —  C2 —  C1 — H
    |    |    |    |     |     |
    H    H    H    H     H     H
```

b. Pentane has 5 carbons and a methyl group has 1 carbon.

```
                        H
                        |
                    H — C — H
    H    H    H         |     H
    |    |    |         |     |
H — C5 — C4 — C3 —      C2 —  C1 — H
    |    |    |         |     |
    H    H    |         H     H
              |
          H — C — H
              |
              H
```

c. Nonane has 9 carbons, a methyl group has 1 carbon and chloro is Cl.

```
              H
              |
          H — C — H
    H    Cl   |    H    H    H    H    H    H
    |    |    |    |    |    |    |    |    |
H — C1 — C2 — C3 — C4 — C5 — C6 — C7 — C8 — C9 — H
    |    |    |    |    |    |    |    |    |
    H    H    H    H    H    H    H    H    H
```

d. Decane has 10 carbons, a methyl group has 1 carbon, bromo is Br and chloro is Cl.

```
                   H              H
                   |              |
               H — C — H      H — C — H
    H    H    H    |    H    H    |    H    H    H
    |    |    |    |    |    |    |    |    |    |
H — C1 — C2 — C3 — C4 — C5 — C6 — C7 — C8 — C9 — C10 — H
    |    |    |    |    |    |    |    |    |    |
    H    Br   H    Cl   |    H    H    H    H    H
                        |
                    H — C — H
                        |
                        H
```

9.73 Draw complete structural formulas corresponding to each of the following modified molecular formulas.

a. $CH_3CH_2CHCHCH_2Cl$

b. $CH_3CH{=}CHCH_2CHO$

c. $CH_2{=}CHCH_2CH_2CH_2OH$

d. $CH_3CH_2(CH_2)_3CH_3$

Analysis/Plan: The complete structural formula clearly shows all bonds. We need to rewrite these, being sure to give each carbon 4 bonds, each oxygen 2 bonds and chlorine 1 bond.

Work:

a.

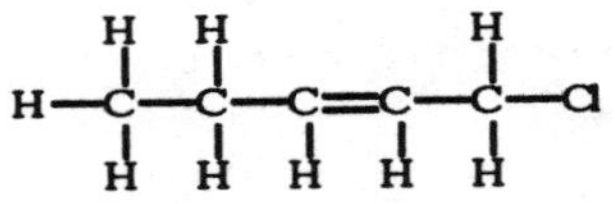

```
    H   H           H
    |   |           |
H — C — C — C = C — C — Cl
    |   |   |   |   |
    H   H   H   H   H
```

b.

```
    H   H   H   H     O
    |   |   |   |    //
H — C — C = C — C — C
    |           |    \
    H           H     H
```

c.

```
H       H   H   H   H
 \      |   |   |   |
  C  =  C — C — C — C — O — H
 /          |   |   |
H           H   H   H
```

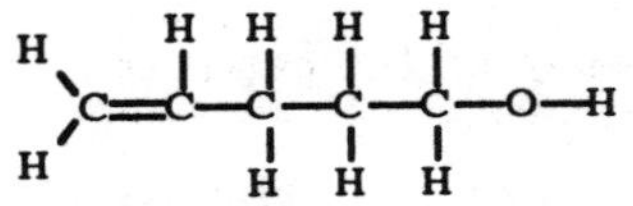

d.

```
    H   H   H   H   H   H
    |   |   |   |   |   |
H — C — C — C — C — C — C — H
    |   |   |   |   |   |
    H   H   H   H   H   H
```

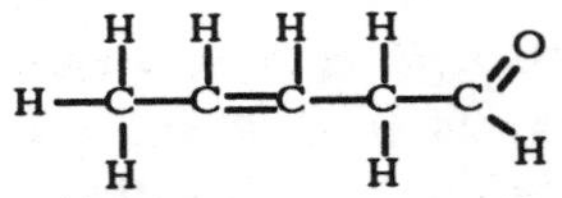

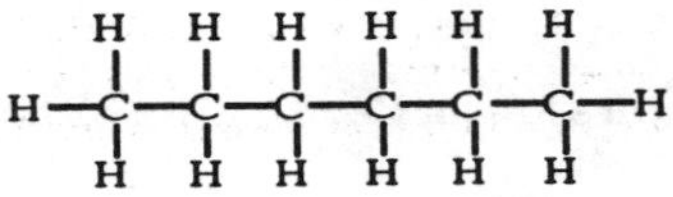

10 Thermodynamics

10.1 What is the difference between thermodynamics and thermochemistry?

Thermochemistry is limited to the study of heat in chemical reactions. Thermodynamics considers work, disorder and spontaneity in addition to heat.

10.3 What is the most common type of work in chemical processes?

The most common form of work in chemical processes is pressure/volume work.

10.5 Define heat. What units are commonly used to measure heat?

Heat is the form of energy which moves from one substance to another because of temperature differences between them. It is a sum of the kinetic energies of all the particles in a substance. The units commonly used to measure heat are calories and joules.

10.7 The phrase, "states of matter", is sometimes used to refer to solid, liquid, and gas. Distinguish between this usage and the usage of the term "state" in thermodynamics.

"States of matter" simply describes the phase of the system; whether it is solid, liquid or gas. The state of a system in thermodynamics includes not only the phase, but the temperature, the composition, the volume, the pressure and all other parameters which determine the nature of the system. Two systems do not have the same thermodynamic state unless **all** of their measurable properties are the same.

10.9 What is Hess's Law, and how is it related to the fact that enthalpy is a state function?

Hess's Law states that the heat (enthalpy) of a process is the same whether the process is carried out in one step or in a series of steps. This is a direct consequence of the fact that enthalpy is a state function. The change in any state function depends only on the difference between the initial and final states, not on the pathway by which the process is carried out.

10.11 State the first law of thermodynamics, in words rather than as an equation.

The change in the internal energy of a system is the sum of the heat applied to or removed from the system and the work done by or on the system.

10.13 Suppose the work in some process is given as –200 J. What meaning should be attached to the negative sign?

The negative sign indicates that work has been done on the surroundings by the system.

10.15 Give an algebraic statement of the first law of thermodynamics.

$\Delta E = q + w$

10.17 If a reaction is characterized by $\Delta H_{rxn} = -500\ kJ\ mol^{-1}$, does it absorb heat from the surroundings or release heat to them?

The negative sign indicates that the reaction releases heat to the surroundings.

10.19 What property of systems is described by entropy?

Entropy is the measure of the disorder or randomness of a system.

10.21 What is meant by the term "standard state"?

The standard state of a substance is its most stable form at 1 atmosphere pressure and a specified temperature, usually 25 °C.

10.23 Show that it is a consequence of the definition of "free energy of formation" that $\Delta G_f^\circ = 0$ for any element in its standard state.

In a formation reaction, substances are formed from elements in their standard states. Therefore, the free energy of formation of an element would be the process of making that element from itself. Because there is no change in the substance in the system and because "standard state" is specified (i.e. temperature, pressure, etc. is set), the change in all properties is zero. ΔG_f° must also be zero.

10.25 What is the relationship between free energy change, spontaneity, and equilibrium?

If ΔG is negative, the process is spontaneous. If ΔG is zero, the system is at equilibrium.

10.27 How much work is involved in moving an object a distance of 5.00 meters against an opposing force of 3.30 N?

Analysis:
Target: The work done in a process. (? J) =
Knowns: The force applied and the distance moved.
Relationships: Work = (force)(distance). 1 J ↔ 1 N•m.
Plan: Use the equation for work directly.
Work:

$$? \text{ J} = \text{work} = (3.30 \text{ N})(5.00 \text{ m}) \times \frac{1 \text{ J}}{1 \text{ N}\cdot\text{m}} = \mathbf{16.5 \text{ J}}$$

10.29 How far can an object be moved against a force of 0.753 N if 100 J is available?

Analysis:
Target: The distance an object can be moved. (? m) =
Knowns: The force applied and the energy available to use as work.
Relationships: Work = (force)(distance). 1 J ↔ 1 N•m.
Plan: Solve the equation for distance, insert the known values and calculate.
Work:

$$? \text{ m} = \text{distance} = \frac{\text{work}}{\text{force}} = \frac{100 \text{ J} \times \frac{1 \text{ N}\cdot\text{m}}{1 \text{ J}}}{0.753 \text{ N}} = \mathbf{133 \text{ m}}$$

10.31 Calculate the work required to move a 150-lb person up one flight of stairs (9.0 feet). The force of gravity is 9.8 N kg^{-1}.

Analysis:
Target: The work done in a process. (? J) =
Knowns: The weight of the person and the distance to be moved.

Relationships: Work = (force)(distance). 1 J ↔ 1 N • m. Other necessary equivalence statements are 1 lb ↔ 0.45359 kg, 1 in ↔ 2.54 cm, 100 cm = 1 m, 1 ft ↔ 12 in, 1 kJ ↔ 1000 J.

Plan: The total force must be determined before the equation can be used. The height must be converted to appropriate units.

Step 3. Use the equation to determine the work.
Step 2. Convert the height to meters using the appropriate unit factors.
Step 1. Convert the weight to total force using the appropriate unit factors.

Work:

Step 1. Convert the weight to total force using the appropriate unit factors.

force = [mass (kg)]•[force per kg ($N\ kg^{-1}$)]

$$\text{force} = 150\text{ lb} \times \frac{0.45359\text{ kg}}{\text{lb}} \times \frac{9.8\text{ N}}{\text{kg}} = 6.7 \times 10^2\text{ N}$$

Step 2. Convert the height to meters using the appropriate unit factors.

$$\text{height} = 9.0\text{ ft} \times \frac{12\text{ in}}{1\text{ ft}} \times \frac{2.54\text{ cm}}{1\text{ in}} \times \frac{1\text{ m}}{100\text{ cm}} = 2.7\text{ m}$$

Step 3. Use the equation to determine the work.

$$?\text{ J} = \text{work} = (6.7 \times 10^2\text{ N})(2.7\text{ m}) \times \frac{1\text{ J}}{1\text{ N}\cdot\text{m}} \times \frac{1\text{ kJ}}{1000\text{ J}} = \mathbf{1.8\ kJ}$$

10.33 One liter (1 kg) of water acquires 0.57 kJ of kinetic energy when it drops over a 190-foot waterfall (the height of Niagara Falls). If all this energy is converted to heat, by how much does the temperature of the water rise? The specific heat of water is 4.184 J g^{-1} K^{-1}. (It is said that Joule, while honeymooning in the Alps, attempted an experimental measurement of this effect.)

Analysis:

Target: The temperature change of the water. (? °C) =

Knowns: The mass of the water, the specific heat of the water and the energy given to the water in the fall.

Relationships: The relationship $q = mc_s\Delta T$ applies. The magnitude of K is the same as that of °C.

Plan: Rearrange the equation to solve for ΔT and calculate.

Work:

$$?\ ^\circ\text{C} = \Delta T = \frac{q}{mc_s} = \frac{0.57\text{ kJ}}{(1\text{ kg})(4.184\text{ J g}^{-1}\ ^\circ\text{C}^{-1})} = \mathbf{0.14\ ^\circ C}$$

10.35 What is the change in internal energy of a system if 100 J heat is added and the system does 50 J work on the surroundings?

Analysis:

Target: The change in internal energy, ΔE. (? J) =

Knowns: The heat, q, and the work, w.

Relationships: The first law of thermodynamics, ΔE = q + w. If heat is added, the sign of q is positive and if the system does work, the sign of w is negative.

Plan: Insert the known values, with the proper signs, into the first law and calculate.

Work:

? J = ΔE = q + w = +100 J + (-50 J) = **+50 J**

10.37 A gas is compressed by the expenditure of 50 J work, and at the same time 50 J heat is added. What is ΔE for the gas?

Analysis:

Target: The change in internal energy, ΔE. (? J) =

Knowns: The heat, q, and the work, w.

Relationships: The first law of thermodynamics, $\Delta E = q + w$. If heat is added, the sign of q is positive and if work is done on the system, as it must be to compress the gas, the sign of w is positive.

Plan: Insert the values, with the proper signs, into the first law and calculate.

Work:

$$? J = \Delta E = q + w = +50 \text{ J} + (+50 \text{ J}) = \mathbf{+100 \text{ J}}$$

10.39 A gas expands against a pressure of 50 kPa, in the process increasing its volume by 0.015 m^3. What is ΔE of the gas?

Analysis:

Target: The change in internal energy, ΔE. (? J) =

Known: The volume change against a constant opposing pressure..

Relationships: The first law of thermodynamics, $\Delta E = q + w$. Because heat is not mentioned, we must assume that q is zero. Work = $-P\Delta V$. Because the volume increases, $\Delta V > 0$. The appropriate equivalence statements are 1 kPa $\leftrightarrow$ 1000 Pa and 1 J $\leftrightarrow$ 1 Pa m^3.

Plan: Find the work, then insert its value into the first law and calculate.

Work:

$$\text{work} = -P\Delta V = -(50 \text{ kPa}) \times \frac{1000 \text{ Pa}}{1 \text{ kPa}} \times (0.015 \text{ m}^3) \times \frac{1 \text{ J}}{1 \text{ Pa m}^3} = -7.5 \times 10^2 \text{ J}$$

$$? J = \Delta E = q + w = 0 + (-7.5 \times 10^2 \text{ J}) = \mathbf{-7.5 \times 10^2 \text{ J}}$$

10.41 Use data from Appendix G to determine ΔS°_{rxn} when acetaldehyde ($S^\circ = 38.3 \text{ J K}^{-1}$) burns according to the reaction $CH_3CHO(\ell) + 3\,O_2(g) \longrightarrow 2\,CO_2(g) + 2\,H_2O(\ell)$.

Analysis:

Target: The ΔS°_{rxn} of a process. (? J K^{-1}) =

Knowns: The balanced equation which provides the stoichiometric coefficients is given. S° must be looked up in Appendix G.

Relationship: $\Delta S^\circ_{rxn} = \Sigma[nS^\circ(\text{products})] - \Sigma[nS^\circ(\text{reactants})]$

Plan: Look up the necessary S° values and use the equation directly.

Work:

$$? \text{ J K}^{-1} = \Delta S^\circ_{rxn} = \Sigma[nS^\circ(\text{products})] - \Sigma[nS^\circ(\text{reactants})]$$

$$? \text{ J K}^{-1} = \Delta S^\circ_{rxn} = 2(S^\circ CO_2(g)) + 2(S^\circ H_2O(\ell)) - [1(S^\circ CH_3CHO(\ell)) + 3(S^\circ O_2(g))]$$

$$? \text{ J K}^{-1} = \Delta S^\circ_{rxn} = 2(213.74) + 2(69.91) - [1(38.3) + 3(205.138)]$$

$$? \text{ J K}^{-1} = \Delta S^\circ_{rxn} = \mathbf{-86.4 \text{ J K}^{-1}}$$

10.43 Use data from Appendix G to calculate the change in entropy when 10.0 g of HgO(s) decomposes, under standard conditions, according to $2\,HgO(s) \longrightarrow 2\,Hg(\ell) + O_2(g)$.

Analysis:

Target: The change in entropy of a process. (? J K^{-1}) =

Knowns: The mass of HgO which decomposes and the balanced equation which provides the stoichiometric coefficients are given. S° must be looked up in Appendix G. The molar mass of HgO is an implicit known.

Relationships: $\Delta S^\circ_{rxn} = \Sigma[nS^\circ(products)] - \Sigma[nS^\circ(reactants)]$

Plan: If we knew ΔS°_{rxn}, then we could use stoichiometric techniques to determine the entropy change.

Step 2. Use the molar mass and stoichiometric techniques to find the entropy change. The change found is for 2 moles of HgO decomposing.

Step 1. Look up the necessary S° values and use the equation directly to determine ΔS°_{rxn}.

Work:

Step 1. Find ΔS°_{rxn} for the process.

$? \text{ J K}^{-1} = \Delta S^\circ_{rxn} = \Sigma[nS^\circ(products)] - \Sigma[nS^\circ(reactants)]$

$? \text{ J K}^{-1} = \Delta S^\circ_{rxn} = 2(S^\circ Hg(\ell)) + 1(S^\circ O_2(g)) - 2(S^\circ HgO(s))$

$? \text{ J K}^{-1} = \Delta S^\circ_{rxn} = 2(76.02) + 1(205.138) - 2(70.29)$

$? \text{ J K}^{-1} = \Delta S^\circ_{rxn} = 216.60 \text{ J K}^{-1}$ (for 2 moles of HgO(s) decomposing)

Step 2. Use stoichiometric techniques to find the entropy change.
The molar mass of HgO is 1(200.59 Hg) + 1(15.9994 O) = 216.59 g mol^{-1}.

$$? \text{ J K}^{-1} = 10.0 \text{ g HgO} \times \frac{1 \text{ mol HgO}}{216.59 \text{ g HgO}} \times \frac{216.60 \text{ J K}^{-1}}{2 \text{ mol HgO}} = \mathbf{5.00 \text{ J K}^{-1}}$$

10.45 The standard entropy of reaction for $PCl_3(g) + Cl_2 \longrightarrow PCl_5(g)$ is -170.27 J K^{-1}. What is the standard entropy of $PCl_5(g)$?

Analysis:

Target: The standard entropy of PCl_5. (? J K^{-1}) =

Knowns: The balanced equation, which provides stoichiometric coefficients, and the standard entropy of a reaction are given. S° values for the other substances must be looked up in Appendix G.

Relationship: $\Delta S^\circ_{rxn} = \Sigma[nS^\circ(products)] - \Sigma[nS^\circ(reactants)]$

Plan: Rearrange the equation to solve for $S^\circ(PCl_5(g))$.

Work:

$\Delta S^\circ_{rxn} = \Sigma[nS^\circ(products)] - \Sigma[nS^\circ(reactants)]$

$\Delta S^\circ_{rxn} = 1(S^\circ PCl_5(g)) - [1S^\circ PCl_3(g)) + 1(S^\circ Cl_2(g))]$

$-170.27 = 1(S^\circ PCl_5(g)) - [1(311.78) + 1(223.066)]$

$? \text{ J K}^{-1} = 1(S^\circ PCl_5(g)) = -170.27 + 311.78 + 223.066 = \mathbf{+364.58 \text{ J K}^{-1}}$

10.47 Use data from Appendix G to calculate ΔG°_{rxn} for the thermite reaction,
$2 Al(s) + Fe_2O_3(s) \longrightarrow 2 Fe(s) + Al_2O_3(s)$.

Analysis:

Target: The ΔG°_{rxn} for a process. (? kJ) =

Knowns: The balanced equation, which provides stoichiometric coefficients, is given.

$\Delta G_f°$ values are available in Appendix G.

Relationship: $\Delta G^\circ_{rxn} = \Sigma[n\Delta G_f°(\text{products})] - \Sigma[n\Delta G_f°(\text{reactants})]$

Plan: Look up the appropriate values and use the equation directly.

Work:

$? \text{ kJ} = \Delta G^\circ_{rxn} = \Sigma[n\Delta G_f°(\text{products})] - \Sigma[n\Delta G_f°(\text{reactants})]$

$? \text{ kJ} = \Delta G^\circ_{rxn} = 2(\Delta G^\circ_f \text{ Fe(s)}) + 1(\Delta G^\circ_f \text{ Al}_2\text{O}_3\text{(s)}) - [2(\Delta G^\circ_f \text{ Al(s)}) + 1(\Delta G^\circ_f \text{ Fe}_2\text{O}_3\text{(s)})]$

$? \text{ kJ} = \Delta G^\circ_{rxn} = 2(0) + 1(-1582.3) - [2(0) + 1(-742.2)]$

$? \text{ kJ} = \Delta G^\circ_{rxn} = \mathbf{-840.1\ kJ}$

10.49 Use data from Appendix G to calculate ΔG°_f for SnO_2(s), given that, for $SnO_2(s) + 2\ CO(g) \longrightarrow 2\ CO_2(g) + Sn(s)$, $\Delta G^\circ_{rxn} = 5.2$ kJ.

Analysis:

Target: The ΔG°_f for SnO_2(s). (? kJ mol^{-1}) =

Knowns: The balanced equation, which provides stoichiometric coefficients, and the ΔG°_{rxn} of a reaction are given. ΔG°_f values for the other substances must be looked up in Appendix G.

Relationship: $\Delta G^\circ_{rxn} = \Sigma[n\Delta G^\circ_f(\text{products})] - \Sigma[n\Delta G^\circ_f(\text{reactants})]$

Plan: Rearrange the equation and insert known values to solve for $\Delta G^\circ_f\ SnO_2$(s).

Work:

$\Delta G^\circ_{rxn} = \Sigma[n\Delta G°(\text{products})] - \Sigma[n\Delta G°(\text{reactants})]$

$\Delta G^\circ_{rxn} = 2(\Delta G^\circ_f \text{ CO}_2\text{(g)}) + 1(\Delta G^\circ_f \text{ Sn(s)}) - [2(\Delta G^\circ_f \text{ CO})) + 1(\Delta G^\circ_f \text{ SnO}_2\text{(s)})]$

$+ 5.2 \text{ kJ} = 2(-394.359) + 1(0) - [2(-137.168) + 1(\Delta G^\circ_f \text{ SnO}_2\text{(s)})]$

$1(\Delta G^\circ_f \text{ SnO}_2\text{(s)}) = 2(-394.359) + 1(0) - [2(-137.168)] - 5.2 = -\ 519.6 \text{ kJ}$

$? \text{ kJ mol}^{-1} = \Delta G_f° \text{SnO}_2\text{(s)} = \dfrac{-519.6 \text{ kJ}}{1 \text{ mol}} = \mathbf{519.6\ kJ\ mol^{-1}}$

10.51 For the reaction $PbO(s) + C(s) \longrightarrow Pb(s) + CO(g)$, $\Delta H^\circ_{rxn} = +106.8$ kJ and $\Delta S^\circ_{rxn} = +188.0$ J K^{-1}. What is the value of ΔG°_{rxn}? Do not use values of ΔG°_f from Appendix G.

Analysis:

Target: The ΔG°_{rxn} for the process. (? kJ) =

Knowns: The ΔH°_{rxn} and ΔS°_{rxn} for the process are given.

Relationships: $\Delta G^\circ_{rxn} = \Delta H^\circ_{rxn} - T\Delta S^\circ_{rxn}$. The standard temperature is 298.15 K. The equivalence statement 1 kJ ↔ 1000 J is needed.

Plan: Use the equation directly, using the factor-label method to convert J to kJ.

Work:

$? \text{ kJ} = \Delta G^\circ_{rxn} = \Delta H^\circ_{rxn} - T\Delta S^\circ_{rxn} = +106.8 \text{ kJ} - \left((298.15 \text{ K})(+188.0 \text{ J K}^{-1}) \times \dfrac{1 \text{ kJ}}{1000 \text{ J}}\right)$

? kJ = +106.8 kJ - 56.05 kJ = **+50.7 kJ**

10.53 The standard entropy of the reaction $Pb(s) + 3/2\ O_2(g) \longrightarrow PbO(s) + SO_2(g)$ is -81.99 J K^{-1}, and ΔG°_{rxn} is -389.4 kJ. Calculate the value of the standard enthalpy of reaction, without using values of ΔH°_f from Appendix G.

Analysis:

Target: The ΔH°_{rxn} for the process. (? kJ) =

Knowns: The ΔG°_{rxn} and ΔS°_{rxn} for the process are given.

Relationships: $\Delta G^\circ_{rxn} = \Delta H^\circ_{rxn} - T\Delta S^\circ_{rxn}$. The standard temperature is 298.15 K. The equivalence statement 1 kJ ↔ 1000 J is needed.

Plan: Rearrange the equation to solve for ΔH°_{rxn}. Use the factor-label method to convert J to kJ.

Work:

$\Delta G^\circ_{rxn} = \Delta H^\circ_{rxn} - T\Delta S^\circ_{rxn}$ therefore, $\Delta H^\circ_{rxn} = \Delta G^\circ_{rxn} + T\Delta S^\circ_{rxn}$

$$? \text{ kJ} = \Delta H^\circ_{rxn} = -389.4 \text{ kJ} + \left((298.15 \text{ K})(-81.99 \text{ J K}^{-1}) \times \frac{1 \text{ kJ}}{1000 \text{ J}}\right)$$

? kJ = -389.4 kJ - 24.44 kJ = **-413.8 kJ**

10.55 If the enthalpy change for a reaction is 24.9 kJ mol^{-1} and the entropy change is 75.0 J mol^{-1} K^{-1} at 298 K, what is the free energy change at 298 K? At 100 K? [Assume that neither ΔH nor ΔS changes with temperature. For most reactions, this assumption is only approximately valid.]

Analysis:

Target: The ΔG°_{rxn} for the process at two different temperatures. (? kJ) =

Knowns: The ΔH°_{rxn} and ΔS°_{rxn} for the process are given.

Relationships: $\Delta G^\circ_{rxn} = \Delta H^\circ_{rxn} - T\Delta S^\circ_{rxn}$. The equivalence statement 1 kJ ↔ 1000 J is needed.

Plan: Use the equation directly, using the factor-label method to convert J to kJ.

Work:

At 298 K:

$$? \text{ kJ} = \Delta G^\circ_{rxn} = \Delta H^\circ_{rxn} - T\Delta S^\circ_{rxn} = +24.9 \text{ kJ mol}^{-1} - \left((298 \text{ K})(75.0 \text{ J mol}^{-1} \text{ K}^{-1}) \times \frac{1 \text{ kJ}}{1000 \text{ J}}\right)$$

? kJ = +24.9 kJ mol^{-1} - 22.4 kJ mol^{-1} = **2.6 kJ mol^{-1}**

At 100 K:

$$? \text{ kJ} = \Delta G^\circ_{rxn} = \Delta H^\circ_{rxn} - T\Delta S^\circ_{rxn} = +24.9 \text{ kJ mol}^{-1} - \left((100 \text{ K})(75 \text{ J mol}^{-1} \text{ K}^{-1}) \times \frac{1 \text{ kJ}}{1000 \text{ J}}\right)$$

? kJ = +24.9 kJ mol^{-1} - 7.50 kJ mol^{-1} = **17.4 kJ mol^{-1}**

10.57 For each of the following processes, determine the sign of ΔS and explain your reasoning.

a. Antifreeze is added to the water in an automobile radiator.

b. A film of ice forms on a puddle of water.

a. ΔS should be **positive**. This is because mixing different substances increases the disorder and, therefore, the entropy of the system.
b. ΔS should be **negative**. As a liquid changes to a solid, the molecules lose their mobility and become fixed in position. This increases the order, decreasing the entropy of the system.

10.59 Suppose a process occurs at 25 °C with ΔH = 100 kJ and ΔS = 100 J K^{-1}. Is the reaction spontaneous at this temperature? At what temperature will the reaction be in equilibrium?

Analysis:
Target: The spontaneity of a process and its equilibrium temperature.
Knowns: A process is spontaneous if $\Delta G^\circ_{rxn} < 0$ and is in equilibrium if $\Delta G_{rxn}^\circ = 0$. The ΔH°_{rxn} and ΔS°_{rxn} for the process are given.
Relationships: $\Delta G^\circ_{rxn} = \Delta H^\circ_{rxn} - T\Delta S^\circ_{rxn}$. The equivalence statements 1 kJ ↔ 1000 J and K ↔ 273/15 + °C are needed.
Plan: Use the equation directly, using the factor-label method to convert J to kJ and °C to K, to determine ΔG°_{rxn}. Rearrange the same relationship to determine the temperature that causes ΔG°_{rxn} to be equal to zero.
Work:
At 25 °C, K = 273.15 + 25 = 298 K.

$$\Delta G^\circ_{rxn} = \Delta H^\circ_{rxn} - T\Delta S^\circ_{rxn} = +100 \text{ kJ} - \left((298 \text{ K})(100 \text{ J K}^{-1}) \times \frac{1 \text{ kJ}}{1000 \text{ J}}\right)$$

$$\Delta G^\circ_{rxn} = +100 \text{ kJ} - 29.8 \text{ kJ} = +70 \text{ kJ}$$

Because $\Delta G^\circ_{rxn} > 0$, the process is **not spontaneous** at 25 °C.

To estimate the equilibrium temperature, set $\Delta G^\circ_{rxn} = 0$ and rearrange to solve for T.

$$0 = \Delta G^\circ_{rxn} = \Delta H^\circ_{rxn} - T\Delta S^\circ_{rxn} \text{ therefore, } \Delta H^\circ_{rxn} = T\Delta S^\circ_{rxn}$$

$$T = \frac{\Delta H^\circ_{rxn}}{\Delta S^\circ_{rxn}} = \frac{100 \text{ kJ}}{100 \text{ J K}^{-1} \times \frac{1 \text{ kJ}}{1000 \text{ J}}} = \mathbf{1000 \text{ K}}$$

10.61 Suppose a reaction is at equilibrium at 298 K.
a. For which of the possible combinations of algebraic signs of ΔS and ΔH will the reaction be spontaneous in the forward direction when the temperature is increased?
b. For which of the combinations of signs will the reaction become spontaneous if the temperature is decreased?

a. If the process is in equilibrium at 298 K, then $\Delta G_{rxn} = 0$. Because the free energy $\Delta G_{rxn} = \Delta H_{rxn} - T\Delta S_{rxn}$, ΔH_{rxn} and ΔS_{rxn} must have the same signs. (Temperature is in Kelvin and **cannot** be negative.) To be spontaneous, ΔG_{rxn} must be negative. If an increase in temperature is to make ΔG_{rxn} negative, **ΔS_{rxn} must be positive** and **ΔH_{rxn} must be positive.** A larger T makes the magnitude of the $T\Delta S_{rxn}$ term larger than that of the ΔH_{rxn} term.
b. Use the same reasoning as in part b. To make a lower temperature cause the process to become spontaneous, **ΔS_{rxn} must be negative** and **ΔH_{rxn} must be negative**. A smaller T makes the magnitude of the $T\Delta S_{rxn}$ term less than that of the ΔH_{rxn} term.

11 Liquids and Solids

11.1 Define, and distinguish between, fluidity and viscosity.

They are opposites. Fluidity is the ability to flow, while viscosity is the resistance to flow.

11.3 Gases are easily compressed, while the volume of a liquid can be decreased only by application of great pressure. Explain.

The distance between gas molecules is large compared to the size of the molecules, therefore, there is a considerable amount of "empty" space into which the molecules can be compressed with only moderate pressure increases. In liquids, the molecules are touching. The liquid volume can only be decreased by decreasing the actual molecular size. This requires far more pressure than is required to decrease gas volume.

11.5 What is the relationship between surface tension and intermolecular forces?

Surface tension arises because a liquid is more attracted to itself than to the surrounding phases. Higher intermolecular forces give rise to greater surface tension.

11.7 What evidence indicates that attractive intermolecular forces exist?

The fact that gases can be condensed when they are compressed or cooled indicates that there is some attraction that causes the molecules to stick together.

11.9 Name five different types of cohesive force that can exist between the particles of a substance. Which of these can be properly called an intermolecular force?

The five major cohesive forces are: ionic bonds, covalent bonds, London forces, dipole-dipole forces, and hydrogen bonding. Because ionic bonds are between ions and covalent bonds are between atoms, they are not intermolecular forces. The others are all intermolecular forces.

11.11 What is the weakest force that exists between the particles of a substance in a condensed state? Give evidence to support your opinion.

The weakest force is the dipole-dipole force because it is zero in any nonpolar molecule. While London forces are weak compared to existing dipoles, they are **always** present in a substance.

11.13 What is an induced dipole? How do induced dipoles arise? With which intermolecular force are induced dipoles associated?

An induced dipole is a dipole that is present only in the presence of an electric field. The fields, which can be produced by another atom, molecule, or ion, cause distortion in the electron cloud, giving rise to a temporary charge imbalance (a temporary dipole). Induced dipoles are associated with London (dispersion) forces.

11.15 Under what circumstances can hydrogen bonding take place?

Hydrogen bonding occurs when a hydrogen atom, covalently bonded to a nitrogen, oxygen or hydrogen atom, is attracted to another nearby nitrogen, oxygen or fluorine atom.

11.17 Which of the following molecules can participate in hydrogen bonding?
a. CH_3OH b. CH_3F c. CH_3OCH_3 d. NF_3

Analysis/Plan: Using previously learned techniques, we need to draw the structural formula of each compound to determine if hydrogen is bonded to the proper atoms (N, O, or F) to give rise to hydrogen bonding.
Work: The structural formulas are:

a. $H_3C{-}\ddot{O}{-}H$ b. $H_3C{-}\ddot{F}:$ c. $H_3C{-}\ddot{O}{-}CH_3$ d. $:\ddot{F}{-}\ddot{N}(\ddot{F}:){-}\ddot{F}:$

a. This species has H bonded to O therefore, it **can hydrogen bond**.
b. Although there is F in the molecule, the H atoms are all bonded to C. This molecule is **not** capable of hydrogen bonding.
c. Although there is O in the molecule, the H atoms are all bonded to C. This molecule is **not** capable of hydrogen bonding.
d. There are no H atoms in this molecule. It is **not** capable of hydrogen bonding.

11.19 Which of the following molecules can participate in hydrogen bonding?
a. PH_3 b. HNO_3 c. GeH_4 d. $HOCl$

Analysis/Plan: We need to draw the structural formula, using previously learned techniques, of each compound to determine if hydrogen is bonded to the proper atoms (N, O, or F) to give rise to hydrogen bonding.
Work: The structural formulas are:

a. $H{-}\ddot{P}(H){-}H$ b. $:\ddot{O}{-}N(=\ddot{O}:){-}\ddot{O}{-}H$ c. $H{-}Ge(H)(H){-}H$ d. $H{-}\ddot{O}{-}\ddot{Cl}:$

a. H is not bonded to N, O or F. Hydrogen bonding is **not** possible.
b. H is bonded to O. This molecule **can hydrogen bond**.
c. H is not bonded to N, O or F. Hydrogen bonding is **not** possible.
d. H is bonded to O. This molecule **can hydrogen bond**.

11.21 Ethanol and dimethyl ether are structural isomers of C_2H_6O. Suggest a reason why the boiling point of one is about 100 °C greater than the other.

$H_3C{-}CH_2{-}O{-}H$ $H_3C{-}O{-}CH_3$

ethanol, bp 78 °C dimethyl ether, bp –25 °C

In ethanol, one of the H atoms is bonded to O, therefore, this molecule is capable of hydrogen bonding. The dimethyl ether, which has no H atoms bonded to its O, cannot hydrogen bond. Hydrogen bonding is a stronger intermolecular force than London forces, giving rise to a higher boiling point for the ethanol.

11.23 Which of the following pairs of liquids would you expect to have the higher boiling point?
a. CH_3CH_3, CH_3Cl b. CH_4, SiH_4
c. CH_3Cl, CH_3OH d. $CH_3CH_2CH_3$, $CH_3CH_2CH_2CH_3$

Analysis/Plan: We need to study the molecules to see which forces are involved. In general, higher boiling points are a result of higher intermolecular forces. Permanent dipole-dipole forces are stronger than London forces and hydrogen bonding is a stronger force than dipole-dipole interactions. If other forces involved are the same, the substance with the higher molar mass has higher London forces and, therefore, a higher boiling point.

Work:

a. CH_3CH_3 is nonpolar; it has only London forces. CH_3Cl is a polar molecule; it has dipole-dipole attractions in addition to London forces. It also has a higher molar mass. With the additional intermolecular forces, **CH_3Cl** has the higher boiling point.

b. Both compounds are nonpolar and have only London forces. The molar mass of **SiH_4** is greater, so it has higher London forces and a higher boiling point.

c. CH_3OH is capable of hydrogen bonding, while CH_3Cl has weaker dipole-dipole forces. **CH_3OH** has the higher boiling point because of its higher intermolecular forces.

d. Both compounds are nonpolar and have only London forces. The molar mass of **$CH_3CH_2CH_2CH_3$** is greater, so it has higher London forces and a higher boiling point.

11.25 Define the terms "sublimation", "evaporation", and "condensation".

Sublimation is the change of a solid directly to the vapor state.
Evaporation is the change of a liquid to the vapor state.
Condensation is the change of a vapor to the liquid state.

11.27 Do liquids always boil at the normal boiling point? Explain.

The boiling point is the temperature at which the vapor pressure of the liquid is equal to the applied pressure. If the liquid is not subjected to normal pressure (1 atm), it will not boil at the normal boiling point. Instead, if the pressure is higher, it will boil at a higher temperature than normal. If the pressure is lower, it will boil at a lower temperature than normal.

11.29 Distinguish between condensation and deposition.

Condensation is the change of vapor to the liquid state. Deposition is the change of vapor to the solid state.

11.31 Calculate the partial pressure of water vapor in the atmosphere when the temperature is 15 °C and the relative humidity is 65%.

Analysis:

Target: A partial pressure of water. (? torr) =

Knowns: The temperature and the relative humidity. The equilibrium partial pressure can be found in Table 11.3 .

Relationship: $RH = \frac{P}{P_{eq}} \times 100$, where RH is the relative humidity, P is the partial pressure and P_{eq} is the equilibrium partial pressure of water at the given temperature.

Plan: Two steps are needed to solve this problem.

Step 2. Rearrange the equation to solve for the partial pressure, insert the values and calculate.

Step 1. Look up the equilibrium partial pressure.

Work:

Step 1. The equilibrium partial pressure at 15 °C is 12.8 torr.

Step 2. $RH = \frac{P}{P_{eq}} \times 100$, therefore,

$$? \text{ torr} = P = \frac{(RH)(P_{eq})}{100} = \frac{(65\%)(12.8 \text{ torr})}{100} = \mathbf{8.3 \text{ torr}}$$

11.33 If the temperature is 50 °F and the partial pressure of atmospheric water vapor is 5.9 torr, what is the relative humidity?

Analysis:

Target: The relative humidity. (? %) =

Knowns: The temperature and the partial pressure of water. The equilibrium partial pressure can be found in Table 11.3 .

Relationships: RH = $\frac{P}{P_{eq}}$ x 100, where RH is the relative humidity, P is the partial pressure and P_{eq} is the equilibrium partial pressure of water at the given temperature. The equivalence statement °C ↔ 5/9(°F - 32) is also needed.

Plan: If we knew the equilibrium partial pressure, we could use the equation to solve for the relative humidity. Table 11.3 is listed by °C so the following steps are needed.

Step 3. Use the equation directly to calculate the relative humidity.
Step 2. Look up the equilibrium partial pressure.
Step 1. Convert the temperature to °C.

Work:

Step 1. °C = 5/9(°F - 32) = 5/9(50 - 32) = 5/9(18) = 10 °C

Step 2. The equilibrium partial pressure at 10 °C is 9.2 torr.

Step 3. $? \% = RH = \frac{P}{P_{eq}} \times 100 = \frac{5.9 \text{ torr}}{9.2 \text{ torr}} \times 100 = \mathbf{64\%}$

11.35 Calculate the dew point of air at 86 °F, whose relative humidity is 40%.

Analysis:

Target: The dew point. (? °C) =

Knowns: The temperature and the relative humidity of air. The equilibrium partial pressure can be found in Table 11.3.

Relationships: RH = $\frac{P}{P_{eq}}$ x 100, where RH is the relative humidity, P is the partial pressure and P_{eq} is the equilibrium partial pressure of water at the given temperature. The equivalence statement °C ↔ 5/9(°F - 32) is needed. The dew point is the temperature at which water begins to condense because the relative humidity is 100%.

Plan: If we knew the partial pressure, we could use the table to determine the temperature at which this would be equal to the equilibrium partial pressure. Table 11.3 is listed by °C, so the following steps are needed.

Step 4. Use Table 11.3 to determine the temperature of the dew point.
Step 3. Rearrange the equation to solve for the partial pressure, insert the values and calculate.
Step 2. Look up the equilibrium partial pressure at the initial temperature.
Step 1. Convert the temperature to °C.

Work:

Step 1. °C = 5/9(°F - 32) = 5/9(86 - 32) = 5/9(54) = 30 °C

Step 2. At 30 °C, the equilibrium partial pressure is 31.8 torr.

Step 3. RH = $\frac{P}{P_{eq}}$ x 100, therefore,

$$P = \frac{(RH)(P_{eq})}{100} = \frac{(40\%)(31.8 \text{ torr})}{100} = 13 \text{ torr}$$

Step 4. According to Table 11.3, water has an equilibrium partial pressure of 12.8 torr at 15 °C. The dew point of this air mass is, therefore, about **15 °C**.

11.37 What is the relationship between boiling point, heat of vaporization, and intermolecular forces?

Both the boiling point and the heat of vaporization increase with increasing intermolecular forces because the molecules are more strongly attracted to each other and harder to separate.

11.39 Are there any substances for which the entropy of vaporization is a negative quantity? Explain.

Entropy of vaporization can **never** be negative. Entropy is a measure of disorder and disorder **always** increases as a liquid is converted to a vapor.

11.41 The heat of vaporization of chloroform ($CHCl_3$) is 31.4 kJ mol^{-1}. Calculate the heat required to vaporize a 256 mg sample.

Analysis:

Target: The heat required to vaporize a sample. (? kJ) =

Knowns: The formula of the compound, the heat of vaporization and the mass of the sample. The molar mass of the sample is an implicit known.

Relationships: The amount of heat required is proportional to the amount of substance. The molar mass provides the equivalence statement g ↔ mol. The equivalence statement is 1 mg ↔ 10^{-3} g is also needed.

Plan: Use the factor-label method to work this problem.

Work:

Molar mass of $CHCl_3$ = 1(12.011 C) + 1(1.0079 H) + 3(35.453 Cl) = 119.378 g mol^{-1}.

$$? \text{ kJ} = 256 \text{ mg } CHCl_3 \times \frac{10^{-3} \text{ g}}{1 \text{ mg}} \times \frac{1 \text{ mol } CHCl_3}{119.373 \text{ g } CHCl_3} \times \frac{31.4 \text{ kJ}}{1 \text{ mol } CHCl_3} = \mathbf{0.0673 \text{ kJ}}$$

11.43 The heat of sublimation of phenol (C_6H_5OH) is 67.8 kJ mol^{-1}. Calculate the heat required to sublime a 10.0 g sample.

Analysis:

Target: The heat required to sublime a sample. (? kJ) =

Knowns: The formula of the compound, the heat of sublimation and the mass of the sample. The molar mass of the sample is an implicit known.

Relationships: The amount of heat required is proportional to the amount of substance. The molar mass provides the equivalence statement g ↔ mol.

Plan: Use the factor-label method to work this problem.

Work:

Molar mass C_6H_5OH = 6(12.011 C) + 6(1.0079 H) + 1(15.9994 O) = 94.113 g mol^{-1}.

$$? \text{ kJ} = 10.0 \text{ g } C_6H_5OH \times \frac{1 \text{ mol } C_6H_5OH}{94.113 \text{ g } C_6H_5OH} \times \frac{67.8 \text{ kJ}}{1 \text{ mol } C_6H_5OH} = \mathbf{7.20 \text{ kJ}}$$

11.45 If 8.50 kJ is required to sublime 15.0 g of solid benzene (C_6H_6), what is the molar heat of sublimation?

Analysis:

Target: The heat of sublimation of a sample. (? kJ mol^{-1}) =

Knowns: The formula of the compound, the heat required to sublime a sample and the mass of the sample. The molar mass of the sample is an implicit known.

Relationships: The amount of heat required is proportional to the amount of substance. The molar mass provides the equivalence statement g ↔ mol.

Plan: Use the factor-label method to find the number of moles of C_6H_6. Divide the heat required by the number of moles to determine the heat of sublimation.

Work:

Molar mass C_6H_6 = 6(12.011 C) + 6(1.0079 H) = 78.114 g mol^{-1}.

$$? \text{ mol} = 15.0 \text{ g } C_6H_6 \times \frac{1 \text{ mol } C_6H_6}{78.114 \text{ g } C_6H_6} = 0.192 \text{ mol } C_6H_6$$

$$? \text{ kJ mol}^{-1} = \frac{8.50 \text{ kJ}}{0.192 \text{ mol } C_6H_6} = \mathbf{44.2 \text{ kJ mol}^{-1}}$$

11.47 Express the heat of vaporization of tribromomethane ($CHBr_3$) in units of J mL^{-1}. The density is 2.89 g mL^{-1} and the heat of vaporization is 40.5 kJ mol^{-1}.

Analysis:

Target: The heat of vaporization. (? J mL^{-1}) =

Knowns: The formula of the compound, the heat of vaporization and the density of the sample. The molar mass of the sample is an implicit known.

Relationships: The amount of heat required is proportional to the amount of substance. The equivalence statement 1 kJ ↔ 1000 J is needed. The molar mass provides the equivalence statement g ↔ mol.

Plan: We know the heat released when 1 mol of $CHBr_3$ is vaporized. We can use the factor-label method to find the volume in mL of 1 mol of CH_3Br. Appropriate steps are:

Step 2. Use the factor-label method to convert the heat of vaporization to J mL^{-1}.

Step 1. Use the factor-label method to find the volume of 1.00 mol of $CHBr_3$.

Work:

Step 1.

Molar mass $CHBr_3$ = 1(12.011 C) + 1(1.0079 H) + 3(79.904 Br) = 252.731 g mol^{-1}.

$$? \text{ mL} = 1.00 \text{ mol } CHBr_3 \times \frac{252.731 \text{ g } CHBr_3}{1 \text{ mol } CHBr_3} \times \frac{1 \text{ mL}}{2.89 \text{ g } CHBr_3} = 87.5 \text{ mL}$$

$$? \text{ J mL}^{-1} = \frac{40.5 \text{ kJ} \times \frac{1000 \text{ J}}{1 \text{ kJ}}}{1 \text{ mol}} \times \frac{1 \text{ mol}}{87.5 \text{ mL}} = \mathbf{463 \text{ J mL}^{-1}}$$

11.49 The heat of vaporization of carbon oxysulfide (COS, also called carbonyl sulfide) is 20.9 kJ mol^{-1} at its normal boiling point of -50 °C. Calculate its entropy of vaporization.

Analysis:

Target: An entropy of vaporization (? J mol^{-1} K^{-1}) =

Knowns: The heat of vaporization and the normal boiling point of the sample.

Relationships: Vaporization is an equilibrium process. At the boiling point, $\Delta G_{vap} = 0$. Therefore, because $\Delta H_{vap} - T\Delta S_{vap} = \Delta G$, $\Delta H_{vap} - T\Delta S_{vap} = 0$ at the boiling point. Necessary equivalence statements are K ↔ 273.15 + °C and 1 kJ ↔ 1000 J.

Plan: Convert the temperature to Kelvin. Rearrange the equation to solve for the entropy of vaporization, using the factor-label method to convert to joules.

Work:

T = 273.15 + °C = 273.15 + (-50 °C) = 223 K

$\Delta H_{vap} - T\Delta S_{vap} = 0$, therefore, $\Delta H_{vap} = T\Delta S_{vap}$

$$? \text{ J mol}^{-1} \text{ K}^{-1} = \Delta S_{vap} = \frac{\Delta H_{vap}}{T} = \frac{20.9 \text{ kJ mol}^{-1} \times \frac{1000 \text{ J}}{1 \text{ kJ}}}{223 \text{ K}} = \mathbf{93.7 \text{ J mol}^{-1} \text{ K}^{-1}}$$

11.51 The heat of fusion of *cis*-2-pentene is 7.12 kJ mol^{-1} at its melting point (-139 °C). Calculate its entropy of fusion.

Analysis:

Target: An entropy of fusion (? J mol^{-1} K^{-1}) =

Knowns: The heat of fusion and the normal melting point of the sample.

Relationships: Melting is an equilibrium process. At the melting point, $\Delta G_{fus} = 0$. Therefore, because $\Delta H_{fus} - T\Delta S_{fus} = \Delta G$, $\Delta H_{fus} - T\Delta S_{fus} = 0$ at the melting point. Necessary equivalence statements are K ↔ 273.15 + °C and 1 kJ ↔ 1000 J.

Plan: Convert the temperature to Kelvin. Rearrange the equation to solve for the entropy of fusion, using the factor-label method to convert to joules.

Work:

$T = 273.15 + °C = 273.15 + (-139°C) = 134\ K$

$\Delta H_{fus} - T\Delta S_{fus} = 0$, therefore, $\Delta H_{fus} = T\Delta S_{fus}$

$$?\ J\ mol^{-1}\ K^{-1} = \Delta S_{fus} = \frac{\Delta H_{fus}}{T} = \frac{7.12\ kJ\ mol^{-1} \times \frac{1000\ J}{1\ kJ}}{134\ K} = 53.1\ J\ mol^{-1}\ K^{-1}$$

11.53 What is a phase diagram?

The phase diagram of a substance is a plot of its vapor pressure versus temperature. A phase diagram enables us to determine the state of the substance under an infinite number of conditions. Regions of solid, liquid and gas are separated by lines composed of P/T points at which equilibrium between two phases exists.

11.55 What name is given to the temperature and pressure below which a substance cannot exist in the liquid state?

This point is the triple point.

11.57 Refer to the phase diagram for CO_2 (Figure 11.28) and describe a sample of CO_2 under the following conditions.

a. -50 °C, 7 atm b. -60 °C, 7 atm

a. Carbon dioxide is a liquid under this set of conditions.
b. Carbon dioxide is a solid under this set of conditions.

11.59 Refer to the phase diagram of water (Figure 11.27) and describe any changes that occur as H_2O is heated from -20 to +120 °C at a pressure of 2×10^{-3} atm.

Water is a solid at -20 °C at this pressure. As the water is heated, it begins to sublime at slightly less than 0 °C. The temperature remains constant until sublimation is complete. After the water is completely converted to vapor, the temperature of the vapor begins to rise.

11.61 On a sketch of the phase diagram for water, draw a rectangle with sides parallel to the axes of the diagram. The rectangle must enclose the triple point, and can be of any size as long as it does not also enclose the normal melting point or boiling point. Describe any changes that occur as a sample is taken around this rectangle, in a clockwise direction beginning at the upper left corner.

As we move from 1 to 2, the solid melts at constant temperature. After the solid has been completely melted, the liquid becomes hotter. Eventually the water begins to boil. After the water has completely vaporized, the vapor gets hotter. As we move from 2 to 3, the vapor expands as the pressure decreases. As we move from 3 to 4, the vapor first cools and then deposits as solid ice at constant temperature. The ice then gets colder. As we move from 4 back to 1, there is not much change because the solid is not very compressible.

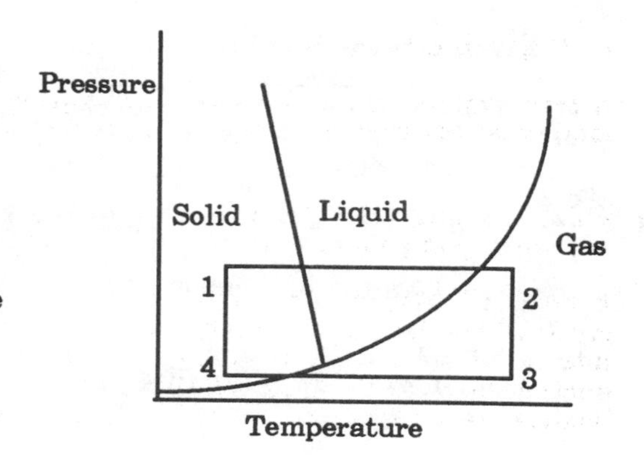

11.63 What is the difference between an amorphous solid and a glass.

In any amorphous solid, the particles are randomly arranged. A glass is a specific type of amorphous solid. Glasses are covalently bonded macromolecules, which lack the crystalline order characteristic of most solids. Lack of a definite crystal structure causes glasses to soften gradually when heated, rather than having a distinct, sharp melting point.

11.65 Compare some of the properties of crystalline and amorphous substances.

Crystalline substances have particles arranged in regular geometric patterns. This causes them to form regularly-shaped macroscopic crystals and to have sharp melting points, both of which can be used to characterize the substance. Amorphous solids have no such regular pattern of particles. They take on many different shapes and liquify over a wide temperature range.

11.67 What type of solid (molecular, metallic, and so on) would you expect each of the following substances to form?

a. MgF_2 b. $SrCl_2$ c. $CaSO_4$ d. CH_3COOH

Analysis/Plan: The formula of the compound indicates the type of solid that a substance will make. Ionic compounds make ionic solids. Metals make metallic solids. Covalent substances and noble gas atoms make molecular solids. A few compounds such as diamond make the covalently bonded macromolecules known as covalent solids. Analyze the formula to determine the class of the substance.

Work:

a. MgF_2 is an ionic compound composed of Mg^{2+} and F^- ions. It makes an **ionic solid**.

b $SrCl_2$ is an ionic compound composed of Sr^{2+} and Cl^- ions. It makes an **ionic solid**.

c. $CaSO_4$ is an ionic compound composed of Ca^{2+} and SO_4^{2-} ions. It makes an **ionic solid**.

d. CH_3COOH is a covalent compound. It makes a **molecular solid**.

11.69 What type of solid would you expect each of the following substances to form?

a. Xe b. KCl c. CH_2Cl_2 d. NH_4F

Analysis/Plan: Proceed as in Problem 11.67.

Work:

a. Xe is a noble gas. It makes a **molecular solid**.

b. KCl is an ionic compound composed of K^+ and Cl^- ions. It makes an **ionic solid**.

c. CH_2Cl_2 is a covalent compound. It makes a **molecular solid**.

d. NH_4F is an ionic compound composed of NH_4^+ and F^- ions. It makes an **ionic solid**.

11.71 Distinguish between a crystal *lattice* and a crystal *structure*.

A crystal structure is the actual arrangement of particles in a crystal. A crystal lattice is the mathematical way of modeling the particles in the crystal.

11.73 What is a unit cell? If you were walking around inside a crystal lattice, would you be able to tell when you passed from one unit cell to the next? If so, how? If not, why not?

A unit cell represents the repeating structural unit of a crystal, the "building block" of the crystal. Lattice points define the corners of the cell and other lattice points may be on the interior of the cell. You would not be able to tell when you had walked from one portion to another because the unit cell is a mathematical model. It does not correspond to real boundaries between the actual particles making up the crystal.

11.75 What is the Bragg equation?

The Bragg equation is used to determine the distance between planes in a crystal from the x-ray diffraction pattern. The Bragg equation is $2d \sin \theta = n\lambda$. In this equation, d is the distance between planes, θ is the angle of reflection, n is an integer number, and λ is the wavelength of light.

11.77 In sodium chloride, the distance between one set of successive layers of ions is 282 pm. Assuming $n = 1$, what is the angle of diffraction of x-rays of wavelength 0.070 nm?

Analysis:
Target: An angle of diffraction. (? °) =
Knowns: The distance between the layers, the wavelength of the x-rays, and $n = 1$.
Relationships: The Bragg equation, $2d \sin \theta = n\lambda$, relates the parameters. Necessary equivalence statements are 1 pm ↔ 10^{-12} m and 1 nm ↔ 10^{-9} m.
Plan: Rearrange the Bragg equation to solve for $\sin \theta$, using appropriate factors to convert distances to the same units. Determine θ from its sin (use $\sin^{-1} \theta$).
Work:

$$2d \sin \theta = n\lambda, \text{ therefore, } \sin \theta = \frac{n\lambda}{2d} = \frac{(1)(0.070 \text{ nm}) \times \dfrac{10^{-9} \text{ m}}{1 \text{ nm}}}{2(282 \text{ pm}) \times \dfrac{10^{-12} \text{ m}}{1 \text{ pm}}} = 0.12$$

? ° = θ = **7.1 °**

11.79 X-rays of wavelength 0.105 nm are diffracted by a metallic crystal. If the diffraction angle is 14.6°, what is the distance between atomic planes in the metal? (Assume $n = 1$.)

Analysis:
Target: The distance between layers. (? pm) =
Knowns: The angle of diffraction, the wavelength of the x-rays, and $n = 1$.
Relationships: The Bragg equation, $2d \sin \theta = n\lambda$, relates the parameters. Necessary equivalence statements are 1 pm ↔ 10^{-12} m and 1 nm = 10^{-9} m.
Plan: Rearrange the Bragg equation to solve for d, using appropriate factors to convert distances to the proper units.
Work:
$2d \sin \theta = n\lambda$

$$? \text{ pm} = d = \frac{n\lambda}{2\sin\theta} = \frac{(1)(0.105 \text{ nm}) \times \dfrac{10^{-9} \text{ m}}{1 \text{ nm}}}{2(\sin 14.6)} \times \frac{1 \text{ pm}}{10^{-12} \text{ m}} = \mathbf{208 \text{ pm}}$$

11.81 Copper, whose atomic radius is 127.8 pm, forms face-centered crystals. What is the length of an edge of the unit cell?

Analysis:

Target: The edge length of the unit cell. (? pm) =

Knowns: The crystal form and the atomic radius.

Relationships: Figure 11.48 in the text shows the face-centered cube. The geometrical relationship derived from this shape and presented in Section 11.8 is $r = \frac{\ell\sqrt{2}}{4}$, where ℓ is the edge length and r is the radius.

Plan: Rearrange the relationship to solve for ℓ, insert the known values and calculate.

Work:

$$r = \frac{\ell\sqrt{2}}{4}, \text{ therefore, } 4\,r = \ell\sqrt{2}$$

$$? \text{ pm} = \ell = \frac{4r}{\sqrt{2}} = \frac{4(127.8 \text{ pm})}{\sqrt{2}} = \mathbf{361.5\ pm}$$

11.83 Niobium forms body-centered cubic crystals. Calculate the length of the unit cell, given that the atomic radius is 142.9 pm.

Analysis:

Target: The edge length of the unit cell. (? pm) =

Knowns: The crystal form and the atomic radius.

Relationships: Figure 11.48 in the text shows the body-centered cube. The geometrical relationship derived from this shape and presented in Section 11.8 is $r = \frac{\ell\sqrt{3}}{4}$, where ℓ is the edge length and r is the radius.

Plan: Rearrange the relationship to solve for ℓ, insert the known values and calculate.

Work:

$$r = \frac{\ell\sqrt{3}}{4}, \text{ therefore, } 4\,r = \ell\sqrt{3}$$

$$? \text{ pm} = \ell = \frac{4r}{\sqrt{3}} = \frac{4(142.9 \text{ pm})}{\sqrt{3}} = \mathbf{330.0\ pm}$$

11.85 Aluminum forms face-centered cubic crystals with a unit cell length of 405.2 pm. What is the radius of an aluminum atom?

Analysis:

Target: The radius of the atom. (? pm) =

Knowns: The crystal form and the edge length of the unit cell.

Relationships: Figure 11.48 in the text shows the face-centered cube. The geometrical relationship derived from this shape and presented in Section 11.8 is $r = \frac{\ell\sqrt{2}}{4}$, where ℓ is the edge length and r is the radius.

Plan: Use the relationship directly, insert the known values and calculate.

Work:

$$? \text{ pm} = r = \frac{\ell\sqrt{2}}{4} = \frac{(405.2 \text{ pm})\sqrt{2}}{4} = \mathbf{143.3\ pm}$$

11.87 Vanadium forms body-centered cubic crystals with a unit cell length of 302.8 pm. What is the radius of a vanadium atom?

Analysis:

Target: The radius of a vanadium atom. (? pm) =

Knowns: The crystal form and the edge length of the unit cell.

Relationships: Figure 11.48 in the text shows the body-centered cube. The geometrical relationship derived from this shape and presented in Section 11.8 is $r = \frac{\ell\sqrt{3}}{4}$, where ℓ is the edge length and r is the radius.

Plan: Use the relationship directly, insert the known values and calculate.

Work:

$$? \text{ pm} = r = \frac{\ell\sqrt{3}}{4} = \frac{(302.8 \text{ pm})\sqrt{3}}{4} = \mathbf{131.1 \text{ pm}}$$

11.89 Cobalt, whose atomic radius is 125.3 pm, forms cubic crystals with a unit cell length of 354.4 pm. Which type of cubic crystals are these?

Analysis:

Target: The crystal type.

Knowns: The atomic radius and the edge length of the unit cell.

Relationships: Figure 11.48 in the text shows the three cubic forms. The geometrical relationships derived from the three shapes and presented in Section 11.8 are $r = \frac{\ell\sqrt{3}}{4}$ for a body-centered cube, $r = \frac{\ell\sqrt{2}}{4}$ for a face-centered cube, and $r = \frac{\ell}{2} = \frac{\ell\sqrt{4}}{4}$ for a simple cube, where ℓ is the edge length and r is the radius.

Plan: The general form of each is $r = \frac{\ell\sqrt{x}}{4}$. Rearrange the relationship, insert values and solve for x to determine the crystal form.

Work:

$$r = \frac{\ell\sqrt{x}}{4}, \text{ therefore, } 4r = \ell\sqrt{x} \text{ and } \sqrt{x} = \frac{4r}{\ell} = \frac{4(125.3 \text{ pm})}{354.4 \text{ pm}} = 1.414$$

$x = (1.414)^2 = 2.000$, therefore, this is a **face-centered cubic crystal structure.**

11.91 What is a liquid crystal? Describe the cohesive forces in liquid crystals.

A liquid crystal is a substance whose particles have a regular arrangement (similar to that in a crystal) in an least one dimension. It also displays randomness, characteristic of a liquid, in the other dimension(s). Typically liquid crystals have a rod-like shape which gives them cohesive forces similar to a liquid, but causes the cohesive forces to be different in different directions. The stronger cohesive forces in one direction give rise to the crystal-like structure in that direction, while the weaker forces in the other direction allow the randomness associated with the liquid state.

11.93 Are there any circumstances in which the freezing and melting points of a substance can be different? Explain.

Under normal circumstances, freezing and melting are equilibrium processes and must occur at the same temperature. In some instances, supercooling is possible if the liquid does not have a nucleation site to cause initiation of crystal formation. When supercooling occurs, the substance begins freezing below its normal freezing/melting point. Once solidification begins, the temperature rises to the normal freezing point and continues at that temperature until solidification is complete.

11.95 Name as many states of matter as you can think of, other than solid, liquid, and vapor.

The text mentions supercritical fluids and liquid crystals. Plasmas and neutron stars are also distinct states.

11.97 Explain the relationship between thermal and electrical conductivity in metals.

Both types of conductivity depend on the movement of electrons in the metal. A metal which is a better conductor of electricity is also a better thermal conductor.

11.99 The entropy of sublimation of carbon dioxide is 54.4 J mol^{-1} K^{-1} at its normal sublimation temperature. Is this an exception to Trouton's rule?

Trouton's rule states that ΔS^{o}_{vap} of a **liquid** is approximately 87 J^{-1} mol^{-1} K^{-1}. Because solid carbon dioxide is not a liquid, Trouton's rule is not expected to apply.

11.101 Refer to the phase diagram for CO_2 (Figure 11.28) and determine the maximum temperature at which solid CO_2 can exist when the pressure is 1.00 atm.

Solid carbon dioxide cannot exist above -78 °C at 1.00 atm.

11.103 Draw the phase diagram of SO_2 from the following information: triple point, -76 °C, 1.25 torr; critical point, 157 °C, 78 atm; normal boiling point, -10 °C, normal melting point, -72.7 °C.

This drawing is not drawn to scale, but has been expanded to show the general shape of the phase diagram.

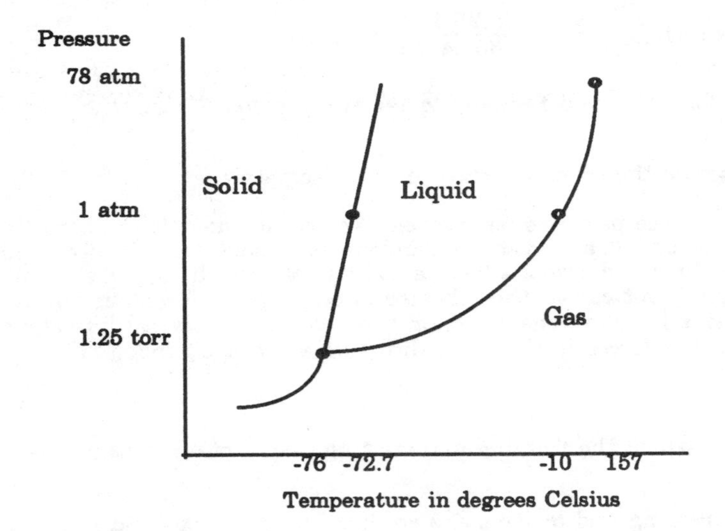

11.105 Why are ionic solids poor conductors of electricity?

To conduct electricity, mobility of particles is needed. In a solid, the electrons are held in position in ionic bonds. Without electron movement, electricity cannot be conducted.

12 Solutions

12.1 Differentiate among the terms "mixture", "solution", and "colloid".

A mixture is any combination of two or more substances; therefore, it can be either heterogeneous (with regions of differing appearance) or homogeneous (uniform in appearance). If the mixture is homogeneous, it is called a solution, and has a dissolved component with a particle size of less than 10 nm (far too small to be seen). In a colloid, particles are suspended in a dispersion medium and have sizes in the range of 10 nm to 1000 nm (still too small to be seen). The larger particle size enables some colloids to exhibit the Tyndall effect.

12.3 What factors influence the enthalpy of solution?

The enthalpy of solution depends on the relative strengths of the interparticle attractions of the solute for itself, of the solvent for itself and of the solute for the solvent.

12.5 Is it possible for a solution to be saturated and yet not in equilibrium with pure solute?

Yes, this is possible. For example, if a saturated solution in equilibrium with a solid is filtered, the solution is still saturated (no more solute can dissolve), yet it is not in equilibrium because there is no solid in contact with the solution.

12.7 Define the terms "deposition", "crystallization", and "precipitation".

Deposition is the formation of a solid from a vapor. Crystallization is the formation of a solid from a solution. Precipitation is also the formation of a solid from a solution, however, it is caused by a chemical reaction, often initiated by mixing solutions of dissolved substances.

12.9 If the definition of a saturated solution is "a solution that can dissolve no more solute," how is it possible for a supersaturated solution to exist?

A supersaturated solution contains more than the equilibrium amount of solute. It cannot be made by dissolving additional solute in a saturated solution. Instead, supersaturated solutions are usually made by preparing a solution at one temperature and slowly cooling the solution. The "extra" solute often stays dissolved until the system is disturbed. Disturbance causes precipitation of the additional solute until the solution is merely saturated. (Most solutes are more soluble at higher temperatures. If the solute is more soluble at a lower temperature, then a supersaturated solution could be prepared by slowly raising the temperature.)

12.11 What is an ideal solution?

An ideal solution is one which obeys Raoult's law. Most ideal solutions are composed of substances whose chemical and physical properties are very similar to each other.

12.13 To what systems does Henry's law apply?

Henry's law applies to solutions composed of a gaseous solute dissolved in a liquid solvent.

12.15 Define the term "colligative property".

A colligative property is any property that depends upon the concentration of solute and not upon the nature of the solute.

12.17 If a small amount of ethanol (bp 78 °C) is dissolved in water, would you expect the boiling point of the solution to be greater or less than 100 °C?

The boiling point would be expected to be less than 100 °C. Solutions comprised of two volatile components usually have boiling points that are between the boiling points of the substances.

12.19 Osmosis is a spontaneous process, and therefore involves either an increase in entropy, a decrease in enthalpy, or both. Which do you think is the more important factor? Explain.

The enthalpy change caused by adding more solvent to a dilute solution is usually very small, and is a minor factor. When dilution occurs, the solute is dispersed over a larger volume, substantially increasing the amount of disorder. Consequently, increase in entropy is the dominant effect in osmosis.

12.21 What is an activity coefficient?

An activity coefficient is the ratio of the activity (effective concentration) of a dissolved species to the actual concentration.

12.23 How might you define a colloid in terms of particle size?

In a colloid, the dispersed substance has particle sizes in the range of 10 nm to 1000 nm.

12.25 Account for the fact that ethanol (C_2H_5OH) is completely miscible with water.

Ethanol has a nonpolar end, but it also has a polar end that is capable of hydrogen bonding with water. Because the polar portion is a substantial fraction of the molecule, its interaction with water is sufficient to make the ethanol miscible in all proportions.

12.27 Classify each of the following ionic substances as soluble, slightly soluble, or insoluble in water.
a. barium hydroxide
b. silver carbonate
c. potassium carbonate
d. silver acetate.

Analysis/Plan: Use and learn Table 12.3 in the text which details the solubility rules.
Work:
a. $Ba(OH)_2$ is an exception to the normal insolubility of hydroxides. It is **soluble**.
b. Silver carbonate is **insoluble**, because most carbonates are insoluble.
c. Potassium is a 1A metal; therefore, potassium carbonate is **soluble**.
d. All acetates are soluble; therefore, silver acetate is **soluble**.

12.29 Classify each of the following ionic substances as soluble, slightly soluble, or insoluble in water.
a. magnesium sulfate
b. magnesium sulfide
c. calcium fluoride
d. ammonium chloride

Analysis/Plan: Use and learn Table 12.3 in the text which details the solubility rules.
Work:
a. Magnesium sulfate is **soluble**, because most sulfates are soluble.
b. Magnesium sulfide is **insoluble**, because most sulfides are insoluble.
c. Calcium fluoride is **insoluble**, because most fluorides are insoluble.
d. All ammonium compounds are soluble; therefore, ammonium chloride is **soluble**.

12.31 In each of the following, both compounds are water-soluble. Predict whether a precipitate will form when solutions of the two are mixed, and if so, identify the compound that precipitates.
a. NH_4Cl, $Ba(NO_3)_2$
b. $Ba(NO_3)_2$, Na_2SO_4
c. NH_4NO_3, $AgNO_3$

Analysis/Plan: Determine the compounds that would be formed if the substances exchanged anions. Use Table 12.3 in the text which details the solubility rules to determine if either of the new combinations is insoluble and will precipitate. Be sure to consider charges to write the correct formula of any precipitate.

Work:

a. The ions are NH_4^+, NO_3^-, Ba^{2+}, and Cl^-. The possible products are NH_4NO_3 and $BaCl_2$. Both combinations are soluble; therefore, **no precipitate forms.**

b. The ions are Ba^{2+}, NO_3^- Na^+, and SO_4^{2-}. The possible products are $BaSO_4$ and $NaNO_3$. $BaSO_4$ is insoluble; therefore, **a precipitate of $BaSO_4$ forms.**

c. Both substances contain the nitrate ion. There is no change when the anions are exchanged; therefore, **no precipitate forms.**

12.33 In each of the following, both compounds are water-soluble. Predict whether a precipitate will form when solutions of the two are mixed, and if so, identify the compound that precipitates.

a. KCN, $Pb(NO_3)_2$ b. $Mg(NO_3)_2$, Na_2S c. NaOH, KBr

Analysis/Plan: Determine the compounds that would occur if the substances exchanged anions. Use Table 12.3 in the text which details the solubility rules to determine if either of the new combinations is insoluble and will precipitate. Be sure to consider charges to write the correct formula of any precipitate.

Work:

a. The ions are K^+, CN^-, Pb^{2+}, and NO_3^-. The possible products are KNO_3 and $Pb(CN)_2$. $Pb(CN)_2$ is insoluble; therefore, **a precipitate of $Pb(CN)_2$ forms.**

b. The ions are Mg^{2+}, NO_3^-, Na^+, and S^{2-}. The possible products are MgS and $NaNO_3$. MgS is insoluble; therefore, **a precipitate of MgS forms.**

c. The ions are Na^+, OH^-, K^+, and Br^-. The possible products are NaBr and KOH. Both compounds are soluble; therefore, **no precipitate forms.**

12.35 If 2.5 moles of O_2 is mixed with 6.3 moles of N_2, what is the composition of the mixture? Express the answer as mole fraction and mole percent of each component.

Analysis:

Target: A mole fraction and a mole percent. (? X) = and (? X%) =

Knowns: The number of moles of each component.

Relationships: The mole fraction of component A, X_A, is equal to $\frac{mol_A}{mol_{total}}$ and the mole percent A, X_A%, is equal to $\frac{mol_A}{mol_{total}} \times 100$.

Plan: Find the total number of moles, insert the values into the relationships and calculate.

Work:

The total number of moles is 2.5 mol O_2 + 6.3 mol N_2 = 8.8 mol.

$$? X_{O_2} = \frac{2.5 \text{ mol } O_2}{8.8 \text{ mol total}} = \mathbf{0.28} \qquad ? X_{O_2}\% = \frac{2.5 \text{ mol } O_2}{8.8 \text{ mol total}} \times 100 = \mathbf{28\%}$$

$$? X_{N_2} = \frac{6.3 \text{ mol } N_2}{8.8 \text{ mol total}} = \mathbf{0.72} \qquad ? X_{N_2}\% = \frac{6.3 \text{ mol } N_2}{8.8 \text{ mol total}} \times 100 = \mathbf{72\%}$$

12.37 What is the mole fraction of carbon tetrachloride in a mixture of 5.00 g CCl_4 and 250 g C_6H_6?

Analysis:

Target: A mole fraction (? X) =

Knowns: The mass of each compound. The molar masses are implicit knowns.

Relationships: The mole fraction of component A, X_A, is equal to $\frac{mol_A}{mol_{total}}$. The molar mass gives the equivalence statement g ↔ mol.

Plan: Use the factor-label method to find the number of moles of each component. Add to find the total number of moles. Use the definition of mole fraction and calculate.

Work:

Molar Mass CCl_4 = 1(12.011 C) + 4(35.453 Cl) = 153.823 g mol^{-1}
Molar Mass C_6H_6 = 6(12.011 C) + 6(1.0079 H) = 78.113 g mol^{-1}

$$? \text{ mol } CCl_4 = 5.00 \text{ g } CCl_4 \times \frac{1 \text{ mol } CCl_4}{153.823 \text{ g } CCl_4} = 0.0325 \text{ mol } CCl_4$$

$$? \text{ mol } C_6H_6 = 250 \text{ g } C_6H_6 \times \frac{1 \text{ mol } C_6H_6}{78.113 \text{ g } C_6H_6} = 3.20 \text{ mol } C_6H_6$$

Total number of moles = 0.0325 mol CCl_4 + 3.20 mol C_6H_6 = 3.23 mol total

$$? X_{CCl_4} = \frac{0.0325 \text{ mol } CCl_4}{3.23 \text{ mol total}} = \mathbf{0.0101}$$

12.39 A certain mixture of NaCl and NaI is 1.00% (w/w) NaI. What is the mole% NaI?

Analysis:

Target: A mole percent. (? X%) =
Knowns: The mass percent NaI. The molar masses are implicit knowns.
Relationships: The mole percent of component A, $X_A\%$, is equal to $\frac{mol_A}{mol_{total}} \times 100$. The mass percent A is equal to $\frac{mass_A}{mass\ total} \times 100$. The molar mass gives the equivalence statement g ↔ mol.

Plan: If we knew the mass of each substance, we could proceed as in Problem 12.37.

Step 3. Insert the values into the definition of mole percent and calculate.
Step 2. Use the factor-label method to determine the number of moles of each component and add to determine the total number of moles.
Step 1. Assume that 100.00 g of the solution is available and determine the mass of each component in the solution. Masses are additive.

Work:

Step 1. Determine the mass of each component.

$$100.00 \text{ g solution} \times \frac{1.00 \text{ g NaI}}{100 \text{ g solution}} = 1.00 \text{ g NaI}$$

100.00 g solution - 1.00 g NaI = 99.00 g NaCl

Step 2. Determine the number of moles of each component and the total number of moles.
Molar Mass NaI = 1(22.98977 Na) + 1(126.9045 I) = 149.8942 g mol^{-1}
Molar Mass NaCl = 1(22.98977 Na) + 1(35.453 Cl) = 58.443 g mol^{-1}

$$? \text{ mol NaI} = 1.00 \text{ g NaI} \times \frac{1 \text{ mol NaI}}{149.8942 \text{ g NaI}} = 6.67 \times 10^{-3} \text{ mol NaI}$$

$$? \text{ mol NaCl} = 99.00 \text{ g NaCl} \times \frac{1 \text{ mol NaCl}}{58.443 \text{ g NaCl}} = 1.694 \text{ mol NaCl}$$

Total number of moles = 6.67×10^{-3} mol NaI + 1.694 mol NaCl = 1.701 mol total

$$? X_{NaI} = \frac{6.67 \times 10^{-3} \text{ mol NaI}}{1.701 \text{ mol total}} \times 100 = \mathbf{0.392 \text{ mol } \%}$$

12.41 Calculate the molality of a solution made by dissolving 1.00 g NaCl in 100.0 g H_2O.

Analysis:
Target: A molality. (? m) =
Knowns: The mass of both the solute and the solvent. The molar mass of NaCl is an implicit known.
Relationships: The molality, m, of a solute in solution is equal to $\frac{\text{mol solute}}{\text{kg solvent}}$. The molar mass gives the equivalence statement g ↔ mol. The equivalence statement 1 kg ↔ 1000 g is also needed.
Plan: The following steps will yield the answer.
Step 3. Insert the values into the definition of molality and calculate.
Step 2. Use the factor-label method to determine the number of moles of NaCl
Step 1. Use the factor-label method to determine the mass of solvent in kg.
Work:
Step 1. Use the factor-label method to determine the mass of solvent in kg.

$$? \text{ kg water} = 100 \text{ g } H_2O \times \frac{1 \text{ kg } H_2O}{1000 \text{ g } H_2O} = 0.100 \text{ kg } H_2O$$

Step 2. Use the factor-label method to determine the number of moles of NaCl.
Molar Mass NaCl = 1(22.98977 Na) + 1(35.453 Cl) = 58.443 g mol^{-1}

$$? \text{ mol NaCl} = 1.00 \text{ g NaCl} \times \frac{1 \text{ mol NaCl}}{58.443 \text{ g NaCl}} = 1.71 \times 10^{-2} \text{ mol NaCl}$$

Step 3. Insert the values into the relationship and solve for the molality.

$$? m \text{ NaCl} = \frac{1.71 \times 10^{-2} \text{ mol NaCl}}{0.100 \text{ kg } H_2O} = \mathbf{0.171}\ m \text{ NaCl}$$

12.43 An aqueous solution contains 2.07% (by mass) Na_2S. What is the molality of the solute?

Analysis:
Target: A molality. (? m) =
Knowns: The mass percent of the solute. The molar mass of Na_2S is an implicit known.
Relationships: The molality, m, of a solute in solution is equal to $\frac{\text{mol solute}}{\text{kg solvent}}$. The mass percent A is equal to $\frac{\text{mass}_A}{\text{mass total}} \times 100$. The molar mass gives the equivalence statement g ↔ mol. The equivalence statement 1 kg ↔ 1000 g is needed.
Plan: If we knew the masses of the solute and the solvent, we could proceed as in Problem 12.41. We can find those masses from the mass percent.
Step 4. Insert the values into the definition of molality and calculate.
Step 3. Use the factor-label method to determine the number of moles of NaCl.
Step 2. Use the factor-label method to determine the mass of solvent in kg.
Step 1. Assume 100.00 g solution and determine the masses of the solute and the solvent, using the factor-label method. Remember that masses are additive.
Work:
Step 1. Assume 100.00 g solution and determine the masses of the two components.

$$? \text{ g } Na_2S = 100.00 \text{ g solution} \times \frac{2.07 \text{ g } Na_2S}{100 \text{ g solution}} = 2.07 \text{ g } Na_2S$$

$$? \text{ g } H_2O = 100.00 \text{ g solution} - 2.07 \text{ g } Na_2S = 97.93 \text{ g } H_2O$$

Step 2. Use the factor-label method to determine the mass of solvent in kg.

$$? \text{ kg water} = 97.93 \text{ g } H_2O \times \frac{1 \text{ kg } H_2O}{1000 \text{ g } H_2O} = 0.09793 \text{ kg } H_2O$$

Step 3. Use the factor-label method to determine the number of moles of Na_2S.
Molar Mass Na_2S = 2(22.98977 Na) + 1(32.06 S) = 78.04 g mol^{-1}

$$? \text{ mol } Na_2S = 2.07 \text{ g } Na_2S \times \frac{1 \text{ mol } Na_2S}{78.04 \text{ g } Na_2S} = 2.65 \times 10^{-2} \text{ mol } Na_2S$$

Step 4. Insert the values into the definition and calculate.

$$? \, m \; Na_2S = \frac{2.65 \times 10^{-2} \text{ mol } Na_2S}{0.09793 \text{ kg } H_2O} = \mathbf{0.271} \; m \; \mathbf{Na_2S}$$

12.45 How much sucrose ($C_{12}H_{22}O_{11}$) must be dissolved in 75.0 g H_2O to produce a 0.100 *m* solution?

Analysis:
Target: A mass of sucrose. (? g) =
Knowns: The molality of the sucrose solution and the mass of the solvent. The molar mass of sucrose is an implicit known.
Relationships: The molality, *m*, of a solute in solution is equal to $\frac{\text{mol solute}}{\text{kg solvent}}$. The molar mass gives the equivalence statement g ↔ mol. The equivalence statement 1 kg ↔ 1000 g is also needed.
Plan: If we knew the number of moles of sucrose, we could find the mass. We can rearrange the definition of molality to solve for the number of moles of sucrose.
Step 2. Use the factor-label method to find the mass of sucrose.
Step 1. Rearrange the definition of molality to determine the number of moles of sucrose, using the factor-label method to convert the mass of water to kg.
Work:
Step 1. Find the number of moles of sucrose.

$$m \text{ sucrose} = \frac{\text{mol sucrose}}{\text{kg } H_2O}\text{; therefore, mol sucrose} = (m \text{ sucrose})(\text{kg } H_2O)$$

$$? \text{ mol sucrose} = \frac{0.100 \text{ mol sucrose}}{1 \text{ kg } H_2O} \times 75.0 \text{ g } H_2O \times \frac{1 \text{ kg } H_2O}{1000 \text{ g } H_2O} = 7.50 \times 10^{-3} \text{ mol}$$

Step 2. Use the factor-label method to determine the mass of sucrose.
Molar Mass $C_{12}H_{22}O_{11}$ = 12(12.011 C) + 22(1.0079 H) + 11(15.9994 O) = 342.299 g mol^{-1}

$$? \text{ g sucrose} = 7.50 \times 10^{-3} \text{ mol sucrose} \times \frac{342.299 \text{ g sucrose}}{1 \text{ mol sucrose}} = \mathbf{2.57 \text{ g sucrose}}$$

12.47 The vapor pressure of octane (C_8H_{18}) is 390 torr at 100 °C, while octadecane ($C_{18}H_{38}$) has a negligibly small vapor pressure at that temperature. What is the vapor pressure of a solution of 10.0 g octadecane in 500 g octane at 100 °C.

Analysis:
Target: A solution vapor pressure. (? torr) =
Knowns: The mass of the solute, the mass of the solvent and the vapor pressure of pure octane. The molar masses are implicit knowns.
Relationships: Raoult's Law, $P_{solution} = X_{solvent}P°_{solvent}$, is needed. The mole fraction of a solvent in a solution, $X_{solvent}$, is equal to $\frac{\text{mol solvent}}{\text{mol total}}$. The molar mass gives the equivalence statement g ↔ mol.

Plan:

Step 2. Insert the values into the definition of vapor pressure and calculate.
Step 1. Use the factor-label method to find the number of moles of each component and add to determine the total number of moles.

Work:

Step 1. Determine the number of moles of each component and the total number of moles.
Molar Mass C_8H_{18} = 8(12.011 C) + 18(1.0079 H) = 114.230 g mol^{-1}
Molar Mass $C_{18}H_{38}$ = 18(12.011 C) + 38(1.0079 H) = 254.498 g mol^{-1}

$$? \text{ mol } C_8H_{18} = 500 \text{ g } C_8H_{18} \times \frac{1 \text{ mol } C_8H_{18}}{114.230 \text{ g } C_8H_{18}} = 4.38 \text{ mol } C_8H_{18}$$

$$? \text{ mol } C_{18}H_{36} = 10.0 \text{ g } C_{18}H_{36} \times \frac{1 \text{ mol } C_{18}H_{36}}{252.498 \text{ g } C_{18}H_{36}} = 3.93 \times 10^{-2} \text{ mol } C_{18}H_{36}$$

Total number of moles = 4.38 mol C_8H_{18} + 3.93 x10^{-2} mol $C_{18}H_{36}$ = 4.42 mol total

Step 2. Insert the values into Raoult's law and solve for the vapor pressure.

$$? \text{ torr} = P_{solution} = \left(\frac{4.38 \text{ mol } C_{18}H_{18}}{4.42 \text{ mol total}}\right)(390 \text{ torr}) = \mathbf{387 \text{ torr}}$$

12.49 The normal boiling point of acetone (C_3H_6O) is 56.5 °C, at which temperature the equilibrium vapor pressure of cyclohexanone ($C_6H_{10}O$) is 24.0 torr. Calculate the vapor pressure (at 56.5 °C) of a cyclohexanone/acetone mixture having 75 mole% cyclohexanone.

Analysis:

Target: A solution vapor pressure. (? torr) =
Knowns: The normal boiling point of acetone, the vapor pressure of cyclohexanone and the mole percent of the mixture.
Relationships: Raoult's Law, $P_{solution} = X_AP°_A + X_BP°_B$, for a solution with two volatile components is needed. At the normal boiling point, the vapor pressure is equal to 760 torr (1 atm). The mole fraction can be found from the mole percent. Mole fractions must add to 1.00.

Plan: Find the mole fraction of each component, then substitute the values into Raoult's law and calculate.

Work:

Mole fraction cyclohexanone = 75%/100 = 0.75. Mole fraction acetone = 1.00 - 0.75 = 0.25.

? torr = $P_{solution}$ =

(0.75 cyclohexanone)(24.0 torr cyclohexanone) + (0.25 acetone)(760 torr acetone)

= **208 torr**

12.51 At 30 °C, the vapor pressure of 1-bromobutane (C_4H_9Br) is 51.3 torr while that of 1-chlorobutane (C_4H_9Cl) is 129.9 torr. What is the vapor pressure of a mixture of these two halobutanes that is 50% (by mass) 1-bromobutane?

Analysis:

Target: A solution vapor pressure. (? torr) =
Knowns: The mass percent of the mixture and the vapor pressures of both pure components are known. The molar masses are implicit knowns.
Relationships: Raoult's Law, $P_{solution} = X_AP°_A + X_BP°_B$, for a solution with two volatile components is needed. The mole fraction of component A, X_A, is equal to $\frac{mol_A}{mol_{total}}$. The

mass percent A is equal to $\frac{mass_A}{mass\ total}$ x 100. Masses are additive. The molar mass gives the equivalence statement g ↔ mol.

Plan: If we knew the mole fraction of each substance, we could proceed as in Problem 12.49.

Step 3. Insert the values into Raoult's law and calculate.

Step 2. Use the factor-label method to determine the number of moles of each component and add to determine the total number of moles.

Step 1. Assume that 100.00 g of the solution is available and determine the mass of each component in the solution.

Work:

Step 1. Determine the mass of each component.

$$100.00 \text{ g solution} \times \frac{50.0 \text{ g } C_4H_9Br}{100 \text{ g solution}} = 50.0 \text{ g } C_4H_9Br$$

100.0 g solution - 50.0 g C_4H_9Br = 50.0 g C_4H_9Cl

Step 2. Determine the number of moles of each component and the total number of moles.
Molar Mass C_4H_9Br = 4(12.011 C) + 9(1.0079 H) + 1(79.904 Br) = 137.019 g mol^{-1}
Molar Mass C_4H_9Cl = 4(12.011 C) + 9(1.0079 H) + 1(35.453 Cl) = 92.568 g mol^{-1}

$$? \text{ mol } C_4H_9Br = 50.0 \text{ g } C_4H_9Br \times \frac{1 \text{ mol } C_4H_9Br}{137.019 \text{ g } C_4H_9Br} = 0.365 \text{ mol } C_4H_9Br$$

$$? \text{ mol } C_4H_9Cl = 50.0 \text{ g } C_4H_9Cl \times \frac{1 \text{ mol } C_4H_9Cl}{92.568 \text{ g } C_4H_9Cl} = 0.540 \text{ mol } C_4H_9Cl$$

Total number of moles = 0.365 mol C_4H_9Br + 0.540 mol C_4H_9Cl = 0.905 mol total

Step 3. Determine the vapor pressure of the solution from Raoult's law.

? torr = $P_{solution}$ =

$$\left(\frac{0.365 \text{ mol } C_4H_9Br}{0.905 \text{ mol total}}\right)(51.3 \text{ torr}) + \left(\frac{0.540 \text{ mol } C_4H_9Cl}{0.905 \text{ mol total}}\right)(129.9 \text{ torr}) = \mathbf{98.2 \text{ torr}}$$

12.53 The vapor pressure of ethyl acetate ($C_4H_8O_2$) is 276.1 torr at 50.0 °C, at which temperature the vapor pressure of propyl acetate ($C_5H_{10}O_2$) is 110.3 torr. What is the partial pressure of each component above a mixture of 2.00 mol propyl acetate and 1.00 mol ethyl acetate? What is the composition of the vapor?

Analysis:

Target: The vapor pressure of each component and the composition of the vapor.
(? torr) = and (? X(g)) =

Knowns: The number of moles and the vapor pressures of both pure components are known.

Relationships: The vapor pressure of each component is given by Raoult's Law, $P_{solution} = X_{solvent}P°_{solvent}$. The mole fraction of component A, X_A, is equal to $\frac{mol_A}{mol_{total}}$. From the gas law chapter, we known that the mole fraction in a vapor is related to its partial pressure and the total pressure $\left(X_A = \frac{P_A}{P_{total}}\right)$.

Plan:

Step 2. Add the vapor pressures of the two components to find the total pressure and use the relationship to find the mole fraction of each component in the gas.

Step 1. Find the total number of moles and use Raoult's law to find the vapor pressure of each component.

Work:

Step 1. Find the vapor pressure of each component.
The total number of moles is 2.00 mol $C_5H_{10}O_2$ + 1.00 mol $C_4H_8O_2$ = 3.00 mol.

$$? \text{ torr} = \text{VP } C_5H_{10}O_2 = \left(\frac{2.00 \text{ mol } C_5H_{10}O_2}{3.00 \text{ mol total}}\right)(110.3 \text{ torr}) = \mathbf{73.5 \text{ torr}}$$

$$? \text{ torr} = \text{VP } C_4H_8O_2 = \left(\frac{1.00 \text{ mol } C_4H_8O_2}{3.00 \text{ mol total}}\right)(276.1 \text{ torr}) = \mathbf{92.0 \text{ torr}}$$

Step 2. Find the mole fraction in the vapor.
The total vapor pressure is 73.5 torr $C_4H_8O_2$ + 92.0 torr $C_5H_{10}O_2$ = 165.5 torr total

$$? X = \text{mol fraction } C_5H_{10}O_2(g) = \frac{73.5 \text{ torr } C_5H_{10}O_2}{165.5 \text{ torr total}} = \mathbf{0.444}$$

$$? X = \text{mol fraction } C_4H_8O_2(g) = \frac{92.0 \text{ torr } C_4H_8O_2}{165.5 \text{ torr total}} = \mathbf{0.556}$$

12.55 For carbon dioxide, the Henry's law constant is 0.039 mol L^{-1} atm^{-1} at room temperature. What is the maximum molarity of CO_2 in an aqueous solution in equilibrium with CO_2 at 5.00 atm?

Analysis:
Target: The maximum molarity of CO_2 in the solution. (? mol L^{-1}) =
Knowns: The Henry's law constant and the pressure of CO_2 in the atmosphere.
Relationship: Henry's law is $S = k_h \cdot P$, where S is the solubility of the solute, k_h is its Henry's law constant at the given temperature and P is the partial pressure of the solute above the solution.
Plan: Insert the values into Henry's law and solve for the molarity.
Work:

$$? \text{ mol } L^{-1} = S = k_h \cdot P = (0.039 \text{ mol } L^{-1} \text{ atm}^{-1})(5.00 \text{ atm}) = \mathbf{0.20 \text{ mol } L^{-1} CO_2}$$

12.57 Assume that a soft drink is carbonated during manufacture by allowing the drink to come to equilibrium with 4.00 atm CO_2 at 0 °C, at which temperature the Henry's law constant is 0.077 mol L^{-1} atm^{-1}. What is the CO_2 concentration in a soft drink?

Analysis:
Target: The molarity of CO_2 in a soft drink. (? mol L^{-1}) =
Knowns: The Henry's law constant and the pressure of CO_2 in the atmosphere.
Relationship: Henry's law is $S = k_h \cdot P$, where S is the solubility of the solute, k_h is its Henry's law constant at the given temperature and P is the partial pressure of the solute above the solution.
Plan: Insert the values into Henry's law and solve for the molarity.
Work:

$$? \text{ mol } L^{-1} = S = k_h \cdot P = (0.077 \text{ mol } L^{-1} \text{ atm}^{-1})(4.00 \text{ atm}) = \mathbf{0.31 \text{ mol } L^{-1} CO_2}$$

12.59 If 0.256 g argon will dissolve in 1 liter of hot water when the argon pressure is 5.00 atm, what is the Henry's law constant at this temperature?

Analysis:
Target: The Henry's law constant (? mol L^{-1} atm^{-1}) =
Knowns: The mass of argon which dissolves and the pressure of Ar in the atmosphere. The molar mass of argon is an implicit known.

Relationships: Henry's law is $S = k_h \cdot P$, where S is the solubility of the solute, k_h is its Henry's law constant at the given temperature and P is the partial pressure of the solute above the solution. The molar mass gives the equivalence statement g ↔ mol.

Plan: This can be solved in a two step process.

Step 2. Rearrange the Henry's law equation to solve for the constant, then insert the known values and calculate.

Step 1. Find the molarity of argon gas, using the factor-label method.

Work:

Step 1. Find the molarity of the argon in the solution.

Molar Mass Ar = 39.948 g mol^{-1}

$$? \text{ mol L}^{-1} \text{ Ar} = \frac{0.256 \text{ g Ar}}{1 \text{ L}} \times \frac{1 \text{ mol Ar}}{39.948 \text{ g Ar}} = 6.41 \times 10^{-3} \text{ mol L}^{-1} \text{ Ar}$$

Step 2. Rearrange Henry's law, insert values and calculate.

$$? \text{ mol L}^{-1} \text{ atm}^{-1} = S = k_h \cdot P\text{; therefore, } k_h = \frac{S}{P} = \frac{6.41 \times 10^{-3} \text{ mol L}^{-1}}{5.00 \text{ atm}} =$$

$$\mathbf{1.28 \times 10^{-3} \text{ mol L}^{-1} \text{ atm}^{-1}}$$

12.61 Calculate the freezing point of a 0.46 *m* aqueous solution of ethylene glycol, a non-electrolyte.

Analysis:

Target: A solution freezing point. (? °C) =

Knowns: The molality of an aqueous solution. The normal freezing point and the freezing point depression constant can be looked up in Table 12.5 and are equal to 0.00 °C and 1.86 °C kg mol^{-1}, respectively.

Relationships: $[T_{f(solution)} - T_{f(pure)}] = \Delta T_f = -i \cdot K_f \cdot m$, where K_f is the freezing point depression constant of the solvent, i is the van't Hoff factor, and *m* is the molality of the solute in the solution, with units of mol kg^{-1}. For a nonelectrolyte, i = 1.

Plan: Rearrange the relationship to solve for $T_{f(solution)}$, insert the known values and calculate.

Work:

$[T_{f(solution)} - T_{f(pure)}] = -i \cdot K_f \cdot m$; therefore, $T_{f(solution)} = T_{f(pure)} - i \cdot K_f \cdot m$

$$? \text{ °C} = 0.00 \text{ °C} - [(1)(1.86 \text{ °C kg mol}^{-1})(0.46 \text{ mol kg}^{-1})] = \mathbf{-0.86 \text{ °C}}$$

12.63 What is the molality of an aqueous solution that freezes at -0.53 °C?

Analysis:

Target: The molality of an aqueous solution. (? mol kg^{-1}) =

Knowns: The freezing point of the solution. The normal freezing point and the freezing point depression constant can be looked up in Table 12.5 and are equal to 0.00 °C and 1.86 °C kg mol^{-1}, respectively.

Relationships: $[T_{f(solution)} - T_{f(pure)}] = \Delta T_f = -i \cdot K_f \cdot m$, where K_f is the freezing point depression constant of the solvent, i is the van't Hoff factor, and *m* is the molality of the solute in the solution, with units of mol kg^{-1}.

Plan: With no solute specified, we must assume that the solute is a nonelectrolyte with i = 1. Rearrange the relationship to solve for *m*, insert the known values and calculate.

Work:

$$[T_{f(solution)} - T_{f(pure)}] = -i \cdot K_f \cdot m\text{; therefore, } m = \frac{[T_{f(solution)} - T_{f(pure)}]}{-i \cdot K_f}$$

$$? \text{ mol kg}^{-1} = \frac{-0.53 \text{ °C} - 0.00 \text{ °C}}{(1)(-1.86 \text{ °C kg mol}^{-1})} = \mathbf{0.28 \text{ mol kg}^{-1}} = \mathbf{0.28}\ \boldsymbol{m}$$

12.65 What are the freezing points of the following aqueous solutions?
a. 0.25 *m* sucrose   b. 0.50 *m* $SrBr_2$

Analysis:
Target: A solution freezing point. (? °C) =
Knowns: The molality of an aqueous solution. The normal freezing point and the freezing point depression constant can be looked up in Table 12.5 and are equal to 0.00 °C and 1.86 °C kg mol^{-1}, respectively.
Relationships: $[T_{f(solution)} - T_{f(pure)}] = -i \cdot K_f \cdot m$, where K_f is the freezing point depression constant of the solvent, i is the van't Hoff factor and *m* is the molality of the solute in the solution, with units of mol kg^{-1}.
Plan: Rearrange the relationship to solve for $T_{f(solution)}$, insert the known values, and calculate.
Work:
For both, $[T_{f(solution)} - T_{f(pure)}] = -i \cdot K_f \cdot m$; therefore, $T_{f(solution)} = T_{f(pure)} - i \cdot K_f \cdot m$
a. Sucrose is a nonelectrolyte so i = 1.

$$? \text{ °C} = 0.00 \text{ °C} - (1)(1.86 \text{ °C kg mol}^{-1})(0.25 \text{ mol kg}^{-1}) = \mathbf{-0.47 \text{ °C}}$$

b. $SrBr_2$ is an ionic compound which dissociates to yield Sr^{2+} and 2 Br^-; therefore, i = 3.

$$? \text{ °C} = 0.00 \text{ °C} - (3)(1.86 \text{ °C kg mol}^{-1})(0.50 \text{ mol kg}^{-1}) = \mathbf{-2.8 \text{ °C}}$$

12.67 Calculate the freezing point of a 1.00% (by mass) solution of bromobenzene (C_6H_5Br) in
a. benzene   b. naphthalene

Analysis:
Target: A solution freezing point. (? °C) =
Knowns: The mass percent of the solute. Normal freezing points and freezing point depression constants can be looked up in Table 12.5. For benzene, they are 5.5 °C and 5.12 °C kg mol^{-1}, respectively. For naphthalene, they are 80.2 °C and 6.8 °C kg mol^{-1}, respectively. The molar mass of C_6H_5Br is an implicit known.
Relationships: $[T_{f(solution)} - T_{f(pure)}] = -i \cdot K_f \cdot m$, where K_f is the freezing point depression constant of the solvent, i is the van't Hoff factor and *m* is the molality of the solute in the solution, with units of mol kg^{-1}. The mass percent A is equal to $\frac{mass_A}{mass\ total} \times 100$. The molar mass gives the equivalence statement g ↔ mol. The equivalence statement 1 kg ↔ 1000 g is needed.
Plan: If we knew the masses of the solute and the solvent, we could determine the molality and proceed as in Problem 12.61. We can find those masses from the mass percent.
Step 5. Rearrange the relationship to solve for $T_{f(solution)}$, insert the known values and solve for the freezing point
Step 4. Insert the values into the relationship and solve for the molality.
Step 3. Use the factor-label method to determine the number of moles of NaCl.
Step 2. Use the factor-label method to determine the mass of solvent in kg.
Step 1. Assume 100.00 g solution and determine the masses of the solute and the solvent, using the factor-label method. Remember that masses are additive.
Work: The first four steps are the same for both solvents.
Step 1. Assume 100.00 g solution and determine the masses of the two components.

$$? \text{ g } C_6H_5Br = 100.00 \text{ g solution} \times \frac{1.00 \text{ g } C_6H_5Br}{100 \text{ g solution}} = 1.00 \text{ g } C_6H_5Br$$

$$? \text{ g solvent} = 100.00 \text{ g solution} - 1.00 \text{g } C_6H_5Br = 99.00 \text{ g solvent}$$

Step 2. Use the factor-label method to determine the mass of solvent in kg.

$$? \text{ kg solvent} = 99.00 \text{ g solvent} \times \frac{1 \text{ kg solvent}}{1000 \text{ g solvent}} = 0.09900 \text{ kg solvent}$$

Step 3. Use the factor-label method to determine the number of moles of C_6H_5Br.
Molar Mass C_6H_5Br = 6(12.011 C) + 5(1.0079 H) + 1(79.904 Br) = 157.010 g mol^{-1}

$$? \text{ mol } C_6H_5Br = 1.00 \text{ g } C_6H_5Br \times \frac{1 \text{ mol } C_6H_5Br}{157.010 \text{ g } C_6H_5Br} = 6.37 \times 10^{-3} \text{ mol } C_6H_5Br$$

Step 4. Insert the values into the relationship and solve for the molality.

$$? \text{ mol kg}^{-1}\ C_6H_5Br = m = \frac{6.37 \times 10^{-3} \text{ mol } C_6H_5Br}{0.09900 \text{ kg solvent}} = 6.43 \times 10^{-2} \text{ mol kg}^{-1}\ C_6H_5Br$$

Step 5. Solve for the freezing point.
For both, $[T_{f(solution)} - T_{f(pure)}] = -i \cdot K_f \cdot m$; therefore, $T_{f(solution)} = T_{f(pure)} - i \cdot K_f \cdot m$

C_6H_5Br is a nonelectrolyte so i = 1.

a. For benzene, $T_{f(pure)}$ = 5.5 °C and K_f = 5.12 °C kg mol^{-1}

$$? \text{ °C} = 5.5 \text{ °C} - (1)(5.12 \text{ °C kg mol}^{-1})(6.43 \times 10^{-2} \text{ mol kg}^{-1}) = \mathbf{5.2\ °C}$$

b. For naphthalene, $T_{f(pure)}$ = 80.2 °C and K_f = 6.8 °C kg mol^{-1}

$$? \text{ °C} = 80.2 \text{ °C} - (1)(6.8 \text{ °C kg mol}^{-1})(6.43 \times 10^{-2} \text{ mol kg}^{-1}) = \mathbf{79.8\ °C}$$

12.69 Suppose 1.0 g of a 50% (by mass) mixture of $MgCl_2$ and $BaCl_2$ is dissolved in 50.0 g H_2O. Calculate the freezing point of the solution.

Analysis:
Target: A solution freezing point. (? °C) =
Knowns: The total mass and the mass percent of each solute and the mass of solvent. The normal freezing point and freezing point depression constant can be looked up in Table 12.5 and are equal to 0.00 °C and 1.86 °C kg mol^{-1}, respectively. The molar masses of the solutes are implicit knowns.
Relationships: $[T_{f(solution)} - T_{f(pure)}] = -i \cdot K_f \cdot m$, where K_f is the freezing point depression constant of the solvent, i is the van't Hoff factor and m is the molality of the solute in the solution, with units of mol kg^{-1}. The mass percent A is equal to $\frac{mass_A}{mass\ total} \times 100$. The molar mass gives the equivalence statement g ↔ mol. The equivalence statement 1 kg ↔ 1000 g is needed.
Plan: If we knew the masses of the solutes, we could determine the molality of each and proceed as in Problem 12.61. We can find those masses from the mass percent.
Step 5. Rearrange the relationship to solve for $T_{f(solution)}$, insert the known values, and solve for the freezing point
Step 4. Insert the values into the relationship and solve for the molality of each solute.
Step 3. Use the factor-label method to determine the number of moles of each solute.
Step 2. Use the factor-label method to determine the mass of solvent in kg.
Step 1. Determine the masses of the solute, using the factor-label method. Remember that masses are additive.
Work:
Step 1. Determine the mass of each solute from its mass percent.

$$? \text{ g } MgCl_2 = 1.0 \text{ g mixture} \times \frac{50 \text{ g } MgCl_2}{100 \text{ g mixture}} = 0.50 \text{ g } MgCl_2$$

$$? \text{ g } BaCl_2 = 1.0 \text{ g mixture} \times \frac{50 \text{ g } BaCl_2}{100 \text{ g mixture}} = 0.50 \text{ g } BaCl_2$$

Step 2. Use the factor-label method to determine the mass of water in kg.

$$? \text{ kg water} = 50.0 \text{ g } H_2O \times \frac{1 \text{ kg } H_2O}{1000 \text{ g } H_2O} = 0.0500 \text{ kg water}$$

Step 3. Use the factor-label method to determine the number of moles of each solute.
Molar Mass $MgCl_2$ = 1(24.305 Mg) + 2(35.453 Cl) = 95.211 g mol^{-1}
Molar Mass $BaCl_2$ = 1(137.33) + 2(35.453 Cl) = 208.24g mol^{-1}

$$? \text{ mol } MgCl_2 = 0.50 \text{ g } MgCl_2 \times \frac{1 \text{ mol } MgCl_2}{95.211 \text{ g } MgCl_2} = 0.0053 \text{ mol } MgCl_2$$

$$? \text{ mol } BaCl_2 = 0.50 \text{ g } BaCl_2 \times \frac{1 \text{ mol } BaCl_2}{208.24 \text{ g } BaCl_2} = 0.0024 \text{ mol } BaCl_2$$

Step 4. Insert the values into the relationship and solve for the molality.

$$? \, m \, MgCl_2 = \frac{0.0053 \text{ mol } MgCl_2}{0.0500 \text{ kg } H_2O} = 0.11 \, m \, MgCl_2$$

$$? \, m \, BaCl_2 = \frac{0.0024 \text{ mol } BaCl_2}{0.0500 \text{ kg } H_2O} = 0.048 \, m \, BaCl_2$$

Step 5. Solve for the freezing point.
For both, $[T_{f(solution)} - T_{f(pure)}] = -i \cdot K_f \cdot m$; therefore, $T_{f(solution)} = T_{f(pure)} - i \cdot K_f \cdot m$

Both solutes dissociate to make a 2+ cation and 2 Cl^- ions; therefore, i = 3. The total molality is 0.11 m $MgCl_2$ + 0.048 m $BaCl_2$ = 0.16 m = 0.16 mol kg^{-1}.

$$? \text{ °C} = 0.000 \text{ °C} - (3)(1.86 \text{ °C kg mol}^{-1})(0.16 \text{ mol kg}^{-1}) = \mathbf{-0.85 \text{ °C}}$$

(Unrounded values were used in subsequent steps throughout the calculation.)

12.71 Dissolution of 0.500 g of a mixture of NaCl and CsBr in 100.0 g H_2O is found to decrease the freezing point by 0.125 °C. Calculate the percent (by mass) of NaCl in the mixture.

Analysis:
Target: The mass percent of NaCl in a mixture. (? %) =
Knowns: The total mass of solute and the mass of solvent. The change in freezing point, ΔT_f, is given. The freezing point depression constant can be looked up in Table 12.5 and is 1.86 °C kg mol^{-1}. The molar masses of the solutes are implicit knowns.
Relationships: $\Delta T_f = -i \cdot K_f \cdot m$, where K_f is the freezing point depression constant of the solvent, i is the van't Hoff factor and m is the molality of the solute in the solution, with units of mol kg^{-1}. The mass percent A is equal to $\frac{mass_A}{\text{mass total}} \times 100$. The molar mass gives the equivalence statement g ↔ mol. The equivalence statement 1 kg ↔ 1000 g is needed.
Plan: If we knew the mass of NaCl, we could determine mass percent. The change in freezing point enables us to find the molality. Use of the mass of solvent with the molality enables us to find the total number of moles. Algebra can be used to determine the mass of each solute.
Step 4. Use the definition of mass percent to determine the mass percent NaCl.
Step 3. Let the mass of NaCl be **y**; then the mass of CsBr is **0.500 - y**. Use the known molar masses to set up expressions for the number of moles of each solute. Add the terms, set them equal to the number of moles found from the molality and solve for **y**.

Step 2. Use the factor-label method to determine the number of moles of solute.
Step 1. Rearrange the relationship to solve for the molality, insert values and calculate.

Work:

Step 1. Rearrange the relationship to solve for the molality, insert values and calculate.

NaCl dissociates to form Na^+ and Cl^- and CsBr dissociates to form Cs^+ and Br^-. Therefore, for both solutes i = 2.

$$\Delta T_f = -i \cdot K_f \cdot m;\ \text{therefore,}\ m = \frac{\Delta T_f}{-i \cdot K_f} = \frac{-0.125\ °C}{-2(1.86\ °C\ kg\ mol^{-1})} = 3.36 \times 10^{-2}\ mol\ kg^{-1}$$

Step 2. Use the factor-label method to determine the number of moles of solute.

$$?\ mol\ solute = \frac{3.36 \times 10^{-2}\ mol\ solute}{kg\ H_2O} \times 100.0\ g\ H_2O \times \frac{1\ kg\ H_2O}{1000\ g\ H_2O} = 0.00336\ mol$$

Step 3. Do the necessary algebra.

Molar Mass NaCl = 1(22.98977 Na) + 1(35.453 Cl) = 58.443 g mol^{-1}
Molar Mass CsBr = 1(132.9054 Cs) + 1(79.904 Br) = 212.809 g mol^{-1}

$$\left(y\ g\ NaCl \times \frac{1\ mol\ NaCl}{58.443\ g\ NaCl}\right) + \left((0.500 - y)\ g\ CsBr \times \frac{1\ mol\ CsBr}{212.809\ g\ CsBr}\right) = 0.00336\ mol$$

$$1.71 \times 10^{-2}\, y + 2.35 \times 10^{-3} - 4.70 \times 10^{-3}\, y = 0.00336 = 3.36 \times 10^{-3}$$

$$1.24 \times 10^{-2}\, y = 3.36 \times 10^{-3} - 2.35 \times 10^{-3} = 1.01 \times 10^{-3}$$

$$y = \frac{1.01 \times 10^{-3}}{1.24 \times 10^{-2}} = 8.15 \times 10^{-2}\ g\ NaCl$$

Step 4. Use the definition of mass percent to determine the mass percent NaCl.

$$\%\ NaCl = \frac{mass\ NaCl}{mass\ mixture} \times 100 = \frac{8.15 \times 10^{-2}\ g\ NaCl}{0.500\ g\ mixture} \times 100 = \mathbf{16.3\%\ NaCl}$$

12.73 0.331 g of an unknown crystalline substance is ground together with 10.0 g camphor, and the resulting powder is found to melt to a clear liquid at 168.0 °C. Given that K_f for camphor is 40 °C kg mol^{-1}, and that pure camphor melts at 176.0 °C, what is the molar mass of the unknown? (Assume it to be a molecular compound.)

Analysis:

Target: The molar mass of an unknown. (? g mol^{-1}) =

Knowns: The mass of solute, the mass of solvent, the solution freezing point, the freezing point of pure camphor and the freezing point depression constant are given.

Relationships: $[T_{f(solution)} - T_{f(pure)}] = -i \cdot K_f \cdot m$, where K_f is the freezing point depression constant of the solvent, i is the van't Hoff factor and m is the molality of the solute in the solution, with units of mol kg^{-1}. The equivalence statement 1 kg ↔ 1000 g is needed.

Plan: We can use the units of molar mass to plan our steps.

Step 3. Calculate the molar mass from the solute mass and number of moles.
Step 2. Determine the number of moles of solute from the molality and solvent mass.
Step 1. Calculate the molality from the known data and given relationship.

Work:

Step 1. Rearrange the relationship to solve for the molality, insert values and calculate. Assume the solute is a nonelectrolyte because it is molecular, so i = 1.

$$[T_{f(solution)} - T_{f(pure)}] = -i \cdot K_f \cdot m;\ \text{therefore,}\ m = \frac{[T_{f(solution)} - T_{f(pure)}]}{-i \cdot K_f}$$

$$? \text{ mol kg}^{-1} = m = \frac{168.0 \text{ °C} - 176.0 \text{ °C}}{-(1)(40 \text{ °C kg mol}^{-1})} = 0.20 \text{ mol kg}^{-1}$$

Step 2. Use the factor-label method to determine the number of moles of solute.

$$? \text{ mol solute} = \frac{0.20 \text{ mol solute}}{\text{kg camphor}} \times 10.0 \text{ g camphor} \times \frac{1 \text{ kg camphor}}{1000 \text{ g camphor}} = 2.0 \times 10^{-3} \text{ mol}$$

Step 3. Determine the molar mass.

$$? \text{ g mol}^{-1} = \frac{0.331 \text{ g}}{2.0 \times 10^{-3} \text{ mol}} = \mathbf{1.7 \times 10^2 \text{ g mol}^{-1}}$$

12.75 Calculate the boiling point of a solution of 12.3 g urea (NH_2CONH_2) in 105 g H_2O. Urea is a nonvolatile molecular substance.

Analysis:

Target: A solution boiling point. (? °C) =

Knowns: The mass of both the solute and the solvent. The normal boiling point and the boiling point elevation constant can be looked up in Table 12.5 and are equal to 100.00 °C and 0.512 °C kg mol^{-1}, respectively. The molar mass is an implicit known.

Relationships: $[T_{b(solution)} - T_{b(pure)}] = \Delta T_b = i \cdot K_b m$, where K_b is the boiling point elevation constant of the solvent, i is the van't Hoff factor, and m is the molality of the solute in the solution, with units of mol kg^{-1}. For a nonelectrolyte, i = 1.

Plan: If we knew the molality, we could rearrange the relationship and solve for $T_{b(solution)}$.

Step 2. Rearrange the relationship to solve for $T_{b(solution)}$, insert the known values and calculate.

Step 1. Use the factor-label method to determine the molality of the solution.

Work:

Step 1. Find the molality of urea in the solution.

Molar Mass $CO(NH_2)_2$ = 1(12.011 C) + 1(15.9994 O) + 2(14.0067 N) + 4(1.0079 H) = 60.055 g mol^{-1}

$$? \; m \text{ urea} = \frac{12.3 \text{ g urea}}{105 \text{ g } H_2O} \times \frac{1 \text{ mol urea}}{60.055 \text{ g urea}} \times \frac{1000 \text{ g } H_2O}{1 \text{ kg } H_2O} = 1.95 \text{ mol kg}^{-1} \text{ urea}$$

Step 2. Rearrange the equation, insert known values and calculate.

$$[T_{b(solution)} - T_{b(pure)}] = i \cdot K_b \cdot m; \text{ therefore, } T_{b(solution)} = T_{b(pure)} + i \cdot K_b \cdot m$$

$$? \text{ °C} = 100.00 \text{ °C} + (1)(0.512 \text{ °C kg mol}^{-1})(1.95 \text{ mol kg}^{-1}) = \mathbf{101.00 \text{ °C}}$$

12.77 When 125 g of dextrose (grape sugar) is dissolved in 850 g water, the normal boiling point of the resulting solution is found to be 100.42 °C. What is the molar mass of dextrose?

Analysis:

Target: The molar mass of an unknown. (? g mol^{-1}) =

Knowns: The mass of solute, the mass of H_2O, and the solution freezing point are given. The normal boiling point and the boiling point elevation constant can be looked up in Table 12.5 and are equal to 100.00 °C and 0.512 °C kg mol^{-1}, respectively.

Relationships: $[T_{b(solution)} - T_{b(pure)}] = i \cdot K_b \cdot m$, where K_b is the freezing point depression constant of the solvent, i is the van't Hoff factor and m is the molality of the solute in the solution, with units of mol kg^{-1}. The equivalence statement 1 kg ↔ 1000 g is needed.

Plan: We can use the units of molar mass to plan our steps.

Step 3. Calculate the molar mass from the solute mass and number of moles.

Step 2. Determine the number of moles of solute from the molality and solvent mass.

Step 1. Calculate the molality from the known data and given relationship.

Work:

Step 1. Rearrange the relationship to solve for the molality, insert values and calculate. The solute is a nonelectrolyte, so i = 1.

$$[T_{b(solution)} - T_{b(pure)}] = i \cdot K_b \cdot m\text{; therefore, } m = \frac{[T_{b(solution)} - T_{b(pure)}]}{i \cdot K_b}$$

$$? \text{ mol kg}^{-1} = m = \frac{100.42\ ^\circ\text{C} - 100.00\ ^\circ\text{C}}{(1)(0.512\ ^\circ\text{C kg mol}^{-1})} = 0.82 \text{ mol kg}^{-1}$$

Step 2. Use the factor-label method to determine the number of moles of solute.

$$? \text{ mol solute} = \frac{0.82 \text{ mol solute}}{\text{kg } H_2O} \times 850 \text{ g } H_2O \times \frac{1 \text{ kg } H_2O}{1000 \text{ g } H_2O} = 0.70 \text{ mol solute}$$

Step 3. Determine the molar mass.

$$? \text{ g mol}^{-1} = \frac{125 \text{ g}}{0.70 \text{ mol}} = \mathbf{1.8 \times 10^2 \text{ g mol}^{-1}}$$

12.79 The osmotic pressure of a nonvolatile nonelectrolyte was found to be 8.38 atm at 25 °C. What is the concentration of the solution? What would the osmotic pressure of this solution be at 0 °C?

Analysis:

Target: A solution concentration and an osmotic pressure. (? mol L^{-1}) = and (? atm) =

Knowns: The osmotic pressure and the temperature. R = 0.0820578 L atm mol^{-1} K^{-1}.

Relationships: The osmotic pressure, Π = MRT. K ↔ 273.15 + °C

Plan: For the first question, rearrange the equation to determine molarity, convert the temperature to kelvins, insert known values, and calculate. For the second question, use the formula for osmotic pressure directly, employing the molarity found in the first answer, and converting the temperature to kelvins.

Work:

For the first question, Π = MRT; therefore, $M = \frac{\Pi}{RT}$.

$$? \text{ mol L}^{-1} = M = \frac{8.38 \text{ atm}}{(0.0820578 \text{ L atm mol}^{-1} \text{ K}^{-1})((273.15 + 25) \text{ K})} = \mathbf{0.343 \text{ mol L}^{-1}}$$

For the second question, the osmotic pressure can be found directly from Π = MRT.

$$? \text{ atm} = \Pi = (0.343 \text{ mol L}^{-1})(0.0820578 \text{ L atm mol}^{-1} \text{ K}^{-1})[(273.15 + 0) \text{ K}] = \mathbf{7.68 \text{ atm}}$$

12.81 Human hemoglobin has a molar mass of 6.84 x 10^4 g mol^{-1}. How much hemoglobin must be dissolved to produce 1.00 L of an aqueous solution with a room-temperature osmotic pressure of 1.00 torr? What is the height of a water column this pressure can support?

Analysis:

Target: A solute mass and water column height. (? g) = and (? mm H_2O) =

Knowns: The osmotic pressure, the temperature, and the molar mass of the solute. R = 0.0820578 L atm mol^{-1} K^{-1}.

Relationships: Π = MRT. K ↔ 273.15 + °C and 1 atm ↔ 760 torr. From Equation 12-15 in the text, 1 atm ↔ 1.034 x 10^4 mm H_2O. The molar mass gives the equivalence statement g ↔ mol.

Plan: For the first question, rearrange the equation to solve for the molarity. Since no temperature is given, assume a standard temperature of 25 °C. Convert temperature to kelvins and the pressure to atm, insert known values, and calculate. Use the factor-label method to determine the mass of hemoglobin from the molarity. For the second question, use the factor-label method to convert the pressure in torr to mm H_2O.

Work:

For the first question, Π = MRT; therefore, M = $\frac{\Pi}{RT}$.

$$? \text{ mol L}^{-1} = \text{M} = \frac{1.00 \text{ torr} \times \frac{1 \text{ atm}}{760 \text{ torr}}}{(0.0820578 \text{ L atm mol}^{-1} \text{ K}^{-1})((273.15 + 25) \text{ K})} = 5.38 \times 10^{-5} \text{ mol L}^{-1}$$

$$? \text{ g hemoglobin} = 5.38 \times 10^{-5} \text{ mol L}^{-1} \times \frac{6.84 \times 10^{4} \text{ g}}{1 \text{ mol}} \times 1.00 \text{ L} = \mathbf{3.68 \text{ g hemoglobin}}$$

The factor-label method is used for the second question.

$$? \text{ mm H}_2\text{O} = 1.00 \text{ torr} \times \frac{1 \text{ atm}}{760 \text{ torr}} \times \frac{1.034 \times 10^{4} \text{ mm H}_2\text{O}}{1 \text{ atm}} = \mathbf{13.6 \text{ mm H}_2\text{O}}$$

12.83 The sheath protein of the tobacco mosaic virus has a huge molar mass. Dissolution of 10 mg of this protein in 1.00 mL water gives a solution whose osmotic pressure at room temperature is sufficient to support a column of water only 0.062 mm high, which is barely noticeable and hard to distinguish from capillary rise. What is the molar mass of this protein? (Other methods are available for the accurate determination of such high molar masses.)

Analysis:

Target: The molar mass of an unknown. (? g mol^{-1}) = or (? mg $mmol^{-1}$) =

Knowns: The volume of the water, the amount of protein dissolved, the height of the water column, and the temperature. (Room temperature is 25 °C.)
R = 0.0820578 L atm mol^{-1} K^{-1}.

Relationships: Π = MRT. K ↔ 273.15 + °C and 1 atm ↔ 760 torr. From Equation 12-15 in the text, 1 atm ↔ 1.034 x 10^4 mm H_2O.

Plan: We can use the units of molar mass to plan our steps. Assume that the solution volume is the same as the H_2O volume.

Step 3. Calculate the molar mass from the solute mass and number of millimoles.
Step 2. Find the number of millimoles of solute from the molarity and solvent volume.
Step 1. Calculate the molarity from the known data and given relationship, using the factor-label method to make the necessary conversions.

Work:

Step 1. Rearrange the relationship to solve for the molarity, insert values and calculate.

Π = MRT; therefore, M = $\frac{\Pi}{RT}$

$$? \text{ mol L}^{-1} = \text{M} = \frac{0.062 \text{ mm H}_2\text{O} \times \frac{1 \text{ atm}}{1.034 \times 10^{4} \text{ mm H}_2\text{O}}}{(0.0820578 \text{ L atm mol}^{-1} \text{ K}^{-1})((273.15 + 25) \text{ K})} = 2.5 \times 10^{-7} \text{ mol L}^{-1}$$

Step 2. Use the factor-label method to determine the number of millimoles of solute. Note that mol L^{-1} is equivalent to mmol mL^{-1}.

$$? \text{ mmol solute} = \frac{2.5 \times 10^{-7} \text{ mmol}}{\text{mL}} \times 1.00 \text{ mL} = 2.5 \times 10^{-7} \text{ mmol}$$

Step 3. Determine the molar mass.

$$? \text{ g mol}^{-1} = \frac{10 \text{ mg}}{2.5 \times 10^{-7} \text{ mmol}} = \mathbf{4.1 \times 10^{7} \text{ mg mmol}^{-1} = 4.1 \times 10^{7} \text{ g mol}^{-1}}$$

12.85 35.00 g NH_4Cl is heated with 65.0 g H_2O and stirred until all the solid dissolves, at which point the solution is cooled to 30 °C. Eventually some NH_4Cl crystallizes. The crystals are separated, dried, and found to weigh 5.62 g. Express the solubility of NH_4Cl at 30 °C as grams solute per 100 g solution (mass %) and as molality.

Analysis:

Target: A mass percent and a molality of a solution. (? %) = and (? mol kg^{-1}) =

Knowns: The mass of solute originally dissolved in a given mass of solvent and the mass which precipitates on cooling. The molar mass of NH_4Cl is an implicit known.

Relationships: The molality of a solute in solution is equal to $\frac{\text{mol solute}}{\text{kg solvent}}$. The mass percent A is equal to $\frac{\text{mass}_A}{\text{mass total}} \times 100$. The molar mass gives the equivalence statement g ↔ mol. The equivalence statement 1 kg ↔ 1000 g is needed.

Plan: The temperature is extraneous data, not needed to answer these questions. The following steps will give the requested answers.

Step 3. Use the factor-label method to solve for the molality, making conversions as necessary.

Step 2. Add the mass of water to that of the NH_4Cl which stayed in solution to determine the solution mass and then use the definition of mass percent.

Step 1. Determine the mass of NH_4Cl that dissolved in the water at 30 °C.

Work:

Step 1. The mass of the solute which dissolved is found by subtraction.

35.00 g NH_4Cl initially dissolved - 5.62 g NH_4Cl precipitated = 29.38 g NH_4Cl

Step 2. Determine the solution mass and use the definition of mass percent. The solution mass is 29.38 g NH_4Cl + 65.0 g H_2O = 94.4 g solution.

$$? \%\ NH_4Cl = \frac{29.38\ g\ NH_4Cl}{94.4\ g\ solution} \times 100 = \mathbf{31.1\%\ NH_4Cl}$$

Step 3. Find the solution molality using the factor-label method.
Molar Mass NH_4Cl = 1(14.0067 N) + 4(1.0079 H) + 1(35.453 Cl) = 53.491 g mol^{-1}

$$? \text{ mol kg}^{-1} = \frac{29.38\ g\ NH_4Cl}{65.0\ g\ H_2O} \times \frac{1\ mol\ NH_4Cl}{53.491\ g\ NH_4Cl} \times \frac{1000\ g\ H_2O}{1\ kg\ H_2O} = \mathbf{8.45}\ \boldsymbol{m}\ \mathbf{NH_4Cl}$$

12.87 The main component of automotive antifreeze is ethylene glycol ($C_2H_6O_2$). Although the mathematical relationships of the colligative properties are accurate only in dilute solution, Equation (12-13) may be used for rough estimates in concentrated solutions. Estimate the molality of ethylene glycol required to decrease the freezing point of water to 0 °F. Also express the composition as weight percent.

Analysis:

Target: A mass percent and a molality of a solution. (? %) = and (? mol kg^{-1}) =

Knowns: The freezing point of the solution is known. The normal freezing point and the freezing point depression constant can be looked up in Table 12.5 and are equal to 0.00 °C and 1.86 °C kg mol^{-1}, respectively. The molar mass of $C_2H_6O_2$ is an implicit known.

Relationships: $[T_{f(solution)} - T_{f(pure)}] = -i \cdot K_f \cdot m$, where K_f is the freezing point depression constant of the solvent, i is the van't Hoff factor and m is the molality of the solute in the solution, with units of mol kg^{-1}. The mass (weight) percent A is equal to $\frac{\text{mass}_A}{\text{mass total}} \times 100$. The molar mass gives the equivalence statement g ↔ mol. The equivalence statements 1 kg ↔ 1000 g and °C = 5/9(°F - 32) are needed.

Plan: If we convert the temperature to °C, we can rearrange the equation and solve for the necessary molality.

Step 4. Add the mass of ethylene glycol to that of water to find the total solution mass and use the definition of mass percent.
Step 3. Assume 1 kg of solvent and determine the mass of ethylene glycol in that solvent mass, using the factor-label method.
Step 2. Rearrange the equation for freezing point depression to solve for the molality, insert the known values and calculate.
Step 1. Determine the freezing point of the solution in °C.

Work:

Step 1. Determine the freezing point of the solution in °C.

? °C = 5/9(°F - 32) = 5/9(0°F - 32) = 5/9(-32) = -18 °C

Step 2. Determine the molality of the ethylene glycol solution. Assume this is a nonelectrolye (it is), so i = 1.

$$[T_{f(solution)} - T_{f(pure)}] = -i \cdot K_f \cdot m\text{; therefore, } m = \frac{[T_{f(solution)} - T_{f(pure)}]}{-i \cdot K_f}$$

$$?\ \text{mol kg}^{-1} = \frac{-18\ °C - 0.00\ °C}{(1)(-1.86\ °C\ \text{kg mol}^{-1})} = \mathbf{9.6\ mol\ kg^{-1}} = \mathbf{9.6}\ m$$

Step 3. Determine the mass of ethylene glycol in 1 kg H_2O.
Molar Mass $C_2H_6O_2$ = 2(12.011 C) + 6(1.0079 H) + 2(15.9994 O) = 62.068 g mol^{-1}

$$?\ g\ C_2H_6O_2 = \frac{9.6\ \text{mol}\ C_2H_6O}{1\ \text{kg}\ H_2O} \times \frac{62.068\ g\ C_2H_6O_2}{1\ \text{mol}\ C_2H_6O_2} \times 1\ \text{kg}\ H_2O =$$

5.9×10^2 g $C_2H_6O_2$

Step 4. Determine the mass percent ethylene glycol.
There are 5.9×10^2 g ethylene glycol in 1 kg = 1000 g H_2O.

The total solution mass is 5.9×10^2 g $C_2H_6O_2$ + 1000 g H_2O = 1.59×10^3 g solution.

$$?\ \%\ C_2H_6O_2 = \frac{5.9 \times 10^2\ g\ C_2H_6O_2}{1.59 \times 10^3\ g\ \text{solution}} \times 100 = \mathbf{37\%\ C_2H_6O_2}$$

12.89 If 0.28 g of an ionic compound with empirical formula Hg_2SO_4 is dissolved in 100.0 g water, the resulting solution has a freezing point depression of 0.021 °C. How many ions are formed when one formula unit of this substance dissolves?

Analysis:

Target: The number of ions the solute forms. (? i) =
Knowns: The freezing point depression of the solution, the mass of solute and the mass of water are known. The freezing point depression constant can be looked up in Table 12.5 and is 1.86 °C kg mol^{-1}. The molar mass of Hg_2SO_4 is an implicit known.
Relationships: $\Delta T_f = -i \cdot K_f \cdot m$, where K_f is the freezing point depression constant of the solvent, i is the van't Hoff factor which indicates the number of ions formed and m is the molality of the solute, with units of mol kg^{-1}. The molar mass gives the equivalence statement g ↔ mol. The equivalence statement 1 kg ↔ 1000 g is needed.

Plan: If we use the factor-label method to solve for the molality of the solution, we can rearrange the freezing point depression relationship to solve for i, insert the known values and calculate.
Step 2. Use the freezing point depression relationship to solve for i.
Step 1. Use the factor-label method to determine the molality of Hg_2SO_4.

Work:

Step 1. Determine the molality of the Hg_2SO_4 solution.
Molar Mass Hg_2SO_4 = 2(200.59 Hg) + 1(32.06 S) + 4(15.9994 O) = 497.24 g mol^{-1}

$$? \text{ mol kg}^{-1} \text{ Hg}_2\text{SO}_4 = \frac{0.28 \text{ g Hg}_2\text{SO}_4}{100.0 \text{ g H}_2\text{O}} \times \frac{1 \text{ mol Hg}_2\text{SO}_4}{497.24 \text{ g Hg}_2\text{SO}_4} \times \frac{1000 \text{ g}}{1 \text{ kg}} =$$

$$5.6 \times 10^{-3} \text{ mol kg}^{-1} = 5.6 \times 10^{-3}\ m$$

Step 2. Determine i.

$$\Delta T_f = -i \cdot K_f \cdot m; \text{ therefore, } i = \frac{\Delta T_f}{-K_f \cdot m} = \frac{-0.021\ °C}{-(1.86\ °C \text{ kg mol}^{-1})(5.6 \times 10^{-3} \text{ mol kg}^{-1})}$$

i = **2.0** (The two ions formed are Hg_2^{2+} and SO_4^{2-}.)

12.91 Bromobenzene normally boils at 156.43 °C, but when 1.008 g benzoic acid ($C_6H_5CO_2H$) is dissolved in 65.32 g bromobenzene, the solution boils at 157.22 °C. When 2.3 g of an unknown molecular substance is dissolved in 50 g bromobenzene, the solution boils at 157.81 °C. What is the molar mass of the unknown?

Analysis:

Target: The molar mass of an unknown. (? g mol^{-1}) =

Knowns: The mass of benzoic acid, the mass of bromobenzene, the boiling point of pure bromobenzene, and the boiling point of that solution are given. The mass of an unknown solute, the mass of bromobenzene, and the boiling point of the unknown solution are given. The molar mass of benzoic acid is an implicit known.

Relationships: $[T_{b(solution)} - T_{b(pure)}] = i \cdot K_b \cdot m$, where K_b is the boiling point elevation constant of the solvent, i is the van't Hoff factor and m is the molality of the solute in the solution, with units of mol kg^{-1}. The equivalence statement 1 kg ↔ 1000 g is needed. The molar mass gives the equivalence statement g ↔ mol.

Plan: If we knew the boiling point elevation constant, we could find the molality of the unknown and proceed as in Problem 12.77. The benzoic acid solution gives us the information to determine the K_b.

Step 5. Calculate the molar mass from the solute mass and number of moles.
Step 4. Determine the number of moles of solute from the molality and solvent mass.
Step 3. Calculate the molality from the known data and given relationship.
Step 2. Rearrange the boiling point depression expression to solve for K_b, insert the known values and calculate.
Step 1. Use the factor-label method to determine the molality of the benzoic acid solution.

Work:

Step 1. Determine the molality of the benzoic acid solution.

Molar Mass $C_6H_5CO_2H$ = 7(12.011 C) + 6(1.0079 H) + 2(15.9994 O) = 122.123 g mol^{-1}

$$? \text{ mol kg}^{-1} = m = \frac{1.008 \text{ g } C_6H_5CO_2H}{65.32 \text{ g bromobenzene}} \times \frac{1 \text{ mol } C_6H_5CO_2H}{122.123 \text{ g } C_6H_5CO_2H} \times$$

$$\frac{1000 \text{ g bromobenzene}}{1 \text{ kg bromobenzene}} = 0.1264\ m\ C_6H_5CO_2H$$

Step 2. Rearrange the boiling point depression expression to solve for K_b, insert the known values and calculate. Assume that i = 1 for benzoic acid. (It does.)

$$[T_{b(solution)} - T_{b(pure)}] = i \cdot K_b \cdot m; \text{ therefore, } K_b = \frac{[T_{b(solution)} - T_{b(pure)}]}{i \cdot m}$$

$$?\ °C \text{ kg mol}^{-1} = K_b = \frac{157.22\ °C - 156.43\ °C}{(1)(0.1264\ m\ C_6H_5CO_2H)} = 6.25\ °C \text{ kg mol}^{-1}$$

Step 3. Rearrange the relationship to solve for the molality, insert values and calculate. Assume the solute is a nonelectrolyte, so i = 1.

$$[T_{b(solution)} - T_{b(pure)}] = i \cdot K_b \cdot m; \text{ therefore, } m = \frac{[T_{b(solution)} - T_{b(pure)}]}{i \cdot K_b}$$

$$? \text{ mol kg}^{-1} = m = \frac{157.81 \text{ °C} - 156.43 \text{ °C}}{(1)(6.25\text{°C kg mol}^{-1})} = 0.221 \text{ mol kg}^{-1}$$

Step 4. Use the factor-label method to determine the number of moles of solute.

$$? \text{ mol solute} = \frac{0.221 \text{ mol solute}}{\text{kg bromobenzene}} \times 50 \text{ g bromobenzene} \times \frac{1 \text{ kg bromobenzene}}{1000 \text{ g bromobenzene}} =$$

$$1.1 \times 10^{-2} \text{ mol}$$

Step 5. Determine the molar mass.

$$? \text{ g mol}^{-1} = \frac{2.3 \text{ g}}{1.1 \times 10^{-2} \text{ mol}} = \mathbf{2.1 \times 10^2 \text{ g mol}^{-1}}$$

12.93 Glutamine synthetase, an enzyme involved in energy transport in animal metabolism, has a molar mass of 592,000 g mol^{-1}. How many grams of this material must be dissolved to produce one liter of an aqueous solution whose (room temperature) osmotic pressure is 1.5 atm? Is this a reasonable value?

Analysis:

Target: A mass of a solute. (? g) =

Knowns: The molar mass of the solute, the osmotic pressure and the solution volume. The value of R = 0.0820578 L atm mol^{-1} K^{-1}. Room temperature is about 25 °C = 298 K.

Relationships: $\Pi = MRT$. The molar mass gives the equivalence statement g ↔ mol.

Plan: Rearrange the equation to determine M, then use the factor-label method to determine the mass of solute needed.

Work:

$$\Pi = MRT; \text{ therefore, } M = \frac{\Pi}{RT}$$

$$? \text{ mol L}^{-1} = M = \frac{1.5 \text{ atm}}{(0.0820578 \text{ L atm mol}^{-1} \text{ K}^{-1})(298 \text{ K})} = 6.1 \times 10^{-2} \text{ mol L}^{-1}$$

Now use the factor-label method to find the necessary mass of solute.

$$? \text{ g solute} = (6.1 \times 10^{-2} \text{ mol L}^{-1})(1.00 \text{ L}) \times \frac{592000 \text{ g solute}}{1 \text{ mol solute}} = \mathbf{3.6 \times 10^4 \text{ g}}$$

This is not likely because dissolving that much solute to make a liter of solution would give a density of $\frac{3.6 \times 10^4 \text{ g}}{1 \text{ L}} \times \frac{1 \text{ L}}{1000 \text{ mL}} = 36 \text{ g mL}^{-1}$. Because the most dense metal only has a density on the order of 22 g mL^{-1}, it is highly unlikely that a solution with a density this high could be obtained.

13 Equilibrium

13.1 Name three features that describe all equilibrium systems.

Any equilibrium system must be reversible. It must be characterized by equal rates of opposing processes (rate forward = rate reverse). It must be independent of the direction from which the equilibrium was approached.

13.3 What is meant by the term "dynamic equilibrium"?

In a dynamic equilibrium, opposing processes occur at equal rates, resulting in no net change in the amounts of reactants or products.

13.5 Explain the statement, "A system at equilibrium has no memory of how the condition of equilibrium was attained."

This means that for a given amount of material, the same equilibrium state is reached, regardless of whether the material was originally all reactants, all products or a mixture of reactants and products.

13.7 What is meant by the phrase, "solute distribution equilibrium"?

If a substance which is soluble in two immiscible liquids is added to a mixture of the liquids, it will dissolve in each solvent to some extent. The ratio of the different concentrations in the two solvents is the solute distribution equilibrium.

13.9 What can be said about the magnitude of the equilibrium constant in a reaction whose equilibrium lies far to the right? To the left?

If the equilibrium lies far to the right, $K_{eq} \gg 1$. If the equilibrium lies far to the left, $K_{eq} \ll 1$.

13.11 Write the equilibrium expression for each of the following reactions.

a. $2\ HgO(s) \rightleftharpoons 2\ Hg(\ell) + O_2(g)$

b. $C(s) + H_2O(g) \rightleftharpoons CO(g) + H_2(g)$

c. $HCO_2H(aq) + H_2O(\ell) \rightleftharpoons HCO_2^-(aq) + H_3O^+(aq)$

Analysis:

Target: An equilibrium expression. (? K_{eq}) =

Known: The balanced equation and the states of all reactants and products.

Relationship: For a generalized reaction, a A(s)+ b B(g) $\rightleftharpoons$ c C(aq) + d D(g),

$K_{eq} = \dfrac{[C]^c \cdot P_D{}^d}{P_B{}^b}$. Gases are shown in terms of partial pressure and dissolved species are shown in terms of concentrations, raised to the power of their coefficients. Pure solids, pure liquids and solvents in dilute solutions are omitted from the K_{eq} expression.

Plan: Study the equation, apply these rules, and write the K_{eq} expression.

Work:

a. ? $K_{eq} = P_{O_2}$ (HgO, a solid, and Hg, a liquid, are omitted from the expression.)

b. ? $K_{eq} = \dfrac{P_{H_2} \cdot P_{CO}}{P_{H_2O}}$ (C, a solid, is omitted from the K_{eq} expression.)

c. ? $K_{eq} = \dfrac{[HCO_2^-][H_3O^+]}{[HCO_2H]}$ (H_2O, the solvent, is omitted from the K_{eq} expression.)

13.13 Write the equilibrium expression for each of the following reactions.

a. $Cl_2(g) + 2\ Br^-(aq) \rightleftharpoons Br_2(aq) + 2\ Cl^-(aq)$
b. $CuO(s) + H_2(g) \rightleftharpoons Cu(s) + H_2O(g)$
c. $H_2(g) + S(s) \rightleftharpoons H_2S(g)$

Analysis:

Target: An equilibrium expression. (? K_{eq}) =
Known: The balanced equation and the states of all reactants and products.
Relationship: For a generalized reaction, a A(s)+ b B(g) $\rightleftharpoons$ c C(aq) + d D(g), $K_{eq} = \frac{[C]^c \cdot P_D^d}{P_B^b}$. Gases are shown in terms of partial pressure and dissolved species are shown in terms of concentrations, raised to the power of their coefficients. Pure solids, pure liquids and solvents in dilute solutions are omitted from the K_{eq} expression.

Plan: Study the equation, apply these rules, and write the K_{eq} expression.
Work:

a. $? K_{eq} = \frac{[Br_2]\,[Cl^-]^2}{P_{Cl_2}[Br^-]^2}$

b. $? K_{eq} = \frac{P_{H_2O}}{P_{H_2}}$ (CuO and Cu, both solids, are omitted from the K_{eq} expression.)

c. $? K_{eq} = \frac{P_{H_2S}}{P_{H_2}}$ (S, a solid, is omitted from the K_{eq} expression.)

13.15 Identify each reaction in problems 11 and 13 as heterogeneous or homogeneous.

A reaction is heterogeneous if two or more different states are present.
11a, with solid, liquid and gas, is heterogeneous.
11b, with solid and gas, is heterogeneous.
11c, with only aqueous solutions and water is homogeneous.
13a, with gas and aqueous solution, is heterogeneous.
13b, with solid and gas, is heterogeneous.
13c, with solid and gas, is heterogeneous.

13.17 The equilibrium constant for the reaction $COCl_2 \rightleftharpoons CO + Cl_2$ is 6.12×10^{-9} atm. What is the value of the equilibrium constant for the reaction $CO + Cl_2 \rightleftharpoons COCl_2$?

Analysis:

Target: An equilibrium constant for a reaction. (? K_{eq}) =
Known: The K_{eq} for a related reaction.
Relationships: If an equation is reversed, $K_{eq,reverse} = (K_{eq,forward})^{-1}$. If a reaction is multiplied by a factor, n, $K_{eq,n} = (K_{eq,original})^n$.

Plan: Determine the relationship of the new reaction to the original reaction and apply the appropriate relationship or combination of relationships.
Work:

The target reaction is the reverse of the given reaction, $K_{eq,reverse} = (K_{eq,forward})^{-1}$.

$? K_{eq} = (6.12 \times 10^{-9}\ atm)^{-1} = \mathbf{1.63 \times 10^{8}\ atm^{-1}}$

13.19 The equilibrium constant for the reaction $2\ SO_2 + O_2 \rightleftharpoons 2\ SO_3$ is 3.4 atm^{-1} at 1000 K. What is the value of the equilibrium constant for each of the following reactions?

a. $2\ SO_3 \rightleftharpoons 2\ SO_2 + O_2$
b. $SO_2 + 1/2\ O_2 \rightleftharpoons SO_3$

Analysis:

Target: An equilibrium constant for a reaction. (? K_{eq}) =

Known: The K_{eq} for a related reaction.

Relationships: If an equation is reversed, $K_{eq,\ reverse} = (K_{eq,forward})^{-1}$. If a reaction is multiplied by a factor, n, $K_{eq,n} = (K_{eq,\ original})^n$.

Plan: Determine the relationship of the new reaction to the original reaction and apply the appropriate relationship or combination of relationships.

Work:

a. The target reaction is the reverse of the given reaction, $K_{eq,reverse} = (K_{eq,forward})^{-1}$.

$? K_{eq} = (3.4\ atm^{-1})^{-1} = \mathbf{0.29\ atm}$

a. The target reaction is one half of the given reaction, $K_{eq,1/2} = (K_{eq,original})^{1/2}$.

$? K_{eq} = (3.4\ atm^{-1})^{1/2} = \mathbf{1.8\ atm^{-1/2}}$

13.21 The dissociation reaction $COCl_2 \rightleftharpoons CO + Cl_2$ has an equilibrium constant of 6.12×10^{-9} atm at 100 °C. For each of the following mixtures, indicate whether there is equilibrium, or, if not, in which direction reaction will occur.

	P_{COCl_2}, atm	P_{CO}, atm	P_{Cl_2}, atm
a.	5×10^{-4}	2×10^{-4}	3×10^{-2}
b.	5×10^{-6}	3×10^{-5}	0.32
c.	2×10^{-3}	5×10^{-8}	6×10^{-5}

Analysis:

Target: To determine whether or not the reaction is at equilibrium and, if it is not, which direction it must proceed to achieve equilibrium.

Knowns: K_{eq} for the reaction and initial pressures of all species.

Relationships: The mixture is at equilibrium if $Q = K_{eq}$. If $Q > K_{eq}$, then the reaction must shift to the left to make more reactants. If $Q < K_{eq}$, then the reaction must shift to the right to make more products.

Plan: Evaluate Q and compare it to K_{eq}.

Work:

For all combinations, $Q = \dfrac{P_{CO} \cdot P_{Cl_2}}{P_{COCl_2}}$.

a. $Q = \dfrac{(2 \times 10^{-4}\ atm)(3 \times 10^{-2}\ atm)}{(5 \times 10^{-4}\ atm)} = 1 \times 10^{-2}$ atm; $Q > K_{eq}$, shift to the **left**.

b. $Q = \dfrac{(3 \times 10^{-5}\ atm)(0.32\ atm)}{(5 \times 10^{-6}\ atm)} = 2$ atm; $Q > K_{eq}$, shift to the **left**.

c. $Q = \dfrac{(5 \times 10^{-8}\ atm)(6 \times 10^{-5}\ atm)}{(2 \times 10^{-3}\ atm)} = 2 \times 10^{-9}$ atm; $Q < K_{eq}$, shift to the **right**.

13.23 The reaction $CO_2 + H_2 \rightleftharpoons H_2O + CO$ has an equilibrium constant of 0.043 at 600 K. For each of the following mixtures, indicate whether there is equilibrium, or, if not, in which direction reaction will occur.

	P_{CO_2}, atm	P_{H_2}, atm	P_{H_2O}, atm	P_{CO}, atm
a.	0.5	0.5	0.25	0.25
b.	0.25	0.25	0.005	0.005
c.	0.15	0.15	0.15	0.3

Analysis:

Target: To determine whether or not the reaction is at equilibrium and, if it is not, which direction it must proceed to achieve equilibrium.

Knowns: K_{eq} for the reaction and initial concentrations of all species.

Relationships: The mixture is at equilibrium if $Q = K_{eq}$. If $Q > K_{eq}$, then the reaction must shift to the left to make more reactants. If $Q < K_{eq}$, then the reaction must shift to the right to make more products.

Plan: Evaluate Q and compare it to K_{eq}.

Work:

For all combinations, $Q = \dfrac{P_{H_2O} \cdot P_{CO}}{P_{CO_2} \cdot P_{H_2}}$.

a. $Q = \dfrac{(0.25 \text{ atm})(0.25 \text{ atm})}{(0.5 \text{ atm})(0.5 \text{ atm})} = 0.3$; $Q > K_{eq}$, shift to the **left**.

b. $Q = \dfrac{(0.005 \text{ atm})(0.005 \text{ atm})}{(0.25 \text{ atm})(0.25 \text{ atm})} = 4 \times 10^{-4}$; $Q < K_{eq}$, shift to the **right**.

c. $Q = \dfrac{(0.15 \text{ atm})(0.3 \text{ atm})}{(0.15 \text{ atm})(0.15 \text{ atm})} = 2$; $Q > K_{eq}$, shift to the **left**.

13.25 The standard free energy change for the hydrogenation of ethyne, $C_2H_2(g) + H_2(g) \rightleftharpoons C_2H_4(g)$, is -141 kJ at 298 K. Evaluate K_{eq} and comment on its magnitude.

Analysis:

Target: K_{eq} for the reaction. (? K_{eq}) =

Knowns: The standard free energy change, ΔG°_{rxn} and the temperature.
$R = 8.31451 \text{ J mol}^{-1} \text{ K}^{-1}$.

Relationships: $\ln K_{eq} = \dfrac{-\Delta G^{\circ}_{rxn}}{RT}$. To cancel units of "$\text{mol}^{-1}$" in R, we must rewrite ΔG°_{rxn} as -141 kJ mol^{-1}. See unit note in Section 13.3 of the text. The equivalence statement 1 kJ ↔ 1000 J is also needed.

Plan: Insert the known values into the given relationship, converting units as necessary, and calculate. The units of K_{eq} are evaluated by writing and analyzing the K_{eq} expression.

Work:

$-141 \text{ kJ mol}^{-1} \times \dfrac{1000 \text{ J}}{\text{kJ}} = -141000 \text{ J mol}^{-1}$

$K_{eq} = \dfrac{P_{C_2H_4}}{P_{C_2H_2} \cdot P_{H_2}} = \dfrac{\text{atm}}{(\text{atm})(\text{atm})} = \text{atm}^{-1}$

$$\ln K_{eq} = \frac{-(-141{,}000 \text{ J mol}^{-1})}{(8.31451 \text{ J mol}^{-1} \text{ K}^{-1})(298 \text{ K})} = 56.9$$

? $K_{eq} = e^{56.9} =$ $\mathbf{5.2 \times 10^{24}}$ **atm^{-1}**

The huge value of K_{eq} indicates that this process goes essentially to completion.

13.27 The equilibrium constant for the reaction $H_2(g) + I_2(g) \rightleftharpoons 2\,HI(g)$ is 620 at 298 K. Calculate ΔG°_{rxn} at 25 °C.

Analysis:

Target: ΔG°_{rxn} for the reaction at 298 K. (? kJ mol^{-1}) =

Knowns: K_{eq} and the temperature. $R = 8.31451 \text{ J mol}^{-1} \text{ K}^{-1}$.

Relationships: $\Delta G^{\circ}_{rxn} = -RT \ln K_{eq}$. The equivalence statement 1 kJ ↔ 1000J is needed.

Plan: Use the known relationship directly, using the factor-label method to convert to kJ.
Work:

$$? \text{ kJ mol}^{-1} = \Delta G^{\circ}_{rxn} = -RT \ln K_{eq} = [-(8.31451 \text{ J mol}^{-1} \text{ K}^{-1})(298 \text{ K}) \ln (620)] \times \frac{1 \text{ kJ}}{1000 \text{J}} =$$

-15.9 kJ mol^{-1}

13.29 At room temperature, the equilibrium constant for the reaction $CH_4(g) + Cl_2(g) \rightleftharpoons CH_3Cl(g) + HCl(g)$ is 7.42×10^{17}. Calculate ΔG°_{rxn} and $\Delta G^{\circ}_f(CH_3Cl)$, given that $\Delta G^{\circ}_f(CH_4) = -50.72$ kJ mol^{-1} and $\Delta G^{\circ}_f(HCl) = -95.30$ kJ mol^{-1}. Check your answer by looking up $\Delta G^{\circ}_f(CH_3Cl)$ in Appendix G.

Analysis:

Target: ΔG°_{rxn} and $\Delta G^{\circ}_f(CH_3Cl)$ for the reaction at 298.15 K. (? kJ) = and (? kJ mol^{-1}) =

Knowns: The K_{eq}, the temperature, and R = 8.31451 J mol^{-1} K^{-1}. ΔG°_f values for the other substances are given.

Relationships: $\Delta G^{\circ}_{rxn} = -RT \ln K_{eq}$. The equivalence statement 1 kJ ↔ 1000J is needed. $\Delta G^{\circ}_{rxn} = \Sigma[n\Delta G^{\circ}_f(\text{products})] - \Sigma[n\Delta G^{\circ}_f(\text{reactants})]$.

Plan: The two values to be determined require a two-step process.

Step 2. Rearrange $\Delta G^{\circ}_{rxn} = \Sigma[n\Delta G^{\circ}_f(\text{products})] - \Sigma[n\Delta G^{\circ}_f(\text{reactants})]$ to solve for $\Delta G^{\circ}_f(CH_3Cl)$, insert the known values and calculate.

Step 1. Determine ΔG°_{rxn} from the known relationship between ΔG°_{rxn} and K_{eq}, using the factor-label method to convert to kJ.

Work:

Step 1. Determine ΔG°_{rxn} from the relationship $\Delta G^{\circ}_{rxn} = -RT \ln K_{eq}$.

$$? \text{ kJ} = \Delta G^{\circ}_{rxn} = [-(8.31451 \text{ J mol}^{-1} \text{ K}^{-1})(298.15 \text{ K}) \ln (7.42 \times 10^{17})] \times \frac{1 \text{ kJ}}{1000 \text{J}} = \mathbf{-102.0 \text{ kJ}}$$

Step 2. Use the value of ΔG°_{rxn} to determine $\Delta G^{\circ}_f(CH_3Cl)$.

$$\Delta G^{\circ}_{rxn} = \Sigma[nG^{\circ}(\text{products})] - \Sigma[nG^{\circ}(\text{reactants})]$$

$$\Delta G^{\circ}_{rxn} = 1(\Delta G^{\circ}_f CH_3Cl(g)) + 1(\Delta G^{\circ}_f HCl(g)) - [1(\Delta G^{\circ}_f CH_4(g)) + 1(\Delta G^{\circ}_f Cl_2(g))]$$

$$-102.0 \text{ kJ} = 1(\Delta G^{\circ}_f CH_3Cl(g)) + 1(-95.30) - [1(-50.72) + 1(0)]$$

$$1(\Delta G^{\circ}_f CH_3Cl(g)) = -102.0 - (-95.30) - 50.72 + 0 = -70.4 \text{ kJ}$$

$$? \text{ kJ mol}^{-1} = \Delta G^{\circ}_f CH_3Cl(g)) = \frac{-70.4 \text{ kJ}}{1 \text{ mol}} = \mathbf{-57.4 \text{ kJ mol}^{-1}}$$

Appendix H gives a value of -57.37 kJ mol^{-1} for $\Delta G^{\circ}_f CH_3Cl(g)$). Our calculated value differs only in the number significant digits (limited by the value of K_{eq}).

13.31 The standard free energies of formation at 25 °C are 68.2 and -32.8 kJ mol^{-1} for gaseous C_2H_4 and C_2H_6, respectively. Calculate K_{eq} at 298 K for the hydrogenation reaction $C_2H_4(g) + H_2(g) \rightleftharpoons C_2H_6(g)$.

Analysis:

Target: K_{eq} for the reaction. (? K_{eq}) =

Knowns: The temperature of the process and ΔG_f° for C_2H_4 and C_2H_6. Because H_2 is an element, its $\Delta G_f^\circ = 0$.

Relationships: $\Delta G_{rxn}^\circ = \Sigma[n\Delta G_f^\circ(\text{products})] - \Sigma[n\Delta G_f^\circ(\text{reactants})]$ and $\ln K_{eq} = \frac{-\Delta G_{rxn}^\circ}{RT}$. The equivalence statement 1 kJ ↔ 1000 J is needed.

Plan: If we knew ΔG_{rxn}°, we could use the relationship to evaluate K_{eq}.

Step 2. Use the calculated value of ΔG_{rxn}° to determine K_{eq}, converting units as necessary. The units of K_{eq} are determined by writing and analyzing the K_{eq} expression.

Step 1. Use the ΔG_f° values to determine ΔG_{rxn}°.

Work:

Step 1. Determine ΔG_{rxn}° from $\Delta G_{rxn}^\circ = \Sigma[n\Delta G_f^\circ(\text{products})] - \Sigma[n\Delta G_f^\circ(\text{reactants})]$.

$$\Delta G_{rxn}^\circ = 1(\Delta G_f^\circ(C_2H_6(g)) - [1(\Delta G_f^\circ(C_2H_4(g)) + 1(\Delta G_f^\circ(H_2(g))]$$

$$\Delta G_{rxn}^\circ = 1(-32.8) - [1(68.2) + 1(0)] = -101.0 \text{ kJ}$$

Step 2. Determine K_{eq} from ΔG_{rxn}°.

$$-101.0 \text{ kJ} \times \frac{1000 \text{ J}}{\text{kJ}} = -101000 \text{ J} \qquad K_{eq} = \frac{P_{C_2H_6}}{P_{C_2H_4} \cdot P_{H_2}} = \frac{\text{atm}}{(\text{atm})(\text{atm})} = \text{atm}^{-1}$$

$$\ln K_{eq} = \frac{-(-101{,}000 \text{ J mol}^{-1})}{(8.31451 \text{ J mol}^{-1} \text{ K}^{-1})(298 \text{ K})} = 40.76$$

$$? K_{eq} = e^{40.76} = \mathbf{5.06 \times 10^{17} \text{ atm}^{-1}}$$

13.33 The reaction $2\ SO_2 + O_2 \rightleftharpoons 2\ SO_3$ is exothermic. What is the effect on the equilibrium of (a) increasing the temperature, (b) adding SO_3 to the system, (c) adding O_2 to the system, and (d) adding Ar to the system (at constant volume)?

Analysis:

Target: To decide if a reaction shifts to the right or left or does not change when a change is made to the system.

Knowns: Each of the disturbances applied and the balanced chemical equation. Δn_{gas} can be determined from the balanced equation.

Relationships: Le Châtelier's Principle can be used to determine the system response.

Plan: Determine the sign of Δn_{gas}. Apply Le Châtelier's Principle separately to each stress.

Work:

a. An increase in temperature decreases K_{eq} for an exothermic reaction. If K_{eq} decreases, then the system must shift **left** until $Q = K_{eq}$.

b. Adding SO_3 makes $Q > K_{eq}$. The equilibrium will shift **left** until $Q = K_{eq}$.

c. Adding O_2 makes $Q < K_{eq}$. The equilibrium will shift **right** until $Q = K_{eq}$.

d. Because the volume stays constant, there is **no change** because the partial pressures of the reactant and product gases do not change and Q is still equal to K_{eq}.

13.35 For each of the following reactions, tell in which direction the equilibrium shifts on (i) decreasing the volume, and (ii) increasing the temperature.

a. $3\ O_2(g) \rightleftharpoons 2\ O_3(g)$; endothermic
b. $2\ NO_2(g) \rightleftharpoons 2\ NO(g) + O_2(g)$; endothermic

Analysis:
Target: To decide if a reaction shifts to the right or left or does not change when a change is made to the system.
Knowns: Each of the disturbances applied and the balanced chemical equation. Δn_{gas} can be determined from the balanced equation.
Relationship: Le Châtelier's Principle can be used to determine the system response.
Plan: Determine the sign of Δn_{gas}. Apply Le Châtelier's Principle separately to each stress.
Work:
a. (i) A decrease in volume causes the system to respond by decreasing the number of moles of gas. For this system, $\Delta n_{gas} = 2 - (3) = -1$. The right side has fewer moles of gas, so the equilibrium would shift to the **right**.
(ii) An increase in temperature increases K_{eq} for an endothermic reaction. If K_{eq} increases, then the system must shift **right** until $Q = K_{eq}$.
b. (i) A decrease in volume causes the system to respond by decreasing the number of moles of gas. For this system, $\Delta n_{gas} = (2 + 1) - (2) = 1$. The left side has fewer moles of gas, so the equilibrium would shift to the **left**.
(ii) An increase in temperature increases K_{eq} for an endothermic reaction. If K_{eq} increases, then the system must shift **right** until $Q = K_{eq}$.

13.37 What is the effect on the equilibrium $CaCO_3(s) \rightleftharpoons CaO(s) + CO_2(g)$ of adding (a) CO_2, and (b) CaO [careful]?

Analysis:
Target: To decide if a reaction shifts to the right or left or does not change when a change is applied to the system.
Knowns: Each of the disturbances applied and the balanced chemical equation.
Relationship: Le Châtelier's Principle can be used to determine the system response.
Plan: Apply Le Châtelier's Principle separately to each stress.
Work:
a. Adding CO_2 makes $Q > K_{eq}$. The equilibrium will shift **left** until $Q = K_{eq}$.
b. Adding CaO does not change Q relative to K_{eq} because the solid does not appear in the equilibrium expression. Therefore, there is **no shift** when this disturbance is applied.

13.39 The equilibrium constant for the reaction $N_2O_4 \rightleftharpoons 2\ NO_2$ is 0.148 atm at 298 K, and ΔH°_{rxn} for the reaction is 57.2 kJ. Calculate the value of the equilibrium constant at 473 K.

Analysis:
Target: The value of K_{eq} at a different temperature. (? K_{eq}) =
Knowns: K_{eq} at one temperature, the second temperature and the enthalpy of reaction. $R = 8.31451\ J\ mol^{-1}\ K^{-1}$.
Relationships: $\ln\left(\frac{K_2}{K_1}\right) = \frac{\Delta H^\circ_{rxn}}{R}\cdot\left(\frac{1}{T_1} - \frac{1}{T_2}\right)$ The equivalence statement 1 kJ ↔ 1000 J is also needed.
Plan: Solve the relationship for the new K_{eq}, using the factor-label method to make necessary conversions.
Work:
Let $K_1 = 0.148$ atm at $T_1 = 298$ K; then K_2 is the unknown at $T_2 = 473$ K.

$$\Delta H^\circ_{rxn} = 57.2\ kJ\ mol^{-1} \times \frac{1000\ J}{1 kJ} = 57200\ J\ mol^{-1}$$

$$\ln\left(\frac{K_2}{0.148\ atm}\right) = \frac{(57200\ J\ mol^{-1})}{(8.31451\ J\ mol^{-1}\ K^{-1})}\cdot\left(\frac{1}{298\ K} - \frac{1}{473\ K}\right) = 8.54$$

$$\frac{K_2}{0.148\ \text{atm}} = e^{8.54}\text{; therefore, } K_2 = (e^{8.54})(0.148\ \text{atm}) = \mathbf{7.58 \times 10^2\ atm}$$

13.41 The equilibrium constant for the reaction $N_2O_4 \rightleftharpoons 2\ NO_2$ is 0.148 atm at 298 K, and ΔH°_{rxn} for the reaction is 57.2 kJ. At what temperature will the equilibrium constant be 1.52 atm?

Analysis:

Target: The temperature which will give a different K_{eq}. (? K) =

Knowns: K_{eq} at one temperature, a second K_{eq}, and the enthalpy of reaction. R = 8.31451 J mol^{-1} K^{-1}.

Relationships: $\ln\left(\frac{K_2}{K_1}\right) = \frac{\Delta H^\circ_{rxn}}{R} \cdot \left(\frac{1}{T_1} - \frac{1}{T_2}\right)$. The equivalence statement 1 kJ ↔ 1000 J is also needed.

Plan: Solve the relationship for the new temperature, using the factor-label method to make necessary conversions.

Work:

Let K_1 = 0.148 atm at T_1 = 298 K; and K_2 = 1.52 atm at T_2 = unknown

$$\Delta H^\circ_{rxn} = 57.2\ \text{kJ mol}^{-1} \times \frac{1000\ \text{J}}{1\text{kJ}} = 57200\ \text{J mol}^{-1}$$

$$\ln\left(\frac{1.52\ \text{atm}}{0.148\ \text{atm}}\right) = \frac{(57200\ \text{J mol}^{-1})}{(8.31451\ \text{J mol}^{-1}\ \text{K}^{-1})} \cdot \left(\frac{1}{298\ \text{K}} - \frac{1}{T_2}\right)$$

$$\ln(10.3) = 23.1 - \frac{6.88 \times 10^3}{T_2}\text{; therefore, } 2.329 = 23.1 - \frac{6.88 \times 10^3}{T_2}$$

$$2.329 - 23.1 = -\left(\frac{6.88 \times 10^3}{T_2}\right)\text{; and, } -20.8 = -\left(\frac{6.88 \times 10^3\ \text{K}}{T_2}\right)$$

$$T_2(-20.8) = -6.88 \times 10^3\ \text{K}$$

$$?\ \text{K} = T_2 = \frac{-6.88 \times 10^3\ \text{K}}{-20.8} = \mathbf{331\ K}$$

13.43 The equilibrium constant for the reaction $PCl_5 \rightleftharpoons PCl_3 + Cl_2$ is 3.0×10^{-7} atm at 298 K and 7.9×10^{-4} at 373 K. What is the value of ΔH°_{rxn} for this reaction?

Analysis:

Target: The change in enthalpy, ΔH°_{rxn}, of a process. (? kJ mol^{-1}) =

Knowns: K_{eq} at two different known temperatures. R = 8.31451 J mol^{-1} K^{-1}.

Relationships: $\ln\left(\frac{K_2}{K_1}\right) = \frac{\Delta H^\circ_{rxn}}{R} \cdot \left(\frac{1}{T_1} - \frac{1}{T_2}\right)$. The equivalence statement 1 kJ ↔ 1000 J is also needed.

Plan: Solve the relationship for ΔH°_{rxn}, using the factor-label method to make necessary conversions.

Work:

Let K_1 = 3.0×10^{-7} atm at T_1 = 298 K; then K_2 = 7.9×10^{-4} atm at T_2 = 373 K.

$$\ln\left(\frac{7.9 \times 10^{-4}\ \text{atm}}{3.0 \times 10^{-7}\ \text{atm}}\right) = \frac{\Delta H^\circ_{rxn}}{(8.31451\ \text{J mol}^{-1}\ \text{K}^{-1})} \cdot \left(\frac{1}{298\ \text{K}} - \frac{1}{373\ \text{K}}\right)$$

$\ln(2.6 \times 10^3) = 7.88 = \Delta H^\circ_{rxn}(8.2 \times 10^{-5}\ J^{-1}\ mol)$

$$? \text{ kJ mol}^{-1} = \Delta H^\circ_{rxn} = \frac{7.88}{8.2 \times 10^{-5}\ J^{-1}\ mol} \times \frac{1\ kJ}{1000\ J} = \mathbf{+97\ kJ\ mol^{-1}}$$

13.45 At its normal boiling point of 100 °C, the heat of vaporization of water is 40.66 kJ mol^{-1}. What is the equilibrium vapor pressure of water at 50 °C? [*Hint:* Treat the vaporization process as a chemical reaction. Recall that, by definition, the vapor pressure of a liquid at its normal boiling point is 1 atm.]

Analysis:

Target: The value of P_{H_2O} at 50 °C. (? atm) =

Knowns: P_{H_2O} at one temperature, the second temperature and the enthalpy of vaporization. R = 8.31451 J mol^{-1} K^{-1}.

Relationships: $\ln\left(\frac{K_2}{K_1}\right) = \frac{\Delta H^\circ_{rxn}}{R} \cdot \left(\frac{1}{T_1} - \frac{1}{T_2}\right)$. For vaporization $H_2O(\ell) \rightleftharpoons H_2O(g)$; so $K_{eq} = P_{H_2O}$. The equivalence statements 1 kJ ↔ 1000 J and K ↔ 273.15 + °C apply.

Plan: Rearrange the relationship to solve for the new P_{H_2O}, using the factor-label method to make necessary conversions.

Work:

Let K_1 = 1 atm (defined) at T_1 = 100 °C + 273.15 = 373 K.
Then K_2 is the unknown P_{H_2O} at T_2 = 50 °C + 273.15 = 323 K.

$$\Delta H^\circ_{rxn} = 40.66\ kJ\ mol^{-1} \times \frac{1000\ J}{1 kJ} = 40660\ J\ mol^{-1}$$

$$\ln\left(\frac{K_2}{1\ atm}\right) = \frac{(40660\ J\ mol^{-1})}{(8.31451\ J\ mol^{-1}\ K^{-1})} \cdot \left(\frac{1}{373\ K} - \frac{1}{323\ K}\right) = -2.0$$

$$\frac{K_2}{1\ atm} = e^{-2.0}; \text{ therefore, } K_2 = P_{H_2O} = (1\ atm)(e^{-2.0}) = \mathbf{0.13\ atm}$$

13.47 A reaction vessel is filled with H_2 and CO_2, each at a partial pressure of 0.75 atm. Later it is found that the pressure of H_2O is 0.25 atm.

a. What is the extent of the reaction $CO_2 + H_2 \rightleftharpoons CO + H_2O$?
b. What are the partial pressures of the other species present?

Analysis:

Target: The extent of the reaction, **x**, and the partial pressures of all species at equilibrium. (? atm) =

Knowns: The initial pressure of all species (CO and H_2O are zero), the equilibrium partial pressure of H_2O and the balanced chemical equation.

Relationships: The balanced chemical equation tells us the reaction stoichiometry. The extent of the reaction can be found from the initial and equilibrium pressures.

Plan: Problems having a known initial pressures and equilibrium pressures are always solved by the structured approach below to determine the extent of reaction and the K_{eq}.

Step 6. Insert the equilibrium pressures into the K_{eq} expression to obtain the value of K_{eq}. (We do not need Step 6 for this problem, but will require it in subsequent cases.)

Step 5. Use the known value of x to determine the extent of the reaction and equilibrium pressures.

Step 4. Add the change, **x**, to the initial amount to determine the equilibrium amount. This is the "equilibrium" line.

Step 3. Write the changes in the amount, in terms of **x**, the extent of the reaction. The multiplier of **x** is always the same as the stoichiometric coefficient in the balanced equation. This is the "change" line.

Step 2. Write the initial amount (pressure or concentration) of each substance under the formula of the substance. This is the "initial" line.
Step 1. Balance the equation if necessary.

Work:

		CO_2	+ H_2	$\rightleftharpoons$	CO	+	H_2O
Step 1.							
Step 2.	Initial	0.75	0.75		0		0
Step 3.	Change	- x	- x		+ x		+ x
Step 4	Equilibrium	0.75 - x	0.75 - x		0 + x		0 + x

Step 5. The equilibrium pressure of H_2O is 0.25 atm. Therefore, 0 + x = 0.25 atm, and **x**, the extent of the reaction = **0.25 atm**.

Determine the equilibrium partial pressures of the other species.

$? \text{ atm} = P_{CO_2} = P_{H_2} = 0.75 - x = 0.75 - 0.25 = \mathbf{0.50\ atm}$

$? \text{ atm} = P_{CO} = 0 + x = 0 + 0.25 = \mathbf{0.25\ atm}$

13.49 SO_2 and O_2 are put into a reaction vessel and allowed to react to produce SO_3. Initially, the pressure of SO_2 is 0.750 atm and that of O_2 is 0.350 atm. Later the pressure of SO_2 is determined to be 0.500 atm. Calculate the partial pressures of the other species, and the total pressure.

Analysis:

Target: Partial pressures of all species at equilibrium and the total pressure. (? atm) =

Knowns: The initial pressure of all species (SO_3 is zero), the equilibrium partial pressure of SO_2, and the balanced chemical equation.

Relationships: The balanced chemical equation tells us the reaction stoichiometry. The extent of the reaction can be found from the initial and equilibrium pressures. The total pressure is the sum of the partial pressures.

Plan: Use the structured approach as detailed in Problem 13.47. Also, add the partial pressures to determine the total pressure.

Work:

		$2\ SO_2$ +	O_2	$\rightleftharpoons$	$2\ SO_3$
Step 1.					
Step 2.	Initial	0.750	0.350		0
Step 3.	Change	- 2x	- x		+ 2x
Step 4.	Equilibrium	0.750 - 2x	0.350 - x		0 + 2x

Step 5. The equilibrium pressure of SO_2 is **0.500 atm**. Therefore, 0.750 - 2x = 0.500 atm.

Rearranging gives: $2x = 0.750 - 0.500 = 0.250$ atm and $x = \frac{0.250}{2} = 0.125$ atm.

Determine the equilibrium partial pressures of the other species.

$? \text{ atm} = P_{O_2} = 0.350 - x = 0.350 \text{ atm} - 0.125 \text{ atm} = \mathbf{0.225\ atm}$

$? \text{ atm} = P_{SO_3} = 0 + 2x = 0 + 2(0.125 \text{ atm}) = \mathbf{0.250\ atm}$

Determine the total pressure by adding the partial pressures.

$? \text{ atm} = P_{total} = P_{SO_2} + P_{O_2} + P_{SO_3} = 0.500 \text{ atm} + 0.225 \text{ atm} + 0.250 \text{ atm} =$

0.975 atm

13.51 The following equilibrium partial pressures were measured at 750 °C: $P_{H_2} = 0.387$ atm, $P_{CO_2} = 0.152$ atm, $P_{CO} = 0.180$ atm, and $P_{H_2O} = 0.252$ atm. What is the value of the equilibrium constant for the reaction $H_2 + CO_2 \rightleftharpoons CO + H_2O$?

Analysis:
Target: The equilibrium constant for the reaction. (? K_{eq}) =
Knowns: The balanced chemical equation and the equilibrium partial pressures of all species.
Relationship: K_{eq} is equal to the reaction quotient at equilibrium.
Plan: Write the equilibrium expression, insert the partial pressures and calculate.
Work:

$$? K_{eq} = \frac{P_{CO} \cdot P_{H_2O}}{P_{H_2} \cdot P_{CO_2}} = \frac{(0.180 \text{ atm})(0.252 \text{ atm})}{(0.387 \text{ atm})(0.152 \text{ atm})} = \mathbf{0.771}$$

13.53 Initially 0.750 atm, the partial pressure of NO decreased to 0.634 atm after the reaction $2\ NO(g) \rightleftharpoons O_2(g) + N_2(g)$ reached equilibrium. What is the value of the equilibrium constant?

Analysis:
Target: The equilibrium constant for the reaction. (? K_{eq}) =
Knowns: The initial pressure of all species (N_2 and O_2 are zero), the equilibrium partial pressure of NO, and the balanced chemical equation.
Relationships: The balanced chemical equation tells us the reaction stoichiometry. The extent of the reaction can be found from the initial and equilibrium pressures. K_{eq} is equal to the reaction quotient at equilibrium.
Plan: We can follow the structured approach detailed in Problem 13.47.
Work:

Step 1.		2 NO	$\rightleftharpoons$	O_2	+	N_2
Step 2.	Initial	0.750		0		0
Step 3.	Change	- 2x		+ x		+ x
Step 4.	Equilibrium	0.750 - 2x		0 + x		0 + x

Step 5. The equilibrium pressure of NO is 0.634 atm. Therefore, 0.750 - 2x = 0.634 atm.

Rearranging gives: $2x = 0.750 - 0.634 = 0.116$ atm. and $x = \frac{0.116}{2} = 0.0580$ atm.

Determine the equilibrium partial pressures of the other species.

$$? \text{ atm} = P_{O_2} = P_{N_2} = 0 + x = 0 + 0.0580 \text{ atm} = 0.0580 \text{ atm}$$

Step 6. Write the equilibrium expression, insert the partial pressures, and evaluate.

$$? K_{eq} = \frac{P_{O_2} \cdot P_{N_2}}{P_{NO}^2} = \frac{(0.0580 \text{ atm})(0.0580 \text{ atm})}{(0.634 \text{ atm})^2} = \mathbf{8.37 \times 10^{-3}}$$

13.55 If the equilibrium pressure of CO_2 over a mixture of solid CaO and $CaCO_3$ is 0.25 atm, what is the value of the equilibrium constant for the reaction $CaCO_3(s) \rightleftharpoons CaO(s) + CO_2$?

Analysis:
Target: The equilibrium constant for the reaction. (? K_{eq}) =
Knowns: The balanced chemical equation and the equilibrium partial pressure of CO_2.
Relationship: K_{eq} is equal to the reaction quotient at equilibrium. CaO and $CaCO_3$ are solids, so do not appear in the K_{eq} expression.
Plan: Write the equilibrium expression, insert the partial pressure, and evaluate.
Work:

$$? K_{eq} = P_{CO_2} = \mathbf{0.25\ atm}$$

13.57 K_{eq} for the reaction $COCl_2(g) \rightleftharpoons CO(g) + Cl_2(g)$ is 4.25 atm at 400 °C. If a vessel initially contains $COCl_2$ at a pressure of 1.60 atm, what are the partial pressures of all species after equilibrium has been attained?

Analysis:

Target: The partial pressures of all species at equilibrium. (? atm) =

Knowns: The initial pressures of all species (CO and Cl_2 are zero), the K_{eq} for the reaction, and the balanced chemical equation.

Relationships: The balanced chemical equation gives the reaction stoichiometry. K_{eq} is equal to the reaction quotient at equilibrium.

Plan: Problems having a known K_{eq} and initial pressures are always solved by the structured approach below.

Step 8. Check by inserting those equilibrium values into the reaction quotient and calculating. Q must equal K_{eq} at equilibrium.

Step 7. Evaluate the equilibrium amounts by means of the expressions in the equilibrium line.

Step 6. Solve the algebraic expression and obtain the numerical value of **x**.

Step 5. Write the equilibrium expression and insert the proper amounts from the equilibrium line.

Step 4. Add the change, **x**, to the initial amount to determine the equilibrium amount. This is the "equilibrium" line.

Step 3. Write the changes in the amount, in terms of **x**, the unknown extent of the reaction. The multiplier of **x** is always the same as the stoichiometric coefficient in the balanced equation. This is the "change" line.

Step 2. Write the initial amount (pressure or concentration) of each substance under the formula of the substance. This is the "initial" line.

Step 1. Balance the equation if necessary.

Work: Because only reactant is present initially, the reaction must shift right.

		$COCl_2$ $\rightleftharpoons$	CO +	Cl_2
Step 1.		$COCl_2$	CO	Cl_2
Step 2.	Initial	1.60	0	0
Step 3.	Change	- x	+ x	+ x
Step 4.	Equilibrium	1.60 - x	0 + x	0 + x

Step 5. Insert the values into the K_{eq} expression.

$$K_{eq} = \frac{P_{CO} \cdot P_{Cl_2}}{P_{COCl_2}} = \frac{(0 + x)(0 + x)}{1.60 - x} = 4.25$$

Step 6. Solve algebraically.

$$\frac{x^2}{1.60 - x} = 4.25, \quad x^2 = 4.25(1.60 - x) = 6.80 - 4.25x; \text{ and, } x^2 + 4.25x - 6.80 = 0,$$

This is a quadratic equation of the form $Ax^2 + Bx + C = 0$ and may be solved using the following relationship. $x = \frac{-B \pm \sqrt{B^2 - (4)(A)(C)}}{2(A)}$

$$x = \frac{-4.25 \pm \sqrt{4.25^2 - (4)(1)(-6.80)}}{2(1)} \qquad x = 1.24 \text{ atm or } -4.68 \text{ atm.}$$

A negative pressure is not possible, so the correct answer is x = 1.24 atm.

Step 7. Find the equilibrium partial pressures.

? atm = $P_{CO} = P_{Cl_2}$ = x = **1.24 atm**

? atm = P_{COCl_2} = 1.60 atm - x = 1.60 atm - 1.24 atm = **0.36 atm**

Step 8. Check: $Q = \frac{P_{CO} \cdot P_{Cl_2}}{P_{COCl_2}} = \frac{(1.24 \text{ atm})(1.24 \text{ atm})}{0.36 \text{ atm}} = 4.3 \text{ atm}$

13.59 For the reaction $H_2(g) + I_2(g) \rightleftharpoons 2\,HI(g)$ at 450 °C, the equilibrium constant is 65. Calculate the equilibrium pressures of all species in a vessel initially containing 0.500 atm HI.

Analysis:

Target: The partial pressures of all species at equilibrium. (? atm) =

Knowns: The initial pressures of all species (H_2 and I_2 are zero), the K_{eq} for the reaction, and the balanced chemical equation.

Relationships: The balanced chemical equation gives the reaction stoichiometry. K_{eq} is equal to the reaction quotient at equilibrium.

Plan: Use the structured approach detailed in Problem 13.57.

Work: Because only product is present initially, the reaction must shift left.

		H_2	+	I_2	$\rightleftharpoons$	2 HI
Step 1.		H_2	+	I_2		2 HI
Step 2.	Initial	0		0		0.500
Step 3.	Change	+ x		+ x		- 2x
Step 4.	Equilibrium	0 + x		0 + x		0.500 - 2x

Step 5. Insert the values into the K_{eq} expression.

$$K_{eq} = \frac{P_{HI}^2}{P_{H_2} \cdot P_{I_2}} = \frac{(0.500 - 2x)^2}{(0 + x)(0 + x)} = \frac{(0.500 - 2x)^2}{x^2} = 65$$

Step 6. Solve algebraically.

Because the left side is a perfect square, take the square root of each side.

$$\frac{(0.500 - 2x)}{x} = 8.1;\ \ 0.500 - 2x = 8.1\,x;\ \ 0.500 = 8.1x + 2x = 10.1\,x;$$

$$x = \frac{0.500}{10.1} = 0.0497 \text{ atm}$$

Step 7. Find the equilibrium partial pressures.

? atm = $P_{H_2} = P_{I_2} = x =$ **0.0497 atm**

? atm = $P_{HI} = 0.500 \text{ atm} - 2x = 0.500 \text{ atm} - 2(0.0497 \text{ atm}) =$ **0.401 atm**

Step 8. Check: $Q = \frac{P_{HI}^2}{P_{H_2} \cdot P_{I_2}} = \frac{(0.401 \text{ atm})^2}{(0.0497 \text{ atm})(0.0497 \text{ atm})} = 65.1$

13.61 At 600 °C, the equilibrium constant for the reaction $CO(g) + H_2O(g) \rightleftharpoons CO_2(g) + H_2(g)$ is 1.85. If 0.100 atm each of CO and H_2O is placed in a vessel and allowed to come to equilibrium, what are the partial pressures of all species?

Analysis:

Target: The partial pressures of all species at equilibrium. (? atm) =

Knowns: The initial pressures of all species (CO_2 and H_2 are zero), the K_{eq} for the reaction, and the balanced chemical equation.

Relationships: The balanced chemical equation gives the reaction stoichiometry. K_{eq} is equal to the reaction quotient at equilibrium.

Plan: Use the structured approach detailed in Problem 13.57.

Work: Because only reactant is present initially, the reaction must shift right.

		CO	+	H_2O	⇌	CO_2	+	H_2
Step 1.								
Step 2.	Initial	0.100		0.100		0		0
Step 3.	Change	- x		- x		+ x		+ x
Step 4.	Equilibrium	0.100 - x		0.100 - x		0 + x		0 + x

Step 5. Insert the values into the K_{eq} expression.

$$K_{eq} = \frac{P_{CO_2} \cdot P_{H_2}}{P_{CO} \cdot P_{H_2O}} = \frac{(0 + x)(0 + x)}{(0.100 - x)(0.100 - x)} = \frac{x^2}{(0.100 - x)^2} = 1.85$$

Step 6. Solve algebraically.
Because the left side is a perfect square, take the square root of each side.

$$\frac{x}{0.100 - x} = 1.36; \quad x = (0.100 - x)(1.36) = 0.136 - 1.36x; \quad x + 1.36x = 0.136$$

$$2.36x = 0.136; \text{ therefore, } x = \frac{0.136}{2.36} = 0.0576 \text{ atm}$$

Step 7. Find the equilibrium partial pressures.

? atm = $P_{H_2} = P_{CO_2} = x =$ **0.0576 atm**
? atm = $P_{CO} = P_{H_2O} = 0.100 \text{ atm} - x = 0.100 \text{ atm} - 0.0576 \text{ atm} =$ **0.042 atm**

Step 8. Check: $Q = \frac{P_{CO_2} \cdot P_{H_2}}{P_{CO} \cdot P_{H_2O}} = \frac{(0.0576 \text{ atm})(0.0576 \text{ atm})}{(0.042 \text{ atm})(0.042 \text{ atm})} = 1.9$

13.63 Suppose 0.100 atm each of CO, H_2, CO_2, and H_2O is placed in a reaction vessel and allowed to come to equilibrium. Use the value of K_{eq} from Problem 13.61 and calculate the equilibrium pressure of all species.

Analysis:
Target: The partial pressures of all species at equilibrium. (? atm) =
Knowns: The initial pressures of all species, the K_{eq} for the reaction, and the balanced chemical equation.
Relationships: The balanced chemical equation gives the reaction stoichiometry. K_{eq} is equal to the reaction quotient at equilibrium.
Plan: Compare Q to K_{eq} to determine which direction the reaction will shift. Then use the structured approach detailed in Problem 13.57.
Work:

$$Q = \frac{P_{CO_2} \cdot P_{H_2}}{P_{CO} \cdot P_{H_2O}} = \frac{(0.100 \text{ atm})(0.100 \text{ atm})}{(0.100 \text{ atm})(0.100 \text{ atm})} = 1.00$$

$Q < K_{eq}$, so the reaction shifts right.

		CO	+	H_2O	⇌	CO_2	+	H_2
Step 1.								
Step 2.	Initial	0.100		0.100		0.100		0.100
Step 3.	Change	- x		- x		+ x		+ x
Step 4.	Equilibrium	0.100 - x		0.100 - x		0.100 + x		0.100 + x

Step 5. Insert the values into the K_{eq} expression.

$$K_{eq} = \frac{P_{CO_2} \cdot P_{H_2}}{P_{CO} \cdot P_{H_2O}} = \frac{(0.100 + x)(0.100 + x)}{(0.100 - x)(0.100 - x)} = \frac{(0.100 + x)^2}{(0.100 - x)^2} = 1.85$$

Step 6. Solve algebraically.

Because the left side is a perfect square, take the square root of each side.

$$\frac{(0.100 + x)}{0.100 - x} = 1.36; \ 0.100 + x = (0.100 - x)(1.36) = 0.136 - 1.36x;$$

$$x + 1.36x = 0.136 - 0.100; \ 2.36x = 0.036; \ \text{therefore, } x = \frac{0.036}{2.36} = 0.015 \text{ atm}$$

Step 7. Find the equilibrium partial pressures.

? atm = $P_{H_2} = P_{CO_2} = 0.100 + x = 0.100$ atm + 0.015atm = **0.115 atm**
? atm = $P_{CO} = P_{H_2O} = 0.100$ atm - x = 0.100 atm - 0.015 atm= **0.085 atm**

Step 8. Check: $Q = \frac{P_{CO_2} \cdot P_{H_2}}{P_{CO} \cdot P_{H_2O}} = \frac{(0.115 \text{ atm})(0.115 \text{ atm})}{(0.085 \text{ atm})(0.085 \text{ atm})} = 1.8$

13.65 K_{eq} for the reaction $N_2O_4 \rightleftharpoons 2\ NO_2$ is 0.148 atm. What is the equilibrium pressure of NO_2 in a vessel originally containing 0.500 atm N_2O_4?

Analysis:
Target: The partial pressure of NO_2 at equilibrium. (? atm) =
Knowns: The initial pressures of all species (NO_2 is zero), the K_{eq} for the reaction, and the balanced chemical equation.
Relationships: The balanced chemical equation gives the reaction stoichiometry. K_{eq} is equal to the reaction quotient at equilibrium.
Plan: Use the structured approach detailed in Problem 13.57.
Work: Because only reactant is present initially, the reaction must shift right.

Step 1.		N_2O_4 $\rightleftharpoons$	$2\ NO_2$
Step 2.	Initial	0.500	0
Step 3.	Change	- x	+ 2x
Step 4.	Equilibrium	0.500 - x	0 + 2x

Step 5. Insert the values into the K_{eq} expression.

$$K_{eq} = \frac{P_{NO_2}^2}{P_{N_2O_4}} = \frac{(0 + 2x)^2}{(0.500 - x)} = 0.148$$

Step 6. Solve algebraically.

$$\frac{4x^2}{0.500 - x} = 0.148, \ 4x^2 = 0.148(0.500 - x) = 0.0740 - 0.148x$$

$4x^2 + 0.148x - 0.0740 = 0$ (This is a quadratic equation. See Step 6 of Problem 13.57.)

$$x = \frac{-0.148 \pm\sqrt{0.148^2 - (4)(4)(-0.0740)}}{2(4)} \qquad x = 0.119 \text{ atm or } -0.156 \text{ atm}$$

A negative pressure is not possible, so the correct answer is x = 0.119 atm.

Step 7. Find the equilibrium partial pressures.

? atm = P_{NO_2} = 2x = 2(0.119 atm) = **0.238 atm**
? atm = $P_{N_2O_4}$ = 0.500 atm - x = 0.500 atm - 0.119 atm = **0.381 atm**

Step 8. Check: $Q = \frac{P_{NO_2}^2}{P_{N_2O_4}} = \frac{(0.238 \text{ atm})^2}{(0.381 \text{ atm})} = 1.49$ atm

13.67 For the reaction $H_2(g) + Br_2(g) \rightleftharpoons 2\ HBr(g)$, the equilibrium constant is about 3500 at 1800 K. If the equilibrium partial pressure of H_2 is 0.00500 atm and that of Br_2 is 0.0500, what is the partial pressure of HBr and what is the total pressure?

Analysis:
Target: The partial pressure of HBr and the total pressure. (? atm) =
Knowns: The equilibrium pressure of H_2 and Br_2, the K_{eq} for the reaction, and the balanced chemical equation.
Relationships: The balanced chemical equation gives the reaction stoichiometry. K_{eq} is equal to the reaction quotient at equilibrium. The total pressure is equal to the sum of the partial pressures.
Plan: Only the pressure of HBr at equilibrium is unknown. Rearrange the K_{eq} expression to solve for the pressure of HBr and calculate. Add the partial pressures to determine the total pressure.
Work:

$$K_{eq} = \frac{P_{HBr}^2}{P_{H_2} \cdot P_{Br_2}} = \frac{P_{HBr}^2}{(0.00500\ \text{atm})(0.0500\ \text{atm})} = 3500$$

$$P_{HBr}^2 = (3500)(0.00500)(0.0500) = 0.875$$

$$?\ \text{atm} = P_{HBr} = \sqrt{0.875} = \mathbf{0.935\ atm}$$

$$\text{Total pressure} = P_{HBr} + P_{H_2} + P_{Br_2} = 0.935\ \text{atm} + 0.00500\ \text{atm} + 0.0500\ \text{atm} = \mathbf{0.990\ atm}$$

13.69 At 700 °C, K_{eq} is 1.50 atm for the reaction $C(s) + CO_2(g) \rightleftharpoons 2\ CO(g)$. If neither CO_2 nor C is present initially, and the initial pressure of CO is 1.75 atm, what is the equilibrium pressure of CO_2?

Analysis:
Target: The partial pressure of CO_2 at equilibrium. (? atm) =
Knowns: The initial pressures of all species (CO_2 is zero), the K_{eq} for the reaction, and the balanced chemical equation.
Relationships: The balanced chemical equation gives the reaction stoichiometry. K_{eq} is equal to the reaction quotient at equilibrium.
Plan: Use the structured approach detailed in Problem 13.57.
Work: Because only product is present initially, the reaction must shift left.

		C	+	CO_2	$\rightleftharpoons$	2 CO
Step 1.						
Step 2.	Initial	*		0		1.75
Step 3.	Change	*		+ x		- 2x
Step 4.	Equilibrium	*		0 + x		1.75 - 2x

***** **Because carbon is a solid, it does not have a partial pressure.**

Step 5. Insert the values into the K_{eq} expression.

$$K_{eq} = \frac{P_{CO}^2}{P_{CO_2}} = \frac{(1.75 - 2x)^2}{x} = 1.50\ \text{atm}$$

Step 6. Solve algebraically.

$$\frac{3.06 - 7x + 4x^2}{x} = 1.50;\ 3.06 - 7x + 4x^2 = 1.50x;\ 4x^2 - 7x - 1.50x + 3.06 = 0,$$

$4x^2 - 8.50x + 3.06 = 0$ (This is a quadratic equation. See Step 6 of Problem 13.57.)

$$x = \frac{-(-8.50) \pm\sqrt{(-8.50)^2 - (4)(4)(3.06)}}{2(4)} \qquad x = 0.459\ \text{atm or}\ 1.67\ \text{atm}$$

The second root would give a negative pressure for CO (1.76 - 2(1.67)). That is not possible, so the correct answer is x = 0.459 atm.

Step 7. Find the equilibrium partial pressures.

? atm = P_{CO_2} = x = **0.459 atm**
? atm = P_{CO} = 1.75 atm - 2x = 1.75 atm - 2(0.459) atm = 0.83 atm

Step 8. Check: $Q = \frac{P_{CO}^2}{P_{CO_2}} = \frac{(0.83)^2}{0.459} = 1.5$ atm

13.71 The reaction $N_2 + O_2 \rightleftharpoons 2\ NO$ has an equilibrium constant of 0.025 at 2400 K. If a vessel initially contains 0.0100 atm each of N_2 and O_2, what are the equilibrium pressures of all species?

Analysis:
Target: The partial pressures of all species at equilibrium. (? atm) =
Knowns: The initial pressures of all species (NO is zero), the K_{eq} for the reaction, and the balanced chemical equation.
Relationships: The balanced chemical equation gives the reaction stoichiometry. K_{eq} is equal to the reaction quotient at equilibrium.
Plan: Use the structured approach detailed in Problem 13.57.
Work: Because only product is present initially, the reaction must shift left.

		N_2	+	O_2	$\rightleftharpoons$	2 NO
Step 1.						
Step 2.	Initial	0.0100		0.0100		0
Step 3.	Change	- x		- x		+ 2x
Step 4.	Equilibrium	0.0100 - x		0.0100 - x		0 + 2x

Step 5. Insert the values into the K_{eq} expression.

$$K_{eq} = \frac{P_{NO}^2}{P_{N_2} \cdot P_{O_2}} = \frac{(2x)^2}{(0.0100 - x)(0.0100 - x)} = \frac{4x^2}{(0.0100 - x)^2} = 0.025$$

Step 6. Solve algebraically.

Because the left side is a perfect square, take the square root of each side.

$$\frac{(2x)}{0.0100 - x} = 0.16;\ \ 2x = 0.16(0.0100 - x) = 0.0016 - 0.16x;\ \ 2x + 0.16x = 0.0016$$

$$2.16x = 0.0016;\ \ x = \frac{0.0016}{2.16} = 7.4 \times 10^{-4} \text{ atm}$$

Step 7. Find the equilibrium partial pressures.

? atm = $P_{N_2} = P_{O_2}$ = 0.0100 atm - x = 0.0100 atm - 7.4×10^{-4} atm = **9.3×10^{-3} atm**
? atm = P_{NO} = 2x = 2(7.4×10^{-4} atm) = **1.5×10^{-3} atm**

Step 8. Check: $Q = \frac{P_{NO}^2}{P_{N_2} \cdot P_{O_2}} = \frac{(1.5 \times 10^{-3} \text{ atm})^2}{(9.3 \times 10^{-3} \text{ atm})(9.3 \times 10^{-3} \text{ atm})} = 0.026$

13.73 The reaction $N_2 + O_2 \rightleftharpoons 2\ NO$ has an equilibrium constant of 0.025 at 2400 K. If the initial concentrations of N_2 and O_2 are both 0.0100 M (and no NO is present), what are the equilibrium concentrations of all species? Note that this problem is similar to 13.71, but the amounts are given as concentrations rather than as partial pressures.

Analysis:

Target: The concentrations of all species at equilibrium. (? M) =

Knowns: The initial concentrations of all species (NO is zero), the K_{eq} for the reaction, and the balanced chemical equation.

Relationships: The balanced chemical equation gives the reaction stoichiometry. K_{eq} is equal to the reaction quotient at equilibrium.

Plan: Use the structured approach detailed in Problem 13.57. Note that we can use the same value of K_{eq} for concentrations as for pressures, because this K_{eq} is dimensionless.

Work: Because only product is present initially, the reaction must shift left.

		N_2	+	O_2	$\rightleftharpoons$	2 NO
Step 1.						
Step 2.	Initial	0.0100		0.0100		0
Step 3.	Change	- x		- x		+ 2x
Step 4.	Equilibrium	0.0100 - x		0.0100 - x		0 + 2x

Step 5. Insert the values into the K_{eq} expression.

$$K_{eq} = \frac{[NO]^2}{[N_2]\cdot[O_2]} = \frac{(2x)^2}{(0.0100 - x)(0.0100 - x)} = \frac{4x^2}{(0.0100 - x)^2} = 0.025$$

Step 6. Solve algebraically.

Because the left side is a perfect square, take the square root of each side.

$$\frac{(2x)}{0.0100 - x} = 0.16;\ 2x = 0.16(0.0100 - x) = 0.0016 - 0.16x;\ 2x + 0.16x = 0.0016;$$

$$2.16\,x = 0.0016;\quad x = \frac{0.0016}{2.16} = 7.4 \times 10^{-4}\ M$$

Step 7. Find the equilibrium concentrations.

$P_{N_2} = P_{O_2} = 0.0100\ M - x = 0.0100\ M - 7.4 \times 10^{-4}\ M = \mathbf{9.3 \times 10^{-3}\ M}$

$P_{NO} = 2x = 2(7.4 \times 10^{-4}\ M) = \mathbf{1.5 \times 10^{-3}\ M}$

Step 8. Check: $Q = \frac{[NO]^2}{[N_2]\cdot[O_2]} = \frac{(1.5 \times 10^{-3}\ M)^2}{(9.3 \times 10^{-3}\ M)(9.3 \times 10^{-3}\ M)} = 0.026$

13.75 At 25 °C, K_{eq} for the reaction $N_2O_4 \rightleftharpoons 2\ NO_2$ is 0.148 atm. What is the equilibrium concentration of NO_2 in a vessel originally containing 0.500 M N_2O_4? Note that the units of the equilibrium constant are different from the units in which the amounts of the chemical species are given. One way of dealing with this is to convert all concentrations to pressures using the ideal gas law, P = MRT.

Analysis:

Target: The concentration of NO_2 at equilibrium. (? M) =

Knowns: The initial concentrations of all species (NO_2 is zero), the K_{eq} for the reaction, and the balanced chemical equation. R = 0.0820578 L atm mol^{-1} K^{-1}.

Relationships: The balanced chemical equation gives the reaction stoichiometry. K_{eq} is equal to the reaction quotient at equilibrium. K ↔ 273.15 + °C.

Plan: Initially, convert the initial concentration to pressure using the ideal gas law. Then use the structured approach detailed in Problem 13.57. Finally, convert the equilibrium pressure to concentration using the ideal gas law.

Work:

Initial Step: Convert the concentration to pressure.

$P = MRT = (0.500\ M)(0.0820578\ L\ atm\ mol^{-1}\ K^{-1})(273.15 + 25\ °C) = 12.2\ atm$

Because only reactant is present initially, the reaction must shift right.

		$N_2O_4 \rightleftharpoons$	$2\ NO_2$
Step 1.			
Step 2.	Initial	12.2	0
Step 3.	Change	- x	+ 2x
Step 4.	Equilibrium	12.2 - x	0 + 2x

Step 5. Insert the values into the K_{eq} expression.

$$K_{eq} = \frac{(P_{NO_2})^2}{P_{N_2O_4}} = \frac{(0 + 2x)^2}{(12.2 - x)} = 0.148$$

Step 6. Solve algebraically.

$$\frac{4x^2}{12.2 - x} = 0.148;\ 4x^2 = 0.148(12.2 - x) = 1.81 - 0.148x;\ \text{and, } 4x^2 + 0.148x - 1.81 = 0$$

(This is a quadratic equation. See Step 6 of Problem 13.57.)

$$x = \frac{-0.148 \pm\sqrt{(0.148)^2 - (4)(4)(-1.81)}}{2(4)}.$$ Therefore, x = 0.654 atm or -0.691 atm.

A negative pressure is not possible, so the correct answer is x = 0.654 atm.

Step 7. Find the equilibrium partial pressures.

$P_{N_2O_4}$ = 12.2 atm - x = 12.2 atm - 0.654 atm = 11.5 atm
P_{NO_2} = 2x = 2(0.654 atm) = 1.31 atm

Step 8. Check: $Q = \frac{(P_{NO_2})^2}{P_{N_2O_4}} = \frac{(1.31\ \text{atm})^2}{(11.5\ \text{atm})} = 0.149$

Finally. Convert the pressure of NO_2 to concentration using the ideal gas law.

$P = MRT$, therefore, $M = \frac{P}{RT}$.

$$?\ M\ NO_2 = \frac{(1.31\ \text{atm})}{(0.0820578\ \text{L atm mol}^{-1}\ \text{K}^{-1})(273.15 + 25\ °\text{C})} = \mathbf{0.0535\ M}$$

13.77 Suppose $\Delta G^\circ_{rxn} = 0$ for some reaction. What is the value of K_{eq}?

Analysis:
Target: K_{eq} for the reaction. (? K_{eq}) =
Knowns: The standard free energy change, ΔG°_{rxn}. R = 8.31451 J mol^{-1} K^{-1}.

Relationship: $\ln K_{eq} = \frac{\Delta G^\circ_{rxn}}{RT}$.

Plan: Let the value of the temperature be x. Insert the known values into the given relationship and calculate.

Work:

$$\ln K_{eq} = \frac{0\ \text{J mol}^{-1}}{(8.31451\ \text{J mol}^{-1}\ \text{K}^{-1})(x\ \text{K})} = 0$$

$$K_{eq} = e^0 = 1$$

13.79 Suppose SO_2, SO_3, and O_2 are in equilibrium according to $2\ SO_2(g) + O_2(g) \rightleftharpoons 2\ SO_3(g)$ and a quantity of inert gas such as helium is added to the system. What is the effect on the equilibrium if (a) He is added without changing the total pressure? In this case the volume of the vessel must increase as the inert gas is added. (b) He is added without changing the volume of the vessel? In this case the total pressure must rise as He is added.

Analysis:

Target: To decide if a reaction shifts to the right or left or does not change when a change is applied to the system.

Knowns: Each of the disturbances applied and the balanced chemical equation. Δn_{gas} can be determined from the balanced equation.

Relationship: Le Châtelier's Principle can be used to determine the system response.

Plan: Determine the sign of Δn_{gas}. Apply Le Châtelier's Principle separately to each stress.

Work:

a. An increase in volume causes the system to respond by increasing the number of moles of gas. For this system, $\Delta n_{gas} = (2) - (2+ 1) = -1$. The left side has more moles of gas, so the equilibrium would shift to the **left**.

b. Addition of He with volume held constant is not a stress. The partial pressures of the reactant and product gases do not change. Therefore, Q does not change relative to K_{eq} and there is **no shift**.

13.81 At 700 °C, K_{eq} is 1.50 atm for the reaction $C(s) + CO_2(g) \rightleftharpoons 2CO(g)$. Suppose the total gas pressure at equilibrium is 1.00 atm. What are the partial pressures of CO and CO_2?

Analysis:

Target: The partial pressures of all gases at equilibrium. (? atm) =

Knowns: The total pressure at equilibrium, the K_{eq} for the reaction, and the balanced chemical equation.

Relationships: The balanced chemical equation gives the reaction stoichiometry. K_{eq} is equal to the reaction quotient at equilibrium. From Dalton's law, $P_{total} = P_{CO} + P_{CO_2}$.

Plan: Let the pressure of P_{CO} be x, then $P_{CO_2} = P_{total} - x. = 1.00 - x$. Insert these values into the K_{eq} expression and solve for x.

Work:

$$K_{eq} = \frac{P_{CO}^2}{P_{CO_2}} = \frac{(x)^2}{(1.00 - x)} = 1.50 \text{ atm}$$

$x^2 = 1.50(1.00 - x) = 1.50 - 1.50x$; and, $x^2 + 1.50x - 1.50 = 0$

(This is a quadratic equation. See Step 6 of Problem 13.57.)

$$x = \frac{-1.50 \pm\sqrt{(1.50)^2 - (4)(1)(-1.50)}}{2(1)}$$. Therefore, x = 0.686 atm or -2.19 atm.

A negative pressure is not possible, so the correct answer is x = 0.686 atm.

Step 7. Find the equilibrium partial pressures.

$P_{CO} = x =$ **0.686 atm**

$P_{CO_2} = 1.00$ atm - 0.686 = **0.31 atm**

(Note: if we let $P_{CO_2} = x$, then $P_{CO_2} = 0.314$ atm.)

Step 8. Check: $Q = \frac{P_{CO}^2}{P_{CO_2}} = \frac{(0.686\text{ atm})^2}{(0.314\text{ atm})} = 1.50$ atm

14 Chemical Kinetics

14.1 List the three postulates of the collision theory of reaction rates.

In order for a reaction to occur, (1) there must be collision between the reacting species; (2) there must be a sufficient amount of energy in the collision (the activation energy); and, (3) the reacting species must have the proper orientation.

14.3 What specific factors influence the rate of a chemical reaction?

Chemical reactions are influenced by the nature of the reactants (including their physical state), the concentration, the pressure (when gaseous reactants are involved), the temperature and the presence of catalysts or inhibitors. In heterogeneous systems, the surface area and type of surface also influence the rate.

14.5 What is the "order" of a chemical reaction? Is it possible for a reaction to be of 11/2 order? If not, why not?

The order of a chemical reaction is the sum of the exponents of the concentrations in the rate law. It is possible to have fractional orders of a reaction if certain mechanisms are followed.

14.7 What are the units of a reaction rate? What are the units of a rate constant? Do the units of either rate or rate constant depend on reaction order? If so, give examples.

Regardless of the reaction order, reaction **rates** are typically measured in mol L^{-1} s^{-1}; however, the units of the **rate constant** depend on the reaction order. If a reaction is zero-order, the rate constant has units of mol L^{-1} s^{-1}; if it is first-order, the units are s^{-1}; and, if it is second-order, the units are L mol^{-1} s^{-1}.

14.9 What is the numerical relationship between the rate constant of a first-order reaction and its half-life?

For a first order reaction the half-life, $t_{1/2} = \frac{0.693}{k}$, where k is the rate constant.

14.11 Discuss the interpretation of the term "reaction coordinate", being careful to distinguish it from the "extent of reaction" used in equilibrium problems.

The reaction coordinate is a picture of the geometrical configuration of the collision complex as it changes from reactants to products. The extent of a reaction, used in equilibrium discussions, indicates how much of a sample has reacted on the macroscopic scale. Extent of a reaction does not deal with the individual reactions at the molecular level.

14.13 Using a simple model, describe the catalysis of a biochemical reaction by an enzyme.

An enzyme is visualized as having a surface region (called an active site) which has a very specific shape. Reactant molecules must have precisely the correct shape to fit into the active site. If they do, then the enzyme can hold them in the proper orientation to maximize the rate of the process of interest.

14.15 What is the specific meaning of the term, "reaction mechanism"?

A reaction mechanism is one or a series of elementary chemical reactions which describe the actual steps by which reacting species are converted to products.

14.17 Describe the significance of intermediates in reaction mechanisms.

Intermediates are produced and subsequently consumed in the reaction mechanism. They do not appear in the overall reaction. If intermediates can be identified, information is obtained about the actual mechanism by which the chemical process is occurring.

14.19 Identify the overall order and the order with respect to each reactant in the following reactions.

a. $C_2H_5Br + OH^- \rightarrow C_2H_5OH + Br^-$; $r = k[OH^-][C_2H_5Br]$

b. $2\ O_3 \rightarrow 3\ O_2$; $r = k[O_3]$

c. $ClCO_2C_2H_5 \rightarrow CO_2 + ClC_2H_5$; $r = k[ClCO_2C_2H_5]$

d. $HI \rightarrow 1/2\ H_2 + 1/2\ I_2$; $r = k[HI]^2$

Analysis/Plan: The rate law of the reaction is of the form $r = k[\text{reactant}_1]^x[\text{reactant}_2]^y...$ where x, is the order with respect to reactant 1, y is the order with respect to reactant 2, etc. The overall order is the sum of the orders of the individual reactants. Look at the equation and determine the order with respect to a substance from its superscript.

Work:

a. This is **first-order** with respect to C_2H_5OH; **first-order** with respect to OH^-; and, 1+1 = **second-order overall.**

b. This is **first-order** with respect to O_3 and **first-order overall.**

c. This is **first-order** with respect to $ClCO_2C_2H_5$ and **first-order overall.**

d. This is **second-order** with respect to HI and **second-order overall.**

14.21 Write the rate of appearance of each product and the rate of disappearance of each reactant in problem 14.19. For example in (a) the rates are:

C_2H_5Br: $\frac{\Delta[C_2H_5Br]}{\Delta t} = -k[C_2H_5Br][OH^-]$ $\quad$ OH^-: $\frac{\Delta[OH^-]}{\Delta t} = -k[C_2H_5Br][OH^-]$

C_2H_5OH: $+\frac{\Delta[C_2H_5OH]}{\Delta t} = k[C_2H_5Br][OH^-]$ $\quad$ Br^-: $+\frac{\Delta[Br^-]}{\Delta t} = k[C_2H_5Br][OH^-]$

Analysis/Plan: For the generalized reaction a A + b B + ... $\rightarrow$ c C + d D + ..., the rate is given by $r = \frac{-1}{a}\frac{\Delta[A]}{\Delta t} = \frac{-1}{b}\frac{\Delta[B]}{\Delta t} = +\frac{1}{c}\frac{\Delta[C]}{\Delta t} = +\frac{1}{d}\frac{\Delta[D]}{\Delta t}$. Inspect each reaction and apply the proper relationship.

Work:

b. O_3: $-\frac{1}{2}\frac{\Delta[O_3]}{\Delta t} = k[O_3]$ $\quad$ O_2: $+\frac{1}{3}\frac{\Delta[O_2]}{\Delta t} = k[O_3]$

c. $ClCO_2C_2H_5$: $-\frac{\Delta[ClCO_2C_2H_5]}{\Delta t} = k[ClCO_2C_2H_5]$

CO_2: $+\frac{\Delta[CO_2]}{\Delta t} = k[ClCO_2C_2H_5]$ $\quad$ ClC_2H_5: $+\frac{\Delta[ClC_2H_5]}{\Delta t} = k[ClCO_2C_2H_5]$

d. HI: $-\frac{\Delta[HI]}{\Delta t} = k[HI]^2$ $\quad$ H_2: $+2\frac{\Delta[H_2]}{\Delta t} = k[HI]^2$ $\quad$ I_2: $+2\frac{\Delta[I_2]}{\Delta t} = k[HI]^2$

14.23 Assume the rate law for a reaction A $\rightarrow$ B is $r = k[A]^y$. Suppose the concentration of A were doubled. What would be the effect on the rate if y = 0? If y = 1, 2, or 3?

Analysis/Plan: For two experiments in which $[A_1] \neq [A_2]$ and all other concentrations are the same, $\frac{Rate_1}{Rate_2} = \left(\frac{[A_1]}{[A_2]}\right)^y$, where y is the reaction order with respect to A. When the concentration is doubled, $\frac{[A_1]}{[A_2]} = 2$. Apply this equation for each order.

Work:

$y = 0$, then $\frac{Rate_1}{Rate_2} = (2)^0 = 1$. The rate is **unchanged**.

$y = 1$, then $\frac{Rate_1}{Rate_2} = (2)^1 = 2$. The rate is **doubled**.

$y = 2$, then $\frac{Rate_1}{Rate_2} = (2)^2 = 4$. The rate is **quadrupled**.

$y = 3$, then $\frac{Rate_1}{Rate_2} = (2)^3 = 8$. The rate **increases by a factor of 8.**

14.25 For the reaction $2\ NO + 2\ H_2 \longrightarrow N_2 + 2\ H_2O$, the rate law is $r = k[H_2][NO]^2$. What would be the effect on the rate if
a. $[H_2]$ were doubled and [NO] were unchanged?
b. $[H_2]$ and [NO] were both doubled?

Analysis/Plan: For two experiments in which $[A_1] \neq [A_2]$ and all other concentrations are the same, $\frac{Rate_1}{Rate_2} = \left(\frac{[A_1]}{[A_2]}\right)^x$, where x is the reaction order with respect to A. If more than one concentration changes, the change in rate is the product of the two effects. When the concentration is doubled, $\frac{[A_1]}{[A_2]} = 2$. Apply these relationships to determine each order.

Work:

a. This reaction is first-order in $[H_2]$, so $\frac{Rate_1}{Rate_2} = (2)^1 = 2$. The rate **doubles.**

b. From part a, doubling $[H_2]$, doubles the rate. The reaction is second-order in [NO], so $\frac{Rate_1}{Rate_2} = (2)^2 = 4$. The overall **rate increases by a factor of 2 x 4 = 8.**

14.27 For the second-order reaction $2\ NO_2 \longrightarrow 2\ NO + O_2$, $k = 1.8 \times 10^{-8}$ L mol^{-1} s^{-1} at 100 °C. What is the rate when $[NO_2] = 0.15$ M?

Analysis/Plan: Because the reaction is second-order, we know the rate law is $r = k[NO_2]^2$. Use the rate law directly, substituting the concentration of NO_2 into the rate law. Remember that M has units of mol L^{-1}.

Work:

$r = (1.8 \times 10^{-8}\ L\ mol^{-1}\ s^{-1})[0.15\ mol\ L^{-1}]^2 = \mathbf{4.1 \times 10^{-10}\ mol\ L^{-1}s^{-1}}$

14.29 At sufficiently high temperatures, nitric oxide reacts with hydrogen to form water according to $H_2 + NO \longrightarrow H_2O + 1/2\ N_2$.
a. If doubling the concentration of NO causes a fourfold increase in the rate, what is the order of the reaction with respect to NO?
b. If doubling the concentration of H_2 causes a twofold increase in the rate, what is the order with respect to H_2?

c. What is the overall order of the reaction?
d. Write the rate law.

Analysis/Plan: For two experiments in which $[A_1] \neq [A_2]$ and all other concentrations are the same, $\frac{Rate_1}{Rate_2} = \left(\frac{[A_1]}{[A_2]}\right)^x$, where x is the reaction order with respect to A. When the concentration is doubled, $\frac{[A_1]}{[A_2]} = 2$. Apply these relationships to determine each order. Add the orders to determine the overall order. Then write the rate law which is of the form $r = k[H_2]^x[NO]^y$, where x and y are the orders with respect to $[H_2]$ and [NO], respectively.

Work:

a. Because the rate is quadrupled when the concentration is doubled, $\frac{Rate_1}{Rate_2} = (2)^y = 4$. Therefore, y = 2 and the reaction is **second-order** in NO.
b. Because the rate is doubled when the concentration is doubled, $\frac{Rate_1}{Rate_2} = (2)^x = 2$. Therefore, x = 1 and the reaction is **first-order** in H_2.
c. The overall order is 2 + 1 = **third-order**.
d. The rate law for the reaction is $r = \mathbf{k[H_2][NO]^2}$.

14.31 The following data were collected during trial runs of a reaction having stoichiometry $A + B + C \longrightarrow D$.

Run	[A]/M	[B]/M	[C]/M	Initial Rate/mol L^{-1} s^{-1}
1	0.05	0.05	0.05	0.80×10^{-3}
2	0.10	0.05	0.05	1.6×10^{-3}
3	0.05	0.10	0.05	3.2×10^{-3}
4	0.05	0.05	0.10	0.80×10^{-3}

a. Determine the order of the reaction with respect to each reactant, and the overall order.
b. Write the rate expression and give the numerical value and units of the rate constant.

Analysis:

Target: The order with respect to each reactant, the overall order, the rate expression and the units of the rate constant.

Knowns: The rate of the reaction at varying concentrations.

Relationship: For two experiments in which $[A_1] \neq [A_2]$ and all other concentrations are the same, $\frac{Rate_1}{Rate_2} = \left(\frac{[A_1]}{[A_2]}\right)^x$, where x is the reaction order with respect to A. The rate law is of the form $r = k[A]^x[B]^y[C]^z$, where x, y and z are the orders with respect to [A], [B] and [C], respectively.

Plan: The data can be used to determine the orders, the rate law and k.

Step 2. For part (b), rearrange the rate law to solve for k, insert the data from one of the runs, and calculate. Units are determined from the order of the reaction.

Step 1. For part (a), use pairs of experiments in which only one concentration changes. Determine the dependence of the rate on this reactant. Repeat for each reactant in turn.

Work:

a. Between Runs 1 and 2, only [A] changes. $\frac{Rate_2}{Rate_1} = \left(\frac{[A_2]}{[A_1]}\right)^x$, so $\frac{1.6 \times 10^{-3}}{0.80 \times 10^{-3}} = \left(\frac{[0.10]}{[0.05]}\right)^x$

$2.0 = 2^x$; x = 1 and the reaction is **first-order in** [A].

Between Runs 1 and 3, only [B] changes. $\frac{Rate_3}{Rate_1} = \left(\frac{[B_3]}{[B_1]}\right)^y$, so $\frac{3.2 \times 10^{-3}}{0.80 \times 10^{-3}} = \left(\frac{[0.10]}{[0.05]}\right)^y$

$4.0 = 2^y$; y= 2 and the reaction is **second-order in** [B].

Between Runs 1 and 4, only [C] changes. $\frac{Rate_4}{Rate_1} = \left(\frac{[C_4]}{[C_1]}\right)^z$, so $\frac{0.80 \times 10^{-3}}{0.80 \times 10^{-3}} = \left(\frac{[0.10]}{[0.05]}\right)^z$

$1.0 = 2^z$; z = 0 and the reaction is **zero-order in [C]**.

The overall order is 1 + 2 + 0 = 3. The reaction is **third-order overall.**

b. $r = k[A][B]^2$; therefore, $k = \frac{r}{[A][B]^2}$. If we use the data from Run 1:

$$k = \frac{0.80 \times 10^{-3}\ \text{mol L}^{-1}\ \text{s}^{-1}}{(0.05\ \text{mol L}^{-1})(0.05\ \text{mol L}^{-1})^2} = \mathbf{6\ L^2\ mol^{-2}\ s^{-1}}$$

14.33 The following data were collected during trial runs of the reaction having stoichiometry A + B → C.

Run	[A]/M	[B]/M	Initial Rate/mol L^{-1} s^{-1}
1	0.05	0.05	3×10^{-5}
2	0.10	0.05	6×10^{-5}
3	0.10	0.10	1.2×10^{-4}

a. Determine the order of the reaction with respect to each reactant, and the overall order.
b. Write the rate expression and give the numerical value and units of the rate constant.

Analysis:

Target: The order with respect to each reactant, the overall order, the rate expression, and the units of the rate constant.

Known: The rate of the reaction at varying known concentrations.

Relationship: For two experiments in which $[A_1] \neq [A_2]$ and all other concentrations are the same, $\frac{Rate_1}{Rate_2} = \left(\frac{[A_1]}{[A_2]}\right)^x$, where x is the reaction order with respect to A. The rate law is of the form $r = k[A]^x[B]^y$, where x and y are the orders with respect to [A] and [B] respectively.

Plan: The data can be used to determine the orders, the rate law and k.

Step 2. For part (b), rearrange the rate law to solve for k, insert the data from one of the runs, and calculate. Units are determined from the order of the reaction.

Step 1. For part (a), use pairs of experiments in which only one concentration changes. Determine the dependence of the rate on this reactant. Repeat for each reactant in turn.

Work:

a. Between Runs 1 and 2, only [A] changes. $\frac{Rate_2}{Rate_1} = \left(\frac{[A_2]}{[A_1]}\right)^x$, so $\frac{6 \times 10^{-5}}{3 \times 10^{-5}} = \left(\frac{[0.10]}{[0.05]}\right)^x$

$2 = 2^x$; x = 1 and the reaction is **first-order in [A]**.

Between Runs 2 and 3, only [B] changes. $\frac{Rate_3}{Rate_2} = \left(\frac{[B_3]}{[B_2]}\right)^y$, so $\frac{1.2 \times 10^{-4}}{6 \times 10^{-5}} = \left(\frac{[0.10]}{[0.05]}\right)^y$

$2 = 2^y$; y= 1 and the reaction is **first-order in [B]**.

The overall order is 1 + 1 = 2. The reaction is **second-order overall.**

b. $r = k[A][B]$; therefore, $k = \frac{r}{[A][B]}$. If we use the data from Run 3:

$$k = \frac{1.2 \times 10^{-4}\ \text{mol L}^{-1}\ \text{s}^{-1}}{(0.10\ \text{mol L}^{-1})(0.10\ \text{mol L}^{-1})} = \mathbf{1.2 \times 10^{-2}\ L\ mol^{-1}\ s^{-1}}$$

14.35 The half-life for the first-order decomposition of cyclobutane, $C_4H_8 \longrightarrow 2\ C_2H_4$, is 89 s at 498 °C. What is the rate constant at this temperature?

Analysis:

Target: A first order rate constant. (? s^{-1}) =

Knowns: The half-life of the reaction and the temperature.

Relationship: $k = \frac{0.693}{t_{1/2}}$.

Plan: Use the equation for k directly. The temperature is not needed.

Work:

$$? \ s^{-1} = k = \frac{0.693}{t_{1/2}} = \frac{0.693}{89 \ s} = \mathbf{7.8 \times 10^{-3} \ s^{-1}}$$

14.37 The rate constant for the first-order decomposition of chloroethane, $CH_3CH_2Cl \longrightarrow C_2H_2 + HCl$, is 0.13 s^{-1} at 100 °C. What is its half-life?

Analysis:

Target: The half-life of a first order reaction. (? s)

Knowns: The rate constant for the reaction and the temperature.

Relationship: $t_{1/2} = \frac{0.693}{k}$.

Plan: Use the equation for $t_{1/2}$ directly. The temperature is not needed.

Work:

$$? \ s = t_{1/2} = \frac{0.693}{k} = \frac{0.693}{0.13 \ s^{-1}} = \mathbf{5.3 \ s}$$

14.39 If the initial concentration of a sample of cyclobutane is 0.500 M, how much remains after an elapsed time of 5.0 minutes? The rate constant for the first-order decomposition is 0.0078 s^{-1}.

Analysis:

Target: The amount of reactant remaining after a given time. (? M) =

Knowns: The rate constant and the elapsed time of the reaction.

Relationships: $\ln\left(\frac{C}{C_o}\right) = -k \cdot t$, where C_o is the initial concentration, C is the concentration after elapsed time t, and k is the rate constant for the reaction. The equivalence statement 1 min ↔ 60 s is needed.

Plan: Insert the data into the equation and solve for C, using the factor-label method to convert the time to seconds.

Work:

$$\ln\left(\frac{C}{C_o}\right) = -k \cdot t; \text{ therefore, } \ln\left(\frac{C}{0.500 \ M}\right) = -(0.0078 \ s^{-1})(5.0 \ min) \times \frac{60 \ s}{1 \ min} = -2.3$$

$$\frac{C}{0.500 \ M} = e^{-2.3}; \quad C = (0.500 \ M)(e^{-2.3}) = \mathbf{0.048 \ M}$$

14.41 If the initial concentration of a N_2O_5 sample is 0.337 M, how much remains after one day of decomposition at 45 °C? (First order decomposition; rate constant 1.0×10^{-5} s^{-1}.)

Analysis:

Target: The amount of reactant remaining after a given time. (? M) =

Knowns: The rate constant, the temperature, the initial concentration, and the elapsed time of the reaction.

Relationships: $\ln\left(\frac{C}{C_o}\right) = -k \cdot t$, where C_o is the initial concentration, C is the concentration after elapsed time t, and k is the rate constant for the reaction. The equivalence statements 1 min ↔ 60 s, 60 min ↔ 1 hr and 24 hr ↔ 1 day are needed.

Plan: Insert the data into the equation and solve for C, using the factor-label method to convert the time to seconds. The temperature is not needed.

Work:

$\ln\left(\frac{C}{C_o}\right) = -k \cdot t$; therefore,

$$\ln\left(\frac{C}{0.337\text{ M}}\right) = -(1.0 \times 10^{-5}\text{ s}^{-1})(1\text{ day}) \times \frac{24\text{ hr}}{1\text{ day}} \times \frac{60\text{ min}}{1\text{ hr}} \times \frac{60\text{ s}}{1\text{ min}} = -0.86$$

$$\frac{C}{0.337\text{ M}} = e^{-0.86};\ C = (0.337\text{ M})(e^{-0.86}) = \mathbf{0.14\text{ M}}$$

14.43 The half-life of the first-order decomposition of sulfuryl chloride is 3.30×10^4 s at 600 K. If the initial concentration is 0.0250 M, what is the concentration after one hour at this temperature?

Analysis:

Target: The amount of reactant remaining after a given time. (? M) =

Knowns: The half-life, the initial amount of reactant, the temperature, and the elapsed time of the reaction.

Relationships: The rate constant, k, is equal to $\frac{0.693}{t_{1/2}}$ and $\ln\left(\frac{C}{C_o}\right) = -k \cdot t$, where C_o is the initial concentration, C is the concentration after elapsed time t, and k is the rate constant for the reaction. The equivalence statements 1 min ↔ 60 s and 60 min ↔ 1 hr are needed.

Plan: Two steps are required. The temperature is not needed.

Step 2. Insert the data into the second equation and solve for C, using the factor-label method to convert the time to seconds.

Step 1. Use the first equation to find the rate constant.

Work:

Step 1. Find the value of k.

$$k = \frac{0.693}{t_{1/2}} = \frac{0.693}{3.30 \times 10^4\text{ s}} = 2.10 \times 10^{-5}\text{ s}^{-1}$$

Step 2. Solve for the reactant remaining. Assume 1 hr has 3 significant digits.

$\ln\left(\frac{C}{C_o}\right) = -k \cdot t$; therefore,

$$\ln\left(\frac{C}{0.0250\text{ M}}\right) = -(2.10 \times 10^{-5}\text{ s}^{-1})(1\text{ hr}) \times \frac{60\text{ min}}{1\text{ hr}} \times \frac{60\text{ s}}{1\text{ min}} = -0.0756$$

$$\frac{C}{0.0250\text{ M}} = e^{-0.0756};\ C = (0.0250\text{ M})(e^{-0.0756}) = \mathbf{0.0232\text{ M}}$$

14.45 If the concentration of a substance decreases from 0.0125 M to 0.00993 M in 176 s, and the process is known to be first order, what is the value of the rate constant?

Analysis:

Target: The rate constant of a reaction. (? s^{-1}) =

Knowns: The time elapsed between two different known concentrations.

Relationship: $\ln\left(\frac{C}{C_o}\right) = -k \cdot t$, where C_o is the initial concentration, C is the concentration after elapsed time t, and k is the rate constant for the reaction.

Plan: Insert the data into the equation and solve for k.

Work:

$$\ln\left(\frac{C}{C_o}\right) = -k \cdot t; \ \ln\left(\frac{0.00993 \text{ M}}{0.0125 \text{ M}}\right) = -(k)(176 \text{ s}); \ \ln(0.794) = -(k)(176 \text{ s})$$

$$-0.230 = -(k)(176 \text{ s}); \text{ therefore } k = \frac{-0.230}{-176 \text{ s}} = \mathbf{1.31 \times 10^{-3} s^{-1}}$$

14.47 Emission of radioactivity and decay of unstable nuclei follow first-order kinetics. These nuclear processes are not affected by temperature or the state of chemical combination. For example, radium chloride, whether pure or in solution, decays at the same rate as radium metal. For this reason it is customary to write the first-order rate law not in terms of concentration, but in terms of the number of moles of radioactive nuclei. The rate law for the decay of Ra-226 nuclei is $\frac{-\Delta n}{\Delta t} = k \cdot n$, where n is the number of moles and $k = 1.36 \times 10^{-11} \text{ s}^{-1}$.

a. What is the half-life?

b. How much of a sample, originally containing 1.00 mmol, remains after 1000 years?

Analysis:

Target: The half-life and the amount of reactant remaining after a given time. (? yr) = and (? mmol) =

Knowns: The rate constant, the initial amount of reactant, and the elapsed time of the reaction.

Relationships: The half-life is equal to $\frac{0.693}{k}$ and $\ln\left(\frac{n}{n_o}\right) = -k \cdot t$, where n_o is the initial number of moles, n is the number of moles after elapsed time t, and k is the rate constant for the reaction. The equivalence statements 1 min ↔ 60 s, 60 min ↔ 1 hr, 24 hr ↔ 1 day and 365 day ↔ 1 year are needed.

Plan: For both, use the factor-label method to convert the rate constant to yr^{-1}. Use the first equation to find the half-life for part (a). Insert the data into the second equation and solve for n for part (b).

Work:

$$k = 1.36 \times 10^{-11} \text{ s}^{-1} \times \frac{60 \text{ s}}{1 \text{ min}} \times \frac{60 \text{ min}}{1 \text{ hr}} \times \frac{24 \text{ hr}}{1 \text{ day}} \times \frac{365 \text{ day}}{1 \text{ yr}} = 4.29 \times 10^{-4} \text{ yr}^{-1}$$

a. $? \text{ yr} = t_{1/2} = \frac{0.693}{k} = \frac{0.693}{4.29 \times 10^{-4} \text{ yr}^{-1}} = \mathbf{1.62 \times 10^3 \text{ years}}$

b. $\ln\left(\frac{n}{n_o}\right) = -k \cdot t$; therefore,

$$\ln\left(\frac{n}{1.00 \text{ mmol}}\right) = -(4.29 \times 10^{-4} \text{ yr}^{-1})(1000 \text{ yr}) = -0.429$$

$$? \text{ mmol} = \frac{n}{1.00 \text{ mmol}} = e^{-0.429}; \ n = (1.00 \text{ mmol})(e^{-0.429}) = \mathbf{0.651 \text{ mmol}}$$

14.49 How long a time must elapse before the radioactivity of a sample of $^{239}_{94}$Pu decreases to 5.0% of its initial value? This is another way of wording the question, "How much time must elapse for the number of moles remaining to be equal to 5.0% of the original amount?" The half-life of this isotope of plutonium is 24,100 years.

Analysis:

Target: The length of time required to decrease the amount of reactant. (? years) =

Knowns: The half-life and the percentage decrease of the amount of reactant.

Relationships: The rate constant, k, is equal to $\frac{0.693}{t_{1/2}}$ and $\ln\left(\frac{n}{n_o}\right) = -k \cdot t$, where n_o is the initial number of moles, n is the number of moles after elapsed time t, and k is the rate constant for the reaction.

Plan: This can be solved in two steps.

Step 2. Insert the data into the second equation and solve for t.

Step 1. Use the first equation to find the rate constant.

Work:

Step 1. Use the half-life to find k.

$$k = \frac{0.693}{t_{1/2}} = \frac{0.693}{24100 \text{ yr}} = 2.88 \times 10^{-5} \text{ yr}^{-1}$$

Step 2. Use the equation to solve for t.

$$\ln\left(\frac{n}{n_o}\right) = -k \cdot t; \quad \text{To decrease to 5.0\%, } \frac{n}{n_o} = \frac{5.0}{100} = 0.050.$$

$$\ln(0.050) = -3.0 = -(2.88 \times 10^{-5} \text{ yr}^{-1})(t)$$

$$? \text{ yr} = t = \frac{-3.0}{-2.88 \times 10^{-5} \text{ yr}^{-1}} = \mathbf{1.0 \times 10^5 \text{ years}}$$

14.51 Sketch the reaction profile for a reaction that is exothermic by 125 kJ mol^{-1} and has an activation energy of 75 kJ mol^{-1}.

Analysis/Plan: Reaction profiles are drawn with the energy of the reactants represented on the left and the products on the right. If the process is exothermic, the products have lower energy than the reactants. The activation energy of the forward reaction is shown as a "hill" between the reactants and the products. Use these guidelines and sketch.

Work:

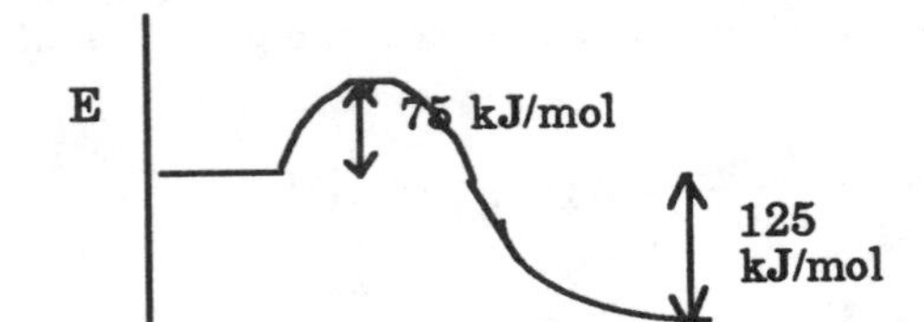

14.53 If an exothermic reaction has $\Delta E_{rxn} = -250$ kJ mol^{-1} and $E_a = 100$ kJ mol^{-1}, what is E_a for the reverse reaction?

Analysis:

Target: The activation energy of a reverse reaction. (? kJ mol^{-1}) =

Knowns: ΔE_{rxn} and the activation energy of the forward reaction.

Relationship: $\Delta E_{rxn} = E_{a,f} - E_{a,r}$.

Plan: Rearrange this relationship to solve for $E_{a,r}$, insert the values and calculate.

Work:

$\Delta E_{rxn} = E_{a,f} - E_{a,r}$; therefore, $E_{a,r} + \Delta E_{rxn} = E_{a,f}$ and $E_{a,r} = E_{a,f} - \Delta E_{rxn}$

$E_{a,r} = 100 \text{ kJ mol}^{-1} - (-250 \text{ kJ mol}^{-1}) = \mathbf{350 \text{ kJ mol}^{-1}}$

14.55 Sketch the reaction profile of a reaction in which $E_{a,f} = E_{a,r}$. Is the reaction exothermic or endothermic?

Analysis/Plan: Reaction profiles are drawn with the energy of the reactants represented on the left and the products on the right. If the process is exothermic, the products have lower energy than the reactants. The activation energy of the forward reaction is shown as a "hill" between the reactants and the products. Use these guidelines and sketch. If the activation energies of the forward and reverse reaction are the same, then **ΔE_{rxn} must be zero.** The reactants and products must have the same energy.
Work:

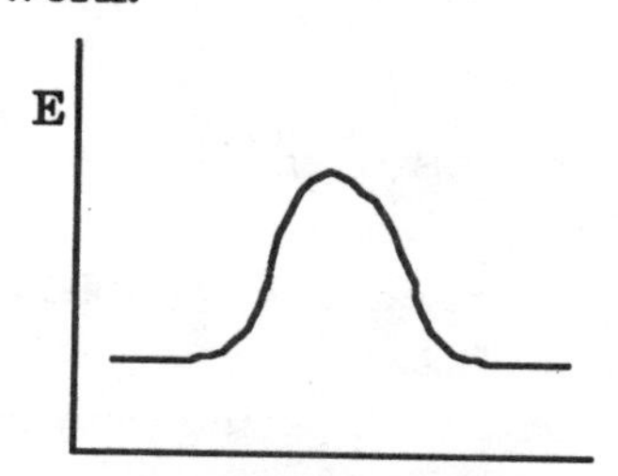

14.57 What is the reverse activation energy of a reaction having $\Delta E_{rxn} = -108$ kJ mol^{-1} and $E_{a,f} = 15$ kJ mol^{-1}?

Analysis:
Target: The activation energy of a reverse reaction. (? kJ mol^{-1}) =
Known: ΔE_{rxn} and the activation energy of the forward reaction.
Relationship: $\Delta E_{rxn} = E_{a,f} - E_{a,r}$.
Plan: Rearrange this relationship to solve for $E_{a,r}$, insert the values and calculate.
Work:

$\Delta E_{rxn} = E_{a,f} - E_{a,r}$; therefore, $E_{a,r} + \Delta E_{rxn} = E_{a,f}$ and $E_{a,r} = E_{a,f} - \Delta E_{rxn}$

$E_{a,r} = 15 \text{ kJ mol}^{-1} - (-108 \text{ kJ mol}^{-1}) = \mathbf{123 \text{ kJ mol}^{-1}}$

14.59 For the reaction $2\,N_2O_5 \longrightarrow 2\,N_2O_4 + O_2$, the rate constant is 2.4×10^{-4} s^{-1} at 293 K and 9.2×10^{-4} s^{-1} at 303 K?
a. Calculate the activation energy?
b. Calculate the frequency factor?

Analysis:
Target: An activation energy and a frequency factor. (? kJ mol^{-1}) = and (? s^{-1}) =
Knowns: The rate constants at two different temperatures. $R = 8.31451$ J mol^{-1} K^{-1}.
Relationships: $\ln\left(\frac{k_2}{k_1}\right) = \frac{E_a}{R} \cdot \left(\frac{1}{T_1} - \frac{1}{T_2}\right)$ and $k = Ae^{-Ea/RT}$. The equivalence statement 1 kJ ↔ 1000 J is needed.
Plan: For part (a), insert the rate constants, their temperatures and R into the first equation and solve for E_a. For part (b), use one of the rate constants, its temperature and the calculated value of E_a in the second relationship to determine A, the pre-exponential factor.
Work:
a. Let $k_1 = 2.4 \times 10^{-4}$ s^{-1} at $T_1 = 293$ K and let $k_2 = 9.2 \times 10^{-4}$ s^{-1} at $T_2 = 303$ K.

$$\ln\left(\frac{9.2 \times 10^{-4} \text{ s}^{-1}}{2.4 \times 10^{-4} \text{ s}^{-1}}\right) = \frac{E_a}{(8.31451 \text{ J mol}^{-1} \text{ K}^{-1})} \cdot \left(\frac{1}{293 \text{ K}} - \frac{1}{303 \text{ K}}\right)$$

$\ln(3.8) = 1.34 = E_a(1.4 \times 10^{-5} \text{ J}^{-1} \text{ mol})$

$$? \text{ kJ mol}^{-1} = E_a = \frac{1.34}{1.4 \times 10^{-5} \text{ J}^{-1} \text{ mol}} = 9.9 \times 10^4 \text{ J mol}^{-1} \times \frac{1 \text{ kJ}}{1000 \text{ J}} = \mathbf{99 \text{ kJ mol}^{-1}}$$

b. Use k_1 and T_1. $k = Ae^{-Ea/RT}$; therefore, $A = \frac{k}{e^{-Ea/RT}} = \frac{2.4 \times 10^{-4} \text{ s}^{-1}}{e^{-(99000)/(8.314)(293)}} = \mathbf{1.1 \times 10^{14} \text{ s}^{-1}}$

14.61 The activation energy of the first-order decomposition of cyclobutane is 260 kJ mol^{-1}, and the rate constant is 6.08×10^{-8} s^{-1} at 600 K. What is the value of the rate constant at 650 K?

Analysis:
Target: The rate constant at a different temperature. (? s^{-1}) =
Knowns: The rate constant at one temperature, the activation energy and the new temperature. The ideal gas constant is R = 8.31451 J mol^{-1} K^{-1}.
Relationships: $\ln\left(\frac{k_2}{k_1}\right) = \frac{E_a}{R} \bullet \left(\frac{1}{T_1} - \frac{1}{T_2}\right)$. The equivalence statement 1 kJ ↔ 1000 J is required.
Plan: Insert the data into the equation, then solve for and evaluate the new rate constant.
Work:
Let $k_1 = 6.08 \times 10^{-8}$ s^{-1} at T_1 = 600 K and let k_2 = ? at T_2 = 650 K.

$$\ln\left(\frac{k_2}{6.08 \times 10^{-8} \text{ s}^{-1}}\right) = \frac{260 \text{ kJ mol}^{-1} \times \frac{1000 \text{ J}}{1 \text{ kJ}}}{(8.31451 \text{ J mol}^{-1} \text{ K}^{-1})} \bullet \left(\frac{1}{600 \text{ K}} - \frac{1}{650 \text{ K}}\right) = 4.0$$

$$\frac{k_2}{6.08 \times 10^{-8} \text{ s}^{-1}} = e^{4.0}; \text{ therefore, } k_2 = (6.08 \times 10^{-8} \text{ s}^{-1})(e^{4.0}) = \mathbf{3.4 \times 10^{-6} \text{ s}^{-1}}$$

14.63 At what temperature does the decomposition of cyclobutane proceed with a rate constant of 6.08×10^{-6} s^{-1}? Use the data from Problem 14.61.

Analysis:
Target: The temperature to give a certain rate constant. (? K) =
Known: The rate constant at one temperature, the activation energy, and the new rate constant. The ideal gas constant is R = 8.31451 J mol^{-1} K^{-1}.
Relationships: $\ln\left(\frac{k_2}{k_1}\right) = \frac{E_a}{R} \bullet \left(\frac{1}{T_1} - \frac{1}{T_2}\right)$. The equivalence statement 1 kJ ↔ 1000 J is needed.
Plan: Insert the known data, then solve for and evaluate the new temperature.
Work:
Let $k_1 = 6.08 \times 10^{-8}$ s^{-1} at T_1 = 600 K and let $k_2 = 6.08 \times 10^{-6}$ s^{-1} at T_2 = ?

$$\ln\left(\frac{6.08 \times 10^{-6} \text{ s}^{-1}}{6.08 \times 10^{-8} \text{ s}^{-1}}\right) = \frac{260 \text{ kJ mol}^{-1} \times \frac{1000 \text{ J}}{1 \text{ kJ}}}{(8.31451 \text{ J mol}^{-1} \text{ K}^{-1})} \bullet \left(\frac{1}{600 \text{ K}} - \frac{1}{T_2}\right)$$

$$\ln(100) = 4.605 = 52.1 - \frac{3.13 \times 10^4 \text{ K}}{T_2}; \quad 4.605 - 52.1 = -\frac{3.13 \times 10^4 \text{ K}}{T_2}$$

$$-47.5 = -\frac{3.13 \times 10^4 \text{ K}}{T_2}; \quad -47.5(T_2) = -3.13 \times 10^4 \text{ K}; \text{ therefore, } T_2 = \frac{-3.13 \times 10^4 \text{ K}}{-47.5} = \mathbf{659 \text{ K}}$$

14.65 For many reactions, the rate constant is approximately doubled if the temperature is increased by 10 °C. This is a good rule of thumb, but it applies only near room temperature and only to those reactions whose activation energy is moderate rather than high or low. What is the activation energy of a reaction whose rate at 35 °C is exactly twice that at 25 °C?

Analysis:

Target: An activation energy. (? kJ mol^{-1}) =

Knowns: The ratio of rates at two different temperatures. R = 8.31451 J mol^{-1} K^{-1}.

Relationships: $\ln\left(\frac{k_2}{k_1}\right) = \frac{E_a}{R} \bullet \left(\frac{1}{T_1} - \frac{1}{T_2}\right)$. If the rate is to be doubled, then the ratio k_2/k_1 = 2. The equivalence statements 1 kJ ↔ 1000 J and K ↔ 273.15 + °C are needed.

Plan: Insert the rate constant ratio, the temperatures and R into the equation and solve for E_a.

Work:

Let k_2/k_1 = 2; then T_1 = 25 °C + 273.15 = 298 K and T_2 = 35 °C + 273.15 = 308 K.

$$\ln(2) = \frac{E_a}{(8.31451 \text{ J mol}^{-1} \text{ K}^{-1})} \bullet \left(\frac{1}{298 \text{ K}} - \frac{1}{308 \text{ K}}\right)$$

$$\ln(2) = 0.693 = E_a(1.3 \times 10^{-5} \text{ J}^{-1} \text{ mol})$$

$$? \text{ kJ mol}^{-1} = E_a = \frac{0.693}{1.3 \times 10^{-5} \text{ J}^{-1} \text{ mol}} = 5.3 \times 10^4 \text{ J mol}^{-1} \times \frac{1 \text{ kJ}}{1000 \text{ J}} = \mathbf{53 \text{ kJ mol}^{-1}}$$

14.67 Write the rate law for each of the following elementary reactions:

a. $H + O_2 \longrightarrow OH + O$ b. $O + O_3 \longrightarrow 2 O_2$ c. $2 Cl + Cl_2 \longrightarrow 2 Cl_2$

Analysis/Plan: In an elementary reaction, the coefficient of a reactant gives the order with respect to that reactant. Use those coefficients to write the rate law which is of the form $r = k[A]^x[B]^y[C]^z$, where x, y and z are the orders with respect to [A], [B] and [C], respectively.

Work:

a. $r = k[H][O_2]$, because the coefficients of H and O_2 are both 1.

b. $r = k[O][O_3]$, because the coefficients of O and O_3 are both 1.

c. $r = k[Cl]^2[Cl_2]$, because the coefficients of Cl is 2 and that of Cl_2 is 1.

14.69 What is the overall reaction corresponding to the following mechanism?

$N_2O_5 \longrightarrow NO_2 + NO_3$

$NO_3 + N_2O_5 \longrightarrow 3 NO_2 + O_2$

Analysis/Plan: To determine the overall reaction, we add the two steps, combining common terms on the same side and deleting like terms on opposite sides.

Work:

$\mathbf{2 N_2O_5 \longrightarrow 4 NO_2 + O_2}$

14.71 The reaction $NO_2(g) + CO(g) \longrightarrow CO_2(g) + NO(g)$ proceeds at low temperatures via the mechanism

a. $NO_2 + NO_2 \longrightarrow NO + NO_3$ (slow)

$NO_3 + CO \longrightarrow NO_2 + CO_2$ (fast)

At high temperatures, on the other hand, the mechanism is the direct transfer of an oxygen atom from NO_2 to CO:

b. $NO_2 + CO \longrightarrow NO + CO_2$

Write the rate laws corresponding to these two mechanisms.

Analysis:

Target: The rate law of the process in which the substances involved in the rate law and their respective orders are to be determined.

Knowns: The steps of the mechanism, identified as fast or slow.

Relationships: The coefficients of substances in the elementary steps give the order with respect to the reactant. The slow step is the rate determining step.

Plan: Identify the rate determining step and write the rate law.

Work:

a. The first step is the slow step. $\mathbf{r = \mathit{k}[NO_2][NO_2] = \mathit{k}[NO_2]^2}$

b. In a one step process, that step is the rate determining step. $\mathbf{r = \mathit{k}[NO_2][CO]}$

14.73 The rate law for the reaction $2\ NO + O_2 \rightarrow 2\ NO_2$ is $r = k[NO]^2[O_2]$. Write a plausible mechanism for the reaction.

Analysis/Plan: We need to write a mechanism in which the coefficient of NO is 2 and that of O_2 is 1 and in which no more than three particles collide.

Work:

A one step mechanism fits those criteria. $\mathbf{2\ NO + O_2 \rightarrow 2\ NO_2}$

14.75 A chemist investigating the first order decomposition of N_2O_5, for which the rate constant is $1.0 \times 10^{-5}\ s^{-1}$ at 45 °C, forgot to record the initial concentration in one experiment. However, the N_2O_5 concentration was 0.025 M after an elapsed time of 5.00×10^4 s. What was the original concentration?

Analysis:

Target: The initial molarity of a reactant. (? M) =

Knowns: The rate constant, the temperature, the final concentration, and the elapsed time of the reaction.

Relationship: $\ln\left(\frac{C}{C_o}\right) = -k \cdot t$, where C_o is the initial concentration, C is the concentration after elapsed time t, and k is the rate constant for the reaction.

Plan: Insert the data into the equation and solve for C_o. The temperature is not needed.

Work:

$\ln\left(\frac{C}{C_o}\right) = -k \cdot t$; therefore,

$$\ln\left(\frac{0.025\ M}{C_o}\right) = -(1.0 \times 10^{-5}\ s^{-1})(5.00 \times 10^4\ s) = -0.50$$

$$\frac{0.025\ M}{C_o} = e^{-0.50};\ 0.025\ M = (C_o)(e^{-0.50});\ C_o = \frac{0.025\ M}{e^{-0.50}} = \mathbf{0.041\ M}$$

14.77 The second-order decomposition, $NO_2 \rightarrow NO + 1/2\ O_2$, has a rate constant equal to $3.1\ L\ mol^{-1}\ s^{-1}$ at 400 °C.

a. How long will it take for the NO_2 concentration to be reduced to 1/2 of its initial value of 0.030 M?

b. Same question as (a), but use an initial concentration of 0.10 M.

Analysis:

Target: The time required for a process to occur. (? s) =

Knowns: The rate constant, the temperature, and the elapsed time of the reaction.

Relationship: For a second order process, $\frac{1}{C} = \frac{1}{C_o} + kt$ where C_o is the initial concentration, C is the concentration after elapsed time t, and k is the rate constant for the reaction.

Plan: Insert the data into the equation and solve for C_o. The temperature is not needed.

Work:

a. If the concentration is reduced to 1/2 its initial value, $C = 1/2C_o = \frac{0.30\text{ M}}{2} = 0.15\text{ M}$.

$$\frac{1}{C} = \frac{1}{C_o} + kt; \quad \frac{1}{0.15\text{ M}} = \frac{1}{0.30\text{ M}} + (3.1\text{ L mol}^{-1}\text{ s}^{-1})(t)$$

$$\frac{1}{0.15\text{ M}} - \frac{1}{0.30\text{ M}} = (33\text{ L mol}^{-1}) = (3.1\text{ L mol}^{-1}\text{ s}^{-1})(t)$$

$$t = \frac{33\text{ L mol}^{-1}}{3.1\text{ L mol s}^{-1}} = \mathbf{11\ s}$$

b. If the concentration is reduced to 1/2 its initial value, $C = 1/2C_o = \frac{0.10\text{ M}}{2} = 0.050\text{ M}$.

$$\frac{1}{C} = \frac{1}{C_o} + kt; \quad \frac{1}{0.050\text{ M}} = \frac{1}{0.10\text{ M}} + (3.1\text{ L mol}^{-1}\text{ s}^{-1})(t)$$

$$\frac{1}{0.050\text{ M}} - \frac{1}{0.10\text{ M}} = (10\text{ L mol}^{-1}) = (3.1\text{ L mol}^{-1}\text{ s}^{-1})(t)$$

$$t = \frac{10\text{ L mol}^{-1}}{3.1\text{ L mol s}^{-1}} = \mathbf{3.2\ s}$$

14.79 For a certain reaction the reverse activation energy is greater than the forward activation energy. Is the reaction exothermic or endothermic?

The energy involved in a reaction is $\Delta E_{rxn} = E_{a,f} - E_{a,r}$. If $E_{a,r}$ is larger than $E_{a,f}$, the sign of ΔE_{rxn} will be negative. Because ΔE_{rxn} and ΔH_{rxn} are usually similar in value, ΔH_{rxn} is also probably negative and, therefore, the reaction is exothermic.

14.81 What is the forward activation energy of a reaction having $\Delta E_{rxn} = -108\text{ kJ mol}^{-1}$ and $E_{a,r} = 15\text{ kJ mol}^{-1}$.

This is impossible. The activation energy of the reverse reaction must be at least as large as the exothermicity. This would give a negative value for $E_{a,f}$ and activation energies are **always** positive.

14.83 Calculate the frequency factor for the decomposition of cyclobutane. Use the data from Problem 14.61.

Analysis:

Target: A frequency factor, A. (? s^{-1}) =

Knowns: The rate constant, the temperature and the activation energy are given. The ideal gas constant is $R = 8.31451\text{ J mol}^{-1}\text{ K}^{-1}$.

Relationships: $k = Ae^{-Ea/RT}$. The equivalence statement 1 kJ ↔ 1000 J is needed.

Plan: Rearrange the equation to solve for A and calculate, using the factor-label method to convert kJ to J.

Work:

$$260\text{ kJ} \times \frac{1000\text{ J}}{1\text{ kJ}} = 260000\text{ J}$$

$k = Ae^{-Ea/RT}$; therefore, $A = \frac{k}{e^{-Ea/RT}} = \frac{6.08 \times 10^{-8} s^{-1}}{e^{-(260000)/(8.314)(600)}}$ = **2.6×10^{15} s^{-1}**

14.85 The exchange reaction between methyl bromide and chlorine has a rate constant of 5.9×10^{-3} L mol^{-1} s^{-1} at 25 °C. If the activation energy is 66.1 kJ mol^{-1}, at what temperature will the rate constant be 5.0×10^{-2} L mol^{-1} s^{-1}?

Analysis:

Target: The temperature to give a certain rate constant. (? K) =

Knowns: The rate constant at one temperature, the activation energy and the new rate constant. The ideal gas constant is R = 8.31451 J mol^{-1} K^{-1}.

Relationships: $\ln\left(\frac{k_2}{k_1}\right) = \frac{E_a}{R} \cdot \left(\frac{1}{T_1} - \frac{1}{T_2}\right)$. The equivalence statements 1 kJ ↔ 1000 J and K = 273.15 + °C are needed.

Plan: Insert the known data, then solve for and evaluate the new temperature.

Work:

Let $k_1 = 5.9 \times 10^{-3}$ L mol^{-1}s^{-1} at T_1 = 25 °C + 273.15 = 298 K.

Let $k_2 = 5.0 \times 10^{-2}$ L mol^{-1} s^{-1} at T_2 = ?

$$\ln\left(\frac{5.0 \times 10^{-2} \text{ L mol}^{-1} \text{ s}^{-1}}{5.9 \times 10^{-3} \text{ L mol}^{-1}\text{s}^{-1}}\right) = \frac{66.1 \text{ kJ mol}^{-1} \times \frac{1000 \text{ J}}{1 \text{ kJ}}}{(8.31451 \text{ J mol}^{-1} \text{ K}^{-1})} \cdot \left(\frac{1}{298 \text{ K}} - \frac{1}{T_2}\right)$$

$$\ln(8.5) = 2.14 = 26.7 - \frac{7.95 \times 10^3}{T_2}; \quad 2.14 - 26.7 = -\frac{7.95 \times 10^3}{T_2}$$

$$-24.6 = -\frac{7.95 \times 10^3}{T_2}; \quad -24.6(T_2) = -7.95 \times 10^3; \text{ therefore, } T_2 = \frac{-7.95 \times 10^3}{-24.6} = \textbf{324 K}$$

14.87 By what factor does the rate of a reaction increase between 0 °C and 100 °C if the activation energy is 100 kJ mol^{-1}?

Analysis:

Target: The rate of increase of a reaction between two temperatures. (? k_2/k_1) =

Knowns: The activation energy and two different temperatures. The ideal gas constant is R = 8.31451 J mol^{-1} K^{-1}. The rate of a reaction is directly proportional to its rate constant.

Relationships: $\ln\left(\frac{k_2}{k_1}\right) = \frac{E_a}{R} \cdot \left(\frac{1}{T_1} - \frac{1}{T_2}\right)$. The equivalence statements 1 kJ ↔ 1000 J and °C + 273.15 ↔ K are needed.

Plan: We want the ratio k_2/k_1. Use the equation to solve for this ratio, first converting the temperatures to Kelvin.

Work:

a. Let k_2 be the rate constant at 100 °C + 273.15 = 373 K.
   Let k_1 be the rate constant at 0 °C + 273.15 = 273 K.

$$\ln\left(\frac{k_2}{k_1}\right) = \frac{100 \text{ kJ mol}^{-1} \times \frac{1000 \text{ J}}{1 \text{ kJ}}}{(8.31451 \text{ J mol}^{-1} \text{ K}^{-1})(273\text{K})(373 \text{ K})} \cdot (373 \text{ K} - 273 \text{ K}) = 11.8$$

$\frac{k_2}{k_1} = e^{11.8} = 1.35 \times 10^5$ **The rate is 1.35×10^5 times faster at 373 K than at 273 K.**

14.89 What is the activation energy of a reaction whose rate constant increases fivefold when the temperature is increased from 50 °C to 100 °C? Can the frequency factor be determined from these data?

Analysis:

Target: An activation energy and a frequency factor. (? kJ mol^{-1}) = and (? s^{-1}) =

Knowns: The ratio of the rate constants, $\frac{k_2}{k_1}$, is 5. The temperatures are also known. The ideal gas constant is R = 8.31451 J mol^{-1} K^{-1}.

Relationships: $\ln\left(\frac{k_2}{k_1}\right) = \frac{E_a}{R} \cdot \left(\frac{1}{T_1} - \frac{1}{T_2}\right)$ and $k = Ae^{-Ea/RT}$. The equivalence statements 1 kJ ↔ 1000 J and K ↔ 273.15 + °C are needed.

Plan: Insert the rate constant ratio, the temperatures and R into the first equation and solve for E_a.

Work:

a. Let T_2 = 100 °C + 273.15 = 373 K and T_1 = 50 °C + 273.15 = 323 K.

$$\ln(5) = \frac{E_a}{(8.31451 \text{ J mol}^{-1} \text{ K}^{-1})} \cdot \left(\frac{1}{323 \text{ K}} - \frac{1}{373 \text{ K}}\right)$$

$$1.609 = E_a(5.0 \times 10^{-5})$$

$$E_a = \frac{1.609}{5.0 \times 10^{-5}} = 3.2 \times 10^4 \text{ J mol}^{-1} \times \frac{1 \text{ kJ}}{1000 \text{ J}} = \mathbf{32 \text{ kJ mol}^{-1}}$$

The frequency factor cannot be determined because we must know the rate constant at a given temperature, not just a ratio of rate constants.

14.91 Sketch the reaction profile for a reaction having E_a = 50 kJ mol^{-1}, and ΔE_{rxn} = -75 kJ mol^{-1}. Suppose that in the presence of a catalyst the activation energy is reduced to 25 kJ mol^{-1}. Sketch the reaction profile of the catalyzed reaction on the same sketch you drew for the uncatalyzed reaction.

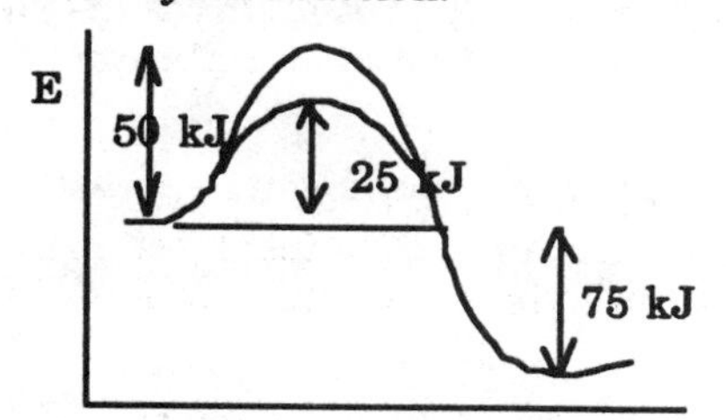

14.93 The mechanism of the acid-catalyzed hydrolysis of acetonitrile is

$CH_3CN + H^+ \rightleftharpoons CH_3CNH^+$ (fast)

$CH_3NH^+ \longrightarrow$ product (slow)

What is the rate law corresponding to this mechanism? [*Hint:* The mechanism is similar to that given on page 627 for the $H_2 + I_2$ reaction.]

Analysis:

Target: The rate law of the process in which the identity of the compounds and their orders are to be determined.

Knowns: The steps of the mechanism, identified as fast or slow.

Relationships: The coefficients of substances in the elementary steps give the order with respect to the reactant. The slow step is the rate determining step.

Plan: Identify the rate determining step and write the rate law.

Work:

The second step is the slow step. From this we would write that $r = k[CH_3CNH^+]$, however, CH_3CNH^+ is an intermediate in the process. and is maintained in a state of equilibrium in the first fast step. The equilibrium has $K_{eq} = \frac{[CH_3CNH^+]}{[CH_3CN][H^+]}$. Rearranging the equilibrium expression gives $[CH_3CNH^+] = K_{eq}[CH_3CN][H^+]$. Substituting this into the rate law gives:

$$r = kK_{eq}[CH_3CN][H^+]$$

14.95 The acid-catalyzed dehydration of ethanol, $CH_2CH_2OH \xrightarrow[H^+]{} CH_2{=}CH_2 + H_2O$, follows the rate law $r = k[CH_3CH_2OH][H^+]$.

a. Write a plausible mechanism for this reaction.
b. Write the rate constant in terms of an equilibrium constant and an elementary-step constant.

Analysis:

Target: A possible mechanism that would agree with the observed rate law.
Knowns: The balanced equation and the observed rate law.
Relationships: The coefficients of substances in the elementary steps give the order with respect to the reactant. The slow step is the rate determining step.

Plan: We need to write a mechanism such that the catalyst $[H^+]$ is in or is derivable from the rate determining step. A mechanism similar to the one in Problem 14.93 is appropriate.

Work:

$$CH_3CH_2OH + H^+ \rightleftharpoons CH_3CH_2OH_2^+ \quad \text{(fast)}$$

$$CH_3CH_2OH_2^+ \longrightarrow CH_2{=}CH_2 + H_2O + H^+ \quad \text{(slow)}$$

This mechanism is justified in the following manner. The second step is the slow step. From this we would write that $r = k[CH_3CH_2OH_2^+]$, however, $CH_3CH_2OH_2^+$ is an intermediate in the process. $CH_3CH_2OH_2^+$ is maintained in a state of equilibrium in the first fast step.

The equilibrium has $K_{eq} = \frac{[CH_3CH_2OH_2^+]}{[CH_3CH_2OH][H^+]}$. Rearranging the equilibrium expression gives $[CH_3CH_2OH_2^+] = K_{eq}[CH_3CH_2OH][H^+]$. Substituting this into the rate law gives:

$$r = kK_{eq}[CH_3CH_2OH][H^+]$$

The rate constant k' is equal to kK_{eq}, where k is the rate constant of the rate determining step, and K_{eq} is for the fast equilibrium step.

15 Acids and Bases

15.1 What is meant by the terms, "Arrhenius acid" and "Arrhenius base"? Give examples of each.

An Arrhenius acid releases H^+ (a proton) when dissolved in water. Examples are HCl, HNO_2, HCN, H_3PO_4 and $HClO_3$. An Arrhenius base releases OH^- when dissolved in water. Examples are KOH, LiOH, $Ba(OH)_2$, CsOH and $Sr(OH)_2$.

15.3 What is the hydronium ion? How does it differ from, say, a hydrated sodium ion?

The hydronium ion is a hydrated hydrogen ion, H_3O^+. In the hydronium ion, a strong covalent bond holds the hydrogen ion to the water molecule. In the hydrated sodium ion, water molecules are attracted to the sodium ion by a weaker ion-dipole attraction.

15.5 What are Brønsted-Lowry acids and bases? Give examples of each.

A Brønsted-Lowry acid is a proton donor. Examples are H_2SO_4, HBr, H_2CO_3, $HC_2O_4^-$ and $HC_2H_3O_2$. A Brønsted-Lowry base is a proton acceptor. NH_3, CH_3NH_2, CN^-, PO_4^{3-} and F^- are examples of Brønsted-Lowry bases.

15.7 Is it possible for a substance to be both an Arrhenius base and a Brønsted-Lowry base? If so, give an example.

No, it is not possible. NaOH is an Arrhenius base because it gives up hydroxide ion. The Brønsted-Lowry base associated with NaOH is OH^-, not the whole NaOH formula unit. NH_3 is a Brønsted-Lowry base because it can accept a proton. It is not an Arrhenius base because it does not have a hydroxide ion in its formula.

15.9 Write the equation of a base ionization in aqueous solution.

The most common weak base is NH_3 which ionizes according to the reaction below.

$$NH_3(aq) + H_2O(\ell) \rightleftharpoons NH_4^+(aq) + OH^-(aq)$$

15.11 What is a conjugate pair? Give examples, including both charged and uncharged acids.

The two members of a conjugate pair differ by one proton. The acid has one more H^+ than does its conjugate base. Examples are H_2SO_4/HSO_4^-, NH_4^+/NH_3, HPO_4^{2-}/PO_4^{3-}, and HF/F^-. In each pair, the acid is shown first.

15.13 What is meant by the leveling effect of water?

Water is a fairly effective base and, thus, enables several acids to ionize 100% when dissolved in it. H_3O^+ is the strongest acid that can exist in aqueous solution because any stronger acid ionizes to produce H_3O^+. Similarly, the strongest base that can exist in aqueous solution is OH^- because any stronger base ionizes to produce OH^-.

15.15 What is meant by autoionization? Write the autoionization reaction for water.

The same molecule serves as both the proton donor and the proton acceptor. The autoionization reaction for water is: $H_2O(\ell) + H_2O(\ell) \rightleftharpoons H_3O^+(aq) + OH^-(aq)$.

15.17 Calculate the molarity of each of the following solutions.
a. 15.18 g NaCl in 100 mL solution
b. 50.0 g H_2SO_4 in 500 mL solution
c. 0.100 g phenol (C_6H_5OH) in 1.00 L solution

Analysis:
Target: The molarity of a solute in a solution. (? mol L^{-1}) =
Knowns: The mass of the solute and the solution volume. The molar mass of the solute is an implicit known.
Relationships: Molarity = $\frac{\text{mol solute}}{\text{volume of solution in L}}$. The molar mass provides the equivalence statement g ↔ mol. The relationship 10^{-3} L ↔ 1 mL also applies.
Plan: Use the definition of molarity with the factor-label method to convert the mass to a number of moles and the volume to liters.
Work:
a. Molar Mass NaCl = 1(22.98977 Na) + 1(35.453 Cl) = 58.443 g mol^{-1}

$$? \text{ mol } L^{-1} = \frac{15.18 \text{ g NaCl}}{100 \text{ mL}} \times \frac{1 \text{ mol NaCl}}{58.443 \text{ g NaCl}} \times \frac{1 \text{ mL}}{10^{-3} \text{ L}} = \mathbf{2.60 \text{ M NaCl}}$$

b. Molar Mass H_2SO_4 = 2(1.0079 H) + 1(32.06 S) + 4(15.9994 O) = 98.07 g mol^{-1}

$$? \text{ mol } L^{-1} = \frac{50.0 \text{ g } H_2SO_4}{500 \text{ mL}} \times \frac{1 \text{ mol } H_2SO_4}{98.07 \text{ g } H_2SO_4} \times \frac{1 \text{ mL}}{10^{-3} \text{ L}} = \mathbf{1.02 \text{ M } H_2SO_4}$$

c. Molar Mass C_6H_5OH = 6(12.011 C) + 6(1.0079 H) + 1(15.9994 O) = 94.113 g mol^{-1}

$$? \text{ mol } L^{-1} = \frac{0.100 \text{ g } C_6H_5OH}{1.00 \text{ L}} \times \frac{1 \text{ mol } C_6H_5OH}{94.113 \text{ g } C_6H_5OH} = \mathbf{1.06 \times 10^{-3} \text{ M } C_6H_5OH}$$

15.19 Calculate the mass of solute required to prepare each of the following solutions.
a. 350 mL of 0.0100 M CsI
b. 50 mL of 3.0 M KOH
c. 500 mL of 0.0100 M glucose ($C_6H_{12}O_6$)

Analysis:
Target: The mass of a solute in a solution. (? g) =
Knowns: The molarity of the solute and the solution volume. The molar mass of the solute is an implicit known.
Relationships: The number of moles is equal to M•V, where M is the molarity, which is $\frac{\text{mol solute}}{\text{volume of solution in L}}$, and V is the volume in liters. The molar mass provides the equivalence statement g ↔ mol. The relationship 10^{-3} L ↔ 1 mL also applies.
Plan: Find the number of moles from the relationship n = MV, using the factor-label method to convert the volume to liters and then to convert the number of moles to grams.
Work:
a. Molar Mass CsI = 1(132.91 Cs) + 1(126.90 I) = 259.81 g mol^{-1}

$$? \text{ g CsI} = \frac{0.0100 \text{ mol CsI}}{1 \text{ L}} \times 350 \text{ mL} \times \frac{10^{-3} \text{ L}}{1 \text{ mL}} \times \frac{259.81 \text{ g CsI}}{1 \text{ mol CsI}} = \mathbf{0.909 \text{ g CsI}}$$

b. Molar Mass KOH = 1(39.0983 K) + 1(15.9994 O) + 1(1.0079 H) = 56.1056 g mol^{-1}

$$? \text{ g} = \frac{3.0 \text{ mol KOH}}{1 \text{ L}} \times 50 \text{ mL} \times \frac{10^{-3} \text{ L}}{1 \text{ mL}} \times \frac{56.1056 \text{ g KOH}}{1 \text{ mol KOH}} = \mathbf{8.4 \text{ g KOH}}$$

c. Molar Mass $C_6H_{12}O_6$ = 6(12.011 C) + 12(1.0079 H) + 6(15.9994 O) = 180.157 g mol^{-1}

$$? \text{ g} = \frac{0.0100 \text{ mol } C_6H_{12}O_6}{1 \text{ L}} \times 500 \text{ mL} \times \frac{10^{-3} \text{ L}}{1 \text{ mL}} \times \frac{180.157 \text{ g } C_6H_{12}O_6}{1 \text{ mol } C_6H_{12}O_6} = \mathbf{0.901 \text{ g } C_6H_{12}O_6}$$

15.21 How much water must be added to 25.0 mL of a 0.100 M solution in order to reduce its concentration to 0.0300 M?

Analysis:
Target: A volume of water. (? mL) =
Knowns: An initial molarity and volume and a final molarity.
Relationships: The relationship $M_{stock}V_{stock} = M_{dilute}V_{dilute}$ applies. (Review Chapter 3 if necessary.) Volumes are additive.
Plan: Rearrange the relationship to solve for the volume of dilute solution. Subtract the initial volume from the volume of dilute solution to determine the volume of water.
Work:

$$V_{dilute} = \frac{M_{stock}V_{stock}}{M_{dilute}} = \frac{(25.0 \text{ m})(0.100 \text{ M})}{(0.0300 \text{ M})} = 83.3 \text{ mL.}$$

? mL H_2O = 83.3 mL dilute solution - 25.0 mL original solution = **58.3 mL H_2O**

15.23 Calculate the pH of the following solutions of strong acids.
a. 5.00×10^{-1} M HCl b. 0.030 M HNO_3 c. 0.75 g L^{-1} $HClO_4$

Analysis:
Target: The pH of solutions of strong acids. (? pH) =
Knowns: The molarity of the solute or the mass of the solute and the solution volume. The molar mass of the solute is an implicit known.
Relationships: pH = $-\log[H_3O^+]$. A strong acid ionizes 100% when dissolved in water to create $[H_3O^+]$ and its conjugate base. Molarity = $\frac{\text{mol solute}}{\text{volume of solution in L}}$. The molar mass provides the equivalence statement g ↔ mol.
Plan: Use the given molarity to determine the pH. Each acid is monoprotic, so the $[H_3O^+]$ is equal to the original molarity of the acid. For part c, first find the molarity from its definition, using the factor-label method to convert the mass to the number of moles.
Work:

a. $[H_3O^+] = 5.00 \times 10^{-1}$ M; so pH = $-\log(5.00 \times 10^{-1}) = -(-0.301) =$ **0.301**

b. $[H_3O^+] = 0.030$ M; so pH = $-\log(0.030) = -(-1.52) =$ **1.52**

c. Molar Mass $HClO_4$ = 1(1.0079 H) + 1(35.453 Cl) + 4(15.9994 O) = 100.458 g mol^{-1}

$$? \text{ mol L}^{-1} = \frac{0.75 \text{ g } HClO_4}{\text{L}} \times \frac{1 \text{ mol } HClO_4}{100.453 \text{ g } HClO_4} = 7.5 \times 10^{-3} \text{ M } HClO_4$$

$[H_3O^+] = 7.5 \times 10^{-3}$ M; so pH = $-\log(7.5 \times 10^{-3}) = -(-2.13) =$ **2.13**

15.25 To what molarity of HCl solution do the following pH values correspond? 3.05, 5.7, 0.0, -1.00, 2.43.

Analysis:
Target: The molarity of several HCl solutions. (? mol L^{-1}) =
Known: HCl is a strong monoprotic acid.
Relationships: $[H_3O^+] = 10^{-pH}$. A strong acid ionizes 100% in water, so the original molarity of the acid is equal to the $[H_3O^+]$ concentration.
Plan: Use the relationship directly.
Work:

pH = 3.05; [HCl] = $[H_3O^+] = 10^{-pH} = 10^{-3.05} =$ **8.9×10^{-4} M HCl**

pH = 5.7; [HCl] = $[H_3O^+]$ = 10^{-pH} = $10^{-5.7}$ = **2 x 10^{-6} M HCl**
pH = 0.0; [HCl] = $[H_3O^+]$ = 10^{-pH} = $10^{-0.0}$ = **1 M HCl**
pH = -1.00; [HCl] = $[H_3O^+]$ = 10^{-pH} = $10^{-(-1.00)}$ = **10. M HCl or 1.0 x 10^1 M HCl**
pH = 2.43; [HCl] = $[H_3O^+]$ = 10^{-pH} = $10^{-2.43}$ = **3.7 x 10^{-3} M HCl**

15.27 Calculate the pOH of solutions having the following values of pH: 4.19, 7.00, 8.25, 12.2.

Analysis/Plan: In aqueous solution at 25°C, pH + pOH = 14.00. Rearrange to solve for the pOH, insert the value of pH and calculate.
Work:
In each case, pOH = 14.00 - pH.
pH = 4.19; so pOH = 14.00 - 4.19 = **9.81**
pH = 7.00; so pOH = 14.00 - 7.00 = **7.00**
pH = 8.25; so pOH = 14.00 - 8.25 = **5.75**
pH = 12.2; so pOH = 14.00 - 12.2 = **1.8**

15.29 Calculate the pH of the following solutions of strong bases.
a. 3 M NaOH b. 1.3 x 10^{-4} M $Sr(OH)_2$ c. 2.5 g L^{-1} LiOH

Analysis:
Target: The pH of solutions of strong bases. (? pH) =
Knowns: The molarity of the solute or the mass of the solute and the solution volume. The molar mass of the solute is an implicit known.
Relationships: pOH = $-\log[OH^-]$. pH + pOH = 14.00. A strong base dissociates 100% when dissolved in water to create OH^- and a cation. Molarity = $\frac{\text{mol solute}}{\text{L solution}}$. The molar mass provides the equivalence statement g ↔ mol.
Plan: Use the given molarity to determine the pOH, considering the stoichiometry of dissociation as needed, then use the relationship pH + pOH = 14.00 to find the pH. For part c, first find the molarity from its definition, using the factor-label method to convert the mass to number of moles.
Work:

a. $$\frac{3 \text{ mol NaOH}}{1 \text{ L}} \times \frac{1 \text{ mol OH}^-}{1 \text{ mol NaOH}} = 3 \text{ M OH}^-$$

pOH = -log(3) = -(0.5) = -0.5; pH = 14.00 - pOH = 14.00 - (-0.5) = **14.5**

b. $$\frac{1.3 \times 10^{-4} \text{ mol Sr(OH)}_2}{1 \text{ L}} \times \frac{2 \text{ mol OH}^-}{1 \text{ mol Sr(OH)}_2} = 2.6 \times 10^{-4} \text{ M OH}^-$$

pOH = -log(2.6 x 10^{-4}) = -(-3.59) = 3.59; pH = 14.00 - pOH = 14.00 - 3.59 = **10.41**

c. Molar Mass LiOH = 1(6.941 Li) + 1(15.9994 O) + 1(1.0079 H) = 23.948 g mol^{-1}

$$? \text{ mol L}^{-1} = \frac{2.5 \text{ g LiOH}}{1 \text{ L}} \times \frac{1 \text{ mol LiOH}}{23.948 \text{ g LiOH}} \times \frac{1 \text{ mol OH}^-}{1 \text{ mol LiOH}} = 0.10 \text{ M OH}^-$$

pOH = -log(0.10) = -(-0.98) = 0.98; pH = 14.00 - pOH = 14.00 - 0.98 = **13.02**

15.31 Calculate the percent ionization in each of the following solutions of weak, monoprotic acids.
a. 0.015 M, $[H_3O^+]$ = 1.0 x 10^{-4} b. 9.95 x 10^{-2} M, $[H_3O^+]$ = 1.5 x 10^{-3}
c. 4.5 x 10^{-3} M, pH = 5.52

Analysis:
Target: A percent ionization. (? % ionization) =
Knowns: The initial concentration of the acid and either its $[H_3O^+]$ or its pH.

Relationships: The percent ionization = $\frac{\text{[ionized form]}}{\text{[initial acid]}}$ x 100. $[H_3O^+] = 10^{-pH}$. H_3O^+ and an anion are the ionized forms of the weak acid.

Plan: Use the formula for percent ionization directly if $[H_3O^+]$ is given. If it is not, find the $[H_3O^+]$ from the pH and then use the formula.

Work:

a. $? \text{ \% ionization} = \frac{1.0 \times 10^{-4}\text{ M}}{0.015\text{ M}} \times 100 = \mathbf{0.67\%}$

b. $? \text{ \% ionization} = \frac{1.5 \times 10^{-3}\text{ M}}{9.95 \times 10^{-2}\text{ M}} \times 100 = \mathbf{1.5\%}$

c. $[H_3O^+] = 10^{-pH} = 10^{-5.52} = 3.0 \times 10^{-6}\text{ M}$

$? \text{ \% ionization} = \frac{3.0 \times 10^{-6}\text{ M}}{4.5 \times 10^{-3}\text{ M}} \times 100 = \mathbf{0.067\%}$

15.33 If a 0.503 M solution of a weak acid is 3.0×10^{-2}% ionized, what is the hydrogen ion concentration? What is the pH?

Analysis:

Target: A hydrogen ion concentration and a pH. (? M) = and (? pH) =

Knowns: The initial concentration of the acid and its percent ionization.

Relationships: The percent ionization = $\frac{\text{[ionized form]}}{\text{[initial acid]}}$ x 100. pH = $-\log[H_3O^+]$. The H_3O^+ and an anion are the ionized forms of the weak acid.

Plan: Rearrange the formula to determine the $[H_3O^+]$. Use this molarity with the definition of pH to determine the pH.

Work:

$\text{\% ionization} = \frac{\text{[ionized form]}}{\text{[initial acid]}} \times 100$; therefore, $\text{[ionized form]} = \frac{\text{(\% ionization)[initial acid]}}{100}$

$? \text{ mol L}^{-1} = [H_3O^+] = \text{[ionized form]} = \frac{(3.0 \times 10^{-2}\text{ \%})(0.503\text{ M})}{100} = \mathbf{1.5 \times 10^{-4}\text{ M } H_3O^+}$

$? \text{ pH} = -\log[H_3O^+] = -\log(1.5 \times 10^{-4}) = -(-3.82) = \mathbf{3.82}$

15.35 Calculate the hydroxide ion concentration and the pH of the following solutions of weak bases, given the percent ionization.

a. 3.6×10^{-2} M (1.5%)

b. 0.78 M (3.2×10^{-3}%)

Analysis:

Target: A hydroxide ion concentration and a pH. (? M) = and (? pH) =

Knowns: The initial concentration of the base and its percent ionization.

Relationships: The percent ionization = $\frac{\text{[ionized form]}}{\text{[initial acid]}}$ x 100. pOH = $-\log[OH^-]$.

pH + pOH = 14.00. OH^- and a cation are the ionized forms of the weak base.

Plan: Rearrange the formula to determine the $[OH^-]$. Use this molarity with the definition to determine the pOH. Then determine the pH from the relationship pH + pOH = 14.00.

Work:

For both a and b, the following relationship is used.

$\text{\% ionization} = \frac{\text{[ionized form]}}{\text{[initial base]}} \times 100$; therefore, $\text{[ionized form]} = \frac{\text{(\% ionization)[initial base]}}{100}$

a. $? \text{ mol L}^{-1} = [OH^-] = \text{[ionized form]} = \frac{(1.5\%)(3.6 \times 10^{-2}\text{ M})}{100} = \mathbf{5.4 \times 10^{-4}\text{ M } OH^-}$

pOH = -log[OH^-] = -log(5.4×10^{-4}) = -(-3.27) = 3.27

? pH = 14.00 - pOH = 14.00 - 3.27 = **10.73**

b. ? mol L^{-1} = [OH^-] = [ionized form] = $\frac{(3.2 \times 10^{-3}\%)(0.78\ M)}{100}$ = **2.5×10^{-5} M OH^-**

pOH = -log[OH] = -log(2.5×10^{-5}) = -(-4.60) = 4.60

? pH = 14.00 - pOH = 14.00 - 4.60 = **9.40**

15.37 Classify each of the following acids as strong or weak: HF, HCl, HBr, HCN, H_2SO_4, HSO_4^-, HNO_3.

Analysis/Plan: The strong acids in Table 15.1 in the text should be memorized. Other acids are assumed to be weak.

Work:

HCl, HBr, HNO_3 and H_2SO_4 are **strong acids**.

HF, HCN, and HSO_4^- are **weak acids.**

15.39 Classify each of the following bases as strong or weak: CsOH, OH^-, H_2O, HCO_3^-, NH_2^-.

Analysis/Plan: All soluble metal hydroxide bases are strong. The other strong bases in Table 15.4 in the text should be memorized. Other bases are assumed to be weak.

Work:

CsOH, OH^- and NH_2^- are **strong bases**.

H_2O and HCO_3^- are **weak bases**.

15.41 Identify all species in the following reactions as either an acid or base, in the Brønsted-Lowry sense.

a. $CN^- + H_2O \rightleftharpoons HCN + OH^-$

b. $HSO_4^- + H_2CO_3 \rightleftharpoons HCO_3^- + H_2SO_4$

c. $H_2C_2O_4 + NO_2^- \rightleftharpoons HNO_2 + HC_2O_4^-$

Analysis/Plan: In Brønsted-Lowry theory, an acid is a proton (H^+) donor and a base is a proton acceptor. Study the equation to determine if the species is giving or receiving a proton to produce its conjugate.

Work:

In each case, conjugates have the same subscript.

a. $CN^- + H_2O \rightleftharpoons HCN + OH^-$
base₁ acid₂ acid₁ base₂

b. $HSO_4^- + H_2CO_3 \rightleftharpoons HCO_3^- + H_2SO_4$
base₁ acid₂ base₂ acid₁

c. $H_2C_2O_4 + NO_2^- \rightleftharpoons HNO_2 + HC_2O_4^-$
acid₁ base₂ acid₂ base₁

15.43 Arrange the species in the reactions of Problem 15.41 as Brønsted-Lowry conjugate pairs.

Analysis/Plan: Conjugates differ by one H^+, with the acid having one more H^+ than its conjugate base. They were designated with subscripts in the answer to Problem 15.41. Conjugate pairs are shown with the acid first.

Work:

a. **HCN/CN^-, H_2O/OH^-**

b. **H_2SO_4/HSO_4^-, H_2CO_3/HCO_3^-**

c. **$H_2C_2O_4/HC_2O_4^-$, HNO_2/NO_2^-**

15.45 What is the conjugate base of each of the following acids?

a. H_2SO_4 b. H_2F^+ c. CH_3OH d. H_3O^+

Analysis:

Target: The conjugate base of a acid.
Known: The formula of the acid.
Relationship: A conjugate base has one less H^+ than its conjugate acid.

Plan: Subtract one H^+ from the formula of the acid to obtain its conjugate base.

Work:

a. Subtracting an H from the formula gives HSO_4. Subtracting 1 from the charge gives 0 - 1 = -1. The conjugate base is **HSO_4^-**.

b. Subtracting an H from the formula gives HF^+. Subtracting 1 from the charge gives 1 - 1 = 0. The conjugate base is **HF**.

c. Subtracting an H from the formula gives CH_3O. Subtracting 1 from the charge gives 0 - 1 = -1. The conjugate base is **CH_3O^-**.

d. Subtracting an H from the formula gives H_2O^+. Subtracting 1 from the charge gives 0 - 1 = -1. The conjugate base is **H_2O**.

15.47 What is the conjugate acid of each of the following bases?

a. H_2SO_4 b. CH_3OH c. I^-

Analysis:

Target: The conjugate acid of an base.
Known: The formula of the base.
Relationship: A conjugate acid has one more H^+ than its conjugate base.

Plan: Add one H^+ to the formula of the base to obtain its conjugate acid.

Work:

a. Adding an H to the formula gives H_3SO_4. Adding 1 to the charge gives 0 + 1 = +1. The conjugate acid is **$H_3SO_4^+$**.

b. Adding an H to the formula gives CH_3OH_2. Adding 1 to the charge gives 0 + 1 = +1. The conjugate acid is **$CH_3OH_2^+$**.

c. Adding an H to the formula gives HI. Adding 1 to the charge gives -1 + 1 = 0. The conjugate acid is **HI**.

15.49 Write the aqueous ionization reaction for each of the following acids.

a. nitrous acid b. hydrobromic acid c. acetic acid

Analysis:

Target: The ionization reaction of an acid in water.
Knowns: The name of the acid. The formula of the acid is an implicit known.
Relationships: An acid will donate a proton to water forming its conjugate base and the hydronium ion. A weak acid will partially ionize, establishing an equilibrium, while a strong acid will ionize almost completely.

Plan: Write the formula of the acid from its name, then show the acid transferring its proton to water to create the conjugates. Use a double arrow for a weak acid and a single arrow for a strong acid.

Work:

a. Nitrous acid is a weak acid, HNO_2. The reaction is:

$$HNO_2 + H_2O \rightleftharpoons H_3O^+ + NO_2^-.$$

b. Hydrobromic acid is a strong acid, HBr. The reaction is:

$$HBr + H_2O \rightarrow H_3O^+ + Br^-.$$

c. Acetic acid is a weak acid, CH_3COOH. The reaction is:

$$CH_3COOH + H_2O \rightleftharpoons H_3O^+ + CH_3COO^-.$$

15.51 Write the aqueous ionization reaction for each of the following bases.
a. nicotine ($C_{10}H_{14}N_2$) b. hydrazine (N_2H_4) c. trimethylamine [$(CH_3)_3N$]

Analysis:
Target: The ionization reaction of a base in water.
Known: The formula of the base.
Relationships: A base will accept a proton from water forming its conjugate acid and the hydroxide ion. All Brønsted-Lowry bases are weak and will partially ionize, establishing an equilibrium.
Plan: Show the base accepting a proton from water to create the conjugates. Use a double arrow to show the base is weak.
Work:

a. $C_{10}H_{14}N_2 + H_2O \rightleftharpoons C_{10}H_{15}N_2^+ + OH^-$

b. $N_2H_4 + H_2O \rightleftharpoons N_2H_5^+ + OH^-$

c. $(CH_3)_3N + H_2O \rightleftharpoons (CH_3)_3NH^+ + OH^-$

15.53 Classify each of the following solutes according to the pH of their aqueous solutions, as less than 7, approximately 7, or greater than 7.
a. KNO_3 b. K_2CO_3 c. NH_4Cl d. Na_2S e. KNH_2

Analysis:
Target: The approximate pH of a salt solution.
Knowns: The formula of the salt. The ions which comprise the salt are implicit knowns. The strong acids and bases must be memorized.
Relationships: The conjugate base of a strong acid has no ability to accept a proton in water, while the conjugate base of a weak acid functions as a base in water. Similarly, the cation of a strong base has no acid or base ability, but the conjugate acid of a weak base functions as an acid, donating a proton to water.
Plan: Determine the ions that make up the salt and decide if they can either accept or donate a proton. If a proton can be donated, the solution will be acidic with pH less than 7. If a proton can be accepted, the solution will be basic with a pH greater than 7. If neither ion has acidic or basic ability in water, the solution will be neutral with a pH of 7.
Work:

a. K^+ is the cation of KOH, a strong base. It cannot donate a proton because it does not have one. It cannot accept a proton because it has no lone pair to which a proton can be transferred. NO_3^- is the conjugate base of the strong acid HNO_3. It has essentially no ability to accept a proton from water. It cannot donate a proton because it does not have one. Because neither ion reacts with water, the solution is neutral with **pH = 7.**

b. K^+ cannot act as an acid or base (part a). CO_3^{2-} is the conjugate base of the weak acid HCO_3^-. It can accept a proton from water, functioning as a base. Because the CO_3^{2-} functions as a base, the solution is basic with a **pH greater than 7**.

c. NH_4^+ is the conjugate acid of the weak base NH_3. It can donate a proton to water. Cl^- is the conjugate base of the strong acid HCl. It has essentially no ability to accept a proton from water. It cannot donate a proton because it does not have one. Because the NH_4^+ can act as an acid, the solution is acidic with a **pH less than 7**.

d. Na^+ is the cation of NaOH, a strong base. It cannot donate a proton because it does not have one. It cannot accept a proton because it has no lone pair to which a proton can be transferred. S^{2-} is the conjugate base of the weak acid HS^-. It can accept a proton from water. Because the S^{2-} acts as a base, the solution is basic with a **pH greater than 7**.

e. K^+ cannot act as an acid or base (part a). NH_2^- is the conjugate base of the very weak acid NH_3. NH_2 is also listed in Table 15.4 as a strong base. It can accept a proton from water. Because the NH_2^- acts as a base, the solution is basic with a **pH greater than 7**.

15.55 Write the equation of the hydrolysis reaction(s) that occur(s) when each of the solutes in Problem 15.53 is dissolved.

Analysis/Plan: Show each reactive species in Problem 15.53 either donating a proton to water or accepting a proton from water in accordance with the discussion in Problem 15.53.

Work:

a. Neither ion has acid/base ability so there is **no reaction**.

b. CO_3^{2-} acts as a base, accepting a proton from water.

$$\mathbf{CO_3^{2-} + H_2O \rightleftharpoons HCO_3^- + OH^-}$$

c. NH_4^+ acts as an acid, donating a proton to water.

$$\mathbf{NH_4^+ + H_2O \rightleftharpoons H_3O^+ + NH_3}$$

d. S^{2-} functions as a base, accepting a proton from water.

$$\mathbf{S^{2-} + H_2O \rightarrow HS^- + OH^-}$$

e. NH_2^- functions as a base, accepting a proton from water.

$$\mathbf{NH_2^- + H_2O \rightarrow NH_3 + OH^-}$$

15.57 For each of the following amphiprotic species, write two different proton transfer reactions with water.

a. HSO_4^- b. HCO_3^-

Analysis/Plan: An amphiprotic species can both accept a proton to act as a base and donate a proton to act as an acid. Write an equation for each process, showing water accepting the proton when the species functions as an acid and donating the proton when the species functions as a base.

Work:

a. HSO_4^- as an acid: $\mathbf{HSO_4^- + H_2O \rightleftharpoons H_3O^+ + SO_4^{2-}}$

HSO_4^- as a base: $\mathbf{HSO_4^- + H_2O \rightleftharpoons H_2SO_4 + OH^-}$

b. HCO_3^- as an acid: $\mathbf{HCO_3^- + H_2O \rightleftharpoons H_3O^+ + CO_3^{2-}}$

HCO_3^- as a base: $\mathbf{HCO_3^- + H_2O \rightleftharpoons H_2CO_3 + OH^-}$

15.59 Which is the stronger acid in each of the following pairs?

a. HCl, HNO_2 b. CH_3COOH, HF

Analysis/Plan: Consult Table 15.2 in the text which lists acids in terms of relative strength. The strong acids should also be memorized.

Work:

a. **HCl** is a stronger acid than HNO_2. (HCl is a strong acid, while HNO_2 is a weak acid.)

b. **HF** is a stronger acid than CH_3COOH.

15.61 Which the stronger base in each of the following pairs?

a. Br^-, OH^- b. NO_2^-, F^-

Analysis/Plan: Consult Table 15.2 in the text which lists acids in terms of relative strength. Note that a stronger acid has a weaker conjugate base.

Work:

a. OH^- is a stronger base that Br^-. (Br^- is the conjugate base of a strong acid.)

b. NO_2^- is a stronger base than F^-.

15.63 Write the equation for the proton transfer equilibrium that occurs between each of the following pairs of reactants. Indicate whether the equilibrium lies to the right or the left.

a. HF, NO_2^- b. HSO_4^-, HCO_3^-

Analysis/Plan: Consult Table 15.2 in the text for relative acid and base strengths. The stronger acid will donate the proton and the stronger base will accept it. The equilibrium will lie to the side of the weaker acid and, thus, the weaker base.

Work:

a. HF is a stronger acid than NO_2^-.

The equilibrium is $HF + NO_2^- \rightleftharpoons HNO_2 + F^-$.

The equilibrium lies to the **right**, because HNO_2 is a weaker acid than HF and F^- is a weaker base than NO_2^-.

b. HSO_4^- is a stronger acid than HCO_3^-.

The equilibrium is $HSO_4^- + HCO_3^- \rightleftharpoons SO_4^{2-} + H_2CO_3$.

The equilibrium lies to the **right**, because SO_4^{2-} is a weaker base than HCO_3^- and H_2CO_3 is a weaker acid than HSO_4^-.

15.65 Write the chemical equations for the stepwise ionization of oxalic acid ($H_2C_2O_4$), a diprotic acid.

Analysis/Plan: In each step of the ionization, the acid donates one proton to water, until no acidic hydrogens remain. Show the proton transfers in turn, with water functioning as the base in the ionization process.

Work:

$$H_2C_2O_4 + H_2O \rightleftharpoons H_3O^+ + HC_2O_4^-$$

$$HC_2O_4^- + H_2O \rightleftharpoons H_3O^+ + C_2O_4^{2-}$$

15.67 Which is the stronger acid, $H_2PO_4^-$ or HPO_4^{2-}? Why

$H_2PO_4^-$ is the stronger acid, because it is harder to remove a proton and its positive charge from the doubly charged HPO_4^{2-} than it is from the singly charged $H_2PO_4^-$.

15.69 Sulfide ion is a dihydric base. Write the equations for its stepwise acceptance of two protons from water.

Analysis/Plan: In each step of the ionization, the base accepts one proton from water. Show the proton transfers in turn, with water functioning as the acid in the ionization process.

Work:

$$S^{2-} + H_2O \rightleftharpoons HS^- + OH^-$$

$HS^- + H_2O \rightleftharpoons H_2S + OH^-$

15.71 Arrange the following binary hydrides in order of increasing acidity, and explain your reasoning: PH_3, NH_3, AsH_3.

Analysis/Plan: The acidity of binary hydrides increases with the size of the central atom within any group. This happens because increasing the bond length makes it easier to break the bond and, therefore, the acid is stronger. Determine the position of the central atom on the chart and apply this logic.
Work:
Radius increases down a group, therefore, acid strengths are **$NH_3 < PH_3 < AsH_3$**.

15.73 In each of the following pairs, identify the stronger acid and explain your reasoning.
a. FCH_2COOH, $F_2CHCOOH$ b. H_2Se, H_2Te c. H_3PO_4, $H_2PO_4^-$

Analysis/Plan: Study the relationship between the two acid formulas and analyze according to the similarities and differences in the structures.
Work:
a. Both of these are oxoacids. **$F_2CHCOOH$ is stronger than FCH_2COOH** because F is more electronegative than H. The additional F atom has a greater inductive effect of withdrawing electron density from the O-H bond and weakening it, making it easier to remove the proton.
b. Both of these are binary acids in the same group. **H_2Te is stronger** because Te has a larger radius than Se. See Problem 15.71.
c. **H_3PO_4 is stronger** because it is harder to remove a proton with its positive charge from a negatively charged species ($H_2PO_4^-$) than from a neutral species (H_3PO_4).

15.75 Arrange the following acids in order of increasing strength, and explain your reasoning: $HClO_2$, $HOCl$, $HOBr$.

Analysis/Plan: Study the relationship between the three acid formulas and analyze according to the similarities and differences in the structures.
Work:
HOCl is stronger than HOBr because Cl is more electronegative than Br. The greater inductive effect weakens the O-H bond and makes it easier to remove the proton in HOCl. Similarly, $HClO_2$ is stronger than HOCl. The additional oxygen pulls electron density from the O-H bond weakening it and making it still easier to remove the proton. The overall order of increasing acid strength is **$HOBr < HOCl < HClO_2$**.

15.77 Why is NH_2^- a stronger base than OH^-? (*Hint:* Explain in terms of the relative strengths of the conjugate acids.)

OH^- is the conjugate base of H_2O and NH_2^- is the conjugate base of NH_3. The acidity of binary hydrides in the same period increases with the electronegativity of the central atom because the H-X bond is more polar and more easily broken. O is more electronegative than N; therefore, H_2O is a stronger acid than NH_3. A stronger acid is more likely to give up a proton; consequently, its conjugate base is less likely to accept a proton and is a weaker base.

15.79 In each of the following reactions, identify each species as a Lewis acid, base, or adduct.
a. $AlCl_3 + Br^- \rightarrow AlCl_3Br^-$
b. $Be(OH)_2(s) + 2\ OH^-(aq) \rightarrow Be(OH)_4^{2-}(aq)$

Analysis/Plan: A Lewis acid has an incomplete or expandable valence shell and accepts a pair of electrons to make a coordinate covalent bond. A Lewis base has a lone pair of electrons that it donates to make the coordinate covalent bond. Draw the Lewis structures and analyze which species is providing the electron pair and which species is accepting it. The adduct is the product of the process.

Work:

a.

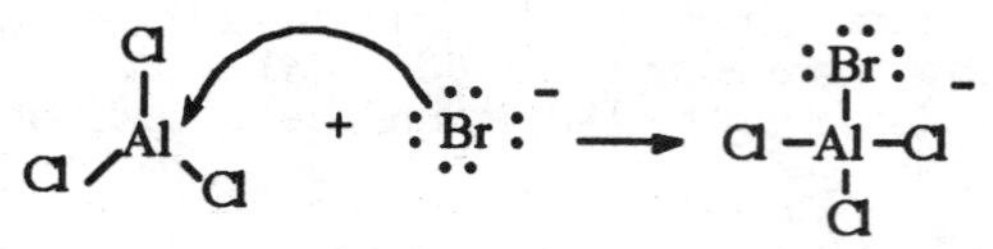

$AlCl_3 + :\ddot{Br}:^{-} \longrightarrow [AlCl_3Br]^{-}$

acid base adduct

b.

$HO{-}Be{-}OH + 2\,:\ddot{O}{-}H^{-} \longrightarrow [Be(OH)_4]^{2-}$

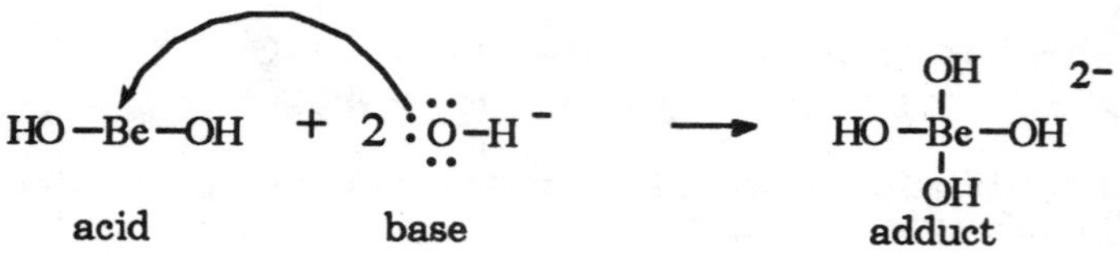

acid base adduct

15.81 What volume of 0.0998 M NaOH is required to reach the equivalence point in a titration of 33.5 mL 0.152 M HClO?

Analysis:

Target: The volume of a base solution. (? mL) = V_b

Knowns: The volume, the molarity and formula of the acid and the formula and molarity of the base.

Relationships: $n_a = n_b \cdot$(stoichiometric mole ratio). $n = M \cdot V$. The formulas of the acid and base enable us to write the balanced chemical equation for neutralization. The stoichiometric mole ratio is given by coefficients in the balanced equation.

Plan: Determine the stoichiometric mole ratio from the balanced chemical equation, and use the relationships $n_a = n_b \cdot$(mol ratio) and $n = M \cdot V$ to find the volume of the base.

Work:

The balanced chemical equation is $NaOH + HClO \longrightarrow NaClO + H_2O$, so the stoichiometric mole ratio is 1 mol acid $\leftrightarrow$ 1 mol base.

$$n_b = n_a\left(\frac{1 \text{ mol base}}{1 \text{ mol acid}}\right); \text{ therefore, } M_bV_b = M_aV_a(1)$$

$$? \text{ mL} = V_b = \frac{M_aV_a}{M_b} = \frac{(0.152 \text{ M})(33.5 \text{ mL})(1)}{0.0998 \text{ M}} = \mathbf{51.0 \text{ mL NaOH solution}}$$

15.83 If 23.07 mL of 0.1000 M NaOH is required in the titration of 16.51 mL aqueous hydrofluoric acid, what is the acid molarity?

Analysis:

Target: The molarity of an acid solution. (? mol L^{-1}) = M_a

Knowns: The name and volume of the acid and the volume, formula and molarity of the base. The name of the acid gives its formula, HF.

Relationships: $n_a = n_b \cdot$(stoichiometric mole ratio). $n = M \cdot V$. The formulas of the acid and base enable us to write the balanced chemical equation for neutralization. The stoichiometric mole ratio is given by coefficients in the balanced equation.

Plan: Determine the stoichiometric mole ratio from the balanced chemical equation, and use the relationships $n_a = n_b \cdot$(mol ratio) and $n = M \cdot V$ to find the molarity of the acid.

Work:

The balanced chemical equation is $NaOH + HF \longrightarrow NaF + H_2O$, so the stoichiometric mole ratio is 1 mol acid ↔ 1 mol base.

$$n_a = n_b\left(\frac{1 \text{ mol acid}}{1 \text{ mol base}}\right); \text{ therefore, } M_aV_a = M_bV_b(1)$$

$$? \text{ mol L}^{-1} = M_a = \frac{M_bV_b}{V_a} = \frac{(0.1000 \text{ M})(23.07 \text{ mL})(1)}{16.51 \text{ mL}} = \mathbf{0.1397 \text{ M HF}}$$

15.85 What volume of 0.2473 M HCl is required to completely neutralize 25.00 mL of 0.0930 M $Ba(OH)_2$?

Analysis:

Target: The volume of an acid solution. (? mL) = V_a

Knowns: The formula and volume of the acid and the volume, formula and molarity of the base.

Relationships: $n_a = n_b \cdot$(stoichiometric mole ratio). $n = M \cdot V$. The formulas of the acid and base enable us to write the balanced chemical equation for neutralization. The stoichiometric mole ratio is given by coefficients in the balanced equation.

Plan: Determine the stoichiometric mole ratio from the balanced chemical equation, and use the relationships $n_a = n_b \cdot$(mol ratio) and $n = M \cdot V$ to find the volume of the base.

Work:

The balanced chemical equation is $Ba(OH)_2 + 2\ HCl \longrightarrow BaCl_2 + 2\ H_2O$, so the stoichiometric mole ratio is 2 mol acid ↔ 1 mol base.

$$n_a = n_b\left(\frac{2 \text{ mol acid}}{1 \text{ mol base}}\right); \text{ therefore, } M_aV_a = M_bV_b(2)$$

$$? \text{ mL} = V_a = \frac{M_bV_b(2)}{M_a} = \frac{(0.0930 \text{ M})(25.0 \text{ mL})(2)}{0.2473 \text{ M}} = \mathbf{18.8 \text{ mL HCl solution}}$$

15.87 What volume of 0.500 M KOH is required to neutralize 5.72 g benzoic acid ($C_6H_5CO_2H$)?

Analysis:

Target: The volume of a base solution. (? mL) = V_b

Knowns: The mass and formula of the acid and the molarity and formula of the base. The molar mass of the acid is an implicit known

Relationships: $n_a = n_b \cdot$(stoichiometric mole ratio). $n = M \cdot V$. The formulas of the acid and base enable us to write the balanced chemical equation for neutralization. The stoichiometric mole ratio is given by coefficients in the balanced equation. The molar mass provides the equivalence statement g ↔ mol. The equivalence statement 10^{-3} L ↔ 1 mL is also needed.

Plan: Use the factor-label method to calculate the number of moles of acid. Determine the stoichiometric mole ratio from the balanced chemical equation, and use the relationships $n_a = n_b \cdot$(mol ratio) and $n = M \cdot V$ to find the volume of the base. Use the factor-label method to convert the volume to milliliters.

Work:

The balanced chemical equation is $KOH + C_6H_5CO_2H \longrightarrow KC_6H_5CO_2 + H_2O$, so the stoichiometric mole ratio is 1 mol acid ↔ 1 mol base.

Molar Mass $C_6H_5CO_2H$ = 7(12.011 C) + 6(1.0079 H) + 2(15.9994 O) = 122.123 g mol^{-1}

$$n_a = 5.72 \text{ g } C_6H_5CO_2H \times \frac{1 \text{ mol } C_6H_5CO_2H}{122.123 \text{ g } C_6H_5CO_2H} = 4.68 \times 10^{-2} \text{ mol } C_6H_5CO_2H$$

$$n_b = n_a\left(\frac{1\ \text{mol base}}{1\ \text{mol acid}}\right);\ \text{therefore, } M_bV_b = n_a(1)$$

$$?\ \text{mL} = V_b = \frac{n_a(1)}{M_b} = \frac{(4.68 \times 10^{-2}\ \text{mol})(1)}{0.500\ \text{M}} = 9.37 \times 10^{-2}\ \text{L} \times \frac{1\ \text{mL}}{10^{-3}\ \text{L}} =$$

93.7 mL KOH solution

15.89 What volume of 0.0200 M HCl is required to neutralize 3.48 g $Ca(OH)_2$?

Analysis:
Target: The volume of an acid solution. (? L) = V_a
Knowns: The mass and formula of the base and the molarity and formula of the acid. The molar mass of the base is an implicit known.
Relationships: $n_a = n_b$•(stoichiometric mole ratio). n = M•V. The formulas of the acid and base enable us to write the balanced chemical equation for neutralization. The stoichiometric mole ratio is given by coefficients in the balanced equation. The molar mass provides the equivalence statement g ↔ mol.
Plan: Use the factor-label method to calculate the number of moles of base. Determine the stoichiometric mole ratio from the balanced chemical equation, and use the relationships $n_a = n_b$•(mol ratio) and n = M•V to find the volume of the base.
Work:
The balanced chemical equation is $Ca(OH)_2 + 2\ HCl \longrightarrow CaCl_2 + 2\ H_2O$, so the stoichiometric mole ratio is 2 mol acid ↔ 1 mol base.
Molar Mass $Ca(OH)_2$ = 1(40.08 Ca) + 2(15.9994 O) + 2(1.0079) = 74.09 g mol^{-1}

$$n_b = 3.48\ \text{g Ca(OH)}_2 \times \frac{1\ \text{mol Ca(OH)}_2}{74.09\ \text{g Ca(OH)}_2} = 4.70 \times 10^{-2}\ \text{mol Ca(OH)}_2$$

$$n_a = n_b\left(\frac{2\ \text{mol acid}}{1\ \text{mol base}}\right);\ \text{therefore, } M_aV_a = n_b(2)$$

$$?\ \text{L} = V_a = \frac{n_b(2)}{M_a} = \frac{(4.70 \times 10^{-2}\ \text{mol})(2)}{0.0200\ \text{M}} = \mathbf{4.70\ L\ HCl\ solution}$$

15.91 What volume of 0.0500 M H_2SO_4 is required to neutralize 0.219 g $Mg(OH)_2$?

Analysis:
Target: The volume of an acid solution. (? mL) = V_a
Knowns: The mass and formula of the base and the molarity and formula of the acid. The molar mass of the base is an implicit known
Relationships: $n_a = n_b$•(stoichiometric mole ratio). n = M•V. The formulas of the acid and base enable us to write the balanced chemical equation for neutralization. The stoichiometric mole ratio is given by coefficients in the balanced equation. The molar mass provides the equivalence statement g ↔ mol. The equivalence statement 10^{-3} L ↔ 1 mL is also needed.
Plan: Use the factor-label method to calculate the number of moles of base. Determine the stoichiometric mole ratio from the balanced chemical equation, and use the relationships $n_a = n_b$•(mol ratio) and n = M•V to find the volume of the base. Use the factor-label method to convert the volume to milliliters.
Work:
The balanced chemical equation is $H_2SO_4 + Mg(OH)_2 \longrightarrow MgSO_4 + 2\ H_2O$, so the stoichiometric mole ratio is 1 mol acid ↔ 1 mol base.
Molar Mass $Mg(OH)_2$ = 1(24.305 Mg) + 2(15.9994 O) + 2(1.0079) = 58.320 g mol^{-1}

$$n_b = 0.219 \text{ g Mg(OH)}_2 \times \frac{1 \text{ mol Mg(OH)}_2}{58.320 \text{ g Mg(OH)}_2} = 3.76 \times 10^{-3} \text{ mol Mg(OH)}_2$$

$$n_a = n_b\left(\frac{1 \text{ mol acid}}{1 \text{ mol base}}\right); \text{ therefore, } M_aV_a = n_b(1)$$

$$? \text{ mL} = V_a = \frac{n_b(1)}{M_a} = \frac{(3.76 \times 10^{-3} \text{ mol})(1)}{0.0500 \text{ M}} = 7.51 \times 10^{-2} \text{ L} \times \frac{1 \text{ mL}}{10^{-3} \text{ L}} =$$

75.1 mL H_2SO_4 solution

15.93 If 19.7 mL of 0.1099 M KOH is required to neutralize a 25.0 mL sample of an acid, and the acid is known to be diprotic, what is its molarity?

Analysis:

Target: The molarity of a diprotic acid (H_2X) solution. (? mol L^{-1}) = M_a

Knowns: The volume and general formula of the acid and the volume, formula and molarity of the base.

Relationships: $n_a = n_b \cdot$(stoichiometric mole ratio). $n = M \cdot V$. The formulas of the acid and base enable us to write the balanced chemical equation for neutralization. The stoichiometric mole ratio is given by coefficients in the balanced equation.

Plan: Determine the stoichiometric mole ratio from the balanced chemical equation, and use the relationships $n_a = n_b \cdot$(mol ratio) and $n = M \cdot V$ to find the molarity of the acid.

Work:

The balanced chemical equation is 2 KOH + $H_2X \longrightarrow K_2X$ + 2 H_2O, so the stoichiometric mole ratio is 1 mol acid ↔ 2 mol base.

$$n_a = n_b\left(\frac{1 \text{ mol acid}}{2 \text{ mol base}}\right), \text{ therefore, } M_aV_a = \frac{M_bV_b}{2}$$

$$? \text{ mol L}^{-1} = M_a = \frac{M_bV_b}{2V_a} = \frac{(0.1099 \text{ M})(19.7 \text{ mL})}{(2)(25.0 \text{ mL})} = \textbf{0.0433 M acid}$$

15.95 It is found that a 0.482 g sample of a pure acid requires 18.3 mL of 0.103 M NaOH to reach the equivalence point in a titration. Assuming that the acid is monoprotic, what is its molar mass?

Analysis:

Target: The molar mass of a monoprotic acid, HX. (? g mol^{-1}) =

Knowns: The volume, formula and molarity of the base and the mass and general formula of the acid.

Relationships: $n_a = n_b \cdot$(stoichiometric mole ratio). $n = M \cdot V$. The formulas of the acid and base enable us to write the balanced chemical equation for neutralization. The stoichiometric mole ratio is given by coefficients in the balanced equation. The equivalence statement 10^{-3} L ↔ 1 mL is also needed. Molar mass has units of g mol^{-1}.

Plan: Determine the stoichiometric mole ratio from the balanced chemical equation, and use the relationships $n_a = n_b \cdot$(mol ratio) and $n = M \cdot V$ to find the number of moles of acid, employing the factor-label method to convert mL to L. Divide the mass of the acid by the number of moles of acid to determine its molar mass.

Work:

The balanced chemical equation is HX + NaOH ⟶ NaX + H_2O, so the stoichiometric mole ratio is 1 mol acid ↔ 1 mol base.

$$n_a = n_b\left(\frac{1 \text{ mol acid}}{1 \text{ mol base}}\right) = M_bV_b(1)$$

$$n_a = (0.103 \text{ mol L}^{-1})(18.3 \text{ mL})(1) \times \frac{10^{-3} \text{ L}}{1 \text{ mL}} = 1.88 \times 10^{-3} \text{ mol acid}$$

$$? \text{ g mol}^{-1} = \frac{0.482 \text{ g acid}}{1.88 \times 10^{-3} \text{ mol acid}} = \mathbf{256 \text{ g mol}^{-1}}$$

15.97 It is found that a 3.52 mg sample of a pure acid requires 6.63 mL of 0.0102 M KOH for complete neutralization. Assuming that the acid is diprotic, what is its molar mass?

Analysis:
Target: The molar mass of a diprotic acid, H_2X. (? g mol^{-1})
Knowns: The volume, formula and molarity of the base and the mass of the acid. The acid is diprotic.
Relationships: $n_a = n_b$•(stoichiometric mole ratio). n = M•V. The formulas of the acid and base enable us to write the balanced chemical equation for neutralization. The stoichiometric mole ratio is given by coefficients in the balanced equation. Molar mass has units of g mol^{-1} or, equivalently, mg $mmol^{-1}$.
Plan: Determine the stoichiometric mole ratio from the balanced chemical equation, and use the relationships $n_a = n_b$•(mol ratio) and n = M•V to find the number of millimoles of acid Divide the mass of the acid by the number of millimoles of acid to find the molar mass.
Work:
The balanced chemical equation is $H_2X + 2\ KOH \longrightarrow K_2X + 2\ H_2O$, so the stoichiometric mole ratio is 1 mol acid ↔ 2 mol base.

$$n_a = n_b\left(\frac{1 \text{ mol acid}}{2 \text{ mol base}}\right) = \frac{M_bV_b}{2}$$

$$n_a = \frac{(0.0102 \text{ mol L}^{-1})(6.63 \text{ mL})}{2} = 3.38 \times 10^{-2} \text{ mmol acid}$$

$$? \text{ mg mmol}^{-1} = \frac{3.52 \text{ mg acid}}{3.38 \times 10^{-2} \text{ mmol acid}} = \mathbf{104 \text{ mg mmol}^{-1} = 104 \text{ g mol}^{-1}}$$

15.99 Write proton-transfer autoionization equations for the following amphiprotic solvents.
a. NH_3 b. NH_2OH c. H_2SO_4

Analysis/Plan: In a proton-transfer autoionization, one solvent molecule serves as the proton donor and another serves as the proton acceptor. Products are the conjugate base and the conjugate acid of the solvent.
Work:
a. $NH_3 + NH_3 \rightleftharpoons NH_4^+ + NH_2^-$
b. $NH_2OH + NH_2OH \rightleftharpoons NH_3OH^+ + NH_2O^-$
c. $H_2SO_4 + H_2SO_4 \rightleftharpoons H_3SO_4^+ + HSO_4^-$

15.101 Two solutions of identical molarity are prepared. One is a weak acid, and the other is a strong acid. Which has the greater pH? Which reacts more vigorously with metallic zinc?

The strong acid will ionize essentially completely, while the weak acid will only partial ionize. Consequently, the $[H_3O^+]$ will be higher in the strong acid. Because pH = $-\log[H_3O^+]$, a higher $[H_3O^+]$ means a lower pH. Therefore, the **weak acid will have a greater pH**. With its higher hydronium concentration the **strong acid will react more vigorously with a metal**.

15.103 All of the following equilibria lie to the right. Use this information to arrange the acids HSO_4^-, HBrO, HCN, HNO_2, and H_2S in order of increasing acid strength.

a. $HSO_4^- + HS^- \rightleftharpoons H_2S + SO_4^{2-}$

b. $HBrO + CN^- \rightleftharpoons BrO^- + HCN$

c. $HSO_4^- + NO_2^- \rightleftharpoons SO_4^{2-} + HNO_2$

d. $H_2S + BrO^- \rightleftharpoons HS^- + HBrO$

e. $HNO_2 + HS^- \rightleftharpoons NO_2^- + H_2S$

Analysis/Plan: The equilibrium lies to the side of the weaker acid and base. Compare each pair in turn and then use those rankings to rank all the acids. If two potential acids are reacted, the stronger acid donates the proton and the weaker acid acts as the base.

Work:

a. HSO_4^- is a stronger acid than H_2S.
b. HBrO is a stronger acid than HCN.
c. HSO_4^- is a stronger acid than HNO_2.
d. H_2S is a stronger acid than HBrO.
e. HNO_2 is a stronger acid than H_2S.

Putting these results together gives: **HCN < HBrO < H_2S < HNO_2 < HSO_4^-.**

15.105 If 39.27 mL of 0.01006 M KOH is needed to reach the first equivalence point in a titration of 25.00 mL H_2SO_4, what is the molarity of the acid?

Analysis:

Target: The molarity of an acid solution. (? mol L^{-1}) = M_a

Knowns: The volume of the acid and the volume, formula and molarity of the base. Only one proton is removed when H_2SO_4 is titrated to the first equivalence point.

Relationships: $n_a = n_b$•(stoichiometric mole ratio). $n = M•V$. The formulas of the acid and base enable us to write the balanced chemical equation for neutralization. The stoichiometric mole ratio is given by coefficients in the balanced equation.

Plan: Determine the stoichiometric mole ratio from the balanced chemical equation, and use the relationships $n_a = n_b$•(mol ratio) and $n = M•V$ to find the molarity of the acid.

Work:

The balanced chemical equation is $KOH + H_2SO_4 \longrightarrow NaHSO_4 + H_2O$, so the stoichiometric mole ratio is 1 mol acid ↔ 1 mol base.

$$n_a = n_b\left(\frac{1 \text{ mol acid}}{1 \text{ mol base}}\right); \text{ therefore, } M_aV_a = M_bV_b(1)$$

$$? \text{ mol L}^{-1} = M_a = \frac{M_bV_b(1)}{V_a} = \frac{(0.01006 \text{ M})(39.27 \text{ mL})}{(25.00 \text{ mL})} = \mathbf{0.01580 \text{ M } H_2SO_4}$$

15.107 How much 0.142 N acid will be required for the titration of 25.0 mL of 0.35 N base?

Analysis:

Target: The volume of an acid. (? mL) =

Knowns: The normality of the acid and the volume and normality of the base.

Relationship: $N_aV_a = N_bV_b$.

Plan: Rearrange the relationship to solve for the volume of the acid and calculate.

Work:

$N_aV_a = N_bV_b$

$$? \text{ mL} = V_a = \frac{N_bV_b}{N_a} = \frac{(0.35 \text{ N})(25.0 \text{ mL})}{0.142 \text{ N}} = \mathbf{62 \text{ mL acid}}$$

15.109 It is found that a 0.191 g sample of a pure acid requires 25.4 mL of 0.103 M NaOH to reach the equivalence point in a titration. What is the equivalent weight of the acid? If the acid is diprotic, what is its molar mass?

Analysis:

Target: The equivalent weight of an acid and the molar mass of the acid if it is diprotic. (? g $equiv^{-1}$) = and (? g mol^{-1}) =

Knowns: The mass of the acid and the volume and normality of the base. If the acid is diprotic, there are two equivalents per mole. (An equivalent weight of an acid is the mass of the acid which can donate one mole of protons.)

Relationships: N_bV_b = equiv. At the equivalence point, the number of equivalents of an acid is equal to the number of equivalents of a base. Equivalent weight has units of g $equiv^{-1}$, while the molar mass has units of g mol^{-1}. The equivalence statement 10^{-3} L ↔ 1 mL is also needed.

Plan: Use the relationships to solve for the number of equivalents of acid, and then divide the mass of acid by the number of equivalents of acid to determine the equivalent weight. Use the factor-label method, and the fact that there are two equivalents per mole, to determine the number of grams in a mole (the molar mass).

Work:

$$? \text{ equivalents} = N_bV_b = (0.103 \text{ eq L}^{-1})(25.4 \text{ mL}) \times \frac{10^{-3} \text{ L}}{\text{mL}} = 2.62 \times 10^{-3} \text{ equivalents}$$

$$? \text{ g equiv}^{-1} = \frac{0.191 \text{ g acid}}{2.62 \times 10^{-3} \text{ equivalents}} = \mathbf{73.0 \text{ g equiv}^{-1}}$$

$$? \text{ g mol}^{-1} = \frac{73.0 \text{ g}}{\text{equivalent}} \times \frac{2 \text{ equivalents}}{1 \text{ mol}} = \mathbf{146 \text{ g mol}^{-1}}$$

16 Ionic Equilibrium: Acids and Bases

16.1 Distinguish between reaction quotient and equilibrium expression.

These expressions have the same format, however, concentrations or partial pressures can have any values in the reaction quotient. In the equilibrium expression, concentrations or partial pressures are specifically those values at equilibrium.

16.3 Is the relationship between acid strength and pK_a value direct or inverse? Explain.

This is an inverse relationship because $pK_a = -\log K_a$. A stronger acid has a higher K_a and, consequently, a lower pK_a.

16.5 Distinguish between equilibrium constant and ionization constant.

There is no real difference between these constants. An ionization constant is simply a specialized form of equilibrium constant for a process in which ionization occurs.

16.7 Derive the general expression $x = \left([-K + \sqrt{K^2 + 4KC}]\right)/2$ for the extent of reaction in an equilibrium ionization of a weak acid or base. C is the initial molarity and K is K_a or K_b. What is the relationship between x and pH in this expression? Modify the expression for x to a form giving the fractional ionization in terms of K and C.

The general ionization reaction for an acid is $HA + H_2O \rightleftharpoons H_3O^+ + A^-$. Using techniques discussed in Chapter 13, we can determine the equilibrium concentrations by looking at the changes in concentration.

Balanced	HA	+	H_2O	$\rightleftharpoons$	H_3O^+	+	A^-
Initial	C				0		0
Change	- x				+ x		+ x
Equilibrium	C - x				x		x

Insert the values into the expression for K_a.

$$K_a = \frac{[H_3O^+][A^-]}{[HA]} = \frac{(x)(x)}{C - x} = \frac{x^2}{C - x}$$

Cross multiplying gives: $x^2 = K_a(C - x)$.
Rearranging yields, $x^2 + K_a x - K_a C = 0$. This is a quadratic equation with a positive root:

$$x = \frac{-K_a + \sqrt{K_a^2 + 4(K_a)(C)}}{2}$$

For acids, $pH = -\log x$ because $x = [H_3O^+]$. Bases would have an analogous solution with K_b and with $x = [OH^-]$. $pOH = -\log x$. Because $pH + pOH = 14.00$, $pH = 14.00 - pOH$. Therefore, $pH = 14.00 - (-\log x) = 14.00 + \log x$, for bases.

Percent ionization $= \frac{[\text{ionized form}]}{[\text{initial}]} \times 100$. Because x = [ionized form] and C = [initial]:

$$\%\ \text{ionization} = \frac{x}{C} = \frac{-K + \sqrt{K^2 + 4(K)(C)}}{2C} \times 100,$$

where $K = K_a$ for an acid or K_b for a base.

16.9 Why is the pH of an aqueous solution of a *very* weak acid, say one with $pK_a > 14$, independent of concentration?

If an acid is too weak, it cannot donate a proton to water and, therefore, cannot change the hydronium concentration. If the hydronium concentration doesn't change, neither does the pH.

16.11 Define hydrolysis, and give some example chemical equations of the hydrolysis of salts.

Hydrolysis is the transfer of a proton between an ion and water. Conjugates of weak acids or bases undergo hydrolysis. For example, NH_4Cl provides NH_4^+, the conjugate acid of the weak base NH_3. (Cl^-, the conjugate base of the strong acid HCl, is a spectator ion.) NH_4^+ undergoes hydrolysis according to the reaction: $NH_4^+ + H_2O \rightleftharpoons H_3O^+ + NH_3$.
Similarly, $NaNO_2$ provides NO_2^-, the conjugate base of the weak acid HNO_2. (Na^+, the cation of the strong base NaOH, is a spectator ion.) NO_2^- undergoes hydrolysis according to the reaction: $NO_2^- + H_2O \rightleftharpoons HNO_2 + OH^-$.

16.13 Distinguish between amphiprotic and amphoteric.

Amphoteric species can react with either acids or bases. Amphiprotic species can also react with either acids or bases, but specifically by donating a proton to a stronger base or by accepting a proton from a stronger acid.

16.15 What is meant by the term "buffer capacity"?

Buffer capacity is the amount of H_3O^+ or OH^- which can be added to a buffer before a significant change in the pH of the solution occurs.

16.17 What is titration?

Titration is the process of adding small measured increments of a strong acid (or base) to a measured amount of a base (or acid) until the sample has been neutralized. By monitoring the amounts, the molarity or molar mass of a sample can be determined.

16.19 Consider the titration of a weak acid with NaOH, and describe the change in composition of the sample as it passes from the undertitrated region, through the equivalence point, and into the overtitrated region.

In the region before the equivalence point (undertitration), the sample is a buffer solution. There are substantial amounts of the acid and its conjugate base both present. At the equivalence point, only the sodium salt of the acid is present. The conjugate base of the weak acid hydrolyzes, causing the pH to be greater than 7 (basic) at the equivalence point. Past the equivalence point (overtitration), the solution is a mixture of excess sodium hydroxide and the sodium salt of the acid.

16.21 What is an indicator? What are the characteristics of a good indicator for an acid/base titration?

An indicator is a weak acid or base, whose acid form has a different color than its basic form. The dominant species determines the color. As the $[H_3O^+]$ of the bulk solution changes, the acid indicator is protonated or deprotonated and the color changes. Because the color changes over a specific pH range that depends on its K_a, an indicator can be used to signal when equivalence of the solution has been achieved. Good indicators are strongly colored in at least one of their forms, if not both, so that relatively tiny amounts are needed to provide a visible color change.

16.23 What is meant by the term "indicator range"?

The indicator range is the range of pH values, usually 1.5 to 2.0 pH units, over which the color change from the protonated to the deprotonated form (or vice versa) is complete.

16.25 In a description of the ionization of a diprotic acid, to what do the symbols, K, K_1, and K_2 refer? Illustrate your answer with chemical and algebraic equations.

K is the overall ionization constant, K_1 is for the ionization of the first proton and K_2 is for the ionization of the second proton. H_2S is a diprotic acid. (See Appendix J in the text for ionization constants. Remember that for a two step process, $K_{overall} = K_1 \times K_2$.)

$$H_2S + H_2O \rightleftharpoons H_3O^+ + HS^- \qquad K_1 = \frac{[H_3O^+][HS^-]}{[H_2S]} = 1.0 \times 10^{-7}$$

$$HS^- + H_2O \rightleftharpoons H_3O^+ + S^{2-} \qquad K_2 = \frac{[H_3O^+][S^{2-}]}{[HS^-]} \sim 10^{-19}$$

$$H_2S + 2\,H_2O \rightleftharpoons 2\,H_3O^+ + S^{2-} \qquad K = \frac{[H_3O^+]^2[S^{2-}]}{[H_2S]} = K_1 \cdot K_2 \sim 10^{-26}$$

16.27 Convert the following acid ionization constants to pK_a's, using the appropriate number of significant figures.

a. fluoroacetic acid, $K_a = 2.2 \times 10^{-3}$
b. benzenesulfonic acid, $K_a = 0.20$
c. butanoic acid, $K_a = 1.51 \times 10^{-5}$
d. *p*-chlorobenzoic acid, $K_a = 1 \times 10^{-4}$

Analysis:
Target: The pK_a of an acid. (? pK_a) =
Knowns: The name of the acid and the acid ionization constant, K_a.
Relationship: $pK_a = -\log K_a$. The pK_a has as many places to the right of the decimal point as there are significant digits in the K_a.
Plan: Use the relationship directly.
Work:
a. ? $pK_a = -\log K_a = -\log (2.2 \times 10^{-3}) = -(-2.66) =$ **2.66**
b. ? $pK_a = -\log K_a = -\log (0.20) = -(-0.70) =$ **0.70**
c. ? $pK_a = -\log K_a = -\log (1.51 \times 10^{-5}) = -(-4.821) =$ **4.821**
d. ? $pK_a = -\log K_a = -\log (1 \times 10^{-4}) = -(-4.0) =$ **4.0**

16.29 Convert the following pK_a values to acid ionization constants, using the appropriate number of significant figures.

a. bromoacetic acid, $pK_a = 2.86$
b. thiophenol, $pK_a = 6.52$
c. picric acid, $pK_a = 0.25$
d. nitrous acid, $pK_a = 3.4$

Analysis:
Target: The K_a of an acid. (? K_a) =
Knowns: The name of the acid and the pK_a.
Relationship: $pK_a = -\log K_a$; therefore, $K_a = 10^{-pK_a}$. The K_a has as many significant digits as there are places to the right of the decimal point in the pK_a.
Plan: Use the second relationship directly.

Work:

a. $? K_a = 10^{-pK_a} = 10^{-2.86} =$ **1.4×10^{-3}**
b. $? K_a = 10^{-pK_a} = 10^{-6.52} =$ **3.0×10^{-7}**
c. $? K_a = 10^{-pK_a} = 10^{-0.25} =$ **0.56**
d. $? K_a = 10^{-pK_a} = 10^{-3.4} =$ **4×10^{-4}**

16.31 Calculate the pH of a 0.375 M solution of hypochlorous acid.

Analysis:
Target: The pH of a solution of weak acid. (? pH) =
Knowns: The molarity of the acid. The K_a is 3.0×10^{-8} (Appendix J).
Relationships: The structured approach to equilibrium problems is appropriate. (Review Chapter 13 if necessary.) $pH = -\log [H_3O^+]$.
Plan: Use the structured approach to equilibrium problems to determine the $[H_3O^+]$ and then use the formula to determine the pH.
Work:

Balanced	$HClO$	+	H_2O	$\rightleftharpoons$	H_3O^+	+	ClO^-
Initial	0.375				0		0
Change	- x				+ x		+ x
Equilibrium	0.375 - x				x		x

Insert the values into the K_a expression.

$$K_a = \frac{[H_3O^+][ClO^-]}{[HClO]} = \frac{(x)(x)}{(0.375 - x)} = 3.0 \times 10^{-8}$$

Solve algebraically. Because K_a is very small, assume $0.375 - x \approx 0.375$.

$$\frac{x^2}{0.375} = 3.0 \times 10^{-8}\text{; therefore, } x^2 = 0.375(3.0 \times 10^{-8}) = 1.1 \times 10^{-8}\text{; } x = \sqrt{1.1 \times 10^{-8}}$$

$$x = [H_3O^+] = 1.1 \times 10^{-4}\ M\text{; } pH = -\log [H_3O^+] = -\log (1.1 \times 10^{-4}) = -(-3.97) = \mathbf{3.97}$$

Check: $\frac{1.1 \times 10^{-4}}{0.375} = 0.00029$; so the approximation that $0.375 - x \approx 0.375$ is valid.

16.33 Calculate $[OH^-]$ and pH in a 0.050 M solution of dimethylamine ($K_b = 5.9 \times 10^{-4}$).

Analysis:
Targets: The $[OH^-]$ and pH of a solution of weak base. (? M) = and (? pH) =
Knowns: The molarity of the base and its K_b. Its formula is $(CH_3)_2NH$ (Appendix J).
Relationships: The structured approach to equilibrium problems is appropriate.
$pOH = -\log [OH^-]$ and $pH + pOH = 14.00$.
Plan: Use the structured approach to equilibrium problems to determine the $[OH^-]$ and then use the other relationships to determine the pH.
Work:

Balanced	$(CH_3)_2NH$	+	H_2O	$\rightleftharpoons$	$(CH_3)_2NH_2^+$	+	OH^-
Initial	0.050				0		0
Change	- x				+ x		+ x
Equilibrium	0.050 - x				x		x

Insert the values into the K_b expression.

$$K_b = \frac{[(CH_3)_2NH_2^+][OH^-]}{[(CH_3)_2NH]} = \frac{(x)(x)}{(0.050 - x)} = 5.9 \times 10^{-4}$$

Solve algebraically. Because K_b is not very small and the concentration is low, we must use the quadratic equation. From Problem 16.7, $x = \frac{-K_b + \sqrt{K_b^2 + 4(K_b)(C)}}{2}$.

$$x = \frac{-5.9 \times 10^{-4} + \sqrt{(5.9 \times 10^{-4})^2 + 4(5.9 \times 10^{-4})(0.050)}}{2}$$

$x = [OH^-] =$ **5.1×10^{-3} M**; pOH = -log $[OH^-]$ = -log (5.1×10^{-3}) = -(-2.29) = 2.29

? pH = 14.00 - pOH = 14.00 - 2.29 = **11.71**

16.35 Calculate the pH and percent ionization in a 0.0750 M solution of HCN.

Analysis:

Targets: The pH and percent ionization of a solution of weak acid. (? pH) = and (% ionization) =

Knowns: The molarity of the acid. The K_a is 4.0×10^{-10} (Appendix J).

Relationships: The structured approach to equilibrium problems is appropriate.

pH = -log $[H_3O^+]$. Percent ionization = $\frac{[\text{ionized form}]}{[\text{initial}]} \times 100$.

Plan: Use the structured approach to equilibrium problems to determine the $[H_3O^+]$ and then use the formula to determine the pH. The $[H_3O^+]$ is the ionized form. The percent ionization can be found from the $[H_3O^+]$ and the initial molarity of the acid.

Work:

Balanced	HCN	+	H_2O	⇌	H_3O^+	+	CN^-
Initial	0.0750				0		0
Change	- x				+ x		+ x
Equilibrium	0.0750 - x				x		x

Insert the values into the K_a expression.

$$K_a = \frac{[H_3O^+][CN^-]}{[HCN]} = \frac{(x)(x)}{(0.0750 - x)} = 4.0 \times 10^{-10}$$

Solve algebraically. Because K_a is very small, assume 0.075 - x ≈ 0.0750.

$\frac{x^2}{0.0750} = 4.0 \times 10^{-10}$; therefore, $x^2 = 0.0750(4.0 \times 10^{-10}) = 3.0 \times 10^{-11}$; $x = \sqrt{3.0 \times 10^{-11}}$

$x = [H_3O^+] = 5.5 \times 10^{-6}$ M; pH = -log $[H_3O^+]$ = -log (5.5×10^{-6}) = -(-5.26) = **5.26**

? % ionization = $\frac{[\text{ionized form}]}{[\text{initial}]} \times 100 = \frac{5.5 \times 10^{-6} \text{ M}}{0.0750 \text{ M}} \times 100 =$ **7.3×10^{-3}%**

Since the % ionization is less than 5%, the approximation that 0.0750 - x ≈ 0.0750 is valid.

16.37 What is the percent ionization in a 0.100 M solution of formic acid? What is the pH?

Analysis:

Targets: The pH and percent ionization of a solution of weak acid. (? pH) = and (% ionization) =

Knowns: The molarity of the acid. The K_a and formula are 1.8×10^{-4} and HCOOH, respectively (Appendix J).
Relationships: The structured approach to equilibrium problems is appropriate.
$pH = -\log [H_3O^+]$. Percent ionization $= \frac{[\text{ionized form}]}{[\text{initial}]} \times 100$.

Plan: Use the structured approach to equilibrium problems to determine the $[H_3O^+]$ and then use the formula to determine the pH. The $[H_3O^+]$ is the ionized form. The percent ionization can be found from the $[H_3O^+]$ and the initial molarity of the acid.

Work:

Balanced	HCOOH	+ H_2O	$\rightleftharpoons$	H_3O^+	+	$HCOO^-$
Initial	0.100			0		0
Change	- x			+ x		+ x
Equilibrium	0.100 - x			x		x

Insert the values into the K_a expression.

$$K_a = \frac{[H_3O^+][HCOO^-]}{[HCOOH]} = \frac{(x)(x)}{(0.100 - x)} = 1.8 \times 10^{-4}$$

Solve algebraically. Because K_a is small, assume $0.100 - x \approx 0.100$.

$$\frac{x^2}{0.100} = 1.8 \times 10^{-4}; \text{ therefore, } x^2 = 0.100(1.8 \times 10^{-4}) = 1.0 \times 10^{-5}; \; x = \sqrt{1.0 \times 10^{-5}}$$

$$x = [H_3O^+] = 4.2 \times 10^{-3}\text{ M}; \; pH = -\log [H_3O^+] = -\log (4.2 \times 10^{-3}) = -(-2.37) = \mathbf{2.37}$$

$$? \% \text{ ionization} = \frac{[\text{ionized form}]}{[\text{initial}]} \times 100 = \frac{4.2 \times 10^{-3}\text{ M}}{0.100\text{ M}} \times 100 = \mathbf{4.2\ \%}$$

Since the % ionization is less than 5%, the approximation that $0.100 - x \approx 0.100$ is valid.

16.39 The pH of a solution of formic acid is 3.00. What was the initial concentration of the acid? What are the equilibrium concentrations of the protonated and deprotonated forms?

Analysis:
Targets: The initial concentration of the acid and the equilibrium concentrations of the protonated and deprotonated forms. (? mol L^{-1}) =
Knowns: The pH of the acid solution. The K_a and formula are in Problem 16.37.
Relationships: The structured approach to equilibrium problems is appropriate.
$pH = -\log [H_3O^+]$, so $[H_3O^+] = 10^{-pH}$.

Plan: Use the structured approach to equilibrium problems and solve algebraically to determine the initial concentration, C, of the acid and the equilibrium concentrations. $x = [H_3O^+]$ and can be found from the pH using the relationship $[H_3O^+] = 10^{-pH}$.

Work:

Balanced	HCOOH	+ H_2O	$\rightleftharpoons$	H_3O^+	+	$HCOO^-$
Initial	C			0		0
Change	- x			+ x		+ x
Equilibrium	C - x			x		x

Insert the values into the K_a expression.

$$K_a = \frac{[H_3O^+][HCO_2^-]}{[HCO_2H]} = \frac{(x)(x)}{(C - x)} = 1.8 \times 10^{-4}$$

Find $x = [H_3O^+]$ from the pH and then solve algebraically for C.

$x = [H_3O^+] = 10^{-pH} = 10^{-3.00} = 1.0 \times 10^{-3}$ M

$\frac{(1.0 \times 10^{-3})^2}{C - 1.0 \times 10^{-3}} = \frac{1.0 \times 10^{-6}}{C - 1.0 \times 10^{-3}} = 1.8 \times 10^{-4}$; so $1.0 \times 10^{-6} = 1.8 \times 10^{-4}(C - 1.0 \times 10^{-3})$

$1.0 \times 10^{-6} = (1.8 \times 10^{-4})C - 1.8 \times 10^{-7}$; $1.0 \times 10^{-6} + 1.8 \times 10^{-7} = (1.8 \times 10^{-4})C$

$1.2 \times 10^{-6} = (1.8 \times 10^{-4})C$; $C = \frac{1.2 \times 10^{-6}}{1.8 \times 10^{-4}} = \mathbf{6.6 \times 10^{-3}\ M}$

The protonated form, HCOOH, = C - x = 6.6×10^{-3} M - 1.0×10^{-3} = **5.6×10^{-3} M.**

The deprotonated form, $HCOO^-$, = x = **1.0×10^{-3} M.**

16.41 The percent ionization in thiophenol is 0.53% when the concentration is 0.0100 M. What is the acid ionization constant of thiophenol?

Analysis:
Target: The acid ionization constant, K_a, of a weak acid. (? K_a) =
Knowns: The molarity and the percent ionization of the acid. The formula is not given, but may be represented by HA.
Relationships: The structured approach to equilibrium problems is appropriate.

Percent ionization = $\frac{[\text{ionized form}]}{[\text{initial}]} \times 100$.

Plan: Use the structured approach to equilibrium problems to write an algebraic expression for the K_a. Then use the definition of percent ionization to determine the equilibrium concentration of $[H_3O^+]$, which is the ionized form. Substitute that value into the K_a expression to determine the K_a.

Work:

Balanced	HA	+	H_2O	$\rightleftharpoons$	H_3O^+	+	A^-
Initial	0.0100				0		0
Change	- x				+ x		+ x
Equilibrium	0.0100 - x				x		x

Insert the values into the K_a expression.

$$K_a = \frac{[H_3O^+][A^-]}{[HA]} = \frac{(x)(x)}{(0.0100 - x)}$$

Use the percent ionization to find $[H_3O^+]$ = x.

$$\% \text{ ionization} = \frac{[\text{ionized form}]}{[\text{initial}]} \times 100; \text{ therefore, } \frac{[H_3O^+]}{0.0100\ M} \times 100 = 0.53\%$$

$$[H_3O^+] = \frac{(0.53\%)(0.0100\ M)}{100} = 5.3 \times 10^{-5}\ M = x$$

Substitute x into the K_a expression.

$$?\ K_a = \frac{(5.3 \times 10^{-5})^2}{0.0100 - 5.3 \times 10^{-5}} = \mathbf{2.8 \times 10^{-7}}$$

16.43 Fluoroacetic acid is 4.6% ionized in a 1.00 M solution. What is its pK_a?

Analysis:
Target: The pK_a of a weak acid. (? pK_a) =

Knowns: The molarity and the percent ionization of the acid. The formula is not given, but may be represented by HA.
Relationships: The structured approach to equilibrium problems is appropriate.

Percent ionization = $\frac{[\text{ionized form}]}{[\text{initial}]}$ x 100. $pK_a = -\log K_a$

Plan: Use the structured approach to equilibrium problems to write an algebraic expression for the K_a. Then, use the definition of percent ionization to determine the equilibrium concentration of $[H_3O^+]$, which is the ionized form. Substitute that value into the K_a expression to determine the K_a. Finally, use the relationship $pK_a = -\log K_a$.

Work:

Balanced	HA	+	H_2O	$\rightleftharpoons$	H_3O^+	+	A^-
Initial	1.00				0		0
Change	- x				+ x		+ x
Equilibrium	1.00 - x				x		x

Insert the values into the K_a expression.

$$K_a = \frac{[H_3O^+][A^-]}{[HA]} = \frac{(x)(x)}{(1.00 - x)}$$

Use the percent ionization to find $[H_3O^+] = x$.

$$\% \text{ ionization} = \frac{[\text{ionized form}]}{[\text{initial}]} \times 100 = \frac{[H_3O^+]}{1.00 \text{ M}} \times 100 = 4.6\%$$

$$[H_3O^+] = \frac{(4.6\%)(1.00 \text{ M})}{100} = 0.046 \text{ M} = x$$

Substitute x into the K_a expression.

$$K_a = \frac{(0.046)^2}{1.00 - 0.046} = 2.2 \times 10^{-3}$$

$? pK_a = -\log K_a = -\log (2.2 \times 10^{-3}) = -(-2.65) =$ **2.65**

16.45 The pH of a 0.025 M solution of butanoic acid is 3.21. What is the acid ionization constant?

Analysis:
Target: The acid ionization constant, K_a, of a weak acid. (? K_a) =
Knowns: The molarity and the pH of the acid solution. The formula is not given, but may be represented by HA.
Relationships: The structured approach to equilibrium problems is appropriate.
$pH = -\log [H_3O^+]$; therefore, $[H_3O^+] = 10^{-pH}$.

Plan: Use the structured approach to equilibrium problems to write an algebraic expression for the K_a. Then use the definition of pH to determine the equilibrium concentration of $[H_3O^+]$. Substitute that value into the K_a expression to determine the K_a.

Work:

Balanced	HA	+	H_2O	$\rightleftharpoons$	H_3O^+	+	A^-
Initial	0.025				0		0
Change	- x				+ x		+ x
Equilibrium	0.025 - x				x		x

Insert the values into the K_a expression.

$$K_a = \frac{[H_3O^+][A^-]}{[HA]} = \frac{(x)(x)}{(0.025 - x)}$$

Use the pH to find $[H_3O^+]$.

$$[H_3O^+] = 10^{-pH} = 10^{-3.21} = 6.2 \times 10^{-4}\ M = x$$

Substitute x into the K_a expression.

$$?\ K_a = \frac{(6.2 \times 10^{-4})^2}{0.025 - 6.2 \times 10^{-4}} = \mathbf{1.6 \times 10^{-5}}$$

16.47 If the equilibrium concentration of formate ion is 0.0115 M in a 0.750 M solution of formic acid, what is the K_a of formic acid? Remember that the value 0.750 M refers to the total concentration of protonated and deprotonated forms.

Analysis:
Target: The acid ionization constant, K_a, of a weak acid. (? K_a) =
Knowns: The original molarity of the acid and the equilibrium formate concentration. The acid may be represented by HA.
Relationship: The structured approach to equilibrium problems is appropriate.
Plan: Use the structured approach to equilibrium problems to write an algebraic expression for the K_a. Then use the equilibrium formate concentration to determine the K_a.
Work:

Balanced	HA	+	H_2O	$\rightleftharpoons$	H_3O^+	+	A^-
Initial	0.750				0		0
Change	- x				+ x		+ x
Equilibrium	0.750 - x				x		x

Insert the values into the K_a expression.

$$K_a = \frac{[H_3O^+][A^-]}{[HA]} = \frac{(x)(x)}{(0.750 - x)}$$

Note that x is equal to the concentration of the formate ion, 0.0115 M. Substitute that value into the K_a expression.

$$?\ K_a = \frac{(0.0115)^2}{(0.750 - 0.0115)} = \mathbf{1.79 \times 10^{-4}}$$

16.49 The pH of a 5.00×10^{-3} M solution of *N*-methylaniline is 8.27. What is the value of the base ionization constant of *N*-methylaniline?

Analysis:
Target: The base ionization constant, K_b, of a weak base. (? K_b) =
Knowns: The molarity and the pH of the base solution. The formula is not given, but may be represented by B.
Relationships: The structured approach to equilibrium problems is appropriate.
pH + pOH = 14.00. pOH = -log $[OH^-]$; therefore, $[OH^-] = 10^{-pOH}$.
Plan: Use the structured approach to equilibrium problems to write an algebraic expression for the K_b. Then use the definition of pOH and the relationship pH + pOH = 14.00 to determine the equilibrium concentration of $[OH^-]$. Substitute that value into the K_b expression to determine the K_b.
Work:

Balanced	B	+	H_2O	$\rightleftharpoons$	BH^+	+	OH^-
Initial	5.00×10^{-3}				0		0
Change	- x				+ x		+ x
Equilibrium	5.00×10^{-3} - x				x		x

Insert the values into the K_b expression.

$$K_a = \frac{[H_3O^+][A^-]}{[HA]} = \frac{(x)(x)}{(5.00 \times 10^{-3} - x)}$$

Use the pH to find the pOH.

pH + pOH = 14.00; pOH = 14.00 - pH = 14.00 - 8.27 = 5.73.

Now use the pOH to find the $[OH^-]$.

$$[OH^-] = 10^{-pOH} = 10^{-5.73} = 1.9 \times 10^{-6}\ M = x$$

Substitute x into the K_b expression.

$$?\ K_b = \frac{(1.9 \times 10^{-6})^2}{5.00 \times 10^{-3} - 1.9 \times 10^{-6}} = \mathbf{6.9 \times 10^{-10}}$$

16.51 Write the chemical equation for the hydrolysis of KNO_2 and calculate the value of K_b.

Analysis:

Targets: The hydrolysis reaction and the K_b of the base. (? K_b) =

Knowns: KNO_2 is a salt. The K_a of HNO_2 is 4.0×10^{-4} (Appendix J).

Relationships: K^+ has no acid base ability; however, NO_2^- is the conjugate base of a weak acid and can accept a proton from water. For conjugate species, $K_a \cdot K_b = K_w = 1.0 \times 10^{-14}$.

Plan: Write the equation showing the base NO_2^- accepting a proton from water. Find the K_b of the base from the K_a of its conjugate acid, using the relationship $K_a \cdot K_b = K_w$.

Work:

$$\mathbf{NO_2^- + H_2O \rightleftharpoons HNO_2 + OH^-}$$

$K_a \cdot K_b = K_w = 1.0 \times 10^{-14}$; therefore, $K_b = \frac{1.0 \times 10^{-14}}{K_a} = \frac{1.0 \times 10^{-14}}{4.0 \times 10^{-4}} = \mathbf{2.5 \times 10^{-11}}$

16.53 Given that pK_a for HOCl is 7.52, calculate K_b for NaOCl. Write the chemical equation for the hydrolysis reaction.

Analysis:

Targets: The hydrolysis reaction and the K_b of the base. (? K_b) =

Knowns: NaOCl is a salt. The pK_a of HOCl is given.

Relationships: K^+ has no acid base ability; however, OCl^- is the conjugate base of a weak acid and can accept a proton from water. For conjugate species, $K_a \cdot K_b = K_w = 1.0 \times 10^{-14}$. $K_a = 10^{-pK_a}$.

Plan: Write the equation showing the base OCl^- accepting a proton from water. Use the relationship $K_a = 10^{-pK_a}$ to determine the K_a of HOCl. Finally, find the K_b of OCl^- from the K_a of HOCl using the relationship $K_a \cdot K_b = K_w$.

Work:

$$\mathbf{OCl^- + H_2O \rightleftharpoons HOCl + OH^-}$$

$pK_a = 7.52$; therefore, $K_a = 10^{-pK_a} = 10^{-7.52} = 3.0 \times 10^{-8}$

$K_a \cdot K_b = K_w = 1.0 \times 10^{-14}$; therefore, $K_b = \frac{1.0 \times 10^{-14}}{K_a} = \frac{1.0 \times 10^{-14}}{3.0 \times 10^{-8}} = \mathbf{3.3 \times 10^{-7}}$

16.55 Classify each of the following aqueous solutions as acidic, neutral, or basic:
a. hydrazinium bromide
b. potassium sulfate
c. potassium chloride
c. trimethylammonium chloride

Analysis:
Target: The acidity or basicity of a salt solution.
Knowns: The formula of the salt. The ions which comprise the salt are implicit knowns. The strong acids and bases must be memorized. Consult Appendix J for unfamiliar ion names derived from weak acids or bases.
Relationships: The conjugate base of a strong acid has no ability to accept a proton in water, while the conjugate base of a weak acid functions as a base, accepting a proton from water. Similarly, the cation of a strong base has no acid or base ability, but the conjugate acid of a weak base functions as an acid, donating a proton to water.
Plan: Determine the ions that make up the salt and decide if either ion can accept or donate a proton. If a proton can be donated, the solution will be acidic. If a proton can be accepted, the solution will be basic. If neither ion has acidic or basic ability in water, the solution will be neutral.
Work:
a. Hydrazinium, $N_2H_5^+$, is the conjugate acid of the weak base N_2H_4. It can donate a proton to water. Br^- is the conjugate base of the strong acid HBr. It has essentially no ability to accept a proton from water. It cannot donate a proton because it does not have one. Because the $N_2H_5^+$ can act as an acid, the solution is **acidic**.
b. K^+ is the cation of KOH, a strong base. It cannot donate a proton because it does not have one. It cannot accept a proton because it has no lone pair to which a proton can be transferred. SO_4^{2-} is the conjugate base of the weak acid HSO_4^-. It can accept a proton from water, functioning as a base. Because the SO_4^{2-} acts as a base, the solution is **basic**.
c. K^+ cannot act as an acid or base (part b). Cl^- is the conjugate base of the strong acid HCl. It has no ability to accept a proton from water. It cannot donate a proton because it does not have one. Because neither ion has acid/base ability, the solution is **neutral**.
d. Trimethylammonium, $(CH_3)_3NH^+$, is the conjugate acid of the weak base $(CH_3)_3N$. It can donate a proton to water. Cl^- has no acid/base ability (part c). Because the $(CH_3)_3NH^+$ can act as an acid, the solution is **acidic**.

16.57 What is the pH of a 0.0500 M pyridinium chloride?

Analysis:
Target: The pH of a salt solution. (? pH) =
Knowns: The molarity of the salt solution. Chloride has no acid/base ability. The pyridinium ion, $C_5H_5NH^+$, is the conjugate acid of the weak base pyridine, C_5H_5N. The K_b of pyridine can be looked up in Appendix J and is 1.4×10^{-9}.
Relationships: For conjugate species, $K_a \cdot K_b = K_w = 1.0 \times 10^{-14}$. The structured approach to equilibrium problems can be used. $pH = -\log [H_3O^+]$.
Plan: Determine the K_a from the relationship $K_a \cdot K_b = K_w$. Write the equation showing the acid $C_5H_5NH^+$ donating a proton to water. Use the structured approach to equilibrium problems to determine the $[H_3O^+]$ and then find the pH from $pH = -\log [H_3O^+]$.
Work:

$$K_a \cdot K_b = K_w = 1.0 \times 10^{-14}; \text{ therefore, } K_a = \frac{1.0 \times 10^{-14}}{K_b} = \frac{1.0 \times 10^{-14}}{1.4 \times 10^{-9}} = 7.1 \times 10^{-6}$$

	$C_5H_5NH^+$ + H_2O	$\rightleftharpoons$	C_5H_5N +	H_3O^+
Balanced	$C_5H_5NH^+$ + H_2O	$\rightleftharpoons$	C_5H_5N +	H_3O^+
Initial	0.0500		0	0
Change	- x		+ x	+ x
Equilibrium	0.0500 - x		x	x

Insert the values into the K_a expression.

$$K_a = \frac{[H_3O^+][C_5H_5N]}{[C_5H_5NH^+]} = \frac{(x)(x)}{(0.0500 - x)} = 7.1 \times 10^{-6}$$

Solve algebraically. Because K_a is very small, assume $0.0500 - x \approx 0.0500$

$$\frac{x^2}{0.0500} = 7.1 \times 10^{-6}; \text{ therefore, } x^2 = 0.0500(7.1 \times 10^{-6}) = 3.6 \times 10^{-7}; x = \sqrt{3.6 \times 10^{-7}}$$

$x = [H_3O^+] = 6.0 \times 10^{-4}$ M; pH = -log $[H_3O^+]$ = -log (6.0×10^{-4}) = -(-3.22) = **3.22**

Check: $\frac{7.1 \times 10^{-6}\text{ M}}{0.0500\text{ M}} = 1.4 \times 10^{-4}$, so the approximation that $0.0500 - x \approx 0.0500$ is valid.

16.59 What is the pH of 0.0375 M ethylammonium chloride? What is the concentration of ethylamine in this solution?

Analysis:

Targets: The pH of a salt solution and the ethylamine concentration. (? pH) = and (? mol L^{-1}) =

Knowns: The molarity of the salt solution. Chloride ion has no acid/base ability. The ethylammonium ion, $CH_3CH_2NH_3^+$, is the conjugate acid of the weak base ethylamine $CH_3CH_2NH_2$. The K_b of ethylamine is 4.3×10^{-4} (Appendix J).

Relationships: For conjugate species, $K_a \cdot K_b = K_w = 1.0 \times 10^{-14}$. The structured approach to equilibrium problems can be used. pH = -log $[H_3O^+]$.

Plan: Determine the K_a from the relationship $K_a \cdot K_b = K_w$. Write the equation showing the acid, $CH_3CH_2NH_3^+$, donating a proton to water. Use the structured approach to equilibrium problems to determine the $[H_3O^+]$ (which is equal to the ethylamine concentration). Finally, find the pH from pH = -log $[H_3O^+]$.

Work:

$$K_a \cdot K_b = K_w = 1.0 \times 10^{-14}; \text{ therefore, } K_a = \frac{1.0 \times 10^{-14}}{K_b} = \frac{1.0 \times 10^{-14}}{4.3 \times 10^{-4}} = 2.3 \times 10^{-11}$$

Work:

Balanced	$CH_3CH_2NH_3^+$	+	H_2O	⇌	H_3O^+	+	$CH_3CH_2NH_2$
Initial	0.0375				0		0
Change	- x				+ x		+ x
Equilibrium	0.0375 - x				x		x

Insert the values into the K_a expression.

$$K_a = \frac{[H_3O^+][CH_3CH_2NH_2]}{[CH_3CH_2NH_3^+]} = \frac{(x)(x)}{(0.0375 - x)} = 2.3 \times 10^{-11}$$

Solve algebraically. Because K_a is very small, assume $0.0375 - x \approx 0.0375$

$$\frac{x^2}{0.0375} = 2.3 \times 10^{-11}; \text{ therefore, } x^2 = 0.0375(2.3 \times 10^{-11}) = 8.7 \times 10^{-13}; x = \sqrt{8.7 \times 10^{-13}}$$

$x = [H_3O^+] = 9.3 \times 10^{-7}$ M; pH = -log $[H_3O^+]$ = -log (9.3×10^{-7}) = -(-6.03) = **6.03**

x also is equal to $[CH_3CH_2NH_2]$ = **9.3×10^{-7} M**

Check: $\frac{9.3 \times 10^{-7}\text{ M}}{0.0375\text{ M}} = 2.5 \times 10^{-5}$, so the approximation that $0.0375 - x \approx 0.0375$ is valid.

16.61 Given that the pH of a 0.044 M solution of hydroxylammonium chloride is 3.66, calculate K_a for the hydroxylammonium ion and K_b for hydroxylamine.

Analysis:

Targets: The K_a of an acid and the K_b of its conjugate base. (? K_a) = and (? K_b) =

Knowns: The molarity and pH of the salt solution. Chloride has no acid/base ability. The hydroxylammonium ion, NH_3OH^+, is the conjugate acid of the weak base NH_2OH.

Relationships: For conjugate species, $K_a \cdot K_b = K_w = 1.0 \times 10^{-14}$. The structured approach to equilibrium problems can be used. $pH = -\log [H_3O^+]$, so $[H_3O^+] = 10^{-pH}$.

Plan: Write the equation showing the acid NH_3OH^+ donating a proton to water. Use the structured approach to equilibrium problems to determine the algebraic expression for the K_a. Use the pH to determine the $[H_3O^+]$, then insert the values into the K_a expression and solve. Finally, determine K_b from $K_a \cdot K_b = K_w = 1.0 \times 10^{-14}$ and the value of K_a.

Work:

	NH_3OH^+	+	H_2O	$\rightleftharpoons$	H_3O^+	+	NH_2OH
Balanced							
Initial	0.044				0		0
Change	- x				+ x		+ x
Equilibrium	0.044 - x				x		x

Insert the values into the K_a expression.

$$K_a = \frac{[H_3O^+][NH_2OH]}{[NH_3OH^+]} = \frac{(x)(x)}{(0.044 - x)}$$

Find the value of x, which is equal to $[H_3O^+]$, from the pH.

$$x = [H_3O^+] = 10^{-pH} = 10^{-3.66} = 2.2 \times 10^{-4}$$

Insert the value of x and solve for the K_a.

$$K_a = \frac{(2.2 \times 10^{-4})(2.2 \times 10^{-4})}{(0.044 - 2.2 \times 10^{-4})} = 1.1 \times 10^{-6}$$

Find K_b for the relationship that $K_a \cdot K_b = K_w = 1.0 \times 10^{-14}$.

$$K_a \cdot K_b = 1.0 \times 10^{-14}; \text{ therefore, } K_b = \frac{1.0 \times 10^{-14}}{1.1 \times 10^{-6}} = 9.1 \times 10^{-9}$$

16.63 Calculate the equilibrium concentration of CN^- in a solution in which the equilibrium concentrations of HCN and H_3O^+ are 0.125 M and 0.830 M, respectively.

Analysis:

Target: The CN^- concentration in a solution. (? mol L^{-1}) =

Knowns: The equilibrium molarity of HCN and H_3O^+ in the solution. The K_a of HCN is 4.0×10^{-10} (Appendix J).

Relationships: The algebraic expression for the equilibrium can be written from the balanced ionization reaction.

Plan: Use the balanced ionization reaction to write the algebraic expression for the K_a. Rearrange the expression to solve for the cyanide concentration, insert the values of the K_a and the other known equilibrium concentrations and calculate.

Work:

Balanced $\quad HCN + H_2O \rightleftharpoons H_3O^+ + CN^-$

The K_a expression is: $K_a = \dfrac{[H_3O^+][CN^-]}{[HCN]}$

Rearrange to solve for $[CN^-]$, insert known values and calculate.

$$? \text{ M } [CN^-] = \frac{K_a[HCN]}{[H_3O^+]} = \frac{(0.125)(4.0 \times 10^{-10})}{(0.830)} = \mathbf{6.0 \times 10^{-11}\ M}$$

16.65 What is the concentration of trimethylamine if 0.0348 mol trimethylammonium bromide is dissolved in 1.00 L of 0.300 M HBr? K_b for trimethylamine is 6×10^{-5}.

Analysis:

Target: The trimethylamine concentration in a solution. (? mol L^{-1}) =

Knowns: The molarity of HBr, the K_b of trimethylamine, $(CH_3)_3N$, and the number of moles of trimethylammonium bromide dissolved in one liter of the solution. HBr is a strong acid, so bromide ion has no acid/base ability in aqueous solution.

Relationships: The structured approach to equilibrium problems can be used because the equilibrium expression holds regardless of what other solutes might be present. $K_a \bullet K_b = 1.0 \times 10^{-14}$. Because HBr is a strong acid, the initial $[H_3O^+] = 0.300$ M. Molarity has units of mol L^{-1}.

Plan: Determine the molarity of the trimethylammonium ion, $(CH_3)_3NH^+$, from its number of moles and the solution volume. Because the solution is acidic, determine K_a of the $(CH_3)_3NH^+$ ion from K_b of its conjugate base, then use the structured approach to equilibrium problems to determine the algebraic expression for the K_a. Solve the expression for the $(CH_3)_3N$ concentration.

Work:

First, find the molarity of the $(CH_3)_3NH^+$ ion.

$$? \text{ mol } L^{-1} = \frac{0.0348 \text{ mol } (CH_3)_3NH^+}{1.00 \text{ L}} = 0.0348 \text{ M } (CH_3)_3NH^+$$

Determine K_a for the $(CH_3)_3NH^+$ ion.

$$K_a \bullet K_b = 1.0 \times 10^{-14}; \text{ therefore, } K_a = \frac{1.0 \times 10^{-14}}{K_b} = \frac{1.0 \times 10^{-14}}{6 \times 10^{-5}} = 2 \times 10^{-10}$$

Now determine the equilibrium constant expression from the balanced equation.

Balanced	$(CH_3)_3NH^+$	+	H_2O	$\rightleftharpoons$	H_3O^+	+	$(CH_3)_3N$
Initial	0.0348				0.300		0
Change	- x				+ x		+ x
Equilibrium	0.0348 - x				0.300 + x		x

Insert the values into the K_a expression.

$$K_a = \frac{[H_3O^+][(CH_3)_3N]}{[(CH_3)_3NH^+]} = \frac{(0.300 + x)(x)}{(0.0348 - x)} = 2 \times 10^{-10}$$

Solve algebraically. Because K_a is very small, assume $0.0348 - x \approx 0.0348$ and $0.300 + x \approx 0.300$.

$$\frac{(0.300)(x)}{(0.0348)} = 2 \times 10^{-10}; \text{ and } x = [(CH_3)_3N] = \frac{(0.0348)(2 \times 10^{-10})}{(0.300)} = \mathbf{2 \times 10^{-11}\ M}$$

Check: $\frac{2 \times 10^{-11} \text{ M}}{0.0348 \text{ M}} = 6 \times 10^{-10}$, so the approximation that $0.0348 - x \approx 0.0375$ is valid.

16.67 The pK_a of HOCl is 7.52. Calculate the pH of a solution in which [HOCl] = 0.493 M and $[OCl^-] = 0.0872$ M.

Analysis:
Target: The pH of a buffer solution. (? pH) =
Knowns: The pK_a of the acid, the molarity of the acid and the molarity of its conjugate base.

Relationship: $pH = pK_a - \log\left(\frac{[acid]}{[base]}\right)$.

Plan: Use the relationship directly to determine the pH.
Work:

$$? \, pH = pK_a - \log\left(\frac{[acid]}{[base]}\right) = 7.52 - \log\left(\frac{[0.493]}{[0.0872]}\right) = 7.52 - 0.752 = \mathbf{6.77}$$

16.69 What is the pH of a solution, 500 mL of which contains 10.0 g each of acetic acid and sodium acetate?

Analysis:
Target: The pH of a buffer solution. (? pH) =
Knowns: The mass of the acid, the mass of its salt (conjugate base), and the volume of solution are given. The pK_a of the acid is 4.74 (Appendix J). The molar mass of the acid and the molar mass of its salt which provides the conjugate base are implicit knowns.

Relationships: $pH = pK_a - \log\left(\frac{[acid]}{[base]}\right)$. Molarity $= \frac{\text{mol solute}}{\text{L solution}}$. The molar mass provides the equivalence statement g ↔ mol. The equivalence statement 10^{-3} L ↔ 1 mL is also needed.

Plan: If we knew the molarities, we could use the relationship directly to determine the pH.

Step 2. Use the relationship $pH = pK_a - \log\left(\frac{[acid]}{[base]}\right)$ to determine the pH.

Step 1. Use the factor-label method to determine the molarity of the acid and its conjugate base.

Work:

Step 1. Use the factor-label method to determine the molarity of the acid and its conjugate base.
Molar Mass CH_3COOH = 2(12.011 C) + 4(1.0079 H) + 2(15.9994 O) = 60.052 g mol^{-1}
Molar Mass $NaCH_3COOH$ = 1(22.98977 Na) + 2(12.011 C) + 3(1.0079 H) + 2(15.9994 O) = 82.034 g mol^{-1}

$$? \, mol \, L^{-1} \, CH_3COOH = \frac{10.0 \, g \, CH_3COOH}{500 \, mL} \times \frac{1 \, mol \, CH_3COOH}{60.052 \, g \, CH_3COOH} \times \frac{1 \, mL}{10^{-3} \, L} = 0.333 \, M$$

$$? \, mol \, L^{-1} \, NaCH_3COO = \frac{10.0 \, g \, NaCH_3COO}{500 \, mL} \times \frac{1 \, mol \, NaCH_3COO}{82.034 \, g \, NaCH_3COO} \times \frac{1 \, mL}{10^{-3} \, L} = 0.244 \, M$$

Step 2. Use the relationship $pH = pK_a - \log\left(\frac{[acid]}{[base]}\right)$ to determine the pH.

$$? \, pH = pK_a - \log\left(\frac{[acid]}{[base]}\right) = 4.74 - \log\left(\frac{[0.333]}{[0.244]}\right) = 4.74 - 0.135 = \mathbf{4.60}$$

16.71 What concentrations of *o*-ethylbenzoic acid (pK_a = 3.79) and potassium *o*-ethylbenzoate are required to prepare a buffer of pH = 4.00?

Analysis:
Target: The relative molarities of an acid and its salt (conjugate base) needed to prepare a specific buffer.
Knowns: The pK_a of the acid and the desired pH.

Relationship: $pH = pK_a - \log\left(\frac{[acid]}{[base]}\right)$.

Plan: Algebraically rearrange the relationship and solve for the ratio of molarities needed.
Work:

$$pH = pK_a - \log\left(\frac{[acid]}{[base]}\right); \text{ therefore, } pH + \log\left(\frac{[acid]}{[base]}\right) = pK_a \text{ and } \log\left(\frac{[acid]}{[base]}\right) = pK_a - pH$$

$$\log\left(\frac{[acid]}{[base]}\right) = 3.79 - 4.00 = -0.21; \quad \frac{[acid]}{[base]} = 10^{-0.21} = \mathbf{0.62}$$

To prepare the buffer, one would need to use 0.62 mol of acid for every mol of salt used.

16.73 What mass ratio of trifluoroacetic acid (molar mass 114 g mol^{-1}) to sodium trifluoroacetate (molar mass 136 g mol^{-1}) must be added to water in order to prepare a buffer of pH = 1.00?

Analysis:
Target: The relative masses of an acid and its salt (conjugate base) needed to prepare a specific buffer.
Knowns: The desired pH and the molar mass of both the acid and its salt. The pK_a of the acid is 0.23 (Appendix J).

Relationship: $pH = pK_a - \log\left(\frac{[acid]}{[base]}\right)$.

Plan: Algebraically rearrange the relationship and solve for the ratio of molarities needed. Use the factor-label method with the given molar masses to convert the ratios of molarities to the necessary mass ratio. Remember that the salt provides the conjugate base.
Work:

$$pH = pK_a - \log\left(\frac{[acid]}{[base]}\right); \text{ therefore, } pH + \log\left(\frac{[acid]}{[base]}\right) = pK_a \text{ and } \log\left(\frac{[acid]}{[base]}\right) = pK_a - pH$$

$$\log\left(\frac{[acid]}{[base]}\right) = 0.23 - 1.00 = -0.77; \quad \frac{[acid]}{[base]} = 10^{-0.77} = 0.17 = \frac{0.17 \text{ mol acid } L^{-1}}{1.0 \text{ mol base } L^{-1}}$$

$$\frac{0.17 \text{ mol } L^{-1} \text{ acid}}{1.0 \text{ mol } L^{-1} \text{ base}} \times \frac{\frac{114 \text{ g acid}}{1 \text{ mol acid}}}{\frac{136 \text{ g base}}{1 \text{ mol base}}} = \frac{\mathbf{0.14 \text{ g acid}}}{\mathbf{1 \text{ g salt}}}$$

16.75 What acid might be a good choice for a buffer of pH = 5.00? What ratio of molar concentrations of acid/salt would produce this pH?

Analysis:
Target: The relative molarities of an acid and its salt needed to prepare a specific buffer.
Knowns: The desired pH. A buffer works best at a pH near its pK_a.

Relationship: $pH = pK_a - \log\left(\frac{[acid]}{[base]}\right)$.

Plan: Consult Appendix J to find an acid with a pK_a close to 5.00, then, algebraically rearrange the relationship and solve for the ratio of molarities needed.
Work:
Propanoic acid, C_2H_5COOH, with pK_a = 4.89, has the tabulated pK_a closest to 5.00.

$$pH = pK_a - \log\left(\frac{[acid]}{[base]}\right); \text{ therefore, } pH + \log\left(\frac{[acid]}{[base]}\right) = pK_a \text{ and } \log\left(\frac{[acid]}{[base]}\right) = pK_a - pH$$

$$\log\left(\frac{[acid]}{[base]}\right) = 4.89 - 5.00 = -0.11; \quad \frac{[acid]}{[base]} = 10^{-0.11} = \mathbf{0.78}$$

To prepare the buffer, one would need to use 0.78 mol of acid for every mol of salt, which provides the conjugate base, used.

16.77 How many moles KOH must be added to 1.00 L of 0.372 M acetic acid in order to prepare a buffer of pH = 5.00?

Analysis:

Target: The number of moles of base needed to make a buffer at a specific pH.

Knowns: The desired pH, the volume of acid and the molarity of the acid. KOH is a strong base. The pK_a of acetic acid can be looked up in Appendix J and is 4.74.

Relationships: $pH = pK_a - \log\left(\frac{[acid]}{[base]}\right)$. As the KOH is added to the acid, partial neutralization occurs. $n = M \bullet V$.

Plan: Rearrange the relationship and solve for the ratio of molarities needed. Then, use the stoichiometry of the neutralization process and algebra to determine the number of moles of KOH required. Finally, use the factor-label method to find the number of moles.

Work:

$pH = pK_a - \log\left(\frac{[acid]}{[base]}\right)$; therefore, $pH + \log\left(\frac{[acid]}{[base]}\right) = pK_a$ and $\log\left(\frac{[acid]}{[base]}\right) = pK_a - pH$

$\log\left(\frac{[acid]}{[base]}\right) = 4.74 - 5.00 = -0.26$; $\frac{[acid]}{[base]} = 10^{-0.26} = \mathbf{0.55}$

Let x be the number of moles of KOH needed and apply reaction stoichiometry.

Balanced	CH_3COOH	+	KOH	$\rightleftharpoons$	KCH_3COOH	+	H_2O
Initial	0.372 M		x		0		
Change	- x		-x		+ x		
After reaction	0.372 - x		0		x		

Substitute these values into the $\frac{[acid]}{[base]}$ ratio and algebraically solve for x.

$\frac{0.372 - x}{x} = 0.55$; therefore, $0.55x = 0.372 - x$; $x + 0.55\,x = 0.372$ and $1.55\,x = 0.372$

$x = \frac{0.372}{1.55} = 0.24\ mol\ L^{-1}$

The number of moles needed is $0.24\ mol\ L^{-1} \times 1.00\ L$ = **0.24 mol KOH**

16.79 Calculate the pH at the equivalence point of a titration of acetic acid with 0.100 M KOH, assuming that the concentration of the acid is as follows.

a. 1.00 M b. 0.100 M c. 0.0100 M

Analysis:

Target: The pH at the equivalence point of a titration. (? pH) =

Knowns: The initial molarity and name of the weak acid and the molarity and formula of the titrant strong base. The formula of the acid is implicit from its name. The K_a of acetic acid can be looked up in Appendix J and is equal to 1.8×10^{-5}.

Relationships: At the equivalence point in the titration of a weak acid with a strong base, the acid has completely reacted and only the salt is present. The salt will hydrolyze creating a basic solution. $n_a = n_b \bullet$(stoichiometric mole ratio). $n = M \bullet V$.

$K_a \bullet K_b = K_w = 1.0 \times 10^{-14}$ for conjugates. $pOH = -\log[OH^-]$. $pH + pOH = 14.00$.

$M = \frac{mol}{L}$.

Plan: We know that hydrolysis will occur, but first we must find the molarity of the salt at the equivalence point.

Step 3. Use the structured approach to equilibrium problems to determine the $[OH^-]$ at the equivalence point and then use the other relationships to find the pH.
Step 2. Use stoichiometry and the definition of molarity to determine the molarity of the salt at the equivalence point.
Step 1. Find the relative volume of base required to reach the equivalence point, using stoichiometric considerations.

Work:

a. **Step 1.** Find the volume of base needed. Because no volume of acid is given assume it be represented by v. The balanced chemical equation is

$NaOH + CH_3COOH \longrightarrow NaCH_3COO + H_2O$; therefore, the stoichiometric mole ratio is 1 mol acid $\leftrightarrow$ 1 mol base.

$$n_a = n_b\left(\frac{1 \text{ mol acid}}{1 \text{ mol base}}\right); \text{ therefore, } M_aV_a = M_bV_b(1)$$

$$V_b = \frac{M_aV_a}{M_b} = \frac{(1.00\ M)(v)(1)}{0.100\ M} = 10.0v \text{ base}$$

Step 2. Use stoichiometry and the definition of molarity to find the molarity of the $NaCH_3COO$. The total solution volume is v + 10.0v = 11.0v.

$$n_{salt} = n_a\left(\frac{1 \text{ mol salt}}{1 \text{ mol acid}}\right) = M_aV_a(1) = (1.00\ M)(v)(1)$$

$$?\ M \text{ salt} = \frac{\text{mol salt}}{\text{solution volume}} = \frac{(1.00\ M)(v)(1)}{11.0v} = 0.0909\ M$$

Step 3. Use the structured approach to equilibrium problems to determine the $[OH^-]$ at the equivalence point and then use the other relationships to find the pH.

$$K_a \cdot K_b = K_w = 1.0 \times 10^{-14}; \text{ therefore, } K_b = \frac{1.0 \times 10^{-14}}{K_a} = \frac{1.0 \times 10^{-14}}{1.8 \times 10^{-5}} = 5.6 \times 10^{-10}$$

Balanced	CH_3COO^- + H_2O	$\rightleftharpoons$	CH_3COOH	+ OH^-
Initial	0.0909		0	0
Change	- x		+ x	+ x
Equilibrium	0.0909 - x		x	x

Insert the values into the K_b expression.

$$K_b = \frac{[CH_3COOH][OH^-]}{[CH_3COO^-]} = \frac{(x)(x)}{(0.0909 - x)} = 5.6 \times 10^{-10}$$

Solve algebraically. Because K_a is very small, assume $0.0909 - x \approx 0.0909$

$$\frac{x^2}{0.0909} = 5.6 \times 10^{-10}; \text{ therefore, } x^2 = 0.0909(5.6 \times 10^{-10}) = 5.1 \times 10^{-11}; x = \sqrt{5.1 \times 10^{-11}}$$

$$x = [OH^-] = 7.1 \times 10^{-6}\ M; \quad pOH = -\log [OH^-] = -\log (7.1 \times 10^{-6}) = -(-5.15) = 5.15$$

$?\ pH = 14.00 - pOH = 14.00 - 5.15 = $ **8.85**

Check: $\frac{7.1 \times 10^{-5}\ M}{0.0909\ M} = 7.8 \times 10^{-4}$, so the approximation that $0.0909 - x \approx 0.0909$ is valid.

b. A "condensed" version of the steps in part (a) is shown.
Step 1. Find the volume of base.

$$V_b = \frac{M_a V_a}{M_b} = \frac{(0.100\ M)(v)(1)}{0.100\ M} = 1.00v \text{ base}$$

Step 2. Use stoichiometry and the definition of molarity to find the molarity of the $NaCH_3COO$. The total solution volume is v + 1.00v = 2.00v.

$$?\ M \text{ salt} = \frac{\text{mol salt}}{\text{solution volume}} = \frac{(0.100\ M)(v)(1)}{2.00v} = 0.0500\ M$$

Step 3. Use the structured approach to equilibrium problems to determine the $[OH^-]$ at the equivalence point and then use the other relationships to find the pH.

$$K_b = \frac{[CH_3COOH][OH^-]}{[CH_3COO^-]} = \frac{(x)(x)}{(0.0500 - x)} = 5.6 \times 10^{-10}$$

Solve algebraically. Because K_a is very small, assume $0.0500 - x \approx 0.0500$

$$\frac{x^2}{0.0500} = 5.6 \times 10^{-10}; \text{ therefore, } x^2 = 0.0500(5.6 \times 10^{-10}) = 2.8 \times 10^{-11}; x = \sqrt{2.8 \times 10^{-11}}$$

$x = [OH^-] = 5.3 \times 10^{-6}$ M; pOH = -log $[OH^-]$ = -log (5.3×10^{-6}) = -(-5.28) = 5.28

? pH = 14.00 - pOH = 14.00 - 5.28= **8.72**

Check: $\frac{5.3 \times 10^{-6}\ M}{0.0500\ M} = 1.1 \times 10^{-4}$, so the approximation that $0.0909 - x \approx 0.0909$ is valid.

c. A "condensed" version of the steps in part (a) is shown.
Step 1. Find the volume of base.

$$V_b = \frac{M_a V_a}{M_b} = \frac{(0.0100\ M)(v)(1)}{0.100\ M} = 0.100\ v \text{ base}$$

Step 2. Use stoichiometry and the definition of molarity to find the molarity of the $NaCH_3COO$. The total solution volume is v + 0.100 v = 1.100 v.

$$?\ M \text{ salt} = \frac{\text{mol salt}}{\text{solution volume}} = \frac{(0.0100\ M)(v)(1)}{1.100\ v} = 0.00909\ M$$

Step 3. Use the structured approach to equilibrium problems to determine the $[OH^-]$ at the equivalence point and then use the other relationships to find the pH.

$$K_b = \frac{[CH_3COOH][OH^-]}{[CH_3COO^-]} = \frac{(x)(x)}{(0.00909 - x)} = 5.6 \times 10^{-10}$$

Solve algebraically. Because K_a is very small, assume $0.00909 - x \approx 0.00909$

$$\frac{x^2}{0.00909} = 5.6 \times 10^{-10}; \text{ therefore, } x^2 = 0.00909\ (5.6 \times 10^{-10}) = 5.1 \times 10^{-12}; x = \sqrt{5.1 \times 10^{-12}}$$

$x = [OH^-] = 2.6 \times 10^{-6}$ M; pOH = -log $[OH^-]$ = -log (2.6×10^{-6}) = -(-5.65) = 5.65

? pH = 14.00 - pOH = 14.00 - 5.65= **8.35**

Check: $\frac{2.6 \times 10^{-6}\ M}{0.00909\ M} = 2.9 \times 10^{-4}$, so the approximation that $0.0909 - x \approx 0.0909$ is valid.

16.81 Calculate the pH during the titration of 50.0 mL 0.10 M HCl with 0.10 M KOH, after the addition of the following volumes of base.
a. 10.0 mL b. 25.0 mL c. 49.0 mL d. 50.0 mL e. 80.0 mL
Sketch a rough titration curve through these points.

Analysis:
Target: The pH at various points on a titration curve. (? pH) =
Knowns: The volume of the acid, the molarity of the acid, the molarity of the base and the volume of the base at several points. Both the acid and the base are strong.
Relationships: pH = -log $[H_3O^+]$, pOH = -log $[OH^-]$ and pH + pOH = 14.00. The acid and base react according to the reaction $HCl + KOH \rightarrow KCl + H_2O$. Molarity = $\frac{mol}{L}$.

Plan: Because the acid and base are both strong, we only need to know the number of moles of acid and base present and the total volume. The molarity of the excess acid or base determines the concentration of excess $[H_3O^+]$ or $[OH^-]$. Because volumes are given in mL, we can work in millimoles and find molarity from $\frac{mmol}{mL}$.

Work:
The initial pH of the acid solution is -log $[H_3O^+]$ = -log (0.10) = -(-1.00) = 1.00.

In each case, the initial number of millimoles of the acid is 50.0 mL(0.10 M) = 5.0 mmol.

a. After addition of 10.0 mL of base, the number of mmol base is:
10.0 mL(0.10 M) = 1.0 mmol.
The total volume is 10.0 mL + 50.0 mL = 60.0 mL. The number of mmol of excess acid is 5.0 mmol acid - 1.0 mmol base = 4.0 mmol acid.

The molarity of the excess acid is: $\frac{4.0\ mmol}{60.0\ mL} = 0.067$ M.

? pH = -log (0.067) = -(-1.18) = **1.18**

b. After addition of 25.0 mL of base, the number of mmol base is:
25.0 mL(0.10 M) = 2.5 mmol.
The total volume is 25.0 mL + 50.0 mL = 75.0 mL. The number of mmol of excess acid is 5.0 mmol acid - 2.5 mmol base = 2.5 mmol acid.

The molarity of the excess acid is: $\frac{2.5\ mmol}{75.0\ mL} = 0.033$ M.

? pH = -log (0.033) = -(-1.48) = **1.48**

c. After addition of 49.0 mL of base, the number of mmol base is:
49.0 mL(0.10 M) = 4.9 mmol.
The total volume is 49.0 mL + 50.0 mL = 99.0 mL. The number of mmol of excess acid is 5.0 mmol acid - 4.9 mmol base = 0.1 mmol acid.

The molarity of the excess acid is: $\frac{0.1\ mmol}{99.0\ mL} = 0.001$ M.

? pH = -log (0.001) = -(-3.0) = **3.0**

d. After addition of 50.0 mL of base, the number of mmol base is:
50.0 mL(0.10 M) = 5.0 mmol.
This is the equivalence point in the titration, because the number of millimoles of base is equal to the number of millimoles of acid.

Because neither ion of the salt can hydrolyze, the solution is neutral with a pH of **7.00**.

e. After addition of 80.0 mL of base, the number of mmol base is:
80.0 mL(0.10 M) = 8.0 mmol.
The total volume is 80.0 mL + 50.0 mL = 130.0 mL. The number of mmol of excess base is
8.0 mmol base - 5.0 mmol acid = 3.0 mmol base.

The molarity of the excess base is: $\frac{3.0 \text{ mmol}}{130.0 \text{ mL}} = 0.023$ M.

pOH = -log (0.023) = -(-1.64) = 1.64.

? pH = 14.00 - pOH = 14.00 - 1.64 = **12.36**

(Note that extra points have been calculated to better define the curve.)

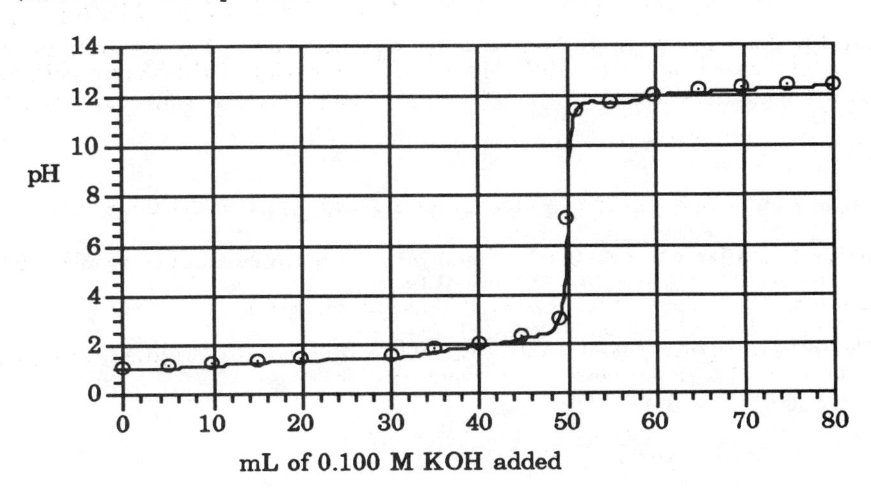

16.83 Calculate the pH during the titration of 50.0 mL 0.10 M KOH with 0.10 M HCl after the addition of the following volumes of acid.
a. 10.0 mL b. 25.0 mL c. 49.0 mL d. 50.0 mL e. 80.0 mL
Sketch a rough titration curve through these points.

Analysis:

Target: The pH at various points on a titration curve. (? pH) =

Knowns: The volume of the base, the molarity of the base, the molarity of the acid and the volume of the acid at several points. Both the acid and the base are strong.

Relationships: pH = -log $[H_3O^+]$, pOH = -log $[OH^-]$ and pH + pOH = 14.00. The acid and base react according to the reaction HCl + KOH $\rightarrow$ KCl + H_2O. Molarity = $\frac{\text{mol}}{\text{L}}$.

Plan: Because the acid and base are both strong, we only need to know the number of moles of acid and base present and the total volume. The molarity of the excess acid or base determines the concentration of excess $[H_3O^+]$ or $[OH^-]$. Because volumes are given in mL, we can work in millimoles and find molarity from $\frac{\text{mmol}}{\text{mL}}$.

Work:

The initial pOH of the base solution is -log $[OH^-]$ = -log (0.10) = -(-1.00) = 1.00.

pH = 14.00 - pOH = 14.00 - 1.00 = 13.00.

In each case, the initial number of millimoles of the base is 50.0 mL(0.10 M) = 5.0 mmol.

a. After addition of 10.0 mL of acid, the number of mmol acid is:

10.0 mL(0.10 M) = 1.0 mmol.
The total volume is 10.0 mL + 50.0 mL = 60.0 mL. The number of mmol of excess base is
5.0 mmol base - 1.0 mmol acid = 4.0 mmol base.

The molarity of the excess base is: $\frac{4.0 \text{ mmol}}{60.0 \text{ mL}}$ = 0.067 M.

pOH = -log (0.067) = -(-1.18) = 1.18

? pH = 14.00 - pOH = 14.00 - 1.18= **12.82**

b. After addition of 25.0 mL of acid, the number of mmol acid is:
25.0 mL(0.10 M) = 2.5 mmol.
The total volume is 25.0 mL + 50.0 mL = 75.0 mL. The number of mmol of excess base is
5.0 mmol base - 2.5 mmol acid = 2.5 mmol base.

The molarity of the excess base is: $\frac{2.5 \text{ mmol}}{75.0 \text{ mL}}$ = 0.033 M.

pOH = -log (0.033) = -(-1.48) = 1.48

? pH = 14.00 - pOH = 14.00 - 1.48 = **12.52**

c. After addition of 49.0 mL of acid, the number of mmol acid is:
49.0 mL(0.10 M) = 4.9 mmol.
The total volume is 49.0 mL + 50.0 mL = 99.0 mL. The number of mmol of excess base is
5.0 mmol base - 4.9 mmol acid = 0.1 mmol base.

The molarity of the excess base is: $\frac{0.1 \text{ mmol}}{99.0 \text{ mL}}$ = 0.001 M.

pOH = -log (0.01) = -(-3.0) = 3.0

? pH = 14.00 - pOH = 14.00 - 3.0 = **11.0**

d. After addition of 50.0 mL of acid, the number of mmol acid is:
50.0 mL(0.10 M) = 5.0 mmol.
This is the equivalence point in the titration, because the number of millimoles of base is equal to the number of millimoles of acid.

Because neither ion of the salt can hydrolyze, the solution is neutral with a pH of **7.00**.

e. After addition of 80.0 mL of acid, the number of mmol acid is:,
80.0 mL(0.10 M) = 8.0 mmol.
The total volume is 80.0 mL + 50.0 mL = 130.0 mL. The number of mmol of excess acid is
8.0 mmol acid - 5.0 mmol base = 3.0 mmol acid.

The molarity of the excess acid is: $\frac{3.0 \text{ mmol}}{130.0 \text{ mL}}$ = 0.023 M.

? pH = -log (0.023) = -(-1.64) = **1.64**.

(Note that this titration is precisely the opposite as that in Problem 16.81--the same molarities volumes and volumes of reactants are used. The resultant titration curve looks like the previous one inverted. As in Problem 16.81, additional points have been calculated to better define the curve.)

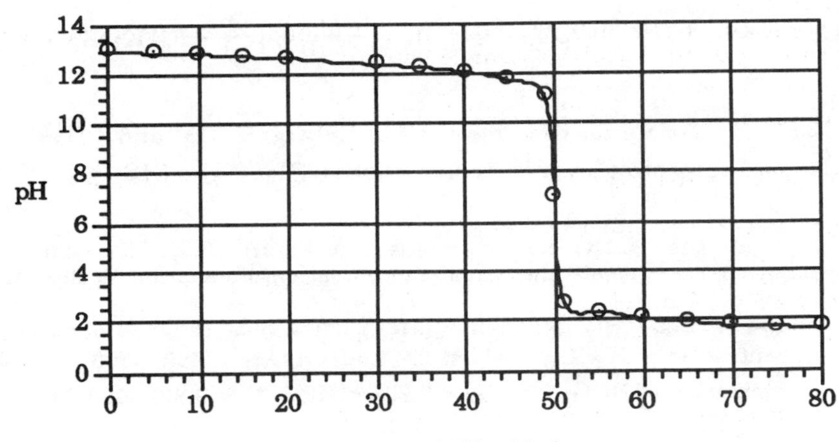

16.85 Write the chemical equations and equilibrium expressions for the successive ionization steps of the following species:
a. H_2SeO_4 (diprotic) b. H_3AsO_4 (triprotic)

Analysis/Plan: Show successive transfers of a proton to water until the acidic hydrogens have all been used.
Work:

a. $H_2SeO_4 + H_2O \rightleftharpoons H_3O^+ + HSeO_4^-$ $K_1 = \dfrac{[H_3O^+][HSeO_4^-]}{[H_2SeO_4]}$

$HSeO_4^- + H_2O \rightleftharpoons H_3O^+ + SeO_4^{2-}$ $K_2 = \dfrac{[H_3O^+][SeO_4^{2-}]}{[HSeO_4^-]}$

b. $H_3AsO_4 + H_2O \rightleftharpoons H_3O^+ + H_2AsO_4^-$ $K_1 = \dfrac{[H_3O^+][H_2AsO_4^-]}{[H_3AsO_4]}$

$H_2AsO_4^- + H_2O \rightleftharpoons H_3O^+ + HAsO_4^{2-}$ $K_2 = \dfrac{[H_3O^+][HAsO_4^{2-}]}{[H_2AsO_4^-]}$

$HAsO_4^{2-} + H_2O \rightleftharpoons H_3O^+ + AsO_4^{3-}$ $K_3 = \dfrac{[H_3O^+][AsO_4^{3-}]}{[HAsO_4^{2-}]}$

16.87 (a) Write the overall chemical equation and equilibrium expression for the complete ionization of H_2CO_3. (b) What is the value of the overall ionization constant?

Analysis/Plan: The overall ionization is the sum of the ionization of the two separate single proton transfers. See Appendix J in the text for ionization constants. Remember that for a two step process, $K_{overall} = K_1 \times K_2$.
Work:

$H_2CO_3 + H_2O \rightleftharpoons H_3O^+ + HCO_3^-$ $K_1 = \dfrac{[H_3O^+][HCO_3^-]}{[H_2CO_3]} = 4.0 \times 10^{-7}$

$HCO_3^- + H_2O \rightleftharpoons H_3O^+ + CO_3^{2-}$ $K_2 = \dfrac{[H_3O^+][CO_3^{2-}]}{[HCO_3^-]} = 4.0 \times 10^{-11}$

$$H_2CO_3 + 2\,H_2O \rightleftharpoons 2\,H_3O^+ + CO_3^{2-} \qquad K = \frac{[H_3O^+]^2[CO_3^{2-}]}{[H_2CO_3]} = K_1 \bullet K_2 = 1.6 \times 10^{-17}$$

16.89 Refer to Figure 16.9 and estimate the concentrations of H_2CO_3, HCO_3^- and CO_3^{2-} when the pH is 7.00 and the total concentration of carbonate (that is, $[H_2CO_3] + [HCO_3^-] + [CO_3^{2-}]$) is 0.0500 M.

Analysis/Plan: Estimate the relative concentrations from **Figure 16.9** in the text. Use the relative concentrations to estimate the actual concentration of each ion in the sample.
Work:

At pH = 7.00, there is essentially no CO_3^{2-}; so $[CO_3^{2-}]$ = **0 M.**
The relative concentration of H_2CO_3 is about 0.2, so $[H_2CO_3]$ = 0.2(0.0500 M) = **0.01 M.**
The relative concentration of HCO_3^- is about 0.8, so $[HCO_3^-]$ = 0.8(0.0500 M) = **0.04 M.**

16.91 Gaseous CO_2 is dissolved in a buffer solution of pH = 4.00, resulting in an equilibrium concentration $[H_2CO_3]$ = 0.0010 M. What are the concentrations of HCO_3^- and CO_3^{2-}?

Analysis:

Target: The concentrations of HCO_3^{2-} and CO_3^{2-} in a buffered solution. (? M) =
Knowns: The molarity of H_2CO_3 and the pH of the solution. The necessary K_a values are shown in Problem 16.87.
Relationships: The equilibrium expressions for each ionization were written in Problem 16.87. $[H_3O^+] = 10^{-pH}$.

Plan: Because the solution is buffered to a specific pH, the equilibrium expressions can be used directly. Insert the known concentrations and algebraically rearrange and solve for the concentration of interest.
Work:

First, determine the $[HCO_3^-]$. Because pH = 4.00, $[H_3O^+] = 10^{-4.00} = 1.0 \times 10^{-4}$.

$$K_1 = \frac{[H_3O^+][HCO_3^-]}{[H_2CO_3]} = \frac{(1.0 \times 10^{-4})[HCO_3^-]}{(0.0010)} = 4.0 \times 10^{-7}$$

$$[HCO_3^-] = \frac{(0.0010)(4.0 \times 10^{-7})}{(1.0 \times 10^{-4})} = \mathbf{4.0 \times 10^{-6}}$$

Now determine the $[CO_3^{2-}]$ from $K_{overall}$.

$$K = \frac{[H_3O^+]^2[CO_3^{2-}]}{[H_2CO_3]} = \frac{(1.0 \times 10^{-4})^2[CO_3^{2-}]}{(0.0010)} = 1.6 \times 10^{-17}$$

$$[CO_3^{2-}] = \frac{(0.0010)(1.6 \times 10^{-17})}{(1.0 \times 10^{-4})^2} = \mathbf{1.6 \times 10^{-12}}$$

16.93 What is the pH of a solution, 1.00 liter of which contains 10.0 g each of Na_2CO_3 and $NaHCO_3$?

Analysis:
Target: The pH of a buffer solution. (? pH) =
Knowns: The two salts provide the acid HCO_3^- and its conjugate base, CO_3^{2-}. The mass of the acid, the mass of its salt, and the volume of solution are given. The pK_a of the acid can be looked up in Appendix J and is equal to 10.40. The molar masses of the acid and its salt are implicit knowns.

Relationships: $pH = pK_a - \log\left(\frac{[acid]}{[base]}\right)$. Molarity $= \frac{mol\ solute}{L\ solution}$. The molar mass provides the equivalence statement g ↔ mol.

Plan: If we knew the molarities, we could use the relationship directly to determine the pH.

Step 2. Use the relationship $pH = pK_a - \log\left(\frac{[acid]}{[base]}\right)$ to determine the pH.

Step 1. Use the factor-label method to determine the molarity of the acid and its salt, which provides the conjugate base.

Work:

Step 1. Use the factor-label method to determine the molarities of the acid and its salt.

Molar Mass Na_2CO_3 = 2(22.98977 Na) + 1(12.011 C) + 3(15.9994 O) = 105.989 g mol^{-1}

Molar Mass $NaHCO_3$ = 1(22.98977 Na) + 1(1.0079 H) + 1(12.011 C) + 3(15.9994 O) = 84.007 g mol^{-1}

$$? \text{ mol L}^{-1}\ Na_2CO_3 = \frac{10.0 \text{ g } Na_2CO_3}{1.00 \text{ L}} \times \frac{1 \text{ mol } Na_2CO_3}{105.989 \text{ g } Na_2CO_3} = 0.0943 \text{ M } Na_2CO_3$$

$$? \text{ mol L}^{-1}\ NaHCO_3 = \frac{10.0 \text{ g } NaHCO_3}{1.00 \text{ L}} \times \frac{1 \text{ mol } NaHCO_3}{84.007 \text{ g } NaHCO_3} = 0.119 \text{ M } NaHCO_3$$

Step 2. Use the relationship $pH = pK_a - \log\left(\frac{[acid]}{[base]}\right)$ to determine the pH.

$$? pH = pK_a - \log\left(\frac{[acid]}{[base]}\right) = 10.40 - \log\left(\frac{[0.119]}{[0.0943]}\right) = 10.40 - 0.101 = \mathbf{10.30}$$

16.95 Calculate the pH if 70 mL 0.100 M KOH is added to 50 mL 0.100 M H_2CO_3. Note that this amount of base is more than enough to convert all H_2CO_3 to HCO_3^-, but not enough to convert all the HCO_3^- to CO_3^{2-}.

Analysis:

Target: The pH of an acid solution to which base has been added. (? pH) =

Knowns: The original volume and molarity of the diprotic weak acid and the volume and molarity of the strong base added. The two salts formed provide the acid, HCO_3^-, and its conjugate base, CO_3^{2-}. The pK_a of the acid HCO_3^- is 10.40 (Appendix J).

Relationships: $pH = pK_a - \log\left(\frac{[acid]}{[base]}\right)$. Molarity $= \frac{mol\ solute}{L\ solution}$. $n = M \bullet V$. The KOH will react in turn with H_2CO_3 and HCO_3^-.

Plan: If we knew the molarities of the ions after the reaction of the strong base, we could use the relationship directly to determine the pH.

Step 3. Use the relationship $pH = pK_a - \log\left(\frac{[acid]}{[base]}\right)$ to determine the pH.

Step 2. Determine the molarity of each of the ions formed from the number of moles and the total solution volume.

Step 1. Write the successive acid/base reactions and use stoichiometric techniques to determine the number of moles of each species present after the KOH has reacted.

Work:

Step 1. Determine the number of millimoles of each species present.

The number of mmol of KOH is 70 mL(0.100 M KOH) = 7.0 mmol KOH.

The number of mmol of H_2CO_3 is 50 mL(0.100 M H_2CO_3) = 5.0 mmol H_2CO_3.

The first balanced acid/base reaction is:

	H_2CO_3	+	KOH $\rightarrow$	$KHCO_3$	+	H_2O
Initial	5.0 mmol		7.0 mmol	0		
Change	-5.0 mmol		-5.0 mmol	+ 5.0 mmol		
After	0 mmol		2.0 mmol	5.0 mmol		

The second balanced acid/base reaction is:

	HCO_3^-	+	KOH $\rightarrow$	K_2CO_3	+	H_2O
Initial	5.0 mmol		2.0 mmol	0		
Change	-2.0 mmol		-2.0 mmol	+2.0 mmol		
After	3.0 mmol		0 mmol	2.0 mmol		

Step 2. Determine the molarity of the ions.
The total volume is 50 mL + 70 mL = 120 mL.

$$? \text{ mol L}^{-1}\ HCO_3^- = \frac{3.0 \text{ mmol}}{120\text{mL}} = 0.025 \text{ M } HCO_3^- \text{ (the acid)}$$

$$? \text{ mol L}^{-1}\ CO_3^{2-} = \frac{2.0 \text{ mmol}}{120\text{mL}} = 0.017 \text{ M } CO_3^{2-} \text{ (the conjugate base)}$$

Step 3. Use the relationship $pH = pK_a - \log\left(\frac{[\text{acid}]}{[\text{base}]}\right)$ to determine the pH.

$$? \text{ pH} = pK_a - \log\left(\frac{[\text{acid}]}{[\text{base}]}\right) = 10.40 - \log\left(\frac{[0.025]}{[0.017]}\right) = 10.40 - 0.18 = \mathbf{10.22}$$

16.97 Show that the error in using the expression $[H_3O^+] = x \approx \sqrt{K_a \cdot C}$ is no more than 5% whenever $x < 0.1 \cdot C$. Hint: assume $x_{approx} = \sqrt{K_a \cdot C}$, show that $x_{true} = \sqrt{[K_a \cdot (C - x)]}$ and then that $(x_{approx}/x_{true}) < 1.05$ when $x < 0.1 \cdot C$.

Analysis/Plan: Use the structured approach to equilibrium problems to set up a general solution for x and then use the hints to complete the proof.
Work:

The general ionization reaction for an acid is $HA + H_2O \rightleftharpoons H_3O^+ + A^-$.

Balanced	HA	+	H_2O	$\rightleftharpoons$	H_3O^+	+	A^-
Initial	C				0		0
Change	- x				+ x		+ x
Equilibrium	C - x				x		x

Insert these values into the equilibrium expression.

$$K_a = \frac{[H_3O^+][A^-]}{[HA]} = \frac{(x)(x)}{C - x} = \frac{x^2}{C - x}$$ Cross multiplying gives: $x^2 = K_a(C - x)$.

$x = \sqrt{K_a(C - x)}$. Let this be x_{true}. If $x = 0.1(C)$, then $x = \sqrt{K_a(C - 0.1C)} = \sqrt{K_a(0.9C)}$

If we make the approximation that $C - x \approx C$, then $x_{approx} = \sqrt{K_a(C)}$.

$$\frac{x_{approx}}{x_{true}} = \frac{\sqrt{K_a(C)}}{\sqrt{K_a(0.9C)}} = \sqrt{\frac{1}{0.9}} = 1.05$$

Because the approximate value is no greater than 1.05 times the true value, the error is no greater than 5% and the approximation is valid.

16.99 The concentration of hydronium ion is 8.0×10^{-5} in a solution of HClO. What is the value of the concentration ratio $[ClO^-]/[HClO]$?

Analysis:

Target: The concentration ratio $[ClO^-]/[HClO]$.
Knowns: The H_3O^+ concentration. The K_a of HClO is 3.0×10^{-8} (Appendix J).
Relationships: The equilibrium expression can be written from the ionization reaction.
Plan: Write the ionization reaction and the equilibrium expression. Insert the values of K_a and $[H_3O^+]$ and algebraically solve for the concentration ratio.

Work:

The equilibrium equation is $HClO + H_2O \rightleftharpoons H_3O^+ + ClO^-$.
Write the equilibrium expression, insert the values and calculate.

$$K_a = \frac{[H_3O^+][ClO^-]}{[HClO]} = 3.0 \times 10^{-8} = \frac{(8.0 \times 10^{-5})[ClO^-]}{[HClO]}$$

$$? \frac{[ClO^-]}{[HClO]} = \frac{3.0 \times 10^{-8}}{8.0 \times 10^{-5}} = \mathbf{3.8 \times 10^{-4}}$$

16.101 The pH of a 0.275 M solution of sodium *p*-methoxyphenolate is 11.85. What is the pK_a of *p*-methoxyphenol?

Analysis:

Targets: The acid ionization constant, K_a, of a weak acid. ($? K_a$) =
Knowns: The molarity and the pH of a solution of its salt (conjugate base). The formula of the base is not given, but may be represented by A^-.
Relationships: The structured approach to equilibrium problems is appropriate.
$pH + pOH = 14.00$. $[OH^-] = 10^{-pOH}$. For conjugate species, $K_a \cdot K_b = 1.0 \times 10^{-14}$. $pK_a = -\log K_a$.

Plan: If we knew the K_b of the base, we could find the K_a and pK_a of its conjugate acid.
Step 2. Use the relationship $K_a \cdot K_b = 1.0 \times 10^{-14}$ to find K_a and from K_a find pK_a.
Step 1. Use the structured approach to equilibrium problems to write an algebraic expression for the K_b. Find the pOH from pH + pOH = 14.00, then use the relationship $[OH^-] = 10^{-pOH}$ to determine the equilibrium concentration of $[OH^-]$. Substitute that value into the K_b expression to determine the K_b.

Work:

Step 1. Find the K_b.

Balanced	A^-	+	H_2O	$\rightleftharpoons$	HA	+	OH^-
Initial	0.275				0		0
Change	- x				+ x		+ x
Equilibrium	0.275 - x				x		x

Insert the values into the K_b expression.

$$K_b = \frac{[HA][OH^-]}{[A^-]} = \frac{(x)(x)}{(0.275 - x)}$$

$$pOH = 14.00 - pH = 14.00 - 11.85 = 2.15$$

$$[OH^-] = 10^{-pOH} = 10^{-2.15} = 7.1 \times 10^{-3} \text{ M}$$

Substitute x (equal to $[OH^-]$) into the K_b expression.

$$K_b = \frac{(7.1 \times 10^{-3})^2}{0.275 - 7.1 \times 10^{-3}} = 1.9 \times 10^{-4}$$

Step 2. Use the relationship $K_a \bullet K_b = 1.0 \times 10^{-14}$ to find K_a. From K_a find the pK_a.

$$K_a = \frac{1.0 \times 10^{-14}}{K_b} = \frac{1.0 \times 10^{-14}}{1.9 \times 10^{-4}} = 5.3 \times 10^{-11}$$

? pKa = $-\log K_a = -\log (5.3 \times 10^{-11}) = -(-10.27) =$ **10.27**

16.103 Calculate the pH of a solution that is 0.0753 M in HOCl and 0.156 M in NaOCl.

Analysis:

Target: The pH of a buffer solution. (? pH) =

Knowns: The molarity of the acid and its conjugate base, in the NaOCl salt. The pK_a of the acid is 7.52 (Appendix J).

Relationship: $pH = pK_a - \log\left(\frac{[acid]}{[base]}\right)$.

Plan: Use the relationship directly to determine the pH.

Work:

$$pH = pK_a - \log\left(\frac{[acid]}{[base]}\right) = 7.52 - \log\left(\frac{[0.0753]}{[0.156]}\right) = 7.52 - (-0.316) = \mathbf{7.84}$$

16.105 Which of the following buffers has the greater resistance to change in pH (buffer capacity)? Explain your answer.

a. [acid] = [salt] = 0.100 M b. [acid] = [salt] = 0.300 M

Combination **b**, with the higher concentrations is a better buffer. This is true because higher concentrations of conjugate species can consume greater quantities of added acid or base.

16.107 Calculate the pH of the solution resulting from mixing equal volumes of 0.10 M HCl with

a. water
b. 0.20 M NaOH
c. 0.10 M sodium acetate
d. 0.20 M sodium acetate
e. 0.10 M NH_3
f. 0.20 M NH_3

Analysis:

Target: The pH of a solution. (? pH) =

Knowns: The molarity and nature of the original substance and the molarity of the added strong acid. From Appendix J, $K_a = 1.8 \times 10^{-5}$ and $pK_a = 4.74$ for acetic acid; $K_b = 1.9 \times 10^{-5}$ and $pK_b = 4.72$ for NH_3.

Relationships: The strong acid will react with the strongest base in the system.

$n = M \bullet V$. $M = \frac{mol}{L}$. $K_a \bullet K_b = 1.0 \times 10^{-14}$ for conjugates. $pH = -\log [H_3O^+]$.

$pH + pOH = 14.00$. $pH = pK_a - \log\left(\frac{[acid]}{[base]}\right)$. $pOH = pK_b - \log\left(\frac{[base]}{[acid]}\right)$.

Plan: Study each system and apply appropriate stoichiometric considerations. Treat the products of the strong acid reaction according to their nature.

Work:

Volumes are not given. In each case assume v be equal to the initial volume. Then, v + v = 2v is equal to the final volume.

a. When the strong acid is added to water, the only result is dilution. The number of moles of acid = $n_a = M \bullet V = (0.10)v = 0.10v$. The molarity of the strong acid is equal to the $[H_3O^+]$ and is $\frac{mol}{L} = \frac{0.10v}{2v} = 0.050$ M.

? pH = $-\log [H_3O^+] = -\log (0.050) = -(-1.30) =$ **1.30**

b. When the strong acid is added to the strong base, reaction stoichiometry must be considered. The number of moles of acid = n_a = M•V = (0.10)v = 0.10v. The number of moles of base = n_b = M•V = (0.20 M)v = 0.20v. The reaction is:

Balanced	HCl	+	NaOH ⟶	NaCl	+	H_2O
Initial	0.10v		0.20v	0		
Change	-0.10v		-0.10v	+ 0.10v		
After	0		0.10v	0.10v		

The molarity of the excess strong base = $\frac{mol}{L} = \frac{0.10v}{2v}$ = 0.050 M.

pOH = -log [OH^-] = -log (0.050) = -(-1.30) = 1.30

? pH = 14.00 - pOH = 14.00 - 1.30 = **12.70**

c. When the strong acid is added to the salt which is the conjugate base of the weak acid CH_3COOH, reaction stoichiometry must be considered. The number of moles of acid = n_a = M•V = (0.10)v = 0.10v. The number of moles of base = n_b = M•V = (0.10 M)v = 0.10v. The reaction is:

Balanced	HCl	+	CH_3COO^- ⟶	CH_3COOH	+	Cl^-
Initial	0.10v		0.10v	0		0
Change	-0.10v		-0.10v	+ 0.10v		+0.10v
After	0		0	0.10v		0.10v

Both the strong acid and weak base are totally consumed. Cl^- has no acid/base ability. The acetic acid produced can be treated by the structured approach to equilibrium. The molarity of the CH_3COOH is $\frac{mol}{L} = \frac{(0.10v)}{2v}$ = 0.050 M.

Balanced	CH_3COOH	+	H_2O	⇌	H_3O^+	+	CH_3COO^-
Initial	0.050				0		0
Change	- x				+ x		+ x
Equilibrium	0.050 - x				x		x

Insert the values into the K_a expression.

$$K_a = \frac{[H_3O^+][CH_3COO^-]}{[CH_3COOH]} = \frac{(x)(x)}{(0.050 - x)} = 1.8 \times 10^{-5}$$

Solve algebraically. Because K_a is very small, assume 0.050 - x ≈ 0.050.

$\frac{x^2}{0.050} = 1.8 \times 10^{-5}$; therefore, $x^2 = 0.050(1.8 \times 10^{-5}) = 9.0 \times 10^{-7}$; $x = \sqrt{9.0 \times 10^{-7}}$

x = [H_3O^+] = 9.5×10^{-4} M; pH = -log [H_3O^+] = -log (9.5×10^{-4}) = -(-3.02) = **3.02**

Check: $\frac{9.5 \times 10^{-4}\ M}{0.050\ M}$ = 0.019, so the approximation that 0.050 -x ≈ 0.050 M is valid.

d. When the strong acid is added to the salt which is the conjugate base of the weak acid CH_3COOH, reaction stoichiometry must be considered. The number of moles of acid = n_a = M•V = (0.10)v = 0.10v. The number of moles of base = n_b = M•V = (0.20 M)v = 0.20v. The reaction is:

Balanced	HCl	+	CH_3COO^-	→	CH_3COOH	+	Cl^-
Initial	0.10v		0.20v		0		0
Change	-0.10v		-0.10v		+ 0.10v		+0.10v
After	0		0.10v		0.10v		0.10v

The Cl^- has no acid/base ability. Both the acid and its conjugate base are present after the reaction of the strong acid, therefore a buffer solution is present and the relationship

$pH = pK_a - \log\left(\frac{[acid]}{[base]}\right)$ applies. $[CH_3COO^-] = [CH_3COOH] = \frac{0.10v}{2v} = 0.050$ M.

$$? \; pH = pK_a - \log\left(\frac{[acid]}{[base]}\right) = 4.74 - \log\left(\frac{[0.050 \text{ M}]}{[0.050 \text{ M}]}\right) = 4.74 - 0 = \mathbf{4.74}$$

e. When the strong acid is added to the weak base, reaction stoichiometry must be considered. The number of moles of acid = n_a = M•V = (0.10)v = 0.10v. The number of moles of base = n_b = M•V = (0.10 M)v = 0.10v. The reaction is:

Balanced	HCl	+	NH_3	→	NH_4^+	+	Cl^-
Initial	0.10v		0.10v		0		0
Change	-0.10v		-0.10v		+ 0.10v		+0.10v
After	0		0		0.10v		0.10v

Both the strong acid and weak base are totally consumed. Cl^- has no acid/base ability. The NH_4^+ produced can be treated by the structured approach to equilibrium. The molarity of the NH_4^+ is $\frac{mol}{L} = \frac{(0.10v)}{2v} = 0.050$ M. Because NH_4^+ is the conjugate acid of the weak base NH_3 ($K_b = 1.8 \times 10^{-5}$), its $K_a = \frac{1.0 \times 10^{-14}}{1.9 \times 10^{-5}} = 5.3 \times 10^{-10}$.

Balanced	NH_4^+	+	H_2O	⇌	H_3O^+	+	NH_3
Initial	0.050				0		0
Change	- x				+ x		+ x
Equilibrium	0.050 - x				x		x

Insert the values into the K_a expression.

$$K_a = \frac{[H_3O^+][NH_3]}{[NH_4^+]} = \frac{(x)(x)}{(0.050 - x)} = 5.3 \times 10^{-10}$$

Solve algebraically. Because K_a is very small, assume $0.050 - x \approx 0.050$.

$$\frac{x^2}{0.050} = 5.3 \times 10^{-10}; \text{ therefore, } x^2 = 0.050(5.3 \times 10^{-10}) = 2.6 \times 10^{-11}; \; x = \sqrt{2.6 \times 10^{-11}}$$

$$x = [H_3O^+] = 5.1 \times 10^{-6} \text{ M}; \; pH = -\log [H_3O^+] = -\log (5.1 \times 10^{-6}) = -(-5.29) = \mathbf{5.29}$$

Check: $\frac{5.1 \times 10^{-6} \text{ M}}{0.050 \text{ M}} = 1.0 \times 10^{-4}$, so the approximation that $0.050 - x \approx 0.050$ M is valid.

f. When the strong acid is added to the weak base, reaction stoichiometry must be considered. The number of moles of acid = n_a = M•V = (0.10)v = 0.10v. The number of moles of base = n_b = M•V = (0.20 M)v = 0.20v. The reaction is:

Balanced	HCl	+	NH_3	→	NH_4^+	+	Cl^-
Initial	0.10v		0.20v		0		0
Change	-0.10v		-0.10v		+ 0.10v		+0.10v
After	0		0.10v		0.10v		0.10v

The Cl^- has no acid/base ability. Both the acid and its conjugate base are present after the reaction of the strong acid, therefore a buffer solution is present and the relationship

$$pOH = pK_b - \log\left(\frac{[base]}{[acid]}\right) \text{ applies. } [NH_3] = [NH_4^+] = \frac{0.10v}{2v} = 0.050 \text{ M.}$$

$$pOH = pK_b - \log\left(\frac{[base]}{[acid]}\right) = 4.72 - \log\left(\frac{[0.050 \text{ M}]}{[0.050 \text{ M}]}\right) = 4.72 - 0 = 4.72$$

$$? \text{ pH} = 14.00 - pOH = 14.00 - 4.72 = \mathbf{9.28}$$

16.109 0.292 g of an unknown acid is dissolved in water and titrated with 0.100 M base. It is noted that the pH is 4.20 after the addition of 11.95 mL base and that the equivalence point is reached after the addition of 23.90 mL base. What are the molar mass and pK_a of the acid? (Assume that the acid is monoprotic.)

Analysis:

Targets: The molar mass of a monoprotic acid, HX and its pK_a. (? g mol^{-1}) = and (? pK_a) =

Knowns: The volume of the base, the molarity of the base and the mass of the acid. Both the acid and the base are monoprotic. The pH after a certain volume of base has been added is also known. n = M•V.

Relationships: $n_a = n_b$•(stoichiometric mole ratio). n = M•V. The equivalence statement 10^{-3} L ↔ 1 mL is also needed. Molar mass has units of g mol^{-1}.

Plan: Determine the stoichiometric mole ratio from the balanced chemical equation, and use the relationships $n_a = n_b$•(mol ratio) and n = M•V to find the number of moles of acid, employing the factor-label method to convert mL to L. Divide the mass of the acid by the number of moles of acid to determine its molar mass. To determine the pK_a, consider the reaction stoichiometry of the titration process.

Work:

The balanced chemical equation is HX + NaOH → NaX + H_2O, so the stoichiometric mole ratio is 1 mol acid ↔ 1 mol base.

$$n_a = n_b\left(\frac{1 \text{ mol acid}}{1 \text{ mol base}}\right) = M_b V_b(1)$$

$$n_a = (0.100 \text{ mol L}^{-1})(23.90 \text{ mL})(1) \times \frac{10^{-3} \text{ L}}{1 \text{ mL}} = 0.00239 \text{ mol acid}$$

$$? \text{ g mol}^{-1} = \frac{0.292 \text{ g acid}}{0.00239 \text{mol acid}} = \mathbf{122 \text{ g mol}^{-1}}$$

To determine the pK_a, consider the reaction of the acid with the base. At pH 4.20, there have been $n_b = M \cdot V = (0.100 \text{ mol L}^{-1}) \times (11.95 \text{ mL}) \times \frac{1 \text{ mL}}{10^{-3} \text{ L}} = 0.00120$ mol base added. The appropriate reaction can be used to determine the relative amounts of the acid and its salt at the volume at which the pH is given.

	HA	+	NaOH	→	NaA	+	H_2O
Initial	0.00239		0.00120		0		0
Change	-0.00120		-0.00120		+ 0.00120		
After	0.00119		0		0.00120		

At this point, a buffer solution exists and the relationship $pH = pK_a - \log\left(\frac{[acid]}{[base]}\right)$ can be used to determine the pK_a of the acid.

Let the solution volume at that point be V, then:

$$[HA] = \frac{0.00119 \text{ mol}}{V} \text{ and } [A^-] = \frac{0.00120 \text{ mol}}{V}.$$

Inserting these into the relationship gives,

$$4.20 = pK_a - \log\left(\frac{\frac{0.00119 \text{ mol}}{V}}{\frac{0.00120 \text{ mol}}{V}}\right) = pK_a - \log(0.992) = pK_a - 0.004$$

Therefore, the $pK_a = 4.20 + 0.004 =$ **4.20**

16.111 Two indicators are available for a titration: phenolphthalein, with a range of 8.0-9.8, and methyl red, with a range of 4.2-6.2. Which would be the better choice for each of the following titrations? Explain your choice.

a. titration of acetic acid with KOH
b. titration of hydrochloric acid with KOH
c. titration of ammonia with HCl
d. titration of sodium cyanide with HCl

Analysis/Plan: Study the system to determine the nature of the species present at the equivalence point. If an acid is present, methyl red should be used since it changes color in the acidic region which coincides with the steeply rising portion of the pH curve at the equivalence point. If a base is present, phenolphthalein should be used because it changes color in the basic region which coincides with the steeply rising portion of the pH curve at the equivalence point. If a neutral species is present, either indicator will be appropriate because the steeply rising portion of the pH curve includes the region on both sides of pH = 7.00.

Work:

a. This is the titration of a weak acid with a strong base. At the equivalence point, the solution will be basic because of the hydrolysis of the acetate ion. Phenolphthalein is an appropriate indicator for this titration.

b. This is the titration of a strong acid with a strong base. Neither ion of the salt formed is capable of hydrolysis. Either indicator is appropriate because the solution is neutral at the equivalence point.

c. This is the titration of a weak base with a strong acid. The ammonium ion hydrolyzes at the equivalence point creating an acidic solution. Methyl red is the appropriate indicator for this titration.

d. This is the titration of a weak base with a strong acid. The acid HCN is formed and causes the solution to be acidic at the equivalence point. Methyl red is the appropriate indicator for this titration.

17 Solubility and Complex Ion Equilibria

17.1 Distinguish between the terms "ion product" and the "solubility product expression".

The ion product, Q, can be evaluated for any concentrations of ions. The solubility product expression, K_{sp}, is specifically for those systems in which ions are in equilibrium with solid.

17.3 Define the terms "solubility" and "molar solubility".

Solubility is defined as the mass of a solute that will dissolve in a specified volume (1 liter or 100 mL) of a solvent. The molar solubility of a solute is the number of moles of the solute which will dissolve in one liter.

17.5 Define the terms, "undersaturated", "saturated", and "supersaturated".

An undersaturated solution contains less than the equilibrium amount of solute. A saturated solution contains the equilibrium amount of solute. A supersaturated solution contains more than the equilibrium amount of the solute. If a supersaturated solution is disturbed, precipitation occurs until the solution is saturated.

17.7 How can the solubility of sparingly soluble hydroxides be manipulated?

The solubility of sparingly soluble hydroxides can be manipulated by adjusting the pH. The equilibrium $H_3O^+ + OH^- \rightleftharpoons 2\,H_2O$ is present in water. By adding acid, the concentration of hydroxide ion can be lowered in accordance with Le Châtelier's Principle, causing the solute to become more soluble. Conversely, adding base increases the hydroxide concentration, decreasing the solubility of the solute.

17.9 What is a complex ion? Give examples.

A complex ion is a metal ion bonded by coordinate covalent bonds to one or more anions or molecules. The anion or molecule provides the electron pair to be shared. Examples are $Fe(CN)_6^{3-}$, $Al(C_2O_4)_2^-$ and $AgCl_2^-$. See Table 17.5 in the text.

17.11 In the term "coordination complex", what does the word "coordination" refer to?

The term coordination refers to the fact that the metal ion in the complex is joined by coordinate covalent bonds to the anions or molecules surrounding it. The anions or molecules serve as the Lewis bases, providing the electron pair. The metal ion serves as the Lewis acid, accepting the electron pair.

17.13 What is meant by the term "qualitative analysis"?

In qualitative analysis, the components in a mixture are identified, but the relative amount of each component is not determined.

17.15 Write the solubility product expression for the following salts.
a. BaF_2 b. $Bi_2(SO_4)_3$ c. CuBr d. $BaCO_3$

Analysis/Plan: The solubility expression is the equilibrium constant expression relating ions in solution to the solid from which they were formed. It is written from the balanced dissociation reaction for the substance. The solvent and the solute each have an activity of 1 and are omitted from the K_{sp} expression. Write the balanced equation for the dissociation reaction and then write the equilibrium expression in the normal manner.

Work:

a. $BaF_2(s) \rightleftharpoons Ba^{2+}(aq) + 2\ F^-(aq)$ $K_{sp} = [Ba^{2+}][F^-]^2$

b. $Bi_2(SO_4)_3(s) \rightleftharpoons 2\ Bi^{3+}(aq) + 3\ SO_4^{2-}(aq)$ $K_{sp} = [Bi^{3+}]^2[SO_4^{2-}]^3$

c. $CuBr(s) \rightleftharpoons Cu^+(aq) + Br^-(aq)$ $K_{sp} = [Cu^+][Br^-]$

d. $BaCO_3(s) \rightleftharpoons Ba^{2+}(aq) + CO_3^{2-}(aq)$ $K_{sp} = [Ba^{2+}][CO_3^{2-}]$

17.17 Calculate the molar solubility of CuCl from its K_{sp}.

Analysis/Plan: The initial molarity of each ion is 0. The solubility is related to the extent of reaction, x. The standard equilibrium procedure can be followed. The K_{sp} for CuCl can be looked up in Appendix H and is 4.2×10^{-8}.

Work:

Balanced	CuCl(s)	$\rightleftharpoons$ $Cu^+(aq)$	+	$Cl^-(aq)$
Initial		0		0
Change		+ x		+ x
Equilibrium		x		x

Insert the values into the K_{sp} expression and solve algebraically.

$$K_{sp} = [Cu^+][Cl^-] = (x)(x) = x^2 = 4.2 \times 10^{-8}$$

$$?\ mol\ L^{-1} = x = \sqrt{4.2 \times 10^{-8}} = \mathbf{2.0 \times 10^{-4}\ mol\ L^{-1}}$$

17.19 Calculate the solubility, in mol L^{-1} and g L^{-1}, of $AuCl_3$.

Analysis/Plan: The solubility is related to the extent of reaction, x. The standard equilibrium procedure can be followed to find the molar solubility. The K_{sp} for $AuCl_3$ is 3.2×10^{-25} (Appendix H). The molar mass of $AuCl_3$ is an implicit known and provides the equivalence statement g ↔ mol. The factor-label method can be used to find the solubility in g L^{-1}.

Work:

Balanced	$AuCl_3(s)$	$\rightleftharpoons$ $Au^{3+}(aq)$	+	$3\ Cl^-(aq)$
Initial		0		0
Change		+ x		+ 3x
Equilibrium		x		3x

Insert the values into the K_{sp} expression.

$$K_{sp} = [Au^{3+}][Cl^-]^3 = (x)(3x)^3 = 27x^4 = 3.2 \times 10^{-25}$$

Solve algebraically.

$$(x)^4 = = \frac{3.2 \times 10^{-25}}{27} = 1.2 \times 10^{-26}$$

$$x = \sqrt[4]{1.2 \times 10^{-26}} = \mathbf{3.3 \times 10^{-7}\ M\ AuCl_3}$$

Molar Mass $AuCl_3$ = 1(196.9665.Au) + 3(35.453 Cl) = 303.326 g mol^{-1}

$$?\ g\ L^{-1} = \frac{3.3 \times 10^{-7}\ mol\ AuCl_3}{L} \times \frac{303.326\ g\ AuCl_3}{1\ mol\ AuCl_3} = \mathbf{1.0 \times 10^{-4}\ g\ L^{-1}\ AuCl_3}$$

17.21 How many grams of PbC_2O_4 will dissolve in 1.00 L of water?

Analysis/Plan: The initial molarity of each ion is 0. The solubility is related to the extent of reaction, x. The standard equilibrium procedure can be followed to find the molar solubility. The K_{sp} for PbC_2O_4 can be looked up in Appendix H and is 3×10^{-11}. The molar mass of PbC_2O_4 is an implicit known and provides the equivalence statement g ↔ mol. The factor-label method can be used to find the mass in 1.00 L of water.

Work:

Balanced	$PbC_2O_4(s)$ ⇌	$Pb^{2+}(aq)$	+	$C_2O_4^{2-}(aq)$
Initial		0		0
Change		+ x		+ x
Equilibrium		x		x

Insert the values into the K_{sp} expression and solve algebraically.

$$K_{sp} = [Pb^{2+}][C_2O_4^{2-}] = (x)(x) = x^2 = 3 \times 10^{-11}$$

$$? \text{ mol L}^{-1} = x = \sqrt{3 \times 10^{-11}} = 5 \times 10^{-6} \text{ mol L}^{-1}\ PbC_2O_4$$

Molar Mass PbC_2O_4 = 1(207.2 Pb) + 2(12.011 C) + 4(15.9994 O) = 295.2 g mol^{-1}

$$? \text{ g} = \frac{5 \times 10^{-6} \text{ mol } PbC_2O_4}{L} \times \frac{295.2 \text{ g } PbC_2O_4}{1 \text{ mol } PbC_2O_4} \times 1.00 \text{ L} = \mathbf{2 \times 10^{-3} g\ PbC_2O_4}$$

17.23 Calculate the solubility, in g L^{-1}, of $AuBr_3$.

Analysis/Plan: The solubility is related to the extent of reaction, x. The standard equilibrium procedure can be followed to find the molar solubility. The K_{sp} for $AuBr_3$ is 4.0×10^{-36} (Appendix H). The molar mass of $AuBr_3$ is an implicit known and provides the equivalence statement g ↔ mol. The factor-label method can be used to find the solubility in g L^{-1}.

Work:

Balanced	$AuBr_3(s)$ ⇌	$Au^{3+}(aq)$	+	$3\ Br^-(aq)$
Initial		0		0
Change		+ x		+ 3x
Equilibrium		x		3x

Insert the values into the K_{sp} expression.

$$K_{sp} = [Au^{3+}][Br^-]^3 = (x)(3x)^3 = 27x^4 = 4.0 \times 10^{-36}$$

Solve algebraically.

$$(x)^4 = \frac{4.0 \times 10^{-36}}{27} = 1.5 \times 10^{-37}$$

$$x = \sqrt[4]{1.5 \times 10^{-37}} = 6.2 \times 10^{-10} \text{ M } AuBr_3$$

Molar Mass $AuBr_3$ = 1(196.9665.Au) + 3(79.904 Br) = 436.679 g mol^{-1}

$$? \text{ g L}^{-1} = \frac{6.2 \times 10^{-10} \text{ mol } AuBr_3}{L} \times \frac{436.679 \text{g } AuBr_3}{1 \text{ mol } AuBr_3} = \mathbf{2.7 \times 10^{-7}\ g\ L^{-1}\ AuBr_3}$$

17.25 It is found that 1.0×10^{-7} moles of AgBr will dissolve in 141 mL of water. What is the value of K_{sp}?

Analysis/Plan: The initial molarity of each ion is 0. The molar solubility is related to the extent of reaction, x, and can be determined, using the factor-label method, from the relationships M = mol L^{-1} and 10^{-3} L ↔ 1 mL. The standard equilibrium procedure can then be followed to find the K_{sp}.

Work:

$$? \text{ mol L}^{-1} = \frac{1.0 \times 10^{-7} \text{ mol}}{141 \text{ mL}} \times \frac{1 \text{ mL}}{10^{-3} \text{ L}} = 7.1 \times 10^{-7} \text{ M} = x, \text{ the molar solubility.}$$

Balanced	$AgBr(s)$	⇌	$Ag^{+}(aq)$	+	$Br^{-}(aq)$
Initial			0		0
Change			+ x		+ x
Equilibrium			x		x

Insert the values into the K_{sp} expression and solve.

$$K_{sp} = [Ag^{+}][Br^{-}] = (x)(x) = x^2 = (7.1 \times 10^{-7})^2 = \mathbf{5.0 \times 10^{-13}}$$

17.27 If 1.55×10^{-4} mol Ag_2SO_4 will dissolve in 10 mL H_2O, what is the value of K_{sp}?

Analysis/Plan: The initial molarity of each ion is 0. The molar solubility is related to the extent of reaction, x, and can be determined, using the factor-label method, from the relationships M = mol L^{-1} and 10^{-3} L ↔ 1 mL. The standard equilibrium procedure can then be followed to find the K_{sp}.

Work:

$$? \text{ mol L}^{-1} = \frac{1.55 \times 10^{-4} \text{ mol}}{10 \text{ mL}} \times \frac{1 \text{ mL}}{10^{-3} \text{ L}} = 1.6 \times 10^{-2} \text{ M} = x, \text{ the molar solubility.}$$

Balanced	$Ag_2SO_4(s)$	⇌	$2\ Ag^{+}(aq)$	+	$SO_4^{2-}(aq)$
Initial			0		0
Change			+ 2x		+ x
Equilibrium			2 x		x

Insert the values into the K_{sp} expression and solve.

$$K_{sp} = [Ag^{+}]^2[SO_4^{2-}] = (2x)^2(x) = 4x^3 = 4(1.6 \times 10^{-2})^3 = \mathbf{1.5 \times 10^{-5}}$$

17.29 What compound (if any) precipitates when equal volumes of 0.0020 M $CaCl_2$ and 0.0040 M Na_2SO_4 are mixed?

Analysis: The possible product salts are $CaSO_4$ and NaCl. NaCl is soluble; so if a precipitate occurs, it will be $CaSO_4$. $CaSO_4$ will be formed if the ion product exceeds the K_{sp}, which is 2.4×10^{-5} (Appendix H).

Plan: Find the concentration of the ions in the solution from the relationships n = M•V and M = mol L^{-1}. Write the dissociation reaction to determine the ion product expression. Insert the concentrations into the ion product expression, calculate and compare to the K_{sp}.

Work:

Let v be the volume of each solution used. The total volume is v + v = 2v.

$$n\ Ca^{2+} = 0.0020 \text{ M} \times v = 0.0020v;\ ? \text{ mol L}^{-1}\ Ca^{2+} = \frac{n}{2v} = \frac{0.0020v}{2v} = 0.0010 \text{ M}$$

$$n\ SO_4^{2-} = 0.0040 \text{ M} \times v = 0.0040v;\ ? \text{ mol L}^{-1}\ SO_4^{2-} = \frac{n}{2v} = \frac{0.0040v}{2v} = 0.0020 \text{ M}$$

The dissociation reaction is: $CaSO_4(s) \rightleftharpoons Ca^{2+}(aq) + SO_4^{2-}(aq)$.

The ion product expression is $Q_{sp} = [Ca^{2+}][SO_4^{2-}] = (0.0010)(0.0020) = 2.0 \times 10^{-6}$.

Because Q_{sp} is less than the K_{sp}, **no precipitation occurs.**

17.31 Will a precipitate form when equal volumes of 0.010 M $BaCl_2$ and 1.5×10^{-4} M NaF are mixed? If so, what compound is it?

Analysis: The possible product salts are BaF_2 and NaCl. NaCl is soluble; so if a precipitate occurs, it will be BaF_2. BaF_2 will be formed if the ion product, Q_{sp}, exceeds the K_{sp}, which is 1.7×10^{-6} (Appendix H).
Plan: Find the concentration of the ions in the solution from the relationships n = M•V and M = mol L^{-1}. Write the dissociation constant to determine the ion product expression. Insert the concentrations into the ion product expression, calculate and compare to the K_{sp}.
Work:
Let v be the volume of each solution used. The total volume is v + v = 2v.

$$n\ Ba^{2+} = 0.010\ M \times v = 0.010v;\ ?\ mol\ L^{-1}\ Ca^{2+} = \frac{n}{2v} = \frac{0.010v}{2v} = 0.010\ M$$

$$n\ F^- = 1.5 \times 10^{-4}\ M \times v = 1.5 \times 10^{4}v;\ ?\ mol\ L^{-1}\ F^- = \frac{n}{2v} = \frac{1.5 \times 10^{-4}v}{2v} = 7.5 \times 10^{-5}\ M$$

The dissociation reaction is: $BaF_2(s) \rightleftharpoons Ba^{2+}(aq) + 2\ F^-(aq)$.
The ion product expression is $Q_{sp} = [Ba^{2+}][F^-]^2 = (0.010)(7.5 \times 10^{-5})^2 = 5.6 \times 10^{-11}$.

Because Q_{sp} is less than the K_{sp}, **no precipitation occurs.**

17.33 How many moles of $Na_2CrO_4(s)$ must be dissolved in 500 mL of 0.010 M $CaCl_2$ in order to cause the appearance of a precipitate? (Assume no change in solution volume on the addition of the solid.)

Analysis: The possible product salts are $CaCrO_4$ and NaCl. NaCl is soluble; so when a precipitate occurs, it will be $CaCrO_4$. $CaCrO_4$ will start to form when the ion product, Q_{sp}, exceeds the K_{sp}, which is 7.1×10^{-4} (Appendix H).
Plan: Write the dissociation reaction to determine the ion product expression. Insert the concentration of Ca^{2+} into the ion product expression and solve for the CrO_4^{2-} concentration needed. Use the relationships n = M•V and 1 mL ↔ 10^{-3} L to find the number of moles of Na_2CrO_4 needed.
Work:
The dissociation reaction is: $CaCrO_4(s) \rightleftharpoons Ca^{2+}(aq) + CrO_4^{2-}(aq)$.

The ion product expression is $Q_{sp} = [Ca^{2+}][CrO_4^{2-}]$.

Precipitation occurs when $Q_{sp} > K_{sp}$. Substituting in values yields:

$$(0.010)[CrO_4^{2-}] > 7.1 \times 10^{-4};\ \text{therefore, } [CrO_4^{2-}] > \frac{7.1 \times 10^{-4}}{0.010} = 7.1 \times 10^{-2}\ M\ CrO_4^{2-}$$

$$?\ mol\ Na_2CrO_4 = M \cdot V = \frac{7.1 \times 10^{-2}\ mol\ CrO_4^{2-}}{L} \times \frac{10^{-3}\ L}{1\ mL} \times 500\ mL =$$

0.036 mol Na_2CrO_4

17.35 If dilute aqueous $AgNO_3$ is added dropwise to a dilute equimolar solution of NaCl and NaBr, what is the composition of the first precipitate to appear?

Analysis/Plan: Possible salts formed are $NaNO_3$, AgBr and AgCl. Because $NaNO_3$ is soluble, either AgBr or AgCl will precipitate first and will begin to precipitate when its ion product exceeds its K_{sp}. Because the Cl^- and the Br^- are equimolar, the salt with the smaller K_{sp} will precipitate first. From Appendix H, the K_{sp} of AgBr is 5.0×10^{-13} and that of AgCl is 1.8×10^{-10}.

Work:

Because AgBr has the smaller K_{sp}, **AgBr will precipitate first.**

17.37 If solid Na_2SO_4 is gradually added to and dissolved in a solution that is 0.010 M in $Ca(NO_3)_2$ and 0.010 M in $Pb(NO_3)_2$, in what order will solid $CaSO_4$ and $PbSO_4$ appear?

Analysis/Plan: Possible salts formed are $NaNO_3$, $CaSO_4$ and $PbSO_4$. Because $NaNO_3$ is soluble, either $CaSO_4$ or $PbSO_4$ will precipitate first and will begin to precipitate when its ion product exceeds its K_{sp}. Because the Ca^{2+} and the Pb^{2+} are equimolar, the salt with the smaller K_{sp} will precipitate first. From Appendix H, the K_{sp} of $CaSO_4$ is 2.4×10^{-5} and that of $PbSO_4$ is 1.6×10^{-8}.

Work:

Because $PbSO_4$ has the smaller K_{sp}, **$PbSO_4$ will precipitate first**, then the **$CaSO_4$ will begin to precipitate**.

17.39 In Problem 17.37, what is the remaining concentration of the ion that formed the first precipitate when the second precipitate just begins to appear?

Analysis/Plan: Write the dissociation reactions to determine the ion product expressions. Use the relationship $Q_{sp} > K_{sp}$ for $CaSO_4$ to determine the molarity of SO_4^{2-} needed to initiate the precipitation of $CaSO_4$. Then, use that value of SO_4^{2-} with the K_{sp} for $PbSO_4$ to determine the molarity of Pb^{2+} remaining in the solution when the $CaSO_4$ begins to precipitate. See Problem 17.37 for the necessary K_{sp} values.

Work:

The dissociation reaction for $CaSO_4$ is: $CaSO_4(s) \rightleftharpoons Ca^{2+}(aq) + SO_4^{2-}(aq)$.

The ion product expression is $Q_{sp} = [Ca^{2+}][SO_4^{2-}]$.

Precipitation occurs when $Q_{sp} > K_{sp}$. Substituting in values yields:

$$(0.010)[SO_4^{2-}] > 2.4 \times 10^{-5}; \text{ therefore, } [SO_4^{2-}] > \frac{2.4 \times 10^{-5}}{0.010} = 2.4 \times 10^{-3} \text{ M } SO_4^{2-}$$

The dissociation reaction for $PbSO_4$ is: $PbSO_4(s) \rightleftharpoons Pb^{2+}(aq) + SO_4^{2-}(aq)$.

The solubility product expression is $K_{sp} = [Pb^{2+}][SO_4^{2-}]$.

Substituting in the value of $[SO_4^{2-}]$ when $CaSO_4$ begins to precipitate gives:

$$[Pb^{2+}](2.4 \times 10^{-3}) = 1.6 \times 10^{-8}; \text{ therefore, } [Pb^{2+}] = \frac{1.6 \times 10^{-8}}{2.4 \times 10^{-3}} = \mathbf{6.7 \times 10^{-6} \text{ M } Pb^{2+}}$$

17.41 What is the maximum concentration of $Ag^+(aq)$ that can exist in 0.0050 M NaBr?

Analysis/Plan: The maximum concentration of Ag^+ which can exist is that in which the ion product is equal to the K_{sp} of AgBr, which is 5.0×10^{-13} (Appendix H). Write the dissociation reaction to determine the ion product expression and then use the structured approach to equilibrium problems. In this case, the initial $[Br^-]$ = 0.0050 M.

Work:

Balanced	$AgBr(s)$	$\rightleftharpoons$	$Ag^+(aq)$	+	$Br^-(aq)$
Initial			0		0.0050
Change			+ x		+ x
Equilibrium			x		0.0050 + x

Insert the values into the K_{sp} expression and solve algebraically.

$$K_{sp} = [Ag^+][Br^-] = (x)(0.0050 + x) = 5.0 \times 10^{-13}$$

Because the K_{sp} is very small, assume that $0.0050 + x \approx 0.0050$

$$(x)(0.0050) = 5.0 \times 10^{-13}; \text{ therefore, } x = \frac{5.0 \times 10^{-13}}{0.0050} = \mathbf{1.0 \times 10^{-10}\ M\ Ag^+}$$

Check: $\dfrac{1.0 \times 10^{-10}\ M}{0.0050\ M} = 2.0 \times 10^{-8}$; the approximation that 0.0050 + x = 0.0050 is valid.

17.43 What is the maximum concentration of $Hg^{2+}(aq)$ that can exist in 0.75 M KBr?

Analysis/Plan: The maximum concentration of Hg^{2+} which can exist is that in which the ion product is equal to the K_{sp} of $HgBr_2$, which is 1×10^{-10} (Appendix H). Write the dissociation reaction to determine the ion product expression and then use the structured approach to equilibrium problems. In this case, the initial $[Br^-]$ = 0.75 M.

Work:

Balanced	$HgBr_2(s)$	$\rightleftharpoons$	$Hg^{2+}(aq)$	+	$2\ Br^-(aq)$
Initial			0		0.75
Change			+ x		+ 2x
Equilibrium			x		0.75+ 2x

Insert the values into the K_{sp} expression and solve algebraically.

$$K_{sp} = [Hg^{2+}][Br^-]^2 = (x)(0.75 + 2x)^2 = 1 \times 10^{-10}$$

Because the K_{sp} is very small, assume that $0.75 + 2x \approx 0.75$

$$(x)(0.75)^2 = 1 \times 10^{-10}; \text{ therefore, } x = \frac{1 \times 10^{-10}}{(0.75)^2} = \mathbf{2 \times 10^{-10}\ M\ Hg^{2+}}$$

Check: $\dfrac{2 \times 10^{-10}\ M}{0.75\ M} = 3 \times 10^{-10}$; the approximation that 0.75 + 2x = 0.75 is valid.

17.45 How many moles of $Mn(NO_3)_2$ will dissolve in 1.00 L of 0.10 M KOH?

Analysis/Plan: KNO_3 is a soluble salt. The maximum concentration of Mn^{2+} which can exist is that in which the ion product is equal to the K_{sp} of $Mn(OH)_2$, which is 2×10^{-13} (Appendix H). Write the dissociation reaction to determine the ion product expression and then use the structured approach to equilibrium problems to determine the molar solubility. In this case, the initial $[OH^-]$ = 0.10 M. To determine the number of moles, use the relationship n = M•V.

Work:

	$Mn(OH)_2(s) \rightleftharpoons$	$Mn^{2+}(aq)$	+	$2\ OH^-(aq)$
Balanced				
Initial		0		0.10
Change		+ x		+ 2x
Equilibrium		x		0.10+ 2x

Insert the values into the K_{sp} expression and solve algebraically.

$$K_{sp} = [Mn^{2+}][OH^-]^2 = (x)(0.10 + 2x)^2 = 2 \times 10^{-13}$$

Because the K_{sp} is small, assume $0.10 + 2x \approx 0.10$.

$$(x)(0.10)^2 = 2 \times 10^{-13}; \text{ therefore, } x = \frac{2 \times 10^{-13}}{(0.10)^2} = 2 \times 10^{-11}\ M$$

$$? \text{ mol} = 2 \times 10^{-11} \text{ mol L}^{-1}\ Mn^{2+} \times 1.00 \text{ L} = \mathbf{2 \times 10^{-11} \text{ mol } Mn(NO_3)_2}$$

Check: $\frac{2 \times 10^{-11}\ M}{0.10\ M} = 2 \times 10^{-10}$; the approximation that $0.10 + 2x \approx 0.10$ is valid.

17.47 What is the solubility (in $g\ L^{-1}$) of Ag_2CrO_4 in 0.0050 M Na_2CrO_4?

Analysis/Plan: The maximum concentration of Ag_2CrO_4 is that in which the ion product is equal to the K_{sp} of Ag_2CrO_4 which is 1.2×10^{-12} (Appendix H). Write the dissociation reaction to determine the ion product expression and then use the structured approach to equilibrium problems to determine the molar solubility. In this case, the initial $[CrO_4^{2-}] = 0.0050$ M. The molar mass of Ag_2CrO_4 gives the equivalence statement g ↔ mol and the factor-label method can be used to determine the solubility in $g\ L^{-1}$.

Work:

	$Ag_2CrO_4(s) \rightleftharpoons$	$2\ Ag^+(aq)$	+	$CrO_4^{2-}(aq)$
Balanced				
Initial		0		0.0050
Change		+ 2x		+ x
Equilibrium		2x		0.0050 + x

Insert the values into the K_{sp} expression and solve algebraically.

$$K_{sp} = [Ag^+]^2[CrO_4^{2-}] = (2x)^2(0.0050 + x) = 1.2 \times 10^{-12}$$

Because the K_{sp} is very small, assume $0.0050 + x \approx 0.0050$

$$(2x)^2(0.0050) = 0.020x^2 = 1.2 \times 10^{-12}; \text{ therefore, } x^2 = \frac{1.2 \times 10^{-12}}{(0.020)} = 6.0 \times 10^{-11}$$

$$x = \sqrt{6.0 \times 10^{-11}} = 7.7 \times 10^{-6} \text{ mol L}^{-1}\ Ag_2CrO_4$$

Molar Mass Ag_2CrO_4 = 2(107.8682 Ag) +1(51.996 Cr) + 4(15.9994 O) = 331.730 g mol^{-1}

$$? \text{ g L}^{-1} = \frac{7.7 \times 10^{-6} \text{ mol } Ag_2CrO_4}{L} \times \frac{331.730 \text{ g } Ag_2CrO_4}{1 \text{ mol } Ag_2CrO_4} = \mathbf{2.6 \times 10^{-3} \text{ g L}^{-1}\ Ag_2CrO_4}$$

17.49 What is the solubility of $SrCO_3$ (in $g\ L^{-1}$) in (a) pure water, and (b) 0.033 M $Sr(NO_3)_2$?

Analysis/Plan: The maximum concentration of $SrCO_3$ is that in which the ion product is equal to the K_{sp} of $SrCO_3$ which is 9.3×10^{-10} (Appendix H). Write the dissociation reaction to determine the ion product expression and then use the structured approach to equilibrium problems to determine the molar solubility. In part (a) the initial concentration of each ion is 0. In part (b), the initial $[Sr^{2+}] = 0.033$ M. The molar mass of $SrCO_3$ gives the equivalence statement g ↔ mole and the factor-label method can be used to determine the solubility in g L^{-1}.

Work:

Molar Mass $SrCO_3$ = 1(87.62 Sr) +1(12.011 C) + 3(15.9994 O) = 147.63 g mol^{-1}

a.

Balanced	$SrCO_3(s)$	⇌	$Sr^{2+}aq)$	+	$CO_3^{2-}(aq)$
Initial			0		0
Change			+ x		+ x
Equilibrium			x		x

Insert the values into the K_{sp} expression and solve algebraically.

$$K_{sp} = [Sr^{2+}][CO_3^{2-}] = (x)(x) = x^2 = 9.3 \times 10^{-10}$$

$$x = \sqrt{9.3 \times 10^{-10}} = 3.0 \times 10^{-5} \text{ mol L}^{-1}\ SrCO_3$$

$$? \text{ g L}^{-1} = \frac{3.0 \times 10^{-5} \text{ mol } SrCO_3}{L} \times \frac{147.63 \text{ g } SrCO_3}{1 \text{ mol } SrCO_3} = \mathbf{4.5 \times 10^{-3} \text{ g L}^{-1}\ SrCO_3}$$

b.

Balanced	$SrCO_3(s)$	⇌	$Sr^{2+}(aq)$	+	$CO_3^{2-}(aq)$
Initial			0.033		0
Change			+ x		+ x
Equilibrium			0.033 + x		x

Insert the values into the K_{sp} expression and solve algebraically.

$$K_{sp} = [Sr^{2+}][CO_3^{2-}] = (0.033 + x)(x) = 9.3 \times 10^{-10}$$

Because the K_{sp} is very small, assume $0.033 + x \approx 0.033$

$$(0.033)(x) = 9.3 \times 10^{-10};\text{ therefore, } x = \frac{9.3 \times 10^{-10}}{(0.033)} = 2.8 \times 10^{-8} \text{ M } SrCO_3$$

$$? \text{ g L}^{-1} = \frac{2.8 \times 10^{-8} \text{ mol } SrCO_3}{L} \times \frac{147.63 \text{ g } SrCO_3}{1 \text{ mol } SrCO_3} = \mathbf{4.2 \times 10^{-6} \text{ g L}^{-1}\ SrCO_3}$$

Check: $\frac{2.8 \times 10^{-8} \text{ M}}{0.033 \text{ M}} = 8.5 \times 10^{-7}$; the approximation that 0.033 + x = 0.033 is valid.

17.51 What is the maximum concentration of Mg^{2+} that can exist in a solution whose pOH is (a) 4.53; (b) 1.70?

Analysis/Plan: The maximum concentration of Mg^{2+} which can exist is that in which the ion product is equal to the K_{sp} of $Mg(OH)_2$ {7.3×10^{-12} (Appendix H)}. Write the dissociation reaction to determine the ion product expression. In each case, the equilibrium $[OH^-]$ can be found from the relationship $[OH^-] = 10^{-pOH}$. Substitute the hydroxide molarity into the ion product expression and solve for the Mg^{2+} concentration.

Work:

For both parts, the balanced dissociation reaction is:

$$Mg(OH)_2(s) \rightleftharpoons Mg^{2+}(aq) + 2\ OH^-(aq).$$

$K_{sp} = [Mg^{2+}][OH^-]^2$; therefore, $[Mg^{2+}] = \frac{K_{sp}}{[OH]^2} = \frac{7.3 \times 10^{-12}}{[OH]^2}$,

a. $[OH^-] = 10^{-4.53} = 3.0 \times 10^{-5}$

$[Mg^{2+}] = \frac{7.3 \times 10^{-12}}{(3.0 \times 10^{-5})^2} = \mathbf{8.4 \times 10^{-3}\ M\ Mg^{2+}}$

b. $[OH^-] = 10^{-1.70} = 0.020$

$[Mg^{2+}] = \frac{7.3 \times 10^{-12}}{(0.020)^2} = \mathbf{1.8 \times 10^{-8} M\ Mg^{2+}}$

17.53 To what value must the pH be adjusted (by addition of HCl) in order to dissolve 5.00 g $Mg(OH)_2$ in a liter of water?

Analysis/Plan: In order to dissolve, the ion product must be no greater than the K_{sp}. Use the molar mass and the factor-label method to determine the potential molarity of the $Mg(OH)_2$ and, therefore, the molarity of Mg^{2+} which will be present at equilibrium. Insert this value into the K_{sp} expression and solve for the maximum value of $[OH^-]$ which can be present. Then use the relationships pOH = $-\log[OH^-]$ and pH = 14.00 - pOH to determine the pH to which the solution must be adjusted. The appropriate K_{sp} and ion product expression were given in Problem 17.51.

Work:

Molar Mass $Mg(OH)_2$ = 1(24.305 Mg) + 2(15.9994 O) + 2(1.0079 H) = 58.320 g mol^{-1}

$? \text{ mol } L^{-1}\ Mg^{2+} = \frac{5.00 \text{ g } Mg(OH)_2}{L} \times \frac{1 \text{ mol } Mg(OH)_2}{58.320 \text{ g } Mg(OH)_2} = 0.0857 \text{ M } Mg^{2+}$

$K_{sp} = [Mg^{2+}][OH^-]^2 = (0.0857)[OH^-]^2 = 7.3 \times 10^{-12}$

$[OH^-]^2 = \frac{7.3 \times 10^{-12}}{0.0857} = 8.5 \times 10^{-11}$; and, $[OH^-] = \sqrt{8.5 \times 10^{-11}} = 9.2 \times 10^{-6}$

$pOH = -\log[OH^-] = -\log(9.2 \times 10^{-6}) = -(-5.03) = 5.03$

$? \text{ pH} = 14.00 - pOH = 14.00 - 5.03 = \mathbf{8.97}$

17.55 Will a precipitate of $Mg(OH)_2$ appear if equal volumes of 1.0×10^{-4} M $MgCl_2$ and 0.010 M NH_3 are mixed? K_b for NH_3 is 1.9×10^{-5}.

Analysis/Plan: A precipitate will form if the ion product exceeds the K_{sp}. We need to find the diluted concentrations of the Mg^{2+} and NH_3 first, from n = M•V and M = mol L^{-1}. Then we must consider the equilibrium of the weak base to determine the concentration of hydroxide that its ionization produces. Those ion concentrations are inserted into the ion product expression, and the value is computed and compared to K_{sp} (Problem 17.51).

Work:

First, find the diluted molarities. Let each initial volume = v, then the total volume is 2v.

$n\ Mg^{2+} = (1.0 \times 10^{-4} \text{ M})v$ and $[Mg^{2+}]_{diluted} = \frac{(1.0 \times 10^{-4} \text{ M})v}{2v} = 5.0 \times 10^{-5} \text{ M}$

$n\ NH_3 = (0.010 \text{ M})v$ and $[NH_3]_{diluted} = \frac{(0.010 \text{ M})v}{2v} = 0.0050 \text{ M}$

Now use the structured approach to equilibrium to determine the $[OH^-]$.

Balanced	NH_3	+	H_2O	$\rightleftharpoons$	NH_4^+	+	OH^-
Initial	0.0050				0		0
Change	- x				+ x		+ x
Equilibrium	0.0050 - x				x		x

Insert the values into the K_b expression.

$$K_b = \frac{[NH_4^+][OH^-]}{[NH_3]} = \frac{(x)(x)}{(0.0050 - x)} = 1.9 \times 10^{-5}$$

Because K_b is not very small, use the quadratic equation. From Problem 16.7,

$$x = [OH^-] = \frac{-K_b + \sqrt{K_b^2 + 4(K_b)(C)}}{2}.$$

$$x = \frac{-1.9 \times 10^{-5} + \sqrt{(1.9 \times 10^{-5})^2 + 4(1.9 \times 10^{-5})(0.0050)}}{2}.$$

$$x = [OH^-] = 3.0 \times 10^{-4}\ M$$

Now compare the ion product to the K_{sp}.

$$Q_{sp} = [Mg^{2+}][OH^-]^2 = (5.0 \times 10^{-5})(3.0 \times 10^{-5})^2 = 4.5 \times 10^{-14} < 7.3 \times 10^{-12}$$

Because the ion product is less than the K_{sp}, **no precipitate forms.**

17.57 A solution contains 0.010 M each of $MgCl_2$ and $CaCl_2$. To what value must the fluoride concentration be adjusted (by addition of solid KF) in order to remove as much Ca^{2+} as possible [in the form of $CaF_2(s)$] while leaving the Mg^{2+} entirely in solution?

Analysis/Plan: The maximum amount of CaF_2 can be precipitated without precipitating MgF_2 if the F^- is increased just until the ion product for MgF_2 is equal to its K_{sp}, which is 6.6 x 10^{-9} (Appendix H). Write the dissociation reaction to determine the solubility product expression. Substitute the molarity of Mg^{2+} into the expression and solve for $[F^-]$.

Work:

The dissociation reaction is $MgF_2(s) \rightleftharpoons Mg^{2+}(aq) + 2F^-(aq)$.

$$[Mg^{2+}][F^-]^2 = (0.010)[F^-]^2 = K_{sp} = 6.6 \times 10^{-9};\ \text{therefore,}\ [F^-]^2 = \frac{6.6 \times 10^{-9}}{0.010} = 6.6 \times 10^{-7}$$

$$?\ [F^-] = \sqrt{6.6 \times 10^{-7}} = \mathbf{8.1 \times 10^{-4}\ M\ F^-}$$

17.59 What is the maximum concentration of Zn^{2+} that can exist in saturated H_2S at pH 3.50?

Analysis/Plan: From Section 17.3 in the text, it is known that $[Zn^{2+}]_{max} = K'_{sp} \cdot 10^{22} \cdot [H_3O^+]^2$. From Appendix H, K'_{sp} for ZnS = 2 x 10^{-24}. Also known is that $[H_3O^+] = 10^{-pH}$. Determine the $[H_3O^+]$ from the pH and then use the given relationship.

Work:

$$[H_3O^+] = 10^{-3.50} = 3.2 \times 10^{-4}\ M$$

$$?\ [Zn^{2+}]_{max} = (2 \times 10^{-24})(10^{22}) \cdot [H_3O^+]^2 = (0.02)(3.2 \times 10^{-4})^2 = \mathbf{2 \times 10^{-9}\ M\ Zn^{2+}}$$

17.61 To what value must the pH of a saturated solution of H_2S be adjusted in order for 0.0050 mg FeS to dissolve in 1.0 L?

Analysis/Plan: In an analogous manner to the discussion presented in Section 17.3 of the text, $[Fe^{2+}]_{max} = K'_{sp} \cdot 10^{22} \cdot [H_3O^+]^2$. For FeS, $K'_{sp} = 6 \times 10^{-18}$. The molar mass of FeS provides the equivalence statement g ↔ mole and it is known that 1 mg ↔ 10^{-3} g. Those relationships can be used with the factor-label method to determine the molarity of Fe^{2+} required. This value can be inserted into the expression for $[Fe^{2+}]$ and then the $[H_3O^+]$ can be determined. Using the relationship pH = $-\log[H_3O^+]$ completes the problem.

Work:

First, find the molarity of FeS which is equal to the desired $[Fe^{2+}]$.

Molar Mass FeS = 1(55.847 Fe) + 1(32.06 S) = 87.91 g mol^{-1}

$$? \text{ mol L}^{-1} = \frac{0.0050 \text{ mg FeS}}{1.0 \text{ L}} \times \frac{10^{-3} \text{ g}}{1 \text{ mg}} \times \frac{1 \text{ mol FeS}}{87.91 \text{ g FeS}} = 5.7 \times 10^{-8} \text{ mol L}^{-1} \text{ Fe}^{2+}$$

Now use the relationship to solve for $[H_3O^+]$.

$$[Fe^{2+}]_{max} = K'_{sp} \cdot 10^{22} \cdot [H_3O^+]^2; \text{ therefore, } [H_3O^+]^2 = \frac{[Fe^{2+}]_{max}}{K'_{sp} \cdot 10^{22}}$$

$$[H_3O^+]^2 = \frac{5.7 \times 10^{-8}}{(6 \times 10^{-18})(10^{22})} = 9 \times 10^{-13}; \text{ and } [H_3O^+] = \sqrt{9 \times 10^{-13}} = 1 \times 10^{-6}$$

$$? \text{ pH} = -\log[H_3O^+] = -\log(1 \times 10^{-6}) = -(-6.0) = \mathbf{6.0}$$

17.63 Write the chemical equation describing the formation of the complex, and the formation constant expression, for each of the following complex ions.

a. $Ag(CN)_2^-$ b. $Cd(NH_3)_4^{2+}$

Analysis/Plan: Study the formula of the ion and write a balanced chemical equation to make the complex ion from metal ions and anions or molecules. Write the equilibrium constant expression in the normal manner.

Work:

a. $Ag^+(aq) + 2\ CN^-(aq) \rightleftharpoons Ag(CN)_2^-(aq)$ $\qquad K_f = \dfrac{[Ag(CN)_2^-]}{[Ag^+][CN^-]^2}$

b. $Cd^{2+}(aq) + 4\ NH_3(aq) \rightleftharpoons Cd(NH_3)_4^{2+}(aq)$ $\qquad K_f = \dfrac{[Cd(NH_3)_4^{2+}]}{[Cd^{2+}][NH_3]^4}$

17.65 Concentrated ammonia is added to a solution of silver nitrate, so that the final value of NH_3 is 0.35 M. Given that K_f for $Ag(NH_3)_2^+$ is 1.6×10^7, what is the value of the ratio $[Ag^+]/[Ag(NH_3)_2^+]$?

Analysis/Plan: Study the formula of the ion and write a balanced chemical equation to make the complex ion from metal ions and anions or molecules. Write the equilibrium constant expression in the normal manner. Insert the values of $[NH_3]$ and the K_f into the equilibrium expression. Rearrange the expression to solve for the ratio and calculate.

Work:

$Ag^+(aq) + 2\ NH_3(aq) \rightleftharpoons Ag(NH_3)_2^+$ $\qquad K_f = \dfrac{[Ag(NH_3)_2^+]}{[Ag^+][NH_3]^2}$

$$\text{Rearranging yields: } \frac{[Ag^+]}{[Ag(NH_3)_2^+]} = \frac{1}{[NH_3]^2 K_f} = \frac{1}{(0.35)^2(1.6 \times 10^7)} = \mathbf{5.1 \times 10^{-7}}$$

17.67 Gaseous ammonia is added to a solution of 0.063 M $AgNO_3$ until the concentration of $NH_3(aq)$ rises to 0.18 M. Given that K_f for $Ag(NH_3)_2^+$ is 1.6×10^7, what are the concentrations of $Ag(NH_3)_2^+$ and Ag^+?

Analysis/Plan: Because the formation constant is so large, assume that essentially all the initial Ag^+, 0.063 M, is tied up in the complex ion. Solve for the $\frac{[Ag^+]}{[Ag(NH_3)_2^+]}$ as in Problem 17.65, then substitute in the concentration of the complex ion to determine the molarity of free Ag^+.

Work:

$[Ag(NH_3)_2^+]$ = **0.063 M $[Ag(NH_3)_2^+]$**

From Problem 17.65, $\frac{[Ag^+]}{[Ag(NH_3)_2^+]} = \frac{1}{[NH_3]^2K_f}$

Inserting the $[NH_3]$ gives $\frac{1}{(0.18)^2(1.6 \times 10^7)} = 1.9 \times 10^{-6} = \frac{[Ag^+]}{[Ag(NH_3)_2^+]}$

$$[Ag^+] = (1.9 \times 10^{-6})[Ag(NH_3)_2^+] = (1.9 \times 10^{-6})(0.063) = \mathbf{1.2 \times 10^{-7}\ M\ Ag^+}$$

17.69 What chemical property distinguishes Analytical Group 1 from Analytical Group 2?

The Group 1 elements have insoluble chloride salts, while the chloride salts of the Group 2 elements are soluble.

17.71 A precipitate from Analytical Group 1 is known to contain no $PbCl_2$, and therefore it is AgCl, Hg_2Cl_2, or a mixture of the two. Suggest a means for determining its composition.

Ag^+ forms a complex ion with NH_3, while Hg_2^{2+} does not. Add ammonia. If the precipitate dissolves completely, it contains only AgCl. If part of the precipitate does not dissolve, then that portion is Hg_2Cl_2. If the solution is reacidified and more precipitate forms, then there was also AgCl present in the sample.

17.73 Lead ion is in Analytical Group 2 as well as Analytical Group 1, because $PbCl_2$ is slightly soluble. Suppose that, in the first step of the analysis procedure, hydrochloric acid is added to a sample until [HCl] rises to 0.15 M. What is the maximum concentration of Pb^{2+} that can be present *without* forming a precipitate?

Analysis/Plan: The maximum concentration of Pb^{2+} is that in which the ion product is equal to the K_{sp} of $PbCl_2$ (1.7×10^{-5}, Appendix H). Write the dissociation reaction to determine the ion product expression. Substitute the value of the chloride concentration into the expression and solve for the maximum value of $[Pb^{2+}]$.

Work:

The dissociation reaction is $PbCl_2(s) \rightleftharpoons Pb^{2+}(aq) + 2\ Cl^-(aq)$

Insert the values into the K_{sp} expression and solve algebraically.

$$K_{sp} = [Pb^{2+}][Cl^-]^2 = [Pb^{2+}](0.15)^2 = 1.7 \times 10^{-5}$$

$$[Pb^{2+}] = \frac{1.7 \times 10^{-5}}{(0.15)^2} = \mathbf{7.6 \times 10^{-4}\ M\ Pb^{2+}}$$

18 Oxidation-Reduction Reactions

18.1 Differentiate among the following terms: combustion, oxidation, reaction with oxygen.

Combustion is a rapid exothermic reaction with oxygen, typically resulting in light as well as heat. Oxidation is any increase in oxidation number, caused by reaction with oxygen or any other oxidizing agent. Reactions with oxygen may be slow, such as gradual rusting, and do not need to be exothermic, such as the reaction with nitrogen to produce NO_2.

18.3 Distinguish between a Lewis base and a reducing agent.

A Lewis base has a lone pair of electrons which it can share with an acceptor agent, without changing its oxidation number. A reducing agent transfers a pair of electrons to another species, resulting in an increase in its oxidation number.

18.5 What is a half-reaction? Can it occur by itself?

A half-reaction is either an oxidation (electrons lost) or a reduction (electrons gained). An oxidation half-reaction must always be paired with a reduction half-reaction. Electrons must be transferred from the reducing agent to the oxidizing agent.

18.7 Give a formal definition of oxidation number.

The oxidation number of an element is the number of electrons that each atom of the element gains (negative oxidation number) or loses (positive oxidation number) in the formation reaction for the compound in which the element appears.

18.9 What is a spectator ion?

A spectator ion is any ion that is not changed in a chemical reaction. It must have the same charge and state on both sides of the chemical equation.

18.11 What is meant by a disproportionation? Give an example.

Disproportionation, also called auto oxidation, occurs when the same species is both oxidized and reduced in the reaction. An example of a disproportionation reaction is:

$$3\ MnO_4^{2-}(aq) + 4\ H^+(aq) \longrightarrow 2\ MnO_4^-(aq) + MnO_2(s) + 2\ H_2O(\ell).$$

18.13 Determine the oxidation number of the metal in each of the following compounds.
a. $SrCl_2$ b. Fe_2O_3 c. $NaNH_2$

Analysis/Plan: All oxidation-number problems should be approached through a ***stepwise*** application of the rules detailed in Table 18.1 of the text. Charges for polyatomic ions must be memorized to assign oxidation numbers in compounds with more than one element not covered by the rules. Rule 1 states that the sum of the oxidation numbers equals 0 in a compound and equals the ion charge in a polyatomic ion.

Work:

a. Rule 2 states that the oxidation number of a Group 2A metal in compounds is always +2; therefore, the oxidation number of Sr is **+2**.

b. Rule 5 states that the oxidation number of oxygen is almost always -2. Iron is not listed as one of the exceptions; therefore, the oxidation number of oxygen is -2.

Atom Type	Number of Atoms	Oxidation Number	Total charge
Fe	2	x	2x
O	3	-2	-6

This is a compound, so: $2x - 6 = 0$; $x = +3$. The oxidation number of Fe is **+3**.

c. Rule 2 states that the oxidation number of a Group 1A metal in compounds is always +1; therefore, the oxidation number of Na is **+1**.

18.15 Determine the oxidation number of all elements in the following species.

a. PCl_5 b. NF_3 c. NO_2
d. N_2O_4 e. BaO_2 f. Na_2SO_3

Analysis/Plan: Follow the approach detailed in Problem 18.13.

Work:

a. Rule 3 states that, in compounds, the halogens have an oxidation number of -1 unless bonded to oxygen or another halogen; therefore, Cl has an oxidation number of **-1**.

Atom Type	Number of Atoms	Oxidation Number	Total charge
P	1	x	x
Cl	5	-1	-5

This is a compound, so: $x - 5 = 0$; $x = +5$. The oxidation number of P is **+5**.

b. Rule 3 states that, in compounds, F always has an oxidation number of -1; therefore, F has an oxidation number of **-1**.

Atom Type	Number of Atoms	Oxidation Number	Total charge
N	1	x	x
F	3	-1	-3

This is a compound, so: $x - 3 = 0$; $x = +3$. The oxidation number of N is **+3**.

c. Rule 3 states that the oxidation number of oxygen is almost always -2. Nitrogen is not one of the exceptions to this rule; therefore, the oxidation number of O is **-2**.

Atom Type	Number of Atoms	Oxidation Number	Total charge
N	1	x	x
O	2	-2	-4

This is a compound, so: $x - 4 = 0$; $x = +4$. The oxidation number of N is **+4**.

d. Rule 3 states that the oxidation number of oxygen is almost always -2. Nitrogen is not one of the exceptions to this rule; therefore, the oxidation number of O is **-2**.

Atom Type	Number of Atoms	Oxidation Number	Total charge
N	2	x	2x
O	4	-2	-8

This is a compound, so: $2x - 8 = 0$; $x = +4$. The oxidation number of N is **+4**.

e. Rule 2 states that the oxidation number of a Group 2A metal in compounds is always +2; therefore, the oxidation number of Ba is **+2**.

Atom Type	Number of Atoms	Oxidation Number	Total charge
Ba	1	+2	+2
O	2	x	2x

This is a compound, so: $2 + 2x = 0$; $x = -1$. The oxidation number of O is **-1**. (This is a peroxide.)

f. Rule 2 states that the oxidation number of a Group 1A metal in compounds is always +1; therefore, the oxidation number of Na is +1. Rule 3 states that the oxidation number of oxygen is almost always -2. This compound is not one of the exceptions to this rule; therefore, the oxidation number of O is **-2**.

Atom Type	Number of Atoms	Oxidation Number	Total charge
Na	2	+1	+2
S	1	x	x
O	3	-2	-6

This is a compound, so: $2 + x - 6 = 0$; $x = +4$. The oxidation number of S is **+4**.

18.17 Determine the oxidation number of all elements in the following species.

a. SF_6 b. XeF_4 c. P_4O_{10}

d. $HAsO_2$ e. NH_4^+ f. CH_3OH

Analysis/Plan: Follow the approach detailed in Problem 18.13.

Work:

a. Rule 3 states that in compounds F always has an oxidation number of -1; therefore, F has an oxidation number of **-1**.

Atom Type	Number of Atoms	Oxidation Number	Total charge
S	1	x	x
F	6	-1	-6

This is a compound, so: $x - 6 = 0$; $x = +6$. The oxidation number of S is **+6**.

b. Rule 3 states that in compounds F always has an oxidation number of -1; therefore, F has an oxidation number of **-1**.

Atom Type	Number of Atoms	Oxidation Number	Total charge
Xe	1	x	x
F	4	-1	-4

This is a compound, so: $x - 4 = 0$; $x = +4$. The oxidation number of Xe is **+4**.

c. Rule 3 states that the oxidation number of oxygen is almost always -2. Phosphorus is not one of the exceptions to this rule; therefore, the oxidation number of O is **-2**.

Atom Type	Number of Atoms	Oxidation Number	Total charge
P	4	x	4x
O	10	-2	-20

This is a compound, so: $4x - 20 = 0$; $x = +5$. The oxidation number of P is **+5**.

d. Rule 4 states that the oxidation number of hydrogen bonded to a nonmetal is always +1; therefore, the oxidation number of H is +1. Rule 3 states that the oxidation number of oxygen is almost always -2. This compound is not one of the exceptions to this rule; therefore, the oxidation number of O is **-2**.

Atom Type	Number of Atoms	Oxidation Number	Total charge
H	1	+1	+1
As	1	x	x
O	2	-2	-4

This is a compound, so: 1 + x - 4 = 0; x = +3. The oxidation number of As is **+3**.

e. Rule 4 states that the oxidation number of hydrogen bonded to a nonmetal is always +1; therefore, the oxidation number of H is **+1**.

Atom Type	Number of Atoms	Oxidation Number	Total charge
H	4	+1	+4
N	1	x	x

This is a polyatomic ion, so: 4 + x - 4 = +1; x = -3. The oxidation number of N is **-3**.

f. Rule 4 states that the oxidation number of hydrogen bonded to a nonmetal is always +1; therefore, the oxidation number of H is **+1**. Rule 3 states that the oxidation number of oxygen is almost always -2. This compound is not one of the exceptions to this rule; therefore, the oxidation number of O is **-2**.

Atom Type	Number of Atoms	Oxidation Number	Total charge
H	4	+1	+4
C	1	x	x
O	1	-2	-2

This is a compound, so: 4 + x - 2 = 0; x = -2. The oxidation number of C is **-2**.

18.19 Which of the following are redox reactions?

a. $PCl_3 + Cl_2 \rightarrow PCl_5$

b. $H_2SO_3 + 2\ KOH \rightarrow K_2SO_3 + 2\ H_2O$

c. $SO_3 + H_2O \rightarrow H_2SO_4$

d. $3\ NO_2 + H_2O \rightarrow 2\ HNO_3 + NO$

e. $Mg + 2\ HCl \rightarrow H_2 + MgCl_2$

Analysis/Plan: A reaction is a redox reaction if oxidation numbers change, indicating that electrons have been transferred. Use the rules detailed in Table 18.1 to assign oxidation numbers to each element. Inspect both sides of the equation to determine if oxidation numbers have changed. (Oxidation numbers are shown above the symbol of the element.)

Work:

+3 -1 0 +5 -1

a. $PCl_3 + Cl_2 \rightarrow PCl_5$

The oxidation number of P changes from +3 to +5 and the oxidation number of Cl changes from 0 to -1. **This is a redox reaction.**

+1 +4-2 +1-2+1 +1+4 -2 +1 -2

b. $H_2SO_3 + 2\ KOH \rightarrow K_2SO_3 + 2\ H_2O$

No oxidation numbers change. **This is not a redox reaction.**

+6-2 +1 -2 +1 +6 -2

c. $SO_3 + H_2O \longrightarrow H_2SO_4$

No oxidation numbers change. **This is not a redox reaction.**

+4 -2 +1 -2 +1+5-2 +2-2

d. $3\ NO_2 + H_2O \longrightarrow 2\ HNO_3 + NO$

The oxidation number of N changes from +4 to +5 and from +4 to +2. **This is a redox reaction** (specifically a disproportionation or auto oxidation reaction).

0 +1 -1 0 +2 -1

e. $Mg + 2\ HCl \longrightarrow H_2 + MgCl_2$

The oxidation number of Mg changes from 0 to +2 and the oxidation number of H changes from +1 to 0. **This is a redox reaction.**

18.21 Which of the following are redox reactions?

a. $Al_2S_3 + 6\ H_2O \longrightarrow 2\ Al(OH)_3 + 3\ H_2S$

b. $Xe + 2\ F_2 \longrightarrow XeF_4$

c. $Br_2 + H_2O \longrightarrow HBr + HBrO$

d. $N_2O_4 \longrightarrow 2\ NO_2$

e. $CaC_2 + 2\ H_2O \longrightarrow Ca(OH)_2 + C_2H_2$

Analysis/Plan: A reaction is a redox reaction if oxidation numbers change, indicating that electrons have been transferred. Use the rules detailed in Table 18.1 to assign oxidation numbers to each element. Inspect both sides of the equation to determine if oxidation numbers have changed. (Oxidation numbers are shown above the symbol of the element.)

Work:

+3 -2 +1 -2 +3 -2+1 +1 -2

a. $Al_2S_3 + 6\ H_2O \longrightarrow 2\ Al(OH)_3 + 3\ H_2S$

No oxidation numbers change. **This is not a redox reaction.**

0 0 +4 -1

b. $Xe + 2\ F_2 \longrightarrow XeF_4$

The oxidation number of Xe changes from 0 to +4 and the oxidation number of F changes from 0 to -1. **This is a redox reaction.**

0 +1 -2 +1 -1 +1 +1-2

c. $Br_2 + H_2O \longrightarrow HBr + HBrO$

The oxidation number of Br changes from +0 to -1 and from 0 to +1. **This is a redox reaction** (specifically a disproportionation or auto oxidation reaction).

+4 -2 +4 -2

d. $N_2O_4 \longrightarrow 2\ NO_2$

No oxidation numbers change. **This is not a redox reaction.**

+2 -1 +1 -2 +2 -2+1 -1 +1

e. $CaC_2 + 2\ H_2O \longrightarrow Ca(OH)_2 + C_2H_2$

No oxidation numbers change. **This is not a redox reaction.**

18.23 For each of the redox reactions in Problem 18.19, identify the species that is oxidized and the species that is reduced.

Analysis/Plan: The species that is oxidized has lost one or more electrons for an increase in oxidation number. The species that is reduced has gained one or more electrons for a decrease in oxidation number. Inspect the oxidation numbers and see which substance is oxidized and which is reduced.

Work:

a. The oxidation number of P increases; therefore, **PCl_3 is oxidized**. The oxidation number of Cl in Cl_2 decreases; therefore, **Cl_2 is reduced**.

d. The oxidation number of N in NO_2 both increases and decreases; therefore, **NO_2 is both oxidized** and **reduced.**

e. The oxidation number of Mg increases; therefore, **Mg is oxidized**. The oxidation number of H in HCl decreases; therefore, **HCl is reduced**.

18.25 For each of the redox reactions in Problem 18.21, identify the oxidant and the reductant.

Analysis/Plan: The species that is the reductant is oxidized and has lost one or more electrons for an increase in oxidation number. The species that is the oxidant is reduced and has gained one or more electrons for a decrease in oxidation number. Inspect the oxidation numbers and see which substance is oxidized and which is reduced.

Work:

b. The oxidation number of Xe increases; therefore, **Xe is the reductant**. The oxidation number of F in F_2 decreases; therefore, **F_2 is the oxidant**.

c. The oxidation number of Br in Br_2 both increases and decreases; therefore, **Br_2 is both the oxidant** and **the reductant**.

18.27 Each of the following is a balanced molecular equation. Write each as an ionic equation, and then, by eliminating the spectator ions, write each as a net ionic equation.

a. $3\ NaClO \longrightarrow NaClO_3 + 2\ NaCl$

b. $3\ Cu(s) + 8\ HNO_3(aq) \longrightarrow 3\ Cu(NO_3)_2(aq) + 2\ NO(g) + 4\ H_2O(\ell)$

Analysis/Relationship: The ionic equation is determined by knowing which species are strong electrolytes, and, therefore, which dissociate upon dissolution. Removal of spectator ions, which are the same on both sides of the equation, yields the *net* ionic equation.

Plan: This requires two steps.

Step 2. Identify and remove the spectator ions from the ionic equation.

Step 1. Rewrite the equation, showing all strong electrolytes as separate ions.

Work:

a. **Step 1.** All of the substances are soluble ionic compounds and, therefore, strong electrolytes. NaClO should be written as Na^+ and ClO^-. $NaClO_3$ should be written as Na^+ and ClO_3^-. NaCl should be written as Na^+ and Cl^-.

$$3[Na^+(aq) + ClO^-(aq)] \longrightarrow Na^+(aq) + ClO_3^-(aq) + 2[Na^+(aq) + Cl^-(aq)]$$

Now the multiplications are carried out. The *ionic equation* is:

$$\mathbf{3\ Na^+(aq) + 3\ ClO^-(aq) \longrightarrow Na^+(aq) + ClO_3^-(aq) + 2\ Na^+(aq) + 2\ Cl^-(aq)}$$

Step 2. Three Na^+ ions appear as both reactant and product and are spectator ions. They are canceled out to yield the ***net ionic*** equation:

$$\mathbf{3\ ClO^-(aq) \longrightarrow ClO_3^-(aq) + 2\ Cl^-(aq)}$$

a. **Step 1.** HNO_3 is a strong acid and should be written as H^+ and NO_3^-. $Cu(NO_3)_2$ is a soluble ionic compound and should be written as Cu^{2+} and 2 NO_3^-. The solid Cu, the gaseous NO and the liquid water should be left in their current forms.

$$3\ Cu(s) + 8[H^+(aq) + NO_3^-(aq)] \longrightarrow 3[Cu^{2+}(aq) + 2\ NO_3^-(aq)] + 2\ NO(g) + 4\ H_2O(\ell)$$

Now the multiplications are carried out. The *ionic equation* is:

$$3\ Cu(s) + 8\ H^+(aq) + 8\ NO_3^-(aq) \longrightarrow 3\ Cu^{2+}(aq) + 6\ NO_3^-(aq) + 2\ NO(g) + 4\ H_2O(\ell)$$

Step 2. Six NO_3^- ions appear as both a reactant and a product and are spectator ions. They are canceled out to yield the *net ionic* equation:

$$3\ Cu(s) + 8\ H^+(aq) + 2\ NO_3^-(aq) \longrightarrow 3\ Cu^{2+}(aq) + 2\ NO(g) + 4\ H_2O(\ell)$$

18.29 Each of the following is a balanced molecular equation. Write each as an ionic equation, and then, by eliminating the spectator ions, write each as a net ionic equation.

a. $8\ HCl(aq) + H_3AsO_4(aq) + 4\ Mg(s) \longrightarrow AsH_3(g) + 4\ MgCl_2(aq) + 4\ H_2O(\ell)$

b. $KNO_2(aq) + 5\ H_2O(\ell) + 2\ Al(s) \longrightarrow 2\ Al(OH)_3(aq) + NH_3(aq) + KOH(aq)$

Analysis/Relationship: See 18.27. The ionic equation is determined by knowing which species are strong electrolytes, and, therefore, which dissociate upon dissolution. Removal of spectator ions, which are the same on both sides of the equation, yields the *net* ionic equation.

Plan: This requires two steps.

Step 2. Identify and remove the spectator ions from the ionic equation.

Step 1. Rewrite the equation, showing all strong electrolytes as separate ions.

Work:

a. **Step 1.** HCl is a strong acid and should be written as H^+ and Cl^-. $MgCl_2$ is a soluble ionic compound and should be written as Mg^{2+} and 2 Cl^-. The weak acid H_3AsO_4, the solid Mg, the gaseous AsH_3 and the liquid water should be left in their current forms.

$$8\ [H^+(aq) + Cl^-(aq)] + H_3AsO_4(aq) + 4\ Mg(s) \longrightarrow AsH_3(g) + 4\ [Mg^{2+}(aq) + 2\ Cl^-(aq)] + 4\ H_2O(\ell)$$

Now the multiplications are carried out. The *ionic equation* is:

$$8\ H^+(aq) + 8\ Cl^-(aq) + H_3AsO_4(aq) + 4\ Mg(s) \longrightarrow AsH_3(g) + 4\ Mg^{2+}(aq) + 8\ Cl^-(aq) + 4\ H_2O(\ell)$$

Step 2. Eight Cl^- ions appear as both a reactant and a product and are spectator ions. They are canceled out to yield the *net ionic* equation:

$$8\ H^+(aq) + H_3AsO_4(aq) + 4\ Mg(s) \longrightarrow AsH_3(g) + 4\ Mg^{2+}(aq) + 4\ H_2O(\ell)$$

b. **Step 1.** KNO_2, $Al(OH)_3$ and KOH are soluble ionic compounds and should be written as K^+ and NO_2^-, Al^{3+} and 3 OH^- and K^+ + OH^-, respectively. The weak base NH_3, the solid Al and the liquid water should be left in their current forms.

$$[K^+(aq) + NO_2^-(aq)] + 5\ H_2O(\ell) + 2\ Al(s) \longrightarrow 2\ [Al^{3+}(aq) + 3\ OH^-(aq)] + NH_3(aq) + [K^+(aq) + OH^-(aq)]$$

Now the multiplications are carried out. The *ionic equation* is:

$$K^+(aq) + NO_2^-(aq) + 5\ H_2O(\ell) + 2\ Al(s) \longrightarrow 2\ Al^{3+}(aq) + 6\ OH^-(aq) + NH_3(aq) + K^+(aq) + OH^-(aq)$$

Step 2. One K^+ ion appears as both a reactant and a product and is a spectator ion. It is canceled out and the product OH^-'s are added to yield the *net ionic* equation:

$$NO_2^-(aq) + 5\ H_2O(\ell) + 2\ Al(s) \longrightarrow 2\ Al^{3+}(aq) + 7\ OH^-(aq) + NH_3(aq)$$

18.31 For each of the following unbalanced redox reactions, write the unbalanced oxidation half-reaction and the unbalanced reduction half-reaction.

a. $UO_2^{2+}(aq) + Sn^{2+}(aq) \rightarrow U^{4+}(aq) + Sn^{4+}(aq)$

b. $Hg(\ell) + H_2SO_4(aq) + Cr_2O_7^{2-}(aq) \rightarrow Hg_2SO_4(s) + Cr^{3+}(aq)$

c. $H_2O_2(aq) + Mn^{2+}(aq) \rightarrow MnO_4^-(aq) + H_2O(\ell)$

d. $C_2O_4^{2-}(aq) + IO_3^-(aq) \rightarrow CO_2(g) + I^-(aq)$

Analysis/Plan: If possible, choose substances that contain elements besides H and O for each of the half-reactions. To determine whether the reaction is the oxidation half-reaction or the reduction half-reaction, the half-reaction must be balanced (see Problem 18.33) or the oxidation numbers must be determined. If electrons are gained (appear as a reactant), it is the reduction half-reaction. If electrons are lost (appear as a product), it is the oxidation half-reaction. Consult Problem 18.33 to designate each half-reaction as the oxidation half-reaction or the reduction half-reaction.

Work:

a. oxidation: $\mathbf{Sn^{2+} \rightarrow Sn^{4+}}$ reduction: $\mathbf{UO_2^{2+} \rightarrow U^{4+}}$

b. oxidation: $\mathbf{Hg \rightarrow Hg_2SO_4}$ reduction: $\mathbf{Cr_2O_7^{2-} \rightarrow Cr^{3+}}$

c. oxidation: $\mathbf{Mn^{2+} \rightarrow MnO_4^-}$ reduction: $\mathbf{H_2O_2 \rightarrow H_2O}$

d. oxidation: $\mathbf{C_2O_4^{2-} \rightarrow CO_2}$ reduction: $\mathbf{IO_3^- \rightarrow I^-}$

18.33 Balance each of the half-reactions obtained as answers to Problem 18.31. Assume that the reactions take place in acidic solution.

Analysis/Plan: Follow the steps detailed in Section 18.3 of the book. They are:

1. Balance all atoms other than H and O.
2. Add H_2O to the side deficient in O to balance O.
3. Add H^+ to the side deficient in H to balance H. This completes the mass balance.
4. Add electrons to the more positive (less negative) side to balance the charge.

Work:

a. 1. Balance the atoms other than O and H.

$Sn^{2+} \rightarrow Sn^{4+}$ $UO_2^{2+} \rightarrow U^{4+}$

2. Add H_2O to the side deficient in O to balance O.

$Sn^{2+} \rightarrow Sn^{4+}$ $UO_2^{2+} \rightarrow U^{4+} + 2\ H_2O$

3. Add H^+ to the side deficient in H to balance H.

$Sn^{2+} \rightarrow Sn^{4+}$ $4\ H^+ + UO_2^{2+} \rightarrow U^{4+} + 2\ H_2O$

4. Add electrons to the more positive (less negative) side to balance the charge.

$\mathbf{Sn^{2+} \rightarrow Sn^{4+} + 2\ e^-}$ $\mathbf{4\ H^+ + UO_2^{2+} + 2\ e^- \rightarrow U^{4+} + 2\ H_2O}$

b. 1. Balance the atoms other than O and H. (Recognize that SO_4^{2-} is the sulfate ion.)

$2\ Hg + SO_4^{2-} \rightarrow Hg_2SO_4$ $Cr_2O_7^{2-} \rightarrow 2\ Cr^{3+}$

2. Add H_2O to the side deficient in O to balance O.

$2\ Hg + SO_4^{2-} \rightarrow Hg_2SO_4$ $Cr_2O_7^{2-} \rightarrow 2\ Cr^{3+} + 7\ H_2O$

3. Add H^+ to the side deficient in H to balance H.

$2\ Hg + SO_4^{2-} \rightarrow Hg_2SO_4$ $14\ H^+ + Cr_2O_7^{2-} \rightarrow 2\ Cr^{3+} + 7H_2O$

4. Add electrons to the more positive (less negative) side to balance the charge.

$\mathbf{2\ Hg + SO_4^{2-} \rightarrow Hg_2SO_4 + 2\ e^-}$

$\mathbf{14\ H^+ + Cr_2O_7^{2-} + 6\ e^- \rightarrow 2\ Cr^{3+} + 7\ H_2O}$

c. 1. Balance the atoms other than O and H.

$Mn^{2+} \rightarrow MnO_4^-$ $\qquad$ $H_2O_2 \rightarrow H_2O$

2. Add H_2O to the side deficient in O to balance O.

$4\ H_2O + Mn^{2+} \rightarrow MnO_4^-$ $\qquad$ $H_2O_2 \rightarrow 2\ H_2O$

3. Add H^+ to the side deficient in H to balance H.

$4\ H_2O + Mn^{2+} \rightarrow MnO_4^- + 8\ H^+$ $\qquad$ $2\ H^+ + H_2O_2 \rightarrow 2\ H_2O$

4. Add electrons to the more positive (less negative) side to balance the charge.

$\mathbf{4\ H_2O + Mn^{2+} \rightarrow MnO_4^- + 8\ H^+ + 5\ e^-}$

$\mathbf{2\ H^+ + H_2O_2 + 2\ e^- \rightarrow 2\ H_2O}$

d. 1. Balance the atoms other than O and H.

$C_2O_4^{2-} \rightarrow 2\ CO_2$ $\qquad$ $IO_3^- \rightarrow I^-$

2. Add H_2O to the side deficient in O to balance O.

$C_2O_4^{2-} \rightarrow 2\ CO_2$ $\qquad$ $IO_3^- \rightarrow I^- + 3\ H_2O$

3. Add H^+ to the side deficient in H to balance H.

$C_2O_4^{2-} \rightarrow 2\ CO_2$ $\qquad$ $6\ H^+ + IO_3^- \rightarrow I^- + 3\ H_2O$

4. Add electrons to the more positive (less negative) side to balance the charge.

$\mathbf{C_2O_4^{2-} \rightarrow 2\ CO_2 + 2\ e^-}$ $\qquad$ $\mathbf{6\ H^+ + IO_3^- + 6\ e^- \rightarrow I^- + 3\ H_2O}$

18.35 Complete the ion-electron balancing procedure begun in Problem 18.33, by writing the overall net ionic equation for each of the reaction in Problem 18.31.

Analysis/Plan: Adjust the coefficients for each half-reaction so that each shows the same number of electrons. This makes the number of electrons lost by the reductant equal the number of electrons gained by the oxidant. Add the adjusted half-reactions, eliminating or combining like terms as appropriate.

Work:

a. Both the oxidation and the reduction half-reactions have 2 e^-. Add the equations:

$Sn^{2+} \rightarrow Sn^{4+} + 2\ e^-$

$UO_2^{2+} + 4\ H^+ + 2\ e^- \rightarrow U^{4+} + 2\ H_2O$

$Sn^{2+} + UO_2^{2+} + 4\ H^+ + 2\ e^- \rightarrow Sn^{4+} + 2\ e^- + U^{4+} + 2\ H_2O$

Only the electrons are common to both sides. The net ionic equation is:

$\mathbf{Sn^{2+}(aq) + UO_2^{2+}(aq) + 4\ H^+(aq) \rightarrow Sn^{4+}(aq) + U^{4+}(aq) + 2\ H_2O(\ell)}$

b. The oxidation half-reaction has 2 e^-, but the reduction half-reaction has 6 e^-. Six is a multiple of 2, so multiply the oxidation half-reaction by 3 and add the equations.

$3[2\ Hg + SO_4^{2-} \rightarrow Hg_2SO_4 + 2\ e^-]$

$14\ H^+ + Cr_2O_7^{2-} + 6\ e^- \rightarrow 2\ Cr^{3+} + 7\ H_2O$

$6\ Hg + 3\ SO_4^{2-} + 14\ H^+ + Cr_2O_7^{2-} + 6\ e^- \rightarrow 3\ Hg_2SO_4 + 6\ e^- + 2\ Cr^{3+} + 7\ H_2O$

Only the electrons are common to both sides. The net ionic equation is:

$\mathbf{6\ Hg(\ell) + 3\ SO_4^{2-}(aq) + 14\ H^+(aq) + Cr_2O_7^{2-}(aq) \rightarrow 3\ Hg_2SO_4(s) + 2\ Cr^{3+}(aq) + 7\ H_2O(\ell)}$

c. The oxidation half-reaction has 5 e^-, but the reduction half-reaction has 2 e^-. Multiply the oxidation half-reaction by 2 and the reduction half-reaction by 5 and add the equations.

$$2[4\ H_2O + Mn^{2+} \longrightarrow MnO_4^- + 8\ H^+ + 5\ e^-]$$

$$5[2\ H^+ + H_2O_2 + 2\ e^- \longrightarrow 2\ H_2O]$$

$$8\ H_2O + 2\ Mn^{2+} + 10\ H^+ + 5\ H_2O_2 + 10\ e^- \longrightarrow 2\ MnO_4^- + 16\ H^+ + 10\ e^- + 10\ H_2O$$

Electrons, H_2O and H^+ are common to both sides. The net ionic equation is:

$$\mathbf{2\ Mn^{2+}(aq) + 5\ H_2O_2(aq) \longrightarrow 2\ MnO_4^-(aq) + 6\ H^+(aq) + 2\ H_2O(\ell)}$$

d. The oxidation half-reaction has 2 e^-, but the reduction half-reaction has 6 e^-. Six is a multiple of 2, so multiply the oxidation half-reaction by 3 and add the equations.

$$3[C_2O_4^{2-} \longrightarrow 2\ CO_2 + 2\ e^-]$$

$$6\ H^+ + IO_3^- + 6\ e^- \longrightarrow I^- + 3\ H_2O$$

$$3\ C_2O_4^{2-} + 6\ H^+ + IO_3^- + 6\ e^- \longrightarrow 6\ CO_2 + 6\ e^- + I^- + 3\ H_2O$$

Only electrons are common to both sides. The net ionic equation is:

$$\mathbf{3\ C_2O_4^{2-}(aq) + 6\ H^+(aq) + IO_3^-(aq) \longrightarrow 6\ CO_2(aq) + I^-(aq) + 3\ H_2O(\ell)}$$

18.37 Balance the following equations by the oxidation number method, assuming that the reaction takes place in acidic solution. Give the number of electrons transferred in the balanced equation.

a. $I_2(s) + H_2S(g) \longrightarrow SO_4^{2-}(aq) + I^-(aq)$

b. $Cl_2(aq) + H_2O(\ell) \longrightarrow ClO^-(aq) + Cl^-(aq)$

Analysis/Plan: The method detailed in the book in Section 18.3 is appropriate. The steps are:

1. Assign oxidation numbers to each element in the reactant and products.
2. Determine the values of the total increase or decrease in oxidation number.
3. Using the least common multiple, adjust the coefficients of the species containing the redox atoms to make the total increase in oxidation number equal the total decrease in oxidation number.
4. Balance any additional species by inspection, adding H_2O and H^+ (because these are in acidic solution) as needed.

Work:

a. 1. Assign oxidation numbers and adjust the coefficient of I^- to balance this redox atom.

$$\overset{0}{I_2} + \overset{+1\ -2}{H_2S} \longrightarrow \overset{+6\ -2}{SO_4^{2-}} + 2\ \overset{-1}{I^-}$$

2. Determine the values of the total increase or decrease in oxidation number.

2(0 → -1) = -2 — 2 electrons gained in reduction

$$\overset{0}{I_2} + \overset{+1\ -2}{H_2S} \longrightarrow \overset{+6\ -2}{SO_4^{2-}} + 2\ \overset{-1}{I^-}$$

1(-2 → 6) = 8 — 8 electrons lost in oxidation

3. Use coefficients to make the number of electrons lost equal the number of electrons gained. The number of electrons transferred is 8.

$$\overset{0}{4\,I_2} + \overset{+1\ -2}{H_2S} \longrightarrow \overset{+6\ -2}{SO_4^{2-}} + \overset{-1}{8\,I^-}$$

−4[2(0 → −1)] = −8 — 8 electrons gained in reduction

1(−2 → 6) = 8 — 8 electrons lost in oxidation

4. Balance the remaining elements by inspection.
First, add H_2O to balance O.

$$4\,I_2 + H_2S + 4\,H_2O \longrightarrow SO_4^{2-} + 8\,I^-$$

Complete the process by adding 10 H^+ to the right to balance the H. This also balances the charge (0 on each side). (8 electrons are transferred in the process.

$$\mathbf{4\,I_2(s) + H_2S(aq) + 4\,H_2O(\ell) \longrightarrow SO_4^{2-}(aq) + 8\,I^-(aq) + 10\,H^+(aq)}$$

b. 1. Notice that we need to begin with 2 Cl_2, one to be oxidized and one to be reduced.
Assign oxidation numbers and adjust the coefficients of ClO^- and Cl^- to balance these redox atoms.

$$\overset{0}{Cl_2} + \overset{0}{Cl_2} \longrightarrow 2\,\overset{+1\ -2}{ClO^-} + 2\,\overset{1}{Cl^-}$$

2. Determine the values of the total increase or decrease in oxidation number.

$$\overset{0}{Cl_2} + \overset{0}{Cl_2} \longrightarrow \overset{+1\ -2}{2ClO^-} + \overset{-1}{2Cl^-}$$

2(0 → +1) = +2 — 2 electrons lost in the oxidation

2(0 → −1) = −2 — 2 electrons gained in the reduction

3. The number of electrons lost is already equal to the number of electrons gained. We can add the like terms (Cl_2) and divide all coefficients by 2. The number of electrons transferred is **2/ 2** = 1.

$$Cl_2 \longrightarrow ClO^- + Cl^-$$

4. Balance the remaining elements by inspection.
First, add H_2O to balance O.

$$H_2O + Cl_2 \longrightarrow ClO^- + Cl^-$$

Complete the process by adding 2 H^+ to the right to balance the H. This also balances the charge (0 on each side).

$$\mathbf{H_2O(\ell) + Cl_2(g) \longrightarrow ClO^-(aq) + Cl^-(aq) + 2\,H^+(aq)}$$

18.39 Each of the following is an unbalanced net ionic equation. Balance each, using any method you choose.

a. $I^-(aq) + H_2O_2(aq) + H^+(aq) \longrightarrow I_2(s) + H_2O(\ell)$

b. $IO_3^-(aq) + Cr(OH)_4^-(aq) \longrightarrow I^-(aq) + CrO_4^{2-}(aq) + H^+(aq) + H_2O(\ell)$

c. $Cu(s) + NO_3^-(aq) + H^+(aq) \longrightarrow NO_2(g) + Cu^{2+}(aq) + H_2O(\ell)$

Analysis/Plan: Because these reactions are all presented as net ionic equations, they can be readily treated by the half-reaction method detailed in the series of Problems 18.31, 18.33 and 18.35. Separate each into two half-reactions first and then follow the rest of the plan. The H^+ in each skeleton equation indicates that each is in acidic solution.

Work:

a. Separate into two half-reactions.

$I^- \longrightarrow I_2$ $\qquad\qquad$ $H_2O_2 \longrightarrow H_2O$

Balance the atoms other than O and H.

$2\ I^- \rightarrow I_2$ $\quad\quad$ $H_2O_2 \rightarrow H_2O$

Add H_2O to the side deficient in O to balance O.

$2\ I^- \rightarrow I_2$ $\quad\quad$ $H_2O_2 \rightarrow 2\ H_2O$

Add H^+ to the side deficient in H to balance H.

$2\ I^- \rightarrow I_2$ $\quad\quad$ $2\ H^+ + H_2O_2 \rightarrow 2\ H_2O$

Add electrons to the more positive (less negative) side to balance the charge.

$2\ I^- \rightarrow I_2 + 2\ e^-$ $\quad\quad$ $2\ H^+ + H_2O_2 + 2\ e^- \rightarrow 2\ H_2O$

Each half-reaction has 2 e^-, so the electrons are already equal. Add the reactions.

$2\ I^- \rightarrow I_2 + 2\ e^-$

$2\ H^+ + H_2O_2 + 2\ e^- \rightarrow 2\ H_2O$

$2\ I^- + 2\ H^+ + H_2O_2 + 2\ e^- \rightarrow I_2 + 2\ e^- + 2\ H_2O$

Only electrons are common to both sides. The net ionic equation is:

$\mathbf{2\ I^-(aq) + 2\ H^+(aq) + H_2O_2(aq) \rightarrow I_2(s) + 2\ H_2O(\ell)}$

b. Separate into two half-reactions.

$IO_3^- \rightarrow I^-$ $\quad\quad$ $Cr(OH)_4^- \rightarrow CrO_4^{2-}$

Balance the atoms other than O and H.

$IO_3^- \rightarrow I^-$ $\quad\quad$ $Cr(OH)_4^- \rightarrow CrO_4^{2-}$

Add H_2O to the side deficient in O to balance O.

$IO_3^- \rightarrow I^- + 3\ H_2O$ $\quad\quad$ $Cr(OH)_4^- \rightarrow CrO_4^{2-}$

Add H^+ to the side deficient in H to balance H.

$6\ H^+ + IO_3^- \rightarrow I^- + 3\ H_2O$ $\quad\quad$ $Cr(OH)_4^- \rightarrow CrO_4^{2-} + 4\ H^+$

Add electrons to the more positive (less negative) side to balance the charge.

$6\ H^+ + IO_3^- + 6\ e^- \rightarrow I^- + 3\ H_2O$ $\quad\quad$ $Cr(OH)_4^- \rightarrow CrO_4^{2-} + 4\ H^+ + 3\ e^-$

The I half-reaction (reduction) has 6 e^-, while the Cr half-reaction (oxidation) has 3 e^-. Multiply the Cr half-reaction by 2 and add the reactions.

$6\ H^+ + IO_3^- + 6\ e^- \rightarrow I^- + 3\ H_2O$

$2\ [Cr(OH)_4^- \rightarrow CrO_4^{2-} + 4\ H^+ + 3\ e^-]$

$6\ H^+ + IO_3^- + 6\ e^- + 2\ Cr(OH)_4^- \rightarrow I^- + 3\ H_2O + 2\ CrO_4^{2-} + 8\ H^+ + 6\ e^-$

H^+ and electrons are common to both sides. The net ionic equation is:

$\mathbf{IO_3^-(aq) + 2\ Cr(OH)_4^-(aq) \rightarrow I^-(aq) + 3\ H_2O(\ell) + 2\ CrO_4^{2-}(aq) + 2\ H^+(aq)}$

c. Separate into two half-reactions.

$Cu \rightarrow Cu^{2+}$ $\quad\quad$ $NO_3^- \rightarrow NO_2$

Balance the atoms other than O and H.

$Cu \rightarrow Cu^{2+}$ $\quad\quad$ $NO_3^- \rightarrow NO_2$

Add H_2O to the side deficient in O to balance O.

$Cu \rightarrow Cu^{2+}$ $\quad\quad$ $NO_3^- \rightarrow NO_2 + H_2O$

Add H^+ to the side deficient in H to balance H.

$Cu \rightarrow Cu^{2+}$ $\quad\quad$ $2\ H^+ + NO_3^- \rightarrow NO_2 + H_2O$

Add electrons to the more positive (less negative) side to balance the charge.

$Cu \rightarrow Cu^{2+} + 2e^-$ $\quad\quad$ $2H^+ + NO_3^- + e^- \rightarrow NO_2 + H_2O$

The Cu half-reaction (oxidation) has 2 e^-, but the N half-reaction (reduction) has 1 e^-. Multiply the N half-reaction by 2 and add the reactions.

$Cu \rightarrow Cu^{2+} + 2e^-$

$2[2H^+ + NO_3^- + 1e^- \rightarrow 1NO_2 + H_2O]$

$Cu + 4H^+ + 2NO_3^- + 2e^- \rightarrow Cu^{2+} + 2NO_2 + 2H_2O + 2e^-$

Only electrons are common to both sides. The net ionic equation is:

$$\mathbf{Cu(s) + 4H^+(aq) + 2NO_3^-(aq) \rightarrow Cu^{2+}(aq) + 2NO_2(g) + 2H_2O(\ell)}$$

18.41 The following equation is easily balanced by the ion-electron method, but, because of uncertainties in oxidation numbers, less easily by the oxidation-number method.

$CN^-(aq) + MnO_4^-(aq) \rightarrow CNO^-(aq) + MnO_2(s)$

The rules given in Table 18.1 do not help in assignment of oxidation numbers in CN^- or CNO^-. The general principle of assigning the bonding electron pairs to the more electronegative atom (C < N < O), however, together with reasonable Lewis structures, leads to an oxidation number of -3 for the N atom in CN^- and -1 for the N atom in CNO^-. Use these assignments to balance the equation by the oxidation-number method, in basic solution.

Analysis/Plan: Use the oxidation number method as detailed in Problem 18.37, using the oxidation number assignments suggested in the problem.

a. 1. Assign the remaining oxidation numbers.

$$\overset{+2\,-3}{CN^-} + \overset{+7\,-2}{MnO_4^-} \rightarrow \overset{+2\,-1\,-2}{CNO^-} + \overset{+4\,-2}{MnO_2}$$

2. Determine the values of the total increase or decrease in oxidation number.

1(-3 → -1) = +2 $\quad$ 2 electrons lost in the oxidation

$$\overset{+2\,-3}{CN^-} + \overset{+7\,-2}{MnO_4^-} \rightarrow \overset{+2\,-1\,-2}{CNO^-} + \overset{+4\,-2}{MnO_2}$$

1(+7 → +4) = -3 $\quad$ 3 electrons gained in the reduction

3. Use coefficients to make the number of electrons lost equal the number of electrons gained. The number of electrons transferred is **6**.

3[1(-3 → -1)] = +6 $\quad$ 6 electrons lost in the oxidation

$$3\overset{+2\,-3}{CN^-} + 2\overset{+7\,-2}{MnO_4^-} \rightarrow 3\overset{+2\,-1\,-2}{CNO^-} + 2\overset{+4\,-2}{MnO_2}$$

2[1(+7 → +4)] = -6 $\quad$ 6 electrons gained in the reduction

4. Balance the remaining elements by inspection.
First, add H_2O to balance O.

$3CN^- + 2MnO_4^- \rightarrow 3CNO^- + 2MnO_2 + H_2O$

Complete the process by adding 2 H^+ to the left side to balance the H. This also balances the charge (-3 on each side).

$2H^+ + 3CN^- + 2MnO_4^- \rightarrow 3CNO^- + 2MnO_2(s) + H_2O$

Convert to basic by adding 2 OH^- to each side.

$2H_2O + 3CN^- + 2MnO_4^- \rightarrow 3CNO^- + 2MnO_2(s) + H_2O + 2OH^-$

Finally, subtract 1 H_2O from each side to delete like terms.

$$\mathbf{H_2O(\ell) + 3CN^-(aq) + 2MnO_4^-(aq) \rightarrow 3CNO^-(aq) + 2MnO_2(s) + 2OH^-(aq)}$$

19 Electrochemistry

19.1 Define electrochemistry. How does its meaning differ from that of redox chemistry?

All electrochemical reactions are redox reactions. In electrochemistry, electricity is either used to drive the reaction or is produced by the reaction. In many redox reactions, electrons are directly transferred between species and no electricity is involved.

19.3 What is the smallest charge found in nature and where is it found?

The smallest charge is $1.6021773 \times 10^{-19}$ C. It is the magnitude of the negative charge found on the electron and the positive charge found on the proton.

19.5 What are the units of electrical potential and work, and how is potential related to work?

The unit of electrical potential is the volt (V), which is equal to a joule per coulomb. The unit of work is the joule (J). 1 J $\leftrightarrow$ 1 V•C.

19.7 What differences, if any exist between the mechanisms of conduction in dissolved sodium chloride and molten sodium chloride?

There is no major difference in the mechanism because, in both cases, ions carry the current. In aqueous sodium chloride, the sodium and chloride ions are surrounded by water molecules.

19.9 What is an electrode process?

An electrode process is a reaction that happens at an electrode surface. Oxidation will occur at one electrode (the anode) and reduction will occur at the other electrode (the cathode).

19.11 What is an electrochemical cell? What are the two types of cell?

An electrochemical cell is any device in which electrochemical reactions occur. In an electrolytic cell, electricity is converted to chemical energy. In a voltaic (galvanic) cell, chemical energy is converted to electrical energy.

19.13 How can the number of electrons transferred in a redox reaction be determined?

The number of electrons transferred in a redox reaction can be determined by monitoring the time of current flow, the magnitude of the current, and the amount of reactant consumed or product made. The current and time measurements enable us to determine the total amount of charge in faradays (1 $\mathscr{F} \leftrightarrow$ 1 mol e^-). The measurements of the amount of reactant or product allow us to determine the number of moles of that species. The number of moles of electrons divided by the number of moles of reactant or product gives the number of moles of electrons transferred in that half-reaction.

19.15 Describe in detail the meaning of the notation Zn(s)| Zn^{2+}(0.3 M) || Cl^- (0.1 M) | AgCl(s) | Ag(s).

This stands for a voltaic cell in which a zinc electrode is immersed in a 0.3 M solution of a zinc salt in one half-cell. The other half-cell consists of a 0.1 M solution of a chloride salt in contact with solid AgCl which coats a silver electrode. The two half-cells are connected with a salt bridge or other porous barrier.

19.17 What is the relationship between $\mathscr{E}_{red}$ and $\mathscr{E}_{ox}$ for a given voltaic cell?

$\mathscr{E}_{red} = -\mathscr{E}_{ox}$

19.19 How is SRP defined and measured?

The SRP is the standard reduction potential of a given half-reaction. It is determined by connecting the cell under standard conditions to the standard hydrogen electrode (SHE) and monitoring the voltage. If the cell of interest functions as the cathode (is the reduction), the measured voltage is its SRP. If the cell of interest functions as the anode (is the oxidation), the SRP is the negative of the measured voltage.

19.21 What is the strongest commonly-occurring oxidant? The weakest?

The strongest commonly-occurring oxidant is $F_2(g)$. The weakest commonly-occurring oxidant is $Li^+(aq)$. This means that Li(s) is the strongest commonly-occurring reductant.

19.23 What are the relationships among cell potential, free energy of reaction, and spontaneity of cell reaction?

$\Delta G = -n\mathscr{F}\mathscr{E}$. ΔG is the free energy, n is the number of moles of electrons transferred in the process, $\mathscr{F}$ is Faraday's constant and $\mathscr{E}$ is the cell potential. If ΔG is negative and, therefore, $\mathscr{E}$ is positive, the process is spontaneous.

19.25 Describe two commonly-used voltaic cells.

The two most common voltaic cells are the lead storage battery, used in cars, boats and other vehicles, and the dry cell, used in flashlights, radios, etc. The lead storage battery consists of six cells, each with one spongy lead electrode (the anode) and another lead electrode coated with PbO_2 (the cathode). Each electrode produces about 2 V, for an overall battery potential of 12 V. The electrodes are immersed in concentrated aqueous sulfuric acid. The dry cell consists of a zinc anode which also serves as the battery container and a graphite cathode surrounded by a moist paste of NH_4Cl and MnO_2. The dry cell develops a cell potential of 1.5 V. See the answers to Problem 19.85a and 19.85b for the reactions occurring in the cells.

19.27 What is a faraday? How is it related to Avogadro's number?

A faraday is the charge carried by one mole of electrons. It is equal to Avogadro's number (6.022137×10^{23}) times the charge on an electron ($1.6021773 \times 10^{-19}$ C) and, therefore, has a value of 96,485.31 C mol^{-1}.

19.29 Calculate the total charge passing through a circuit if a current of 0.25 A flows for 20 seconds. How much work is done by this amount of charge if the potential difference is 10 V?

Analysis/Plan: Because the current and time are known, these questions can be solved by the factor-label method, using the equivalence statements 1 C ↔ 1 A•s and 1 J ↔ 1 V•C. The unit of charge is the coulomb (C) and the unit of work is the joule (J).

Work:

$$? \text{ C} = 0.25 \text{ A} \times \frac{1 \text{ C}}{1 \text{ A}\cdot\text{s}} \times 20 \text{ s} = \mathbf{5.0\ C}$$

$$? \text{ J} = 10 \text{ V} \times \frac{1 \text{ J}}{1 \text{ V}\cdot\text{C}} \times 5.0 \text{ C} = \mathbf{5.0 \times 10^1\ J}$$

19.31 What length of time must elapse for a current of 75 mA to deliver 15 coulombs of charge?

Analysis/Plan: Because the current and charge are known, this question is solved by the factor-label method, using the equivalence statement 1 C ↔ 1 A•s. The equivalence statement 1 mA = 10^{-3} A is also needed.

Work:

$$? \text{ s} = 15 \text{ C} \times \frac{1 \text{ A}\cdot\text{s}}{1 \text{ C}} \times \frac{1 \text{ mA}}{10^{-3} \text{ A}} \times \frac{1}{75 \text{ mA}} = \mathbf{2.0 \times 10^2 \text{ s}}$$

19.33 The watt (W) is the unit that measures electrical power or the rate of electrical work. The familiar measure of electrical work, used by the electrical power industry, is the ***kilowatt-hour***. The relationship, in SI units, is 1 watt•second ↔ 1 joule. How much work is done when a 60-watt light bulb operates for one hour?

Analysis/Plan: This problem can be solved by the factor-label method. Additional equivalence statements required are 60 s ↔ 1 min and 60 min ↔ 1 hr.

Work:

$$? \text{ J} = 60 \text{ watt} \times \frac{1 \text{ J}}{1 \text{ watt}\cdot\text{s}} \times \frac{60 \text{ s}}{1 \text{ min}} \times \frac{60 \text{ min}}{1 \text{ hr}} \times 1.00 \text{ hr} = \mathbf{2.2 \times 10^5 \text{ J}}$$

19.35 A portable electric heater is rated at 1.3 kilowatts. If the voltage is 117 V, what is the current? How much charge (expressed in faradays) passes through the circuit in three hours?

Analysis/Plan: Both parts of this question can be solved by the factor-label method. Necessary equivalence statements are: 1 watt•1 second ↔ 1 joule, 10^3 watt ↔ 1 kwatt, 60 s ↔ 1 min, 60 min ↔ 1 hr, 1 A•s ↔ 1 C, 1 J ↔ 1 V•C and 1 $\mathcal{F}$ ↔ 96,485.31 C.

Work:

$$? \text{ A} = 1.3 \text{ kwatt} \times \frac{10^3 \text{ watt}}{1 \text{ kwatt}} \times \frac{1 \text{ J}}{1 \text{ watt}\cdot\text{s}} \times \frac{1 \text{ V}\cdot\text{C}}{1 \text{ J}} \times \frac{1}{117 \text{ V}} \times \frac{1 \text{ A}\cdot\text{s}}{\text{C}} = \mathbf{11 \text{ A}}$$

$$? \mathcal{F} = 11 \text{ A} \times \frac{1 \text{ C}}{1 \text{ A}\cdot\text{s}} \times \frac{60 \text{ s}}{1 \text{ min}} \times \frac{60 \text{ min}}{1 \text{ hr}} \times 3.00 \text{ hr} \times \frac{1 \mathcal{F}}{96485.31 \text{ C}} = \mathbf{1.2 \mathcal{F}}$$

19.37 Using shorthand notation, describe the cell in Figure 19.9.

Analysis/Plan: Assign oxidation numbers or write the half-reactions to determine which species is oxidized and which is reduced. Remember OIL, oxidation is on the left in the shorthand. A | is used to separate phases and a comma separates two dissolved species in the same half-cell. A || represents a porous barrier, needed when the two half-cells have different dissolved species. Aqueous substances are shown in the middle and solids on the outside. A Pt electrode is used when there is no solid involved in oxidation or reduction.

Work:

In Figure 19.9, the zinc is oxidized from Zn^0 to Zn^{2+}, so is placed on the left. The copper is reduced from Cu^{2+} to Cu, so is placed on the right A porous barrier (salt bridge) is shown in the picture. The metals serve as their own electrodes. The shorthand is:

Zn(s) | Zn^{2+}(aq) || Cu^{2+}(aq) | Cu(s).

19.39 Make pictorial sketches, similar to Figure 19.9, of each of the following cells.

a. Sn(s) | Sn^{2+}(aq) || Ag^+(aq) | Ag(s)

b. Zn(s) | Zn^{2+}(aq) || Fe^{3+}(aq),Fe^{2+}(aq) | Pt

Analysis/Plan: Show the oxidation on the left and the reduction on the right, using the appropriate metals for electrodes and showing the formulas of the dissolved species.

Work:

a.

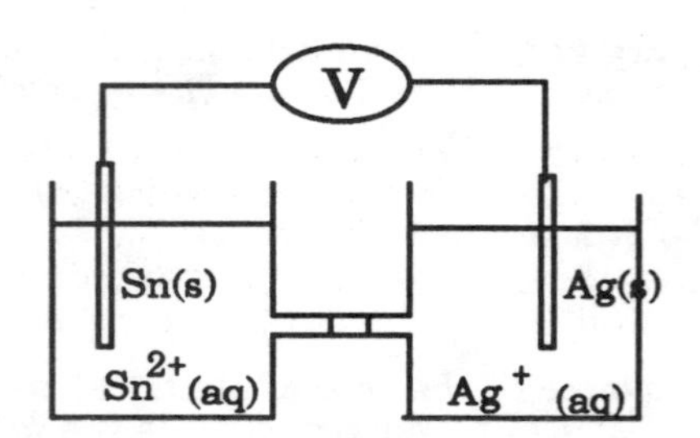

b.

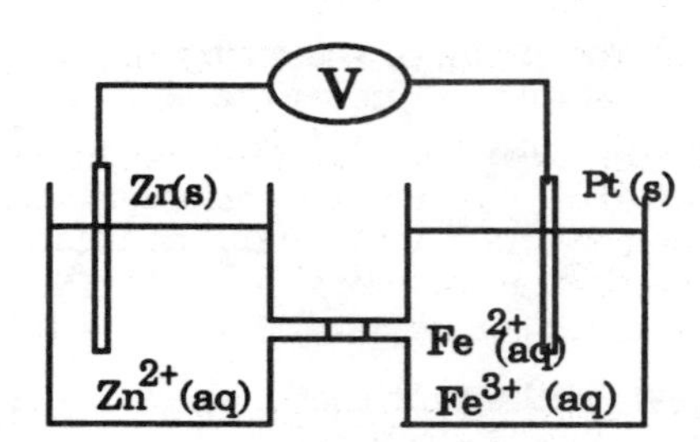

19.41 Write the anode and cathode half-reactions for the cell in Problem 19.37.

Analysis/Plan: Determine the half-reaction by adding electrons to balance the charge. In oxidation, which occurs at the anode, oxidation number increases and electrons are a product. In reduction, which occurs at the cathode, oxidation number decreases and electrons are a reactant.

Work:

Zinc is oxidized. The anode reaction is: $\mathbf{Zn(s) \longrightarrow Zn^{2+}(aq) + 2\ e^-}$.

Copper is reduced. The cathode reaction is: $\mathbf{Cu^{2+}(aq) + 2\ e^- \longrightarrow Cu(s)}$.

19.43 Write the overall cell reaction for each of the cells in Problem 19.39. What is the value of n in each case?

Analysis/Plan: Determine the half-reaction by adding electrons to balance the charge. In oxidation, oxidation number increases and electrons are a product. In reduction, oxidation number decreases and electrons are a reactant. Choose coefficients to make the number of electrons lost equal to the number of electrons gained. Finally, add the equations, deleting or combining like terms.

Work:

a. The oxidation is $Sn(s) \longrightarrow Sn^{2+}(aq) + 2\ e^-$. The reduction is $Ag^+(aq) + 1\ e^- \longrightarrow Ag(s)$.

The lowest common multiple is 2, so n = **2**. Multiply the silver half-reaction by 2 and add, deleting the like terms (only the electrons).

$Sn(s) \longrightarrow Sn^{2+}(aq) + 2\ e^-$

$2[Ag^+(aq) + 1\ e^- \longrightarrow 1\ Ag(s)]$.

$\mathbf{Sn(s) + 2\ Ag^+(aq) \longrightarrow Sn^{2+}(aq) + 2\ Ag(s)}$

b. The oxidation is $Zn(s) \longrightarrow Zn^{2+}(aq) + 2\ e^-$. The reduction is $Fe^{3+}(aq) + 1\ e^- \longrightarrow Fe^{2+}$.

The lowest common multiple is 2, so n = **2**. Multiply the iron half-reaction by 2 and add, deleting the like terms (only the electrons).

$Zn(s) \longrightarrow Zn^{2+}(aq) + 2\ e^-$

$2[Fe^{3+}(aq) + 1\ e^- \longrightarrow Fe^{2+}(aq)]$

$\mathbf{Zn(s) + 2\ Fe^{3+}(aq) \longrightarrow Zn^{2+}(aq) + 2\ Fe^{2+}(aq)}$

19.45 Which of the following reactions occur(s) spontaneously?

a. $2\ MnO_4^-(aq) + 5\ Hg_2^{2+}(aq) + 16\ H^+(aq) \longrightarrow 2\ Mn^{2+}(aq) + 10\ Hg^{2+}(aq) + 8\ H_2O(\ell)$

b. $NiCl_2(aq) + Pb(s) \longrightarrow PbCl_2(g) + Ni(s)$

c. $KClO_4(g) + 2\ HCl(aq) \longrightarrow KClO_3(aq) + Cl_2(g) + H_2O(\ell)$

Analysis/Plan: Write the net ionic equations if necessary, then look up the standard reduction potentials (SRP) of the appropriate species in Appendix I. Determine which reaction is proceeding as the oxidation and determine $\mathscr{E}°_{ox}$ from the relationship $\mathscr{E}°_{ox} = -\mathscr{E}°_{red}$. Calculate the cell potential from $\mathscr{E}°_{cell} = \mathscr{E}°_{red} + \mathscr{E}°_{ox}$. If the $\mathscr{E}°_{cell}$ is negative, the reaction is not spontaneous, but if the $\mathscr{E}°_{cell}$ is positive, the reaction is spontaneous.

Work:

a. The applicable standard reduction potentials, $\mathscr{E}°_{red}$, are:

$8\ H^+(aq) + MnO_4^-(aq) + 5\ e^- \longrightarrow Mn^{2+}(aq) + 4\ H_2O(\ell)$ $\quad \mathscr{E}°_{red} = +1.51$ V

$2\ Hg^{2+}(aq) + 2\ e^- \longrightarrow Hg_2^{2+}(aq)$ $\quad \mathscr{E}°_{red} = +0.920$ V

In the reaction of interest, the mercury half-reaction is the reverse of this and is the oxidation. Therefore, $\mathscr{E}°_{ox} = -0.920$ V.

$\mathscr{E}°_{cell} = \mathscr{E}°_{red} + \mathscr{E}°_{ox} = 1.51\text{ V} + (-0.920\text{ V}) = +0.59\text{ V}$

Because $\mathscr{E}°_{cell} > 0$, **this process is spontaneous**.

b. Both chloride compounds are soluble ionic salts. Cl^- is a spectator ion.

The net ionic equation is: $Ni^{2+}(aq) + Pb(s) \longrightarrow Pb^{2+}(aq) + Ni(s)$.

The applicable standard reduction potentials, $\mathscr{E}°_{red}$, are:

$Ni^{2+}(aq) + 2\ e^- \longrightarrow Ni(s)$ $\quad \mathscr{E}°_{red} = -0.25$ V

$Pb^{2+}(aq) + 2\ e^- \longrightarrow Pb(s)$ $\quad \mathscr{E}°_{red} = -0.126$ V

In the reaction of interest, the lead half-reaction is the reverse of this and is the oxidation. Therefore, $\mathscr{E}°_{ox} = +0.126$ V.

$\mathscr{E}°_{cell} = \mathscr{E}°_{red} + \mathscr{E}°_{ox} = -0.25\text{ V} + (0.126\text{ V}) = -0.12\text{ V}$

Because $\mathscr{E}°_{cell} < 0$, **this process is not spontaneous**.

c. Both potassium compounds are soluble ionic salts. K^+ is a spectator ion.

The net ionic equation is: $ClO_4^-(aq) + 2\ H^+(aq) + Cl^-(aq) \longrightarrow ClO_3^-(aq) + Cl_2(g) + H_2O(\ell)$.

The applicable standard reduction potentials, $\mathscr{E}°_{red}$, are:

$ClO_4^-(aq) + 2\ H^+(aq) + 2\ e^- \longrightarrow ClO_3^-(aq) + H_2O(\ell)$ $\quad \mathscr{E}°_{red} = +1.19$ V

$Cl_2(g) + 2\ e^- \longrightarrow 2\ Cl^-(aq)$ $\quad \mathscr{E}°_{red} = +1.358$ V

In the reaction of interest, the $Cl_2(g)$ reaction is the reverse of this and is the oxidation. Therefore, $\mathscr{E}°_{ox} = -1.358$ V.

$\mathscr{E}°_{cell} = \mathscr{E}°_{red} + \mathscr{E}°_{ox} = 1.19\text{ V} + (-1.358\text{ V}) = -0.17\text{ V}$

Because $\mathscr{E}°_{cell} < 0$, **this process is not spontaneous**.

19.47 Write the equation for the reaction that takes place when the following reagents are mixed. If no reaction takes place, write "NR". Assume standard conditions.

a. $NaF(aq)$, $NaCl(aq)$ b. $Cl_2(g)$, $KBr(aq)$ c. $I_2(s)$, $KBr(aq)$

d. $AgBr(s)$, $Ni(s)$, $H_2O(\ell)$ e. $Cu(s)$, $H^+(aq)$

Analysis/Plan: Write the possible reactions, using coefficients to make the number of electrons lost equal to the number of electrons gained. Look up the standard reduction potentials (SRP) of the appropriate species in Appendix I. Determine which reaction is proceeding as the oxidation and determine $\mathscr{E}°_{ox}$ from the relationship $\mathscr{E}°_{ox} = -\mathscr{E}°_{red}$. Calculate the cell potential from $\mathscr{E}°_{cell} = \mathscr{E}°_{red} + \mathscr{E}°_{ox}$. If the $\mathscr{E}°_{cell}$ is negative, the reaction is not spontaneous, but if the $\mathscr{E}°_{cell}$ is positive, the reaction is spontaneous.

Work:

a. The ions present are Na^+, Cl^- and Br^-. Cl^- and Br^- are in their lowest oxidation states. Because neither can be further reduced, they cannot react with each other. Na^+ can be reduced to Na; however, if either Cl^- or Br^- could reduce Na^+ to Na, the salts NaCl and NaBr would not exist. Therefore, there is no reaction (**NR**).

b. The species present are K^+, Cl_2 and Br^-. K^+ can be reduced to K; however, if Br^- could reduce K^+ to K, the salt KBr would not exist. K^+ is a spectator ion. Br^- can be oxidized to Br_2, and Cl_2 can be reduced to Cl^-. The possible reaction is:

$$Cl_2(g) + 2\ KBr(aq) \longrightarrow 2\ KCl(aq) + Br_2(\ell)$$

The applicable standard reduction potentials, $\mathcal{E}°_{red}$, are:

$Cl_2(g) + 2\,e^- \longrightarrow 2\ Cl^-(aq)$ $\qquad \mathcal{E}°_{red} = +1.358$ V

$Br_2(\ell) + 2\,e^- \longrightarrow 2\ Br^-(aq)$ $\qquad \mathcal{E}°_{red} = +1.066$ V

In the possible reaction, the bromine half-reaction is the reverse of this and is the oxidation. Therefore, $\mathcal{E}°_{ox} = -1.066$ V.

$$\mathcal{E}°_{cell} = \mathcal{E}°_{red} + \mathcal{E}°_{ox} = +1.358\text{ V} - 1.066\text{ V} = +0.292\text{ V}$$

Because $\mathcal{E}°_{cell} > 0$, this process is spontaneous.

The spontaneous reaction is: $\mathbf{Cl_2(g) + 2\ KBr(aq) \longrightarrow 2\ KCl(aq) + Br_2(\ell)}$.

c. The species present are K^+, I_2 and Br^-. K^+ can be reduced to K; however, if Br^- could reduce K^+ to K, the salt KBr would not exist. Br^- can be oxidized to Br_2, and I_2 can be reduced to I^-. The possible reaction is:

$$I_2(g) + 2\ Br^-(aq) \longrightarrow 2\ I^-(aq) + Br_2(\ell)\,.$$

The applicable standard reduction potentials, $\mathcal{E}°_{red}$, are:

$I_2(g) + 2\,e^- \longrightarrow 2\ I^-(aq)$ $\qquad \mathcal{E}°_{red} = +0.535$ V

$Br_2(\ell) + 2\,e^- \longrightarrow 2\ Br^-(aq)$ $\qquad \mathcal{E}°_{red} = +1.066$ V

In the possible reaction, the bromine half-reaction is the reverse of this and is the oxidation. Therefore, $\mathcal{E}°_{ox} = -1.066$ V.

$$\mathcal{E}°_{cell} = \mathcal{E}°_{red} + \mathcal{E}°_{ox} = 0.535\text{ V} - 1.066\text{ V} = -0.531\text{ V}$$

Because $\mathcal{E}°_{cell} < 0$, this process is not spontaneous. Therefore, there is no reaction (**NR**).

d. Assume that water only provides the medium to dissolve the species and, therefore, is neither oxidized or reduced. (This assumption is valid.) The potential redox species are then AgBr which could be reduced to Ag and Br^-, and Ni, which could be oxidized to Ni^{2+}. The possible reaction is:

$$2\ AgBr(s) + Ni(s) \longrightarrow 2\ Ag(s) + 2\ Br^-(aq) + Ni^{2+}(aq)$$

The applicable standard reduction potentials, $\mathcal{E}°_{red}$, are:

$AgBr(s) + 1\,e^- \longrightarrow Ag(s) + Br^-(aq)$ $\qquad \mathcal{E}°_{red} = +0.0713$ V

$Ni^{2+}(aq) + 2\,e^- \longrightarrow Ni(s)$ $\qquad \mathcal{E}°_{red} = -0.25$ V

In the possible reaction, the nickel half-reaction is the reverse of this and is the oxidation. Therefore, $\mathcal{E}°_{ox} = -(-0.25\text{ V}) = +0.25$ V.

$$\mathcal{E}°_{cell} = \mathcal{E}°_{red} + \mathcal{E}°_{ox} = 0.0713\text{ V} + 0.25\text{ V} = +0.32\text{ V}$$

Because $\mathcal{E}°_{cell} > 0$, this process is spontaneous. The reaction is:

$$\mathbf{2\ AgBr(s) + Ni(s) \longrightarrow 2\ Ag(s) + 2\ Br^-(aq) + Ni^{2+}(aq)}$$

e. The species present are Cu, which could be oxidized to Cu^+ or Cu^{2+}, and H^+, which could be reduced to H_2. The possible reactions are:

$2\ Cu(s) + 2\ H^+(aq) \longrightarrow 2\ Cu^+(aq) + H_2(g)$ and

$Cu(s) + 2\ H^+(aq) \longrightarrow Cu^{2+}(aq) + H_2(g)$

The applicable standard reduction potentials, $\mathscr{E}°_{red}$, are:

$2\ H^+(aq) + 2\ e^- \longrightarrow H_2(g)$ $\qquad \mathscr{E}°_{red} = +0.0000$ V (defined)

$Cu^+(aq) + e^- \longrightarrow Cu(s)$ $\qquad \mathscr{E}°_{red} = +0.521$ V

$Cu^{2+}(aq) + 2\ e^- \longrightarrow Cu(s)$ $\qquad \mathscr{E}°_{red} = +0.337$ V

In the first possible reaction, the copper half-reaction is the reverse of this and is the oxidation. Therefore, $\mathscr{E}°_{ox} = -(0.521\ V) = -0.521$ V.

$\mathscr{E}°_{cell} = \mathscr{E}°_{red} + \mathscr{E}°_{ox} = 0.0000\ V + (-0.521\ V) = -0.521$ V

Because $\mathscr{E}°_{cell} < 0$, this process is not spontaneous.

In the second possible reaction, the copper half-reaction is the reverse of this and is the oxidation. Therefore, $\mathscr{E}°_{ox} = -(0.337V) = -0.337$ V.

$\mathscr{E}°_{cell} = \mathscr{E}°_{red} + \mathscr{E}°_{ox} = 0.0000\ V + (-0.337\ V) = -0.337V$

Because $\mathscr{E}°_{cell} < 0$, this process is not spontaneous.

Because neither possible process is spontaneous, there is no reaction (NR).

19.49 Write the equation for any reaction that occurs in the following situations. If no reaction takes place, write "NR."

a. A strip of tin is immersed in $Cu(NO_3)_2(aq)$.

b. A strip of zinc is immersed in $Zn(NO_3)_2(aq)$.

c. Gaseous oxygen is bubbled into water.

Analysis/Plan: Write the possible reactions, using coefficients to make the number of electrons lost equal to the number of electrons gained. Look up the standard reduction potentials (SRP) of the appropriate species in Appendix I. Determine which reaction is proceeding as the oxidation and determine $\mathscr{E}°_{ox}$ from the relationship $\mathscr{E}°_{ox} = -\mathscr{E}°_{red}$. Calculate the cell potential from $\mathscr{E}°_{cell} = \mathscr{E}°_{red} + \mathscr{E}°_{ox}$. If the $\mathscr{E}°_{cell}$ is negative, the reaction is not spontaneous, but if the $\mathscr{E}°_{cell}$ is positive, the reaction is spontaneous.

Work:

a. Assume that the nitrate ion is a spectator ion, because all its reduction reactions require the presence of an acid, H^+. Then, the other species present are Cu^{2+}, which could be reduced to Cu, and Sn, which could be oxidized to Sn^{2+}. The possible reaction is:

$Cu^{2+}(aq) + Sn(s) \longrightarrow Cu(s) + Sn^{2+}(aq)$.

The applicable standard reduction potentials, $\mathscr{E}°_{red}$, are:

$Sn^{2+}(aq) + 2\ e^- \longrightarrow Sn(s)$ $\qquad \mathscr{E}°_{red} = -0.14$ V

$Cu^{2+}(aq) + 2\ e^- \longrightarrow Cu(s)$ $\qquad \mathscr{E}°_{red} = +0.337$ V

In the possible reaction, the tin half-reaction is the reverse of this and is the oxidation. Therefore, $\mathscr{E}°_{ox} = -(-0.14\ V) = +0.14$ V.

$\mathscr{E}°_{cell} = \mathscr{E}°_{red} + \mathscr{E}°_{ox} = 0.337\ V + (0.14\ V) = +0.48$ V

Because $\mathscr{E}°_{cell} > 0$, this process is spontaneous.

The reaction which occurs is: $\mathbf{Cu^{2+}(aq) + Sn(s) \longrightarrow Cu(s) + Sn^{2+}(aq)}$.

b. Assume that the nitrate ion is a spectator ion, because all its reduction reactions require the presence of an acid, H^+. Both oxidation states of zinc are present. If Zn is oxidized and Zn^{2+} is reduced, there is no change in the relative amounts of these species, therefore, there is no reaction (NR).

c. O_2 cannot be oxidized except by F_2, so its only possible reaction is reduction. The only available species to be oxidized is water, which could be oxidized to H_2O_2. Possible half-reactions are:

$O_2(g) + 2\ H_2O(\ell) + 4\ e^- \longrightarrow 4\ OH^-(aq)$ $\mathscr{E}°_{red} = +1.229$ V and

$H_2O_2(aq) + 2\ H^+(aq) + 2\ e^- \longrightarrow 2\ H_2O(\ell)$ $\mathscr{E}°_{red} = +1.77$ V

If the second reaction is reversed, becoming the oxidation, $\mathscr{E}°_{ox} = -1.77$ V. The possible reaction would have $\mathscr{E}°_{cell} = \mathscr{E}°_{red} + \mathscr{E}°_{ox} = 1.229\ V + (-1.77V) = -0.54$ V. Because $\mathscr{E}°_{cell} < 0$, there is no reaction (**NR**).

19.51 Use the position of the reduction half-reaction in Table 19.2 to determine which is the better oxidizing agent (under standard conditions) in each of the following pairs.

a. Br_2, Cl_2 b. $Pb^{2+}(aq)$, $Cd^{2+}(aq)$ c. $KMnO_4(aq)$, $Na_2Cr_2O_7(aq)$
d. $CuCl(aq)$, $CuCl_2(aq)$ e. $Zn^{2+}(aq)$, $Al^{3+}(aq)$ f. $AgCl(s)$, $AgBr(s)$

Analysis/Plan: Eliminate any spectator ions from consideration. The better oxidizing agent is closer to the top left on the SRP table. It is more easily reduced and, therefore, has a higher reduction potential.

Work:

a. $\mathbf{Cl_2}$ ($\mathscr{E}°_{red} = +1.358$ V) is a better oxidizing agent than Br_2 ($\mathscr{E}°_{red} = +1.066$ V).
b. $\mathbf{Pb^{2+}}$ ($\mathscr{E}°_{red} = -0.126$ V) is a better oxidizing agent than Cd^{2+} ($\mathscr{E}°_{red} = -0.430$ V).
c. $\mathbf{KMnO_4}$ (MnO_4^-, $\mathscr{E}°_{red} = +1.51$ V) is a better oxidizing agent than $Na_2Cr_2O_7$ ($Cr_2O_7^{2-}$, $\mathscr{E}°_{red} = +1.33$ V). K^+ and Na^+ are spectator ions.
d. **CuCl** (Cu^+, $\mathscr{E}°_{red} = +0.521$ V) is a better oxidizing agent than $CuCl_2$ (Cu^{2+}, $\mathscr{E}°_{red} = +0.337$ V). Cl^- is a spectator ion.
e. $\mathbf{Zn^{2+}}$ ($\mathscr{E}°_{red} = -0.763$ V) is a better oxidizing agent than Al^{3+} ($\mathscr{E}°_{red} = -1.66$ V).
f. **AgCl(s)** ($\mathscr{E}°_{red} = 0.222$ V) is a better oxidizing agent than AgBr ($\mathscr{E}°_{red} = -0.0713$ V).

19.53 Use the position of the reduction half-reaction in the table of SRP's to decide which is the better reducing agent (under standard conditions) in each of the following pairs.

a. Pb(s), Ni(s) b. $Fe^{2+}(aq)$, $Hg_2^{2+}(aq)$ c. Ag(s), Sn(s)
d. $HNO_2(aq)$, $H_2S(aq)$ e. $CrCl_3(aq)$, $MnSO_4(aq)$ f. Fe(s), $H_2(g)$

Analysis/Plan: The better reducing agent is closer to the bottom right on the SRP table. It is more easily oxidized and, therefore, its has a higher oxidation potential. Remember that $\mathscr{E}°_{ox} = -\mathscr{E}°_{red}$.

Work:

a. **Ni** ($\mathscr{E}°_{ox} = +0.25$ V) is a better reducing agent than Pb($\mathscr{E}°_{ox} = +0.126$V).
b. $\mathbf{Fe^{2+}}$ ($\mathscr{E}°_{ox} = -0.771$ V) is a better reducing agent than Hg_2^{2+} ($\mathscr{E}°_{ox} = -0.920$ V).
c. **Sn** ($\mathscr{E}°_{ox} = +0.14$) is a better reducing agent than Ag ($\mathscr{E}°_{ox} = -0.7994$V).
d. $\mathbf{H_2S}$ ($\mathscr{E}°_{ox} = -0.14$ V) is a better reducing agent than HNO_2 (NO_2^-, $\mathscr{E}°_{ox} = -0.94$V).
e. $\mathbf{CrCl_3}$ (Cr^{3+}, $\mathscr{E}°_{ox} = -1.33$ V) is a better reducing agent than $MnSO_4$ (Mn^{2+}, $\mathscr{E}°_{ox} = -1.51$ V).
f. **Fe** ($Fe \longrightarrow Fe^{2+}$, $\mathscr{E}°_{ox} = +0.44$ V) is a better reducing agent than H_2 ($\mathscr{E}°_{ox} = 0.0000$ V).

19.55 In each of the following pairs, identify the species that is the more easily reduced under standard conditions.

a. Cu^+, Cu^{2+} b. Hg^{2+}, Fe^{3+} c. Al^{3+}, Fe^{3+} d. ClO_4^-, $Cr_2O_7^{2-}$

Analysis/Plan: A substance which is more easily reduced is closer to the top left on the SRP table and, therefore, has a higher reduction potential.

Work:

a. $\mathbf{Cu^+}$ ($\mathscr{E}°_{red} = +0.521$ V) is more easily reduced than Cu^{2+} ($\mathscr{E}°_{red} = +0.337$ V).
b. $\mathbf{Hg^{2+}}$ ($\mathscr{E}°_{red} = +0.920$ V or 0.855 V) is more easily reduced than Fe^{3+} ($\mathscr{E}°_{red} = +0.771$ V).
c. $\mathbf{Fe^{3+}}$ ($\mathscr{E}°_{red} = +0.771$ V) is more easily reduced than Al^{3+} ($\mathscr{E}°_{red} = -1.66$ V).
d. $\mathbf{Cr_2O_7^{2-}}$ ($\mathscr{E}°_{red} = +1.33$ V) is more easily reduced than ClO_4^- ($\mathscr{E}°_{red} = +1.19$ V).

19.57 Calculate the cell potential and identify the anode of the cell in Figure 19.9.

Analysis/Plan: The nature of the possible reactants and products is given in the figures. The spontaneous process must have $\mathscr{E}^\circ_{cell} > 0$. Look up the two appropriate $\mathscr{E}^\circ_{red}$ in Appendix I. The half-reaction with the more positive $\mathscr{E}^\circ_{red}$ will be the reduction and function as the cathode. The half-reaction with the more negative $\mathscr{E}^\circ_{red}$ will reverse, becoming the oxidation, and will function as the anode. The cell potential can be found from the two relationships $\mathscr{E}^\circ_{ox} = -\mathscr{E}^\circ_{red}$ and $\mathscr{E}^\circ_{cell} = \mathscr{E}^\circ_{red} + \mathscr{E}^\circ_{ox}$.

Work:

The two appropriate half-reactions and SRP's are:

$Zn^{2+}(aq) + 2\,e^- \longrightarrow Zn(s)$ $\qquad \mathscr{E}^\circ_{red} = -0.763$ V

$Cu^{2+}(aq) + 2\,e^- \longrightarrow Cu(s)$ $\qquad \mathscr{E}^\circ_{red} = +0.337$ V

The zinc reaction reverses and becomes the oxidation. **Zn** is the anode.

$\mathscr{E}^\circ_{ox} = -(-0.763\text{ V}) = +0.763$ V

$\mathscr{E}^\circ_{cell} = \mathscr{E}^\circ_{red} + \mathscr{E}^\circ_{ox} = +0.337\text{ V} + 0.763\text{ V} = \mathbf{1.100\text{ V}}$

19.59 Calculate the standard potential of the cells consisting of the following electrode pairs. When the cell operates spontaneously under standard conditions, what is the cell reaction? Describe the spontaneously-operating cell in shorthand notation, with the anode on the left.

a. Pt | $Hg^{2+}(aq)$, Hg_2^{2+}; Pt | $I_2(s)$ | $I^-(aq)$

b. $Cu^+(aq)$ | Cu(s); Pt | S(s) | $H_2S(aq)$

c. AgBr(s) | Ag(s); $Zn^{2+}(aq)$ | Zn(s)

d. $NO_3^-(aq)$, $HNO_2(aq)$ | Pt; $ClO_4^-(aq)$, $ClO_3^-(aq)$ | Pt

Analysis/Plan: The spontaneous process must have $\mathscr{E}^\circ_{cell} > 0$. Look up the two appropriate $\mathscr{E}^\circ_{red}$ in Appendix I. The half-reaction with the more positive $\mathscr{E}^\circ_{red}$ will be the reduction and function as the cathode. The half-reaction with the more negative $\mathscr{E}^\circ_{red}$ will reverse, becoming the oxidation, and will function as the anode. The cell potential can be found from the two relationships $\mathscr{E}^\circ_{ox} = -\mathscr{E}^\circ_{red}$ and $\mathscr{E}^\circ_{cell} = \mathscr{E}^\circ_{red} + \mathscr{E}^\circ_{ox}$. In the shorthand, dissolved species are shown in the center and electrodes on the outside. A | separates phases and a || is used for a porous barrier if needed.

Work:

a. The two appropriate half-reactions and SRP's are:

$2\,Hg^{2+}(aq) + 2\,e^- \longrightarrow Hg_2^{2+}(aq)$ $\qquad \mathscr{E}^\circ_{red} = +0.920$ V

$I_2(s) + 2\,e^- \longrightarrow 2\,I^-(aq)$ $\qquad \mathscr{E}^\circ_{red} = +0.535$ V

The iodine reaction reverses and becomes the oxidation at the anode. There are two electrons in each half-reaction, so neither reaction needs to be multiplied.

anode: $2\,I^-(aq) \longrightarrow I_2 + 2\,e^-$ $\qquad \mathscr{E}^\circ_{ox} = -(+0.535\text{V}) = -0.535$ V

cathode: $2\,Hg^{2+}(aq) + 2\,e^- \longrightarrow Hg_2^{2+}(aq)$ $\qquad \mathscr{E}^\circ_{red} = +0.920$ V

net: $\mathbf{2\,I^-(aq) + 2\,Hg^{2+}(aq) \longrightarrow I_2(s) + Hg_2^{2+}(aq)}$

$\mathscr{E}^\circ_{cell} = \mathscr{E}^\circ_{red} + \mathscr{E}^\circ_{ox} = +0.920\text{ V} + (-0.535\text{ V}) = \mathbf{0.385\text{ V}}$

Pt electrodes are needed for both cells. The shorthand for the process is:

Pt | $I_2(s)$ | $I^-(aq)$ || $Hg^{2+}(aq)$, $Hg_2^{2+}(aq)$ | Pt

b. The two appropriate half-reactions and SRP's are:

$Cu^+(aq) + 1\,e^- \longrightarrow Cu(s)$ $\qquad \mathscr{E}^\circ_{red} = +0.521$ V

$S(s) + 2\,H^+(aq) + 2\,e^- \longrightarrow H_2S(aq)$ $\qquad \mathscr{E}^\circ_{red} = +0.14$ V

The sulfur reaction reverses and becomes the oxidation at the anode. The first half-reaction needs to be multiplied by 2 to make the electrons equal.

anode:	$H_2S(aq) \longrightarrow S(s) + 2\,H^+(aq) + 2\,e^-$	$\mathscr{E}°_{ox} = -(+0.14) = -0.14$ V
cathode:	$2\,Cu^+(aq) + 2\,e^- \longrightarrow 2\,Cu(s)$	$\mathscr{E}°_{red} = +0.521$ V
net:	$\mathbf{2\,Cu^+(aq) + H_2S(aq) \longrightarrow S(s) + 2\,H^+(aq) + 2\,Cu(s)}$	

$\mathscr{E}°_{cell} = \mathscr{E}°_{red} + \mathscr{E}°_{ox} = +0.521\text{ V} + (-0.14\text{ V}) = \mathbf{0.38\text{ V}}$

A Pt electrode is needed for the sulfur cell. The shorthand for the process is:
Pt | S(s) | H_2S(aq) || Cu^+(aq) | Cu(s)

c. The two appropriate half-reactions and SRP's are:

$Zn^{2+}(aq) + 2\,e^- \longrightarrow Zn(s)$ $\mathscr{E}°_{red} = -0.763$ V

$AgBr(s) + 1\,e^- \longrightarrow Ag(s) + Br^-(aq)$ $\mathscr{E}°_{red} = +0.0713$ V

The zinc reaction reverses and becomes the oxidation at the anode. The second half-reaction needs to be multiplied by 2 to make the electrons equal.

anode:	$Zn(s) \longrightarrow Zn^{2+}(aq) + 2\,e^-$	$\mathscr{E}°_{ox} = -(-0.763\text{ V}) = +0.763$ V
cathode:	$2\,AgBr(s) + 2\,e^- \longrightarrow 2\,Ag(s) + 2\,Br^-(aq)$	$\mathscr{E}°_{red} = +0.0713$ V
net:	$\mathbf{Zn(s) + 2\,AgBr(s) \longrightarrow Zn^{2+}(aq) + 2\,Ag(s) + 2\,Br^-(aq)}$	

$\mathscr{E}°_{cell} = \mathscr{E}°_{red} + \mathscr{E}°_{ox} = +0.0713\text{ V} + (+0.763\text{ V}) = \mathbf{0.834\text{ V}}$

In each cell, the metal can serve as the electrode. The shorthand for the process is:
Zn(s) | Zn^{2+}(aq) || Br^-(aq) | AgBr(s) | Ag(s)

d. The two appropriate half-reactions and SRP's are:

$NO_3^-(aq) + 3\,H^+(aq) + 2\,e^- \longrightarrow HNO_2(aq) + H_2O(\ell)$ $\mathscr{E}°_{red} = +0.94$ V

$ClO_4^-(aq) + 2\,H^+(aq) + 2\,e^- \longrightarrow ClO_3^-(aq) + H_2O(\ell)$ $\mathscr{E}°_{red} = +1.19$V

The nitrate reaction reverses and becomes the oxidation at the anode. Neither reaction needs to be multiplied because 2 electrons are involved in each half-reaction. Delete like terms after adding.

anode:	$HNO_2(aq) + H_2O(\ell) \longrightarrow NO_3^-(aq) + 3\,H^+(aq) + 2\,e^-$	$\mathscr{E}°_{ox} = -(+0.94\text{ V}) = -0.94$ V
cathode:	$ClO_4^-(aq) + 2\,H^+(aq) + 2\,e^- \longrightarrow ClO_3^-(aq) + H_2O(\ell)$	$\mathscr{E}°_{red} = +1.19$V
net:	$HNO_2(aq) + ClO_4^-(aq) \longrightarrow NO_3^-(aq) + ClO_3^-(aq) + H^+(aq)$	

$\mathscr{E}°_{cell} = \mathscr{E}°_{red} + \mathscr{E}°_{ox} = +1.19\text{ V} + (-0.94\text{ V}) = \mathbf{0.25\text{ V}}$

Pt electrodes are needed for each cell. The shorthand for the process is:
Pt | NO_3^-(aq), HNO_2(aq) || ClO_4^-(aq), ClO_3^-(aq)| Pt

19.61 Calculate the standard potential of the following cells. When the cell operates spontaneously under standard conditions, what is the cell reaction? Does the cell diagram correctly identify the electrode on the left as the anode? If not, write the diagram correctly.

a. Zn(s) | Zn^{2+}(aq) || Sn^{2+}(aq) | Sn(s)

b. Ag(s) |Ag^+(aq) || Cd^{2+}(aq) | Cd(s)

c. Ag(s) | AgCl(s) | HCl(aq) |H_2(g) | Pt

d. Pt | H_2(g) | H^+(aq) || Cl^-(aq) | Cl_2(g) | Pt

Analysis/Plan: The spontaneous process must have $\mathscr{E}°_{cell} > 0$. Look up the two appropriate $\mathscr{E}°_{red}$ in Appendix I. The half-reaction with the more positive $\mathscr{E}°_{red}$ will be the reduction and function as the cathode. The half-reaction with the more negative $\mathscr{E}°_{red}$ will reverse, becoming the oxidation, and will function as the anode. The cell potential can be found from the two relationships $\mathscr{E}°_{ox} = -\mathscr{E}°_{red}$ and $\mathscr{E}°_{cell} = \mathscr{E}°_{red} + \mathscr{E}°_{ox}$. Compare the anode with the shorthand and rewrite the shorthand if it is incorrect.

Work:

a. The two appropriate half-reactions and SRP's are:

$Zn^{2+}(aq) + 2\,e^- \longrightarrow Zn(s)$ $\quad\mathscr{E}°_{red} = -0.763$ V

$Sn^{2+}(aq) + 2\,e^- \longrightarrow Sn(s)$ $\quad\mathscr{E}°_{red} = -0.14$ V

The zinc reaction reverses and becomes the oxidation. Neither reaction needs to be multiplied because 2 electrons are involved in each half-reaction.

anode: $Zn(s) \longrightarrow Zn^{2+}(aq) + 2\,e^-$ $\quad\mathscr{E}°_{ox} = -(-0.763V) = +0.763$ V

cathode: $Sn^{2+}(aq) + 2\,e^- \longrightarrow Sn(s)$ $\quad\mathscr{E}°_{red} = -0.14$ V

net: $\mathbf{Zn(s) + Sn^{2+}(aq) \longrightarrow Zn^{2+}(aq) + Sn(s)}$

$\mathscr{E}°_{cell} = \mathscr{E}°_{red} + \mathscr{E}°_{ox} = +0.763V + (-0.14\ V) = \mathbf{0.62\ V}$

The shorthand diagram is correct.

b. The two appropriate half-reactions and SRP's are:

$Ag^{+}(aq) + 1\,e^- \longrightarrow Ag(s)$ $\quad\mathscr{E}°_{red} = +0.7994$ V

$Cd^{2+}(aq) + 2\,e^- \longrightarrow Cd(s)$ $\quad\mathscr{E}°_{red} = -0.403$ V

The cadmium reaction reverses and becomes the oxidation. The silver reaction needs to be multiplied by two to make the electrons involved in each half-reaction equal.

anode: $Cd(s) \longrightarrow Cd^{2+}(aq) + 2\,e^-$ $\quad\mathscr{E}°_{ox} = -(-0.403\ V) = +0.403V$

cathode: $2\,Ag^{+}(aq) + 2\,e^- \longrightarrow 2\,Ag(s)$ $\quad\mathscr{E}°_{red} = +0.7994$ V

net: $\mathbf{Cd(s) + 2\,Ag^{+}(aq) \longrightarrow Cd^{2+}(aq) + 2\,Ag(s)}$

$\mathscr{E}°_{cell} = \mathscr{E}°_{red} + \mathscr{E}°_{ox} = +0.403V + (0.7994\ V) = \mathbf{1.202\ V}$

The shorthand diagram should be **Cd(s) | Cd^{2+}(aq) ‖ Ag^{+}(aq) | Ag(s).**

c. The two appropriate half-reactions and SRP's are:

$AgCl(s) + 1\,e^- \longrightarrow Ag(s) + Cl^-(aq)$ $\quad\mathscr{E}°_{red} = +0.222$ V

$2\,H^{+}(aq) + 2\,e^- \longrightarrow H_2(g)$ $\quad\mathscr{E}°_{red} = 0.0000$ V

The hydrogen reaction reverses and becomes the oxidation. The silver reaction needs to be multiplied by two to make the electrons involved in each half-reaction equal.

anode: $H_2(g) \longrightarrow 2\,H^{+}(aq) + 2\,e^-$ $\quad\mathscr{E}°_{ox} = -(0.0000\ V) = 0.0000$ V

cathode: $2\,AgCl(s) + 2\,e^- \longrightarrow 2\,Ag(s) + 2\,Cl^-(aq)$ $\quad\mathscr{E}°_{red} = +0.222$ V

net: $\mathbf{H_2(g) + 2\,AgCl(s) \longrightarrow 2\,Ag(s) + 2\,Cl^-(aq) + 2\,H^{+}(aq)}$

$\mathscr{E}°_{cell} = \mathscr{E}°_{red} + \mathscr{E}°_{ox} = +0.222\ V + (0.0000\ V) = \mathbf{0.222\ V}$

The shorthand diagram should be **Pt | H_2(g) | H^{+}(aq) | Cl^-(aq) | AgCl(s) | Ag(s).**

d. The two appropriate half-reactions and SRP's are:

$2\,H^{+}(aq) + 2\,e^- \longrightarrow H_2(g)$ $\quad\mathscr{E}°_{red} = 0.0000$ V

$Cl_2(g) + 2\,e^- \longrightarrow 2\,Cl^-(aq)$ $\quad\mathscr{E}°_{red} = +1.358$ V

The hydrogen reaction reverses and becomes the oxidation. Neither reaction needs to be multiplied, because there are two electrons in each half-reaction.

anode:	$H_2(g) \longrightarrow 2\ H^+(aq) + 2\ e^-$	$\mathscr{E}°_{ox} = -(0.0000\ V) = 0.0000\ V$
cathode:	$Cl_2(g) + 2\ e^- \longrightarrow 2\ Cl^-(aq)$	$\mathscr{E}°_{red} = +1.358\ V$
net:	$\mathbf{H_2(g) + Cl_2(g) \longrightarrow 2\ H^+(aq) + 2\ Cl^-(aq)}$	

$\mathscr{E}°_{cell} = \mathscr{E}°_{red} + \mathscr{E}°_{ox} = +1.358\ V + (0.0000\ V) = \mathbf{1.358\ V}$

The shorthand diagram is correct.

19.63 Write the chemical equation for the spontaneous reaction that occurs when each of the following electrode pairs is coupled in a voltaic cell. Indicate the value of n, the hidden stoichiometric coefficient of the transferred electrons. Write the algebraic expression for the reaction quotient Q.

a. $Pt \mid MnO_4^-(aq),\ Mn^{2+}(aq);\ NO_3^-(aq) \mid NO(g) \mid Pt$
b. $Sn(s) \mid Sn^{2+}(aq);\ Ag^+(aq) \mid Ag(s)$
c. $Zn(s) \mid Zn^{2+}(aq);\ Fe^{3+}(aq),\ Fe^{2+}(aq) \mid Pt$
d. $Pt \mid H_2(1\ atm) \mid H^+(aq);\ Cl^-(aq) \mid AgCl(s) \mid Ag(s)$

Analysis/Plan: The spontaneous process must have $\mathscr{E}°_{cell} > 0$. Look up the two appropriate $\mathscr{E}°_{red}$ in Appendix I. The half-reaction with the more positive $\mathscr{E}°_{red}$ will be the reduction and function as the cathode. The half-reaction with the more negative $\mathscr{E}°_{red}$ will reverse, becoming the oxidation, and will function as the anode. The cell potential can be found from the two relationships $\mathscr{E}°_{ox} = -\mathscr{E}°_{red}$ and $\mathscr{E}°_{cell} = \mathscr{E}°_{red} + \mathscr{E}°_{ox}$. The stoichiometric coefficient n is the number of electrons needed to make the number of electrons lost by the reducing agent equal to the number of electrons gained by the oxidizing agent. Q is written in the normal manner with the products, raised to the powers of their coefficients, divided by the reactants, raised to the power of their coefficients. Gases are expressed in partial pressure. Dissolved species are expressed in concentration. Solids, liquids and the solvent in dilute solutions are omitted from the expression for Q.

Work:

a. The two appropriate half-reactions and SRP's are:

$MnO_4^-(aq) + 8\ H^+(aq) + 5\ e^- \longrightarrow Mn^{2+}(aq) + 4\ H_2O(\ell)$ $\quad \mathscr{E}°_{red} = +1.51\ V$

$NO_3^-(aq) + 4\ H^+(aq) + 3\ e^- \longrightarrow NO(g) + 2\ H_2O(\ell)$ $\quad \mathscr{E}°_{red} = +0.96\ V$

The nitrate reaction reverses and becomes the oxidation. The permanganate reaction needs to be multiplied by 3 and the nitrate reaction by 5 to make the electrons equal. The hidden stoichiometric coefficient $\boldsymbol{n} = \mathbf{15}$. Add the two equations, deleting like terms.

anode:	$5\ NO(g) + 10\ H_2O(\ell) \longrightarrow 5\ NO_3^-(aq) + 20\ H^+(aq) + 15\ e^-$
cathode:	$3\ MnO_4^-(aq) + 24\ H^+(aq) + 15\ e^- \longrightarrow 3\ Mn^{2+}(aq) + 12\ H_2O(\ell)$
net:	$\mathbf{3\ MnO_4^-(aq) + 4\ H^+(aq) + 5\ NO(g) \longrightarrow 3\ Mn^{2+}(aq) + 2\ H_2O(\ell) + 5\ NO_3^-(aq)}$

$\mathscr{E}°_{ox} = -(0.96V) = -0.96\ V \qquad \mathscr{E}°_{cell} = \mathscr{E}°_{red} + \mathscr{E}°_{ox} = +1.51V + (-0.96\ V) = 0.55\ V$

$$\mathbf{Q} = \frac{[\mathbf{Mn^{2+}}]^3[\mathbf{NO_3^-}]^5}{[\mathbf{MnO_4^-}]^3[\mathbf{H^+}]^4\mathbf{P_{NO}}^5}$$

b. The two appropriate half-reactions and SRP's are:

$Sn^{2+}(aq) + 2\ e^- \longrightarrow Sn(s)$ $\quad \mathscr{E}°_{red} = -0.14\ V$

$Ag^+(aq) + 1\ e^- \longrightarrow Ag(s)$ $\quad \mathscr{E}°_{red} = +0.7994\ V$

The tin reaction reverses and becomes the oxidation. The silver reaction needs to be multiplied by 2 to make the electrons equal. The hidden stoichiometric coefficient $\boldsymbol{n} = \mathbf{2}$. Add the two equations, deleting like terms.

anode: $Sn(s) \longrightarrow Sn^{2+}(aq) + 2\,e^-$ $\qquad \mathscr{E}°_{ox} = -(-0.14\ V) = +0.14\ V$

cathode: $2\,Ag^+(aq) + 2\,e^- \longrightarrow 2\,Ag(s)$ $\qquad \mathscr{E}°_{red} = +0.7994\ V$

net: $Sn(s) + 2\,Ag^+(aq) \longrightarrow Sn^{2+}(aq) + 2\,Ag(s)$

$$\mathscr{E}°_{cell} = \mathscr{E}°_{red} + \mathscr{E}°_{ox} = +0.7994\ V + (+0.14V) = 0.94\ V$$

$$Q = \frac{[Sn^{2+}]}{[Ag^+]^2}$$

c. The two appropriate half-reactions and SRP's are:

$Zn^{2+}(aq) + 2\,e^- \longrightarrow Zn(s)$ $\qquad \mathscr{E}°_{red} = -0.763\ V$

$Fe^{3+}(aq) + 1\,e^- \longrightarrow Fe^{2+}(aq)$ $\qquad \mathscr{E}°_{red} = +0.771\ V$

The zinc reaction reverses and becomes the oxidation. The iron reaction needs to be multiplied by 2 to make the electrons equal. The hidden stoichiometric coefficient $n = 2$. Add the two equations, deleting like terms.

anode: $Zn(s) \longrightarrow Zn^{2+}(aq) + 2\,e^-$ $\qquad \mathscr{E}°_{ox} = -(-0.763V) = +0.763\ V$

cathode: $2\,Fe^{3+}(aq) + 2\,e^- \longrightarrow 2\,Fe^{2+}(aq)$ $\qquad \mathscr{E}°_{red} = +0.771\ V$

net: $Zn(s) + 2\,Fe^{3+}(aq) \longrightarrow Zn^{2+}(aq) + 2\,Fe^{2+}(aq)$

$$\mathscr{E}°_{cell} = \mathscr{E}°_{red} + \mathscr{E}°_{ox} = +0.763\ V + (+0.771\ V) = 1.534\ V$$

$$Q = \frac{[Zn^{2+}][Fe^{2+}]^2}{[Fe^{3+}]^2}$$

d. This system was analyzed in 19.61 c. The reaction is:

$H_2(g) + 2\,AgCl(s) \longrightarrow 2\,Ag(s) + 2\,Cl^-(aq) + 2\,H^+(aq)$.

The hidden stoichiometric coefficient $n = 2$. $Q = \dfrac{[H^+]^2[Cl^-]^2}{P_{H_2}}$

19.65 Write the chemical equation for the cell reaction in each of the following cells. Indicate the value of n, the hidden stoichiometric coefficient of the transferred electrons. Write the algebraic expression for the reaction quotient Q.

a. $Zn(s)\,|\,Zn^{2+}(aq)\,\|\,Sn^{2+}(aq)\,|\,Sn(s)$
b. $Ag(s)\,|\,Ag^+(aq)\,\|\,Cd^{2+}(aq)\,|\,Cd(s)$
c. $Ag(s)\,|\,AgCl(s)\,|\,HCl(aq)\,|\,H_2(g)\,|\,Pt$
d. $Pt\,|\,H_2(g)\,|\,H^+(aq)\,\|\,Cl^-(aq)\,|\,Cl_2(g)\,|\,Pt$

Analysis/Plan: The balanced chemical reactions and $\mathscr{E}°$ values for these cells were determined in Problem 19.61. Proceed in exactly the same manner as in Problem 19.63.

Work:

a. The reaction is: $Zn(s) + Sn^{2+}(aq) \longrightarrow Zn^{2+}(aq) + Sn(s)$. $\quad n = 2$.

$$Q = \frac{[Zn^{2+}]}{[Sn^{2+}]}$$

b. The reaction is: $Cd(s) + 2\,Ag^+(aq) \longrightarrow Cd^{2+}(aq) + 2\,Ag(s)$. $\quad n = 2$.

$$Q = \frac{[Cd^{2+}]}{[Ag^+]^2}$$

c. The reaction is: $H_2(g) + 2\,AgCl(s) \longrightarrow 2\,Ag(s) + 2\,Cl^-(aq) + 2\,H^+(aq)$. $\quad n = 2$.

$$Q = \frac{[H^+]^2[Cl^-]^2}{P_{H_2}}$$

d. The reaction is $H_2(g) + Cl_2(g) \longrightarrow 2\,H^+(aq) + 2\,Cl^-(aq)$. $n = 2$.

$$Q = \frac{[H^+]^2[Cl^-]^2}{P_{H_2} \cdot P_{Cl_2}}$$

19.67 Calculate the cell potential of each of the following electrochemical cells at 25°C.

a. $Sn(s)\,|\,Sn^{2+}(4.5 \times 10^{-3}\ M)\,\|\,Ag^+\ (0.100\ M)\,|\,Ag(s)$
b. $Zn(s)\,|\,Zn^{2+}\ (0.500\ M)\,\|\,Fe^{3+}\ (7.2 \times 10^{-6}\ M),\ Fe^{2+}\ (0.15\ M)\,|\,Pt$
c. $Pt\,|\,H_2\ (1\ atm)\,|\,HCl\ (0.00623\ M)\,|\,Cl_2\ (1\ atm)\,|\,Pt$

Analysis: Any cell having a concentration different from 1 M or a partial pressure different from 1 atm must be treated by the Nernst equation ($\mathcal{E} = \mathcal{E}° - \frac{0.0592}{n} \log_{10} Q$). The values of $\mathcal{E}°$ for these cells were determined in Problems 19.61 and 19.63.
Plan: Set up the Nernst equation, insert all known values and evaluate the $\mathcal{E}_{cell}$.
Work:

a. From Problem 19.63b, the equation is: $2\,Ag^+(aq) + Sn(s) \longrightarrow 2\,Ag(s) + Sn^{2+}(aq)$
$\mathcal{E}° = 0.94\ V$ and $n = 2$.

$$\mathcal{E} = 0.94\ V - \frac{0.0592}{2} \log_{10}\frac{[Sn^{2+}]}{[Ag^+]^2} = 0.94 - \frac{0.0592}{2} \log_{10}\left(\frac{4.5 \times 10^{-3}}{(0.100)^2}\right)$$

$$\mathcal{E} = 0.94 - \left(\frac{0.0592}{2}\right)(-0.35) = 0.94 - (-0.010) = \mathbf{0.95\ V}$$

b. From Problem 19.63c, the equation is: $2\,Fe^{3+}(aq) + Zn(s) \longrightarrow Zn^{2+}(aq) + 2\,Fe^{2+}(aq)$
$\mathcal{E}° = 1.534\ V$ and $n = 2$.

$$\mathcal{E} = 1.534\ V - \frac{0.0592}{2} \log_{10}\frac{[Zn^{2+}][Fe^{2+}]^2}{[Fe^{3+}]^2} = 1.534 - \frac{0.0592}{2} \log_{10}\left(\frac{(0.500)(0.15)^2}{(7.2 \times 10^{-6})^2}\right)$$

$$\mathcal{E} = 1.534 - \left(\frac{0.0592}{2}\right)(+8.336) = 1.534 - (0.247) = \mathbf{1.287\ V}$$

c. From Problem 19.61d, the equation is $H_2(g) + Cl_2(g) \longrightarrow 2\,H^+(aq) + 2\,Cl^-(aq)$
$\mathcal{E}° = 1.358\ V$ and $n = 2$.

$$\mathcal{E} = 1.358V - \frac{0.0592}{2} \log_{10}\frac{[H^+]^2[Cl^-]^2}{P_{H_2} \cdot P_{Cl_2}} = 1.358 - \frac{0.0592}{2} \log_{10}\left(\frac{(0.00623)^2(0.00623)^2}{(1)(1)}\right)$$

$$\mathcal{E} = 1.358 - \left(\frac{0.0592}{2}\right)(-8.822) = 1.358 - (-0.261) = \mathbf{1.619\ V}$$

19.69 For each of the following cells, calculate the ratio of concentrations or pressures required to produce the indicated cell potential.

a. $Ni(s)\,|\,Ni^{2+}(aq)\,\|\,Sn^{2+}(aq)\,|\,Sn(s)$; $[Ni^{2+}]/[Sn^{2+}]$ = ? when $\mathcal{E} = +0.27\ V$
b. $Sn(s)\,|\,Sn^{2+}(aq)\,\|\,Ni^{2+}(aq)\,|\,Ni(s)$; $[Sn^{2+}]/[Ni^{2+}]$ = ? when $\mathcal{E} = +0.27\ V$

Analysis: Any cell having a concentration different from 1 M or a partial pressure different from 1 atm must be treated by the Nernst equation ($\mathscr{E} = \mathscr{E}° - \frac{0.0592}{n} \log_{10} Q$). The cell reaction, n and $\mathscr{E}°$ must be determined first.

Plan: Find $\mathscr{E}°$ of the spontaneous cell and n, using techniques described in Problems 19.63 and 19.65. Compare the desired cell to the indicated reaction to determine its $\mathscr{E}°$. Set up the Nernst equation, insert all known values and solve for the requested ratio.

Work:

a. The two appropriate half-reactions and SRP's are:

$Sn^{2+}(aq) + 2\,e^- \longrightarrow Sn(s)$ $\qquad \mathscr{E}°_{red} = -0.14$ V

$Ni^{2+}(aq) + 2\,e^- \longrightarrow Ni(s)$ $\qquad \mathscr{E}°_{red} = -0.25$ V

The nickel reaction reverses and becomes the oxidation. Because there are 2 electrons in each half-reaction, no multiplication is needed. The hidden stoichiometric coefficient $n = 2$. Add the two equations, deleting like terms.

anode: $Ni(s) \longrightarrow Ni^{2+}(aq) + 2\,e^-$ $\qquad \mathscr{E}°_{ox} = -(-0.25V) = +0.25V$

cathode: $Sn^{2+}(aq) + 2\,e^- \longrightarrow Sn(s)$ $\qquad \mathscr{E}°_{red} = -0.14$ V

net: $Ni(s) + Sn^{2+}(aq) \longrightarrow Ni^{2+}(aq) + Sn(s)$

$$\mathscr{E}°_{cell} = \mathscr{E}°_{red} + \mathscr{E}°_{ox} = -0.14\text{ V} + (+0.25\text{ V}) = 0.11\text{ V}$$

In the requested reaction, Ni is the anode, so this $\mathscr{E}°$ applies. Now use the Nernst equation and solve for the requested concentration ratio.

$$0.27\text{ V} = 0.11\text{ V} - \frac{0.0592}{2}\log_{10}\frac{[Ni^{2+}]}{[Sn^{2+}]};\ 0.27\text{ V} - 0.11\text{ V} = 0.16\text{ V} = -\frac{0.0592}{2}\log_{10}\frac{[Ni^{2+}]}{[Sn^{2+}]}$$

$$\log_{10}\frac{[Ni^{2+}]}{[Sn^{2+}]} = \frac{2(0.16)}{-0.0592} = -5.4 \text{ and } \frac{[Ni^{2+}]}{[Sn^{2+}]} = 10^{-5.4} = \mathbf{3.9 \times 10^{-6}}$$

b. The requested cell is the reverse of that in part a. Therefore, $\mathscr{E}° = -0.11$ V. Use the Nernst equation to find the requested concentration ratio.

$$0.27\text{ V} = -0.11\text{ V} - \frac{0.0592}{2}\log_{10}\frac{[Sn^{2+}]}{[Zn^{2+}]};\ 0.27\text{ V} + 0.11\text{ V} = 0.38\text{ V} = -\frac{0.0592}{2}\log_{10}\frac{[Sn^{2+}]}{[Ni^{2+}]}$$

$$\log_{10}\frac{[Sn^{2+}]}{[Ni^{2+}]} = \frac{2(0.38)}{-0.0592} = -12.8 \text{ and } \frac{[Ni^{2+}]}{[Sn^{2+}]} = 10^{-12.8} = \mathbf{1.9 \times 10^{-13}}$$

19.71 For the following cells, calculate the concentration or pressure ratio required to produce the indicated cell potential. Remember that when the cell potential is zero, the cell reaction is at equilibrium.

a. Pt | $Cr^{2+}(aq)$, $Cr^{3+}(aq)$ || H^+ (pH = 3) | H_2 (1.00 atm)| Pt; calculate $[Cr^{2+}]/[Cr^{3+}]$ when $\mathscr{E} = 0$ V.

b. Ag(s) | AgBr(s) || HBr (2.50 M) | $H_2(g)$ | Pt; calculate P_{H_2} when $\mathscr{E} = +0.077$ V.

Analysis: Any cell having a concentration different from 1 M or a partial pressure different from 1 atm must be treated by the Nernst equation ($\mathscr{E} = \mathscr{E}° - \frac{0.0592}{n} \log_{10} Q$). The cell reaction, n and $\mathscr{E}°$ must be determined first.

Plan: Find $\mathscr{E}°$ of the spontaneous cell and n, using techniques described in Problems 19.63 and 19.65. Compare the desired cell to the indicated reaction to determine its $\mathscr{E}°$. Set up the Nernst equation, insert all known values and solve for the requested ratio.

Work:

a. The two appropriate half-reactions and SRP's are:

$Cr^{3+}(aq) + 1\ e^- \longrightarrow Cr^{2+}(aq)$ $\quad \mathscr{E}^\circ_{red} = -0.41\ V$

$2\ H^+(aq) + 2\ e^- \longrightarrow H_2(g)$ $\quad \mathscr{E}^\circ_{red} = 0.0000\ V$

The chromium reaction reverses and becomes the oxidation. Because there are 2 electrons in the hydrogen reaction, the chromium half-reaction must be multiplied by 2. The hidden stoichiometric coefficient $n = 2$. Add the two equations, deleting like terms.

anode: $2\ Cr^{2+}(aq) \longrightarrow 2\ Cr^{3+}(aq) + 2\ e^-$ $\quad \mathscr{E}^\circ_{ox} = -(-0.41\ V) = +\ 0.41\ V$

cathode: $2\ H^+(aq) + 2\ e^- \longrightarrow H_2(g)$ $\quad \mathscr{E}^\circ_{red} = 0.0000\ V$

net: $2\ Cr^{2+}(aq) + 2\ H^+(aq) \longrightarrow 2\ Cr^{3+}(aq) + H_2(g)$

$$\mathscr{E}^\circ_{cell} = \mathscr{E}^\circ_{red} + \mathscr{E}^\circ_{ox} = 0.0000\ V + (+0.41\ V) = 0.41\ V$$

In the requested reaction, chromium(II) is at the anode, so this $\mathscr{E}^\circ$ applies. Now use the Nernst equation and solve for the requested ratio. If the pH = 3, then $[H^+] = 10^{-3}$ M.

$$0\ V = 0.41\ V - \frac{0.0592}{2} \log_{10}\frac{[Cr^{3+}]^2 P_{H_2}}{[Cr^{2+}]^2[H^+]^2};$$

$$0\ V - 0.41\ V = -0.41\ V = -\frac{0592}{2} \log_{10}\frac{[Cr^{3+}]^2(1)}{[Cr^{2+}]^2(10^{-3})^2}$$

$$\log_{10}\frac{[Cr^{3+}]^2(1)}{[Cr^{2+}]^2(10^{-3})^2} = \frac{2(-0.41)}{-0.0592} = 14 \text{ and } \frac{[Cr^{3+}]^2(1)}{[Cr^{2+}]^2(10^{-3})^2} = 10^{14}$$

$$\frac{[Cr^{3+}]^2}{[Cr^{2+}]^2} = \frac{(10^{14})(10^{-3})^2}{1} = 10^8 \text{ and } \frac{[Cr^{3+}]}{[Cr^{2+}]} = \sqrt{10^8} = 10^4$$

$$\frac{[Cr^{2+}]}{[Cr^{3+}]} = \frac{1}{10^4} = 10^{-4}$$

b. The two appropriate half-reactions and SRP's are:

$AgBr(s) + 1\ e^- \longrightarrow Ag(s) + Br^-(aq)$ $\quad \mathscr{E}^\circ_{red} = 0.0713\ V$

$2\ H^+(aq) + 2\ e^- \longrightarrow H_2(g)$ $\quad \mathscr{E}^\circ_{red} = 0.0000\ V$

The hydrogen reaction reverses and becomes the oxidation. Because there are 2 electrons in the hydrogen reaction, the silver half-reaction must be multiplied by 2. The hidden stoichiometric coefficient $n = 2$. Add the two equations, deleting like terms.

anode: $H_2(g) \longrightarrow 2\ H^+(aq) + 2\ e^-$ $\quad \mathscr{E}^\circ_{ox} = 0.0000\ V$

cathode: $2\ AgBr(s) + 2\ e^- \longrightarrow 2\ Ag(s) + 2\ Br^-(aq)$ $\quad \mathscr{E}^\circ_{red} = 0.0713V$

net: $H_2(g) + 2\ AgBr(s) \longrightarrow 2\ H^+(aq) + 2\ Ag(s) + 2\ Br^-(aq)$

$$\mathscr{E}^\circ_{cell} = \mathscr{E}^\circ_{red} + \mathscr{E}^\circ_{ox} = 0.0000\ V + (+0.0713\ V) = 0.0713\ V$$

In the requested reaction, silver is the anode, therefore, the reaction of interest is the **reverse** of this reaction and $\mathscr{E}^\circ = -0.0713$ V. Now use the Nernst equation and solve for the requested pressure.

$$+0.077\ V = -0.0713\ V - \frac{0.0592}{2} \log_{10}\frac{P_{H_2}}{[Br^-]^2[H^+]^2}$$

$$+0.077 \text{ V} - (-0.0713 \text{ V}) = +0.148 \text{ V} = -\frac{0.0592}{2}\log_{10}\frac{P_{H_2}}{[2.50]^2[2.50]^2}$$

$$\log_{10}\frac{P_{H_2}}{[2.50]^2[2.50]^2} = \frac{(0.148)(2)}{-0.0592} = -5.00 \text{ and } \frac{P_{H_2}}{[2.50]^2[2.50]^2} = 10^{-5.00} = 1.0 \times 10^{-5}$$

$$P_{H_2} = (1.0 \times 10^{-5})(2.50)^4 = \mathbf{3.9 \times 10^{-4} \text{ atm}}$$

19.73 For each of the following cations, how many faradays are required for reduction and deposition of one mole of metal at the cathode of an electrolytic cell?
a. Fe^{3+} b. Ag^+ c. Hg_2^{2+}

Analysis/Plan: A Faraday carries the charge of one mole of electrons. Write the reduction equation for converting the ion to the free metal. The number of moles of electrons required for that process is equal to the number of Faradays required.

Work:

a. $Fe^{3+}(aq) + 3\ e^- \longrightarrow Fe(s)$ $\mathcal{F} = 3$

b. $Ag^+(aq) + 1\ e^- \longrightarrow Ag(s)$ $\mathcal{F} = 1$

c. $1/2\ Hg_2^{2+}(aq) + 1\ e^- \longrightarrow Hg(\ell)$ $\mathcal{F} = 1$

19.75 How much charge, expressed in coulombs, is required to reduce each of the following ions to 1.00 g metal?
a. Fe^{3+} b. Ag^+ c. Hg_2^{2+}

Analysis/Plan: In Problem 19.73, we determined the number of Faradays required to reduce one mole of the metal. The molar mass of the metal is an implicit known and provides the equivalence statement g ↔ mol. The equivalence statement 1 mol e^- ↔ 96485.31 C is needed. Use the factor-label method to determine the number of coulombs.

Work:

a. The molar mass of Fe is 55.849 g mol^{-1}.

$$? \text{ C} = 1.00 \text{ g Fe} \times \frac{1 \text{ mol Fe}}{55.849 \text{ g Fe}} \times \frac{3 \text{ mol } e^-}{1 \text{ mol Fe}} \times \frac{96485.31 \text{ C}}{1 \text{ mol } e^-} = \mathbf{5.18 \times 10^3 \text{ C}}$$

b. The molar mass of Ag is 107.8682 g mol^{-1}.

$$? \text{ C} = 1.00 \text{ g Ag} \times \frac{1 \text{ mol Ag}}{107.8682 \text{ g Ag}} \times \frac{1 \text{ mol } e^-}{1 \text{ mol Ag}} \times \frac{96485.31 \text{ C}}{1 \text{ mol } e^-} = \mathbf{8.94 \times 10^2 \text{ C}}$$

c. The molar mass of Hg is 200.59 g mol^{-1}.

$$? \text{ C} = 1.00 \text{ g Hg} \times \frac{1 \text{ mol Hg}}{200.59 \text{ g Hg}} \times \frac{1 \text{ mol } e^-}{1 \text{ mol Hg}} \times \frac{96485.31 \text{ C}}{1 \text{ mol } e^-} = \mathbf{4.81 \times 10^2 \text{ C}}$$

19.77 How many moles of $H_2(g)$ are produced when water is electrolyzed with a current of 0.553 A for a period of 15 min? What volume is occupied by this amount of gas at STP?

Analysis: To determine the number of moles of hydrogen, we must write the electrochemical equation to determine the reaction stoichiometry. The number of moles of electrons depends on the total charge passed through the cell, which can be found from the time and the relationship 1 C•s^{-1} ↔ 1 A. The equivalence statements 1 mol e^- ↔ 96485.31 C,

60 s ↔ 1 min, 1 mol ↔ 22.4 L at STP and 1 mL ↔ 10^{-3} L are needed.

Plan: Write the equation first, then use the factor-label method to determine the number of moles and then the volume. Note that 0.553 A = 0.553 C s^{-1}.

Work:

The reduction equation is: $2\ H^+(aq) + 2\ e^- \longrightarrow H_2(g)$.

$$? \text{ mol } H_2 = \frac{0.553 \text{ C}}{\text{s}} \times \frac{60 \text{ s}}{1 \text{ min}} \times 15.0 \text{ min} \times \frac{1 \text{ mol } e^-}{96485.31 \text{ C}} \times \frac{1 \text{ mol } H_2}{2 \text{ mol } e^-} = \mathbf{2.58 \times 10^{-3} \text{ mol } H_2}$$

$$? \text{ mL } H_2 = 2.58 \times 10^{-3} \text{ mol } H_2 \times \frac{22.4 \text{ L}}{1 \text{ mol } H_2} \times \frac{1 \text{ mL}}{10^{-3} \text{ L}} = \mathbf{57.8 \text{ mL}}$$

19.79 The mass of silver deposited on a spoon during electroplating was 0.634 mg. How much electrical charge passed through the cell?

Analysis: To determine the amount of charge, we must write the electrochemical equation to determine the reaction stoichiometry. The molar mass is an implicit known and provides the equivalence statement g ↔ mol. The equivalence statements 1 mol e^- ↔ 96485.31 C and 1 mg ↔ 10^{-3} g are also needed.

Plan: Write the equation first, then use the factor-label method to determine the number of coulombs required.

Work:

The reduction equation is $Ag^+(aq) + 1\ e^- \longrightarrow Ag(s)$. Molar mass Ag = 107.8682 g mol^{-1}.

$$? \text{ C} = 0.634 \text{ mg Ag} \times \frac{10^{-3} \text{ g Ag}}{1 \text{ mg Ag}} \times \frac{1 \text{ mol Ag}}{107.8682 \text{ g Ag}} \times \frac{1 \text{ mol } e^-}{1 \text{ mol Ag}} \times \frac{96485.31 \text{ C}}{1 \text{ mol } e^-} = \mathbf{0.567 \text{ C}}$$

19.81 How long does an electrolysis cell have to run at a current of 150 A in order to produce 1 pound (454 g) of aluminum?

Analysis: To determine the time required, we must write the electrochemical equation to determine the reaction stoichiometry. The molar mass is an implicit known and provides the equivalence statement g ↔ mol. The equivalence statements 1 A ↔ 1 C•s^{-1}, 1 min ↔ 60 s, 1 hr ↔ 60 min and 1 mol e^- ↔ 96485.31 C are also needed.

Plan: Write the equation first, then use the factor-label method to determine the length of time required. Note that 150 A = 150 C s^{-1}.

Work:

The reduction equation is $Al^{3+}(aq) + 3\ e^- \longrightarrow Al(s)$. Molar mass Al = 26.98154 g mol^{-1}.

$$? \text{ s} = 454 \text{ g Al} \times \frac{1 \text{ mol Al}}{26.98154 \text{ g Al}} \times \frac{3 \text{ mol } e^-}{1 \text{ mol Al}} \times \frac{96485.31 \text{ C}}{1 \text{ mol } e^-} \times \frac{1 \text{ s}}{150 \text{ C}} = \mathbf{3.25 \times 10^4 \text{ s}}$$

$$? \text{ hr} = 3.25 \times 10^4 \text{ s} \times \frac{1 \text{ min}}{60 \text{ s}} \times \frac{1 \text{ hr}}{60 \text{ min}} = \mathbf{9.02 \text{ hr}}$$

19.83 An electrolytic cell contains 50.0 mL of a 0.152 M solution of $FeCl_3$. A current of 0.775 A is passed through the cell, causing deposition of Fe(s) at the cathode. What is the concentration of $Fe^{3+}(aq)$ in the cell after this current has run for 20.0 min?

Analysis: In $FeCl_3$, the iron has a 3+ charge. To determine the concentration, we must write the electrochemical equation to determine the reaction stoichiometry. We can determine the initial number of millimoles of Fe^{3+} from the relationship n = M•V. Concentration can be found from $\frac{mmol}{mL}$. The equivalence statements 1 A ↔ 1 $C \cdot s^{-1}$, 1 min ↔ 60 s, 1 mol e^- ↔ 96485.31 C and 1 mmol ↔ 10^{-3} mol are also needed.

Plan: Use the relationship n = M•V to determine the initial number of millimoles of Fe^{3+}. Write the equation and use the factor-label method to determine the number of millimoles of Fe^{3+} consumed. Subtract that number of millimoles from the original number of millimoles to determine the amount remaining. Divide the number of millimoles remaining by the volume to determine the concentration.

Work:

The initial number of millimoles of $Fe^{3+} = \frac{0.152 \text{ mol}}{\text{L}} \times 50.0 \text{ mL} = 7.60 \text{ mmol } Fe^{3+}$.

The reduction equation is $Fe^{3+}(aq) + 3\ e^- \longrightarrow Fe(s)$.

$$? \text{ mmol } Fe^{3+} \text{ used} = \frac{0.775 \text{ C}}{\text{s}} \times \frac{60 \text{ s}}{1 \text{ min}} \times 20.0 \text{ min} \times \frac{1 \text{ mol } e^-}{96485.31 \text{ C}} \times \frac{1 \text{ mol } Fe^{3+}}{3 \text{ mol } e^-} \times \frac{1 \text{ mmol } Fe^{3+}}{10^{-3} \text{ mol } Fe^{3+}} = 3.21 \text{ mmol } Fe^{3+}$$

$$? \text{ mmol } Fe^{3+} \text{ remaining} = 7.60 \text{ mmol } Fe^{3+} - 3.21 \text{ mmol } Fe^{3+} = 4.39 \text{ mmol } Fe^{3+}$$

$$? \text{ M } Fe^{3+} = \frac{4.39 \text{ mmol } Fe^{3+}}{50.0 \text{ mL}} = \mathbf{0.0877 \text{ M } Fe^{3+}}$$

19.85 Write the anode and cathode half-reactions, and the cell reaction, of the following voltaic cells.
a. dry cell b. lead storage battery (discharge cycle) c. silver cell

See the text for a complete discussion.

a. anode: $Zn(s) \longrightarrow Zn^{2+}(aq) + 2\ e^-$

cathode: $2\ NH_4^+(aq) + 2\ MnO_2(s) + 2\ e^- \longrightarrow Mn_2O_3(s) + H_2O(\ell) + 2\ NH_3(aq)$

overall: $2\ NH_4^+(aq) + 2\ MnO_2(s) + Zn(s) \longrightarrow Mn_2O_3(s) + H_2O(\ell) + 2\ NH_3(aq) + Zn^{2+}(aq)$

b. anode: $Pb(s) + HSO_4^-(aq) \longrightarrow PbSO_4(s) + H^+(aq) + 2\ e^-$

cathode: $PbO_2(s) + 3\ H^+(aq) + HSO_4^-(aq) + 2\ e^- \longrightarrow PbSO_4(s) + 2\ H_2O(\ell)$

overall: $PbO_2(s) + Pb(s) + 2\ H^+ + 2\ HSO_4^-(aq) \longrightarrow 2\ PbSO_4(s) + 2\ H_2O(\ell)$

c. anode: $Zn(s) + 2\ OH^-(aq) \longrightarrow ZnO(s) + H_2O(\ell) + 2\ e^-$

cathode: $AgO(s) + H_2O(\ell) + 2\ e^- \longrightarrow Ag(s) + 2\ OH^-(aq)$

overall: $AgO(s) + Zn(s) \longrightarrow ZnO(s) + Ag(s)$

19.87 Tarnished silver is coated with a patina of $Ag_2S(s)$. This coating can be removed by boiling the silverware in an aluminum pan, with some baking soda or salt added to make the solution conductive. Explain this from the point of view of electrochemistry.

The system operates as a voltaic cell.

anode: $Al(s) \longrightarrow Al^{3+}(aq) + 3\ e^-$

cathode: $Ag_2S(s) + 2\ H^+(aq) + 2\ e^- \longrightarrow 2\ Ag(s) + H_2S(g)$

overall: $2\ Al(s) + 3\ Ag_2S(s) + 6\ H^+(aq) \longrightarrow 2\ Al^{3+}(aq) + 6\ Ag(s) + 3\ H_2S(g)$

19.89 Calculate the standard free energy change of each of the reactions in Problem 19.45. What is the value of the equilibrium constant at 25°C?

Analysis/Plan: The relationships $\Delta G^\circ_{rxn} = -n\mathcal{F}\mathcal{E}^\circ$ and $\frac{\Delta G^\circ_{rxn}}{-RT} = \ln K$ are required to solve this problem. $\mathcal{E}^\circ$ and n were determined in Problem 19.45. The equivalence statements $K \leftrightarrow 273.15 + {}^\circ C$, $1\ kJ \leftrightarrow 10^3\ J$, $1\ V \leftrightarrow \frac{1\ J}{1\ C}$ and $\mathcal{F} \leftrightarrow \frac{96485.31\ C}{1\ mol\ e^-}$ are also needed. The gas constant, R, is equal to $8.31451\ \frac{J}{mol{\cdot}K}$. Insert the known values directly into the relationships, using the factor-label method to convert as necessary.

Work:

In each case, T = 273.15 + 25 °C = 298 K.

a. $n = 10\ e^-$ and $\mathcal{E}^\circ = 0.59\ V$

$$\Delta G^\circ_{rxn} = -n\mathcal{F}\mathcal{E}^\circ = -(10\ mol\ e^-) \times \frac{96485.31\ C}{1\ mol\ e^-} \times 0.59\ V \times \frac{\frac{1\ J}{C}}{V} \times \frac{1\ kJ}{10^3\ J} = \mathbf{-5.7 \times 10^2\ kJ}$$

$$\ln K = \frac{\Delta G^\circ_{rxn}}{-RT} = \frac{-5.7 \times 10^2\ kJ \times \frac{10^3\ J}{1\ kJ}}{(-8.31451\ \frac{J}{mol{\cdot}K} \times 298\ K)} = 230;\ K = e^{230} = 10^{230/2.303} \approx \mathbf{10^{100}}$$

b. $n = 2\ e^-$ and $\mathcal{E}^\circ = -0.12\ V$ (used in calculations as the unrounded 0.124 V)

$$\Delta G^\circ_{rxn} = -n\mathcal{F}\mathcal{E}^\circ = -(2\ mol\ e^-) \times \frac{96485.31\ C}{1\ mol\ e^-} \times (-0.12\ V) \times \frac{\frac{1\ J}{C}}{V} \times \frac{1\ kJ}{10^3\ J} = \mathbf{+24\ kJ}$$

$$\ln K = \frac{\Delta G^\circ_{rxn}}{-RT} = \frac{+24\ kJ \times \frac{10^3\ J}{1\ kJ}}{(-8.31451\ \frac{J}{mol{\cdot}K} \times 298\ K)} = -9.3\ ;\ K = e^{-9.3} = \mathbf{6.4 \times 10^{-5}}$$

c. $n = 2\ e^-$ and $\mathcal{E}^\circ = -0.17\ V$ (used in calculations as the unrounded 0.168 V)

$$\Delta G^\circ_{rxn} = -n\mathcal{F}\mathcal{E}^\circ = -(2\ mol\ e^-) \times \frac{96485.31\ C}{1\ mol\ e^-} \times (-0.17\ V) \times \frac{\frac{1\ J}{C}}{V} \times \frac{1\ kJ}{10^3\ J} = \mathbf{+32\ kJ}$$

$$\ln K = \frac{\Delta G^\circ_{rxn}}{-RT} = \frac{+32\ kJ \times \frac{10^3\ J}{1\ kJ}}{(-8.31451\ \frac{J}{mol{\cdot}K} \times 298\ K)} = -13\ ;\ K = e^{-13} = \mathbf{2.1\ 10^{-6}}$$

19.91 The standard free energies of formation of H^+ and Cl^- are 0.00 and -131 kJ mol^{-1}, respectively. Use these values to calculate the SRP of $Cl_2(g)$, and check your answer against the value in Table 19.2.

Analysis/Plan: The applicable reaction is $H_2(g) + Cl_2(g) \longrightarrow 2\ H^+(aq) + 2\ Cl^-(aq)$. ΔG°_{rxn} can be determined from the relationship $\Delta G^\circ_{rxn} = \Sigma\ n\Delta G^\circ_f\,(products) - \Sigma\ n\Delta G^\circ_f\,(reactants)$. $\mathcal{E}^\circ$ can then be determined from the relationship $\Delta G^\circ_{rxn} = -n\mathcal{F}\mathcal{E}^\circ$. Because the hydrogen reaction is defined as having a potential as 0.000 V whether used as the anode or the cathode, the observed cell voltage will be equal to the SRP of the chlorine half-reaction. The equivalence statements 1 kJ ↔ 10^3 J, 1 V ↔ $\frac{1\ J}{1\ C}$ and $\mathcal{F}$ ↔ $\frac{96485.31\ C}{1\ mol\ e^-}$ are needed.

Work:

First, use ΔG_f^o values to find ΔG_{rxn}^o. The elements $H_2(g)$ and $Cl_2(g)$ are defined as having ΔG_f^o = 0; therefore, G_{rxn}^o = (2)(0.00) + (2)(-131) - (0 + 0) = - 262 kJ mol^{-1}.

Now, use the relationship $\Delta G_{rxn}^o = -n\mathcal{F}\mathcal{E}^o$ to find $\mathcal{E}^o$, with the factor-label method to convert as necessary. The stoichiometric coefficient n = 2 in this process (Problem 19.61d).

$$\mathcal{E}^o = \frac{\Delta G_{rxn}^o}{-n\mathcal{F}} = \left(\frac{\frac{-262\ kJ}{mol} \times \frac{10^3\ J}{1\ kJ}}{(-2\ mol\ e^-)\left(96485.31 \frac{C}{1\ mol\ e^-}\right)} \right) \times \left(\frac{\frac{1\ J}{C}}{V}\right) = +1.36\ V$$

This is the same as the value in the Table (+1.358 V), within rounding.

19.93 Which of Fe^{3+}, Cl^-, and Cu^+ is the most powerful oxidant?

Analysis/Plan: The most powerful oxidant has the highest reduction potential. Look up the reduction potentials and compare them.

Work:

$Fe^{3+}(aq) + 2\ e^- \longrightarrow Fe^{2+}(aq)$ $\qquad \mathcal{E}^o_{red}$ = +0.771 V

$Cl^-(aq)$ is already in the lowest oxidation state for chlorine and cannot be reduced.

$Cu^+(aq) + 1\ e^- \longrightarrow Cu(s)$ $\qquad \mathcal{E}^o_{red}$ = +0.521V

Fe^{3+} is the best oxidant because it has the highest reduction potential.

19.95 Write an algebraic expression of the form $\mathcal{E}$ = A + B•pH for the potential of the cell Zn(s)(1.00 M) || H^+(xM) | H_2(1 atm) | Pt. What are the numerical values of the constants A and B at 25 °C?

Analysis: Any cell having a concentration different from 1 M or a partial pressure different from 1 atm must be treated by the Nernst equation ($\mathcal{E} = \mathcal{E}^o - \frac{0.0592}{n} \log_{10} Q$). The cell reaction, n and $\mathcal{E}^o$ must be determined first. pH = -log[H^+].

Plan: Find $\mathcal{E}^o$ of the spontaneous cell and n, using techniques described in Problems 19.63 and 19.65. Compare the desired cell to the indicated reaction to determine its $\mathcal{E}^o$. Set up the Nernst equation, insert all known values and solve for the general equation.

Work:

The two appropriate half-reactions and SRP's are:

$Zn^{2+}(aq) + 2\ e^- \longrightarrow Zn(s)$ $\qquad \mathcal{E}^o_{red}$ = -0.763V

$2\ H^+(aq) + 2\ e^- \longrightarrow H_2(g)$ $\qquad \mathcal{E}^o_{red}$ = 0.0000 V

The zinc reaction reverses and becomes the oxidation. Because there are 2 electrons in each half-reaction, no multiplication is needed. The hidden stoichiometric coefficient n = 2. Add the two equations, deleting like terms.

anode: $Zn(s) \longrightarrow Zn^{2+}(aq) + 2\ e^-$ $\qquad \mathcal{E}^o_{ox}$ = -(-0.763V) = + 0.763 V

cathode: $2\ H^+(aq) + 2\ e^- \longrightarrow H_2(g)$ $\qquad \mathcal{E}^o_{red}$ = 0.000 V

---

net: $Zn(s) + 2\ H^+(aq) \longrightarrow Zn^{2+}(aq) + H_2(g)$

$\mathcal{E}^o_{cell} = \mathcal{E}^o_{red} + \mathcal{E}^o_{ox}$ = 0.0000 V + (+0.763 V) = 0.763 V

In the requested reaction, Zn is the anode, so this $\mathscr{E}°$ applies. Now use the Nernst equation and evaluate the general relationship.

$$\mathscr{E} = 0.763\text{ V} - \frac{0.0592}{2}\log_{10}\frac{[Zn^{2+}]P_{H_2}}{[H^+]^2} = 0.763\text{ V} - \frac{0.0592}{2}\log_{10}\frac{(1)(1)}{[H^+]^2}$$

The $\log_{10}$ can be reevaluated, using the definition of pH.

$$\log_{10}\left(\frac{1}{[H^+]^2}\right) = -\log[H^+]^2 = -2\log[H^+] = 2(\text{pH})$$

Substituting this value gives: $\mathscr{E} = 0.763\text{ V} - \left(\frac{0.0592}{2}\right)(2\text{ pH}) = \mathscr{E} = \mathbf{0.763\text{ V} - 0.0592 \cdot pH}$

In this equation, **A** = **0.763 V** and **B** = **-0.0592 •pH**.

19.97 **A student in a freshman laboratory performs an electrolysis experiment in which 1.952 g Ag metal is plated out from a $AgNO_3$ solution by running a 0.500 ampere current through the cell for exactly 1 hour. He uses the data to determine the value of the faraday. What is his result, and what is the percent error from the true value of 96,485 C mol^{-1}? Suggest a likely explanation for the discrepancy.**

Analysis: To determine the value of $\mathscr{F}$, we must write the electrochemical equation to determine the reaction stoichiometry. The molar mass is an implicit known and provides the equivalence statement g ↔ mol. The equivalence statements 1 min ↔ 60 s, 1 hr ↔ 60 min and 1 A ↔ 1 C/s, are also needed.

Plan: Write the equation first, then use the factor-label method to determine the number of moles of electrons and the number of coulombs which passed through the cell. Divide the number of coulombs by the number of electrons to evaluate $\mathscr{F}$. Determine the percent error by taking the difference and dividing by the actual value of $\mathscr{F}$.

Work:

The reduction equation is $Ag^+(aq) + 1\,e^- \longrightarrow Ag(s)$. Molar mass Ag = 107.8682 g mol^{-1}.

$$?\text{ mol } e^- = 1.952\text{ g Ag}\ \frac{1\text{ mol Ag}}{107.8682\text{ g Ag}} \times \frac{1\text{ mol } e^-}{1\text{ mol Ag}} = 1.810 \times 10^{-2}\text{ mol } e^-$$

$$?\text{ C} = \frac{0.500\text{ C}}{\text{s}} \times \frac{60\text{ s}}{1\text{ min}} \times \frac{60\text{ min}}{1\text{ hr}} \times 1\text{ hr} = 1.80 \times 10^3\text{ C}$$

$$?\,\mathscr{F} = \frac{1.80 \times 10^3\text{ C}}{1.810 \times 10^{-2}\text{ mol } e^-} = \mathbf{9.95 \times 10^4\ C/mol\ e^-}$$

The difference is 9.95×10^4 C/mol e^- - 96485 C/ mol e^- = 3.0×10^3 C/mol e^-.

The percent difference is $\dfrac{3.0 \times 10^3\text{ C/mol } e^-}{96485.31\text{ C/mol } e^-} \times 100 = 3.0\%$

The difference is probably due to a mass error (too low) or a current error (too high) or a combination of the two effects. The time is usually not a source of error.

20 Organic Compounds

20.1 What is the distinction between organic and inorganic compounds?

Organic compounds always contain carbon. Organic compounds almost always contain hydrogen. Often oxygen, nitrogen, sulfur, phosphorus, or other elements are also present. Organic compounds were originally considered to be produced only by living organisms, but now many have been synthetically produced. Inorganic compounds are the compounds of all other elements. They include some carbon compounds (CO and CO_2) and the salts or acids of the ions CN^-, CO_3^{2-}, etc.

20.3 What is a functional group? Give examples.

A functional group is an atom, group of atoms or a structural feature that confers special characteristic chemical reactivity and physical properties on a compound. Examples are carboxylic acid (-COOH), alcohol (-OH), amine ($-NH_2$) or a triple bond (alkyne, $C\equiv C$).

20.5 What evidence supports the statement, "Resonance confers stability on a molecule?"

Even though aromatic compounds are drawn as though they have three or more double bonds, they don't undergo addition reactions like those of alkenes which may only have one double bond. Instead, resonance causes aromatic compounds to be as unreactive as the alkanes. An additional example of resonance stabilization is provide by the anions of the carboxylic acids which are stabilized by resonance of the $-CO_2^-$ group. The carboxylic acids are much stronger acids than the alcohols which have an $-O^-$ anion, but are not resonance stabilized.

20.7 Identify the two major types of polymer. What feature differentiates them?

Addition polymers are formed by addition reactions at a double bond. The polymer is the only product of the reaction. Condensation polymers are formed in condensation reactions between two different functional groups (acid and alcohol in polyester, for example). Water or some other small molecule is the byproduct of a condensation reaction.

20.9 Is it possible for a nonplanar conformation of cyclopropane to exist? Explain.

The conformation of an organic molecule is defined by its carbon skeleton. Cyclopropane must always be planar because, geometrically, the three carbon atoms must define a plane. The hydrogen atoms do not lie in the same plane as the carbon atoms as they do in benzene.

20.11 Identify each of the following chemical changes as an oxidation, reduction, or neither.

a. methyl ethyl ketone ⟶ 2-butanol

b. $CH_2=CHCH_3 \longrightarrow CH\equiv CCH_3$

c. methylbenzene ⟶ benzoic acid

d. bromomethane ⟶ fluoromethane

Analysis: In organic chemistry, oxidation is often defined as a gain of oxygen or a loss of hydrogen. Similarly, reduction is often defined as the gain of hydrogen or the loss of oxygen. This was illustrated in conjunction with determination of oxidation numbers in Example 20.1 in the text.

Plan: Draw out the structures if necessary, following the naming rules discussed in the text, and determine if either oxygen or hydrogen was gained or lost.

Work:

a. $H_3C-C(=O)-CH_3 \longrightarrow H_3C-CH(OH)-CH_3$ Two hydrogen atoms have been gained. This is a **reduction**.

b. Two hydrogen atoms have been removed. This is an **oxidation**.

c. $C_6H_5-CH_3 \longrightarrow C_6H_5-CO_2H$ Two hydrogen atoms have been removed and two oxygen atoms have been added. This is an **oxidation**.

d. One halogen has been replaced by another. Because both Br and F are more electronegative than carbon, there is no change in oxidation number assignment. This is **neither** an oxidation or a reduction.

20.13 What hybridization of carbon characterizes the bonding in (a) alkanes, (b) alkenes, (c) alkynes, (d) cycloalkanes, and (e) aromatic hydrocarbons?

Analysis: The hybridization of the carbon atom depends on the number of electron regions around it. Review Chapter 9 if necessary.

Plan: Draw a Lewis structure of a characteristic molecule (or look at the structures in the text) and count the number of electron regions around the carbon atoms in that structure.

Work:

a. H_3C-CH_3 All carbon atoms have four electron regions around them, so all carbon atoms are **sp^3** hybridized.

b. 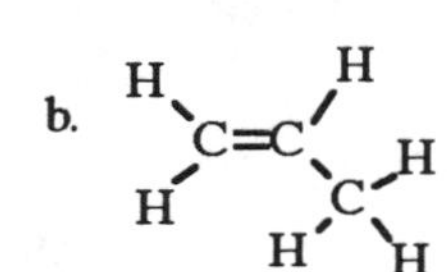

The carbon atoms with the double bond have three electron regions. They are **sp^2** hybridized. The other carbon is **sp^3** hybridized. If this were a longer carbon chain, any additional carbons not involved in the double bond would also be **sp^3** hybridized. Carbons on either side of a double bond are **sp^2** hybridized.

c. $H-C\equiv C-CH_3$ The carbon atoms with the triple bond have two electron regions. They are **sp** hybridized. The other carbon is **sp^3** hybridized. If this were a longer carbon chain, any additional carbons not involved in the triple bond would also be **sp^3** hybridized. Carbons on either side of a triple bond are **sp** hybridized.

d. (cyclohexane, C_6H_{12}) All six carbon atoms have four electron regions around them, so they are all **sp^3** hybridized.

e. 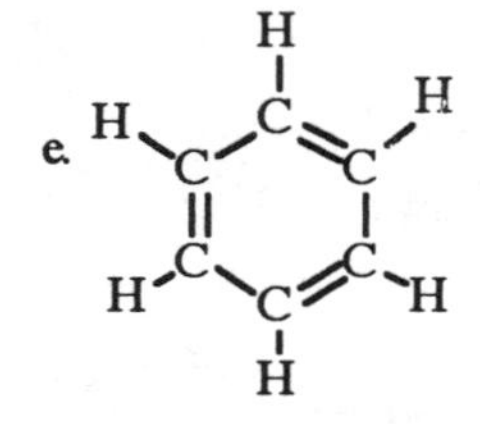

All six carbon atoms have three electron regions around them, so they are all **sp^2** hybridized.

20.15 Draw the structural formula of each of the following.

a. 2,2-dimethylpropane
b. 1,2,4-trichlorobenzene
c. 2-chloro-2-pentene
d. cyclopentene

Analysis/Plan: The IUPAC rules must be learned to draw the structure from the name. Draw the parent carbon chain, then add substituents at the positions indicated by the name.
Work:

a. 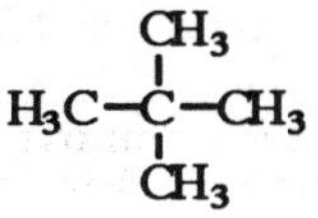

b. 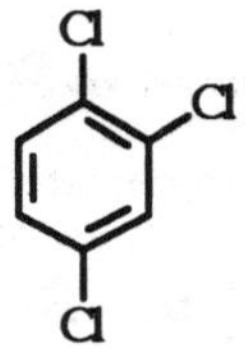

c. 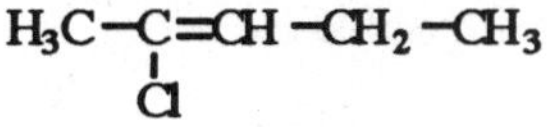

d. 

20.17 Name each of the following.

a. 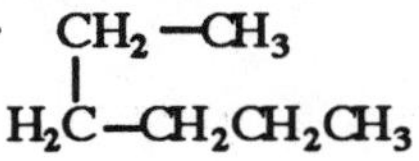

b. 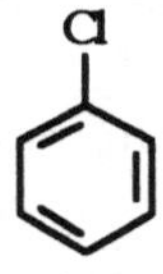

c. 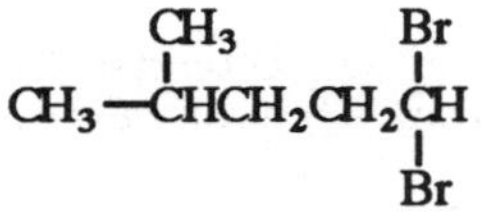

d. 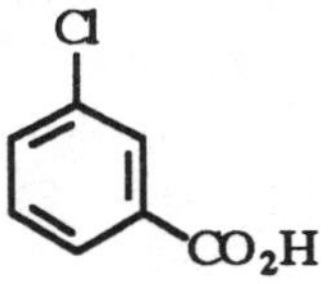

Analysis/Plan: Follow the IUPAC rules as detailed in the text.
a. hexane
b. chlorobenzene
c. 1,1-dibromo-4-methylpentane
d. 3-chlorobenzoic acid

20.19 Which of the following names carries enough information so that a unique structural formula can be drawn?
a. hexane
b. ethane
c. pentenol
d. dichlorobutane
e. cyclopentene
f. dichloromethane

Analysis/Plan: Inspect the name and consider IUPAC rules. If more than two possible structures can be drawn which would fit the name it is not unique.
Work:
a. There is only one straight chain alkane with 6 carbons. This is **unique**.
b. There is only one straight chain alkane with 2 carbons. This is **unique**.
c. The double bond could be between the first and second carbon or between the second and the third carbon. The -OH group could be on any of the five carbons There are many possibilities: 1-penten-1-ol, 1-penten-2-ol, 1-penten-3-ol, 1-penten-4-ol, 1-penten-5-ol, 2-penten-1-ol, 2-penten-2-ol, 2-penten-3-ol, 2-penten-4-ol and 2-penten-5-ol. This name is **not unique**.
d. The two -Cl groups could be on the same carbon (first or second). They could also be on different carbons. Possible compounds that would fit the general description provided by this name are: 1,1-dichlorobutane, 2,2-dichlorobutane, 1,2-dichlorobutane, 1,3-dichlorobutane, 1,4-dichlorobutane and 2,3-dichlorobutane. It is **not unique**.
e. There is only one cyclic ring with 5 carbons and one double bond. This is **unique**.
f. There is only one way in which two chlorines can be bonded to one carbon with two hydrogens. This name is **unique**.

20.21 Explain why "3-methylpropane" is *not* a correct IUPAC name.

Analysis/Plan: Draw the structure which would agree with the incorrect name and determine why it is inappropriate according to IUPAC rules.
Work:

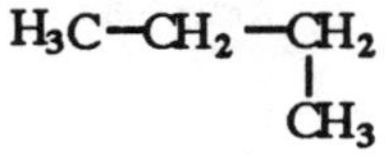

The longest continuous chain has 4 carbons. This should be called **butane**, instead of the name presented in the problem

20.23 Explain why "*cis*-1,2-dichloroethane" is *not* a correct IUPAC name.

Analysis/Plan: Draw the structure which would agree with the incorrect name and determine why it is inappropriate according to IUPAC rules.
Work:

```
   Cl Cl
   |  |
H—C—C—H
   |  |
   H  H
```

There is free rotation about the C-C bond, therefore, there is no restriction of the chlorine atoms to these positions as there is in molecules with rings or double bonds. The IUPAC name is **1,2-dichloroethane**.

20.25 Draw the structural formula corresponding to each of the following.

a. *cis*-CHCl=CHCl
b. $CH_3CH_2C(CH_3)_3$
c. $(CH_3)_3CCH_2CH_2CH{=}CHCH_3$
d. $C(CH_3)_4$

Analysis/Plan: Expand the structure according to the rules discussed in the text, being certain to give each carbon 4 bonds.
Work:

a. 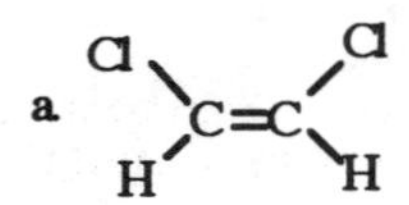

b.
```
                CH3
                |
CH3—CH2—C—CH3
                |
                CH3
```

c.
```
       CH3
       |
CH3—C—CH2—CH2—CH=CH—CH3
       |
       CH3
```

d.
```
       CH3
       |
CH3—C—CH3
       |
       CH3
```

20.27 Name each of the compounds in Problem 20.25.

Analysis/Plan: Use the IUPAC rules discussed in the text.
Work:

a. *cis*-1,2-dichloroethene
b. 2,2-dimethylbutane
c. 6,6-dimethyl-2-heptene
d. 2,2-dimethylpropane

20.29 Draw structural formulas and give names for as many structural and geometrical isomers of C_5H_{10} as you can. [*Hint:* Do not overlook cyclic structures. There are 12 possible isomers.]

Analysis/Plan: Begin with the longest straight chain isomer, then begin to remove $-CH_3$ groups to be used as substituents. Shift them into different positions. Repeat the process with the largest possible ring, also using $-CH_2CH_3$ groups, if necessary.
Work:

$CH_3—CH_2—CH_2—CH{=}CH_2$

1-pentene

```
CH3—CH2     CH3
       \   /
        C=C
       /   \
      H     H
```

cis-2-pentene

```
CH3—CH2     H
       \   /
        C=C
       /   \
      H     CH3
```

trans-2-pentene

```
CH3—CH—CH=CH2
       |
       CH3
```

3-methyl-1-butene

```
CH3—C=CH—CH3
       |
       CH3
```

2-methyl-2-butene

```
CH3—CH2—CH=CH2
             |
             CH3
```

2-methyl-1-butene

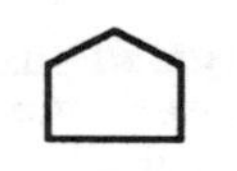

cyclopentane

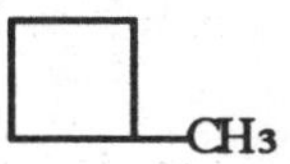

methylcyclobutane 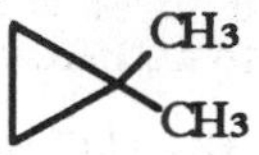 | 1,1-dimethylcyclopropane 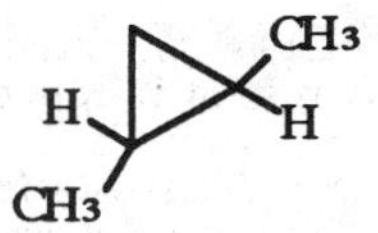 | *trans*-1,2-dimethylcyclopropane

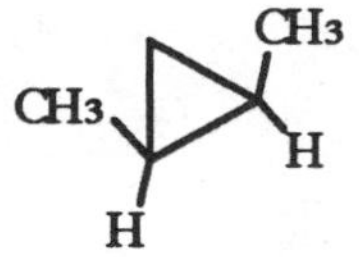

cis-1,2-dimethylcyclopropane 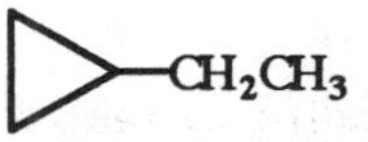 | ethylcyclopropane

20.31 Which of the following compounds can exist as *cis* and *trans* isomers?

a. 1,2-dichloroethane
b. 1,2-dichloroethyne
c. 1,2-dichloroethene
d. 1,1-dichloroethene

Analysis/Plan: To exist as *cis* and *trans* isomers, the substance must have a ring structure or a double bond and the substituent groups must be on different carbons. Analyze the structures and see which fit(s) these criteria.

Work:

a. This has only single bonds, so **cannot** exhibit *cis*/*trans* isomerism.
b. This has a triple bond, so **cannot** exhibit *cis*/*trans* isomerism.
c. This has chlorine substituents on either side of the double bond, so **can** exhibit *cis*/*trans* isomerism.
d. The chlorine substituents are attached to the same carbon on the double bond, so **cannot** exhibit *cis*/*trans* isomerism.

20.33 Draw structures and name isomers of $C_3H_4Cl_2$, other than the two already given in Section 20.1.

Analysis/Plan: Follow the approach used in Problem 20.29. The structures given in Section 20.1 are:

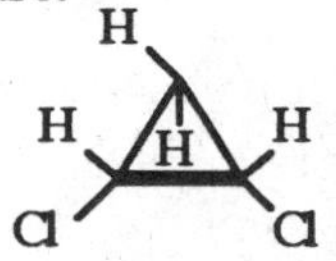

cis-1,2-dichlorocyclopropane 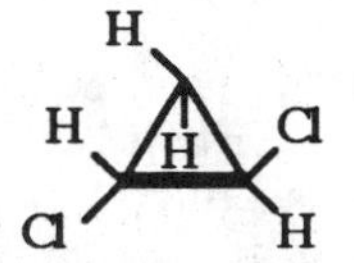 | *trans*-1,2-dichlorocyclopropane

Work:

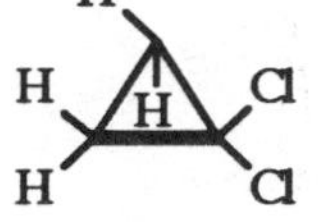

1,1-dichlorocyclopropane 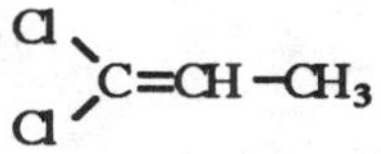 | 1,1-dichloropropene 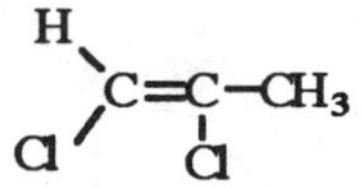 | *cis*-1,2-dichloropropene 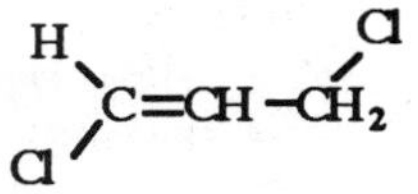 | *trans*-1,3-dichloropropene

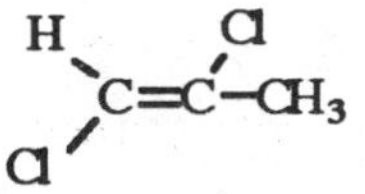

trans-1,2-dichloropropene 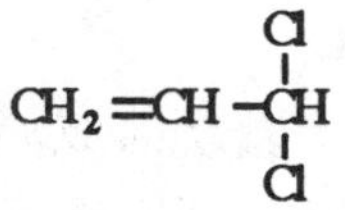 | 3,3-dichloropropene 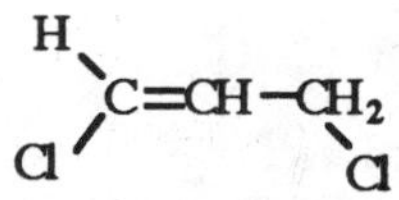 | *cis*-1,3-dichloropropene 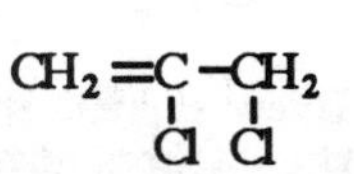 | 2,3-dichloropropene

20.35 Give the name of the class(es) of compounds containing each of the following functional groups.

a. $-CONH_2$ b. $-NHR$ c. $>C=O$

Analysis/Plan: The general functional group structures must be learned.
Work:
a. **amide**   b. **2° amine**   c. **ketones, carboxylic acids, aldehydes, esters, amides**

20.37 Name the classes of compounds containing the following groups.
a. $R-CH_2OH$ b. $RCH(OH)CH_3$ c. $RCHO$ d. $-SO_3H$

Analysis/Plan: The general functional group structures must be learned.
Work:
a. **1° alcohol**   b. **2° alcohol**   c. **aldehyde**   d. **sulfonic acid**

20.39 Draw modified molecular formulas and structural formulas for the following.
a. diethylamine   b. acetic acid   c. acetamide   d. acetone

Analysis/Plan: The general functional group structures and IUPAC naming rules must be learned and applied. To draw the structural formula, expand the modified formula, being sure to give each carbon four bonds.
Work:

a. $(CH_3CH_2)_2NH$ $CH_3-CH_2-N(H)-CH_2-CH_3$

b. CH_3CO_2H $CH_3-C(=O)-OH$

c. CH_3CONH_2 $CH_3-C(=O)-NH_2$

d. $(CH_3)_2CO$ $CH_3-C(=O)-CH_3$

20.41 Draw modified molecular formulas and structural formulas of three isomeric alcohols C_4H_9OH, and label them as 1°, 2°, or 3°.

Analysis/Plan: Follow the general pattern for writing the structure of isomers. A 1° alcohol has the -OH bonded to a carbon bonded to one other carbon, a 2° alcohol has the -OH bonded to a carbon bonded to two other carbons, and a 3° alcohol has the -OH bonded to a carbon bonded to three other carbons.
Work:

1° $CH_3(CH_2)_2CH_2OH$ $CH_3-CH_2-CH_2-CH_2OH$ (1-butanol)

2° $CH_3CH_2CHOHCH_3$ $CH_3-CH_2-CH(OH)-CH_3$ (2-butanol)

3° $(CH_3)_3COH$ $CH_3-C(CH_3)_2-OH$ (2-methyl-2-propanol)

20.43 Draw structural formulas for acetone and 2-butanone, and specify the hybridization of each of the carbon atoms in these molecules. Generalize your results to describe the hybridization of the carbon atoms in any ketone $CH_3(CH_2)_nCOCH_3$.

Analysis/Plan: The names of the compounds determine the structural formula. Count the electron regions about each carbon to determine the hybridization about the carbon atom.
Work:

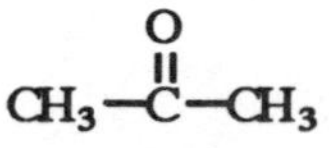

$CH_3-C(=O)-CH_3$ (acetone) $\quad CH_3-CH_2-C(=O)-CH_3$ (2-butanone)

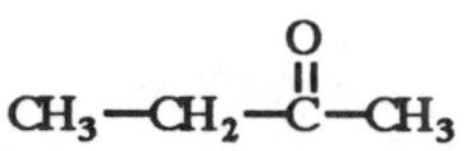

In acetone, the carbon doubly bonded to the oxygen is sp^2 hybridized, while the other two carbons are sp^3 hybridized. In 2-butanone, the carbon doubly bonded to the oxygen is sp^2 hybridized, while the other three carbons are sp^3 hybridized.

The general rule is that for any ketone of the formula $CH_3(CH_2)_nCOCH_3$, the carbon doubly bonded to the oxygen is sp^2 hybridized and the other (n + 2) carbons are sp^3 hybridized.

20.45 How many N-H bonds are there in a secondary amine?

In a secondary amine, the nitrogen is bonded to two carbons. Because nitrogen characteristically makes only three bonds, there can be only **one** N-H bond.

20.47 What is the major organic product expected from each of the following reactions? (Do not try to balance the equations.)

a. $CH_2{=}CH_2 + H_2O \longrightarrow ?$

b. $CH_3(CH_2)_3CH_2OH + KMnO_4(aq) \longrightarrow ?$

c. $CH_3CH_2OH + Na \longrightarrow ?$

d. $CH_3CH_2OH + HCO_2H \longrightarrow ?$

Analysis/Plan: Inspect the nature of the reactants. The general classes of organic reactions must be learned to be able to predict the product.
Work:

a. Reaction at a double bond usually occurs by addition. H adds to one side of the double bond, and OH to the other side, so the product is $\mathbf{CH_3CH_2OH}$.
b. A primary alcohol is being oxidized by a strong oxidizing agent. The product is the carboxylic acid, $\mathbf{CH_3(CH_2)_3COOH}$.
c. Alcohols react with sodium in a similar manner to that of water, producing a strong base that can dissociate in solution. The products are $\mathbf{Na^+}$ and $\mathbf{CH_3CH_2O^-}$.
d. An alcohol and a carboxylic acid react to form an ester, with water as a byproduct. The ester is $\mathbf{HCO_2CH_2CH_3}$.

20.49 What is the major organic product expected from each of the following reactions?

a. 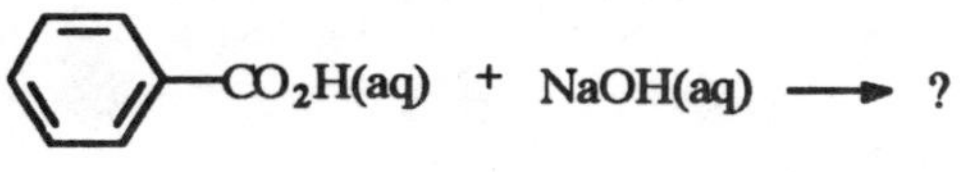 $C_6H_5-CO_2H(aq) + NaOH(aq) \longrightarrow ?$

b. cyclohexanone + $\xrightarrow[H_2O]{NaBH_4}$?

c. 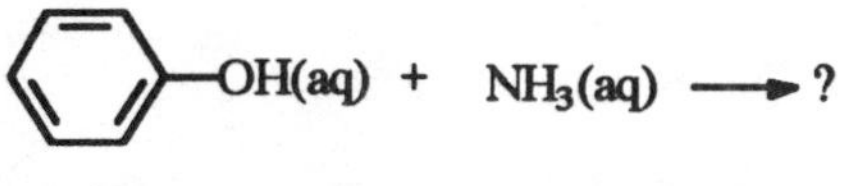 $C_6H_5-OH(aq) + NH_3(aq) \longrightarrow ?$

Analysis/Plan: Inspect the nature of the reactants. The general classes of organic reactions must be learned to be able to predict the product.
Work:

a. The carboxylic acid donates a proton to the strong base, creating its anion. The product is shown at the right, but could also be shown as the sodium salt.

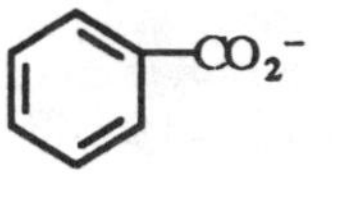

$C_6H_5-CO_2^-$

b. $NaBH_4$ is a potent reducing agent. It converts the cyclohexanone to the alcohol. The product is cyclohexanol.

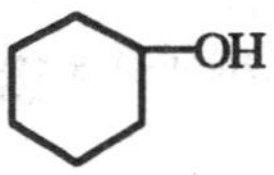

c. Phenol is an acid and can give a proton to the weak base NH_3, creating the phenoxide ion.

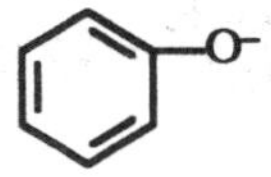

20.51 Calculate the theoretical yield of the organic product if 15 g each of acetaldehyde and $KMnO_4$ react in aqueous solution. The equation was balanced in Problem 20.50 and is:

$$2\ H^+ + 3\ CH_3CHO + 2\ MnO_4^- \longrightarrow H_2O + 3\ CH_3CO_2H + 2\ MnO_2.$$

Analysis:

Target: The mass of the product, acetic acid. (? g CH_3CO_2H) =

Knowns: The mass of CH_3CHO, the mass of $KMnO_4$ and the balanced chemical equation. The molar masses of the reactants and product are implicit knowns.

Relationships: The balanced chemical equation gives mole ratios of all species. The molar masses provide the equivalence statements g ↔ mol.

Plan: We recognize that this is a limiting reactant problem because starting amounts of both reactants are given. Consult the discussion in the plan for Problem 3.55.

Step 2. Compare the amounts of CH_3CO_2H made if each reactant could be totally consumed.

Step 1. Use the standard stoichiometric procedure with the appropriate conversion factors to predict the amount of product that could be made from each reactant.

$$g\ CH_3CHO \longrightarrow mol\ CH_3CHO \longrightarrow mol\ CH_3CO_2H \longrightarrow g\ CH_3CO_2H$$

$$g\ KMnO_4 \longrightarrow mol\ KMnO_4 \longrightarrow mol\ CH_3CO_2H \longrightarrow g\ CH_3CO_2H$$

Work:

Step 1. The SSP for each reactant can be treated as a single 3 step conversion process. The molar masses required are:

CH_3CHO = 2(12.011 C) + 4(1.0079 H) + 1(15.9994 O) = 44.053 g mol^{-1}

$KMnO_4$ = 1(39.0983 K) + 1(54.9380 Mn) + 4(15.9994 O) = 158.0339 g mol^{-1}

CH_3CO_2H = 2(12.011 C) + 4(1.0079 H) +2(15.9994 O) = 60.053 g mol^{-1}

The equivalence statements for moles are:

3 mol CH_3CHO ↔ 3 mol CH_3CO_2H; 2 mol $KMnO_4$ ↔ 3 mol CH_3CO_2H

$$?\ g\ CH_3CO_2H = 15\ g\ CH_3CHO \times \frac{1 mol\ CH_3CHO}{44.053\ g\ CH_3CHO} \times \frac{3\ mol\ CH_3CO_2H}{3\ mol\ CH_3CHO} \times$$

$$\frac{60.053\ g\ CH_3CO_2H}{1\ mol\ CH_3CO_2H} = 21\ g\ CH_3CO_2H$$

$$?\ g\ CH_3CO_2H = 15\ g\ KMnO_4 \times \frac{1\ mol\ KMnO_4}{158.0339\ g\ KMnO_4} \times \frac{3\ mol\ CH_3CO_2H}{2\ mol\ KMnO_4}$$

$$\times \frac{60.053\ g\ CH_3CO_2H}{1\ mol\ CH_3CO_2H} = 8.6\ g\ CH_3CO_2H$$

Step 2. Compare the two amounts of product.

8.6 g is less than 21 g. Only **8.6 g CH_3CO_2H** can be made.

Note that if potassium acetate is isolated instead of the acid, the $KMnO_4$ would still be the limiting reactant. The molar mass of CH_3CO_2K = 2(12.011 C) + 3(1.0079 H) + 2(15.9994 O) + 1(39.0983 K) = 98.143 g mol^{-1}. The theoretical yield of CH_3CO_2K is shown on the next page.

$$? \text{ g } CH_3CO_2K = 15 \text{ g } KMnO_4 \times \frac{1 \text{ mol } KMnO_4}{158.0339 \text{ g } KMnO_4} \times \frac{3 \text{ mol } CH_3CO_2K}{2 \text{ mol } KMnO_4}$$

$$\times \frac{98.143 \text{ g } CH_3CO_2K}{1 \text{ mol } CH_3CO_2K} = 14 \text{ g } CH_3CO_2K$$

20.53 **The examples of condensation polymers given in the text are all copolymers, that is, they contain two different monomers. Is it possible for a single monomer to polymerize so as to form a condensation *homo*polymer? If so, suggest an example. If not, explain why not.**

This is possible if the monomer has different groups capable of reacting at either end. For example, if one end is an alcohol and the other end is a carboxylic acid, a polyester could be made. $HOCH_2CH_2CO_2H \rightarrow [-CH_2CH_2CO_2-]_n$. If one end is an amine and the other end is a carboxylic acid, a polyamide could be formed. $NH_2CH_2CH_2CO_2H \rightarrow [CH_2CH_2CONH-]_n$

20.55 **Nylon is decomposed by acids, while polyethylene is not. Suggest an explanation for this difference in behavior.**

Nylon is a polyamide and is subject to hydrolysis at the amide bonds, in a reaction that is the reverse of the condensation reaction by which it is made. Polyethylene is a long chain alkane and, as such, is quite unreactive under normal conditions.

20.57 **Name two types of polymer you have used today. Give the type (addition or condensation) and the monomer(s).**

See the Tables 20.3 and 20.4 in the text and determine which of these you have used and their nature.

20.59 **What is the percentage (by mass) of chlorine in Saran Wrap?**

Analysis:

Target: The percent by mass chlorine in Saran Wrap. (? % Cl) =

Known: Table 20.4 in the text indicates that the general formula for Saran Wrap is $[CH_2CCl_2]_n$. The molar mass is an implicit known.

Relationship: The percent of an element in a compound is found from the relationship:

$$\frac{\text{mass of the element in one mole of the compound}}{\text{mass of one mole of the compound}} \times 100$$

The percent in the empirical unit is the same as the percent in the whole compound.

Plan: Calculate the molar mass of an empirical unit (CH_2CCl_2) of Saran and then apply the above relationship.

Work:

Calculate the molar mass of CH_2CCl_2.

C	2 mol x 12.011 g mol^{-1}	=	24.022 g C
Cl:	2 mol x 35.453 g mol^{-1}	=	70.906 g Cl
H:	2 mol x 1.0079 g mol^{-1}	=	2.0158 g H
CH_2Cl_2:	1 mol	=	96.944 g CH_2Cl_2

$$? \% \text{ Cl} = \frac{70.906 \text{ g Cl}}{96.944 \text{ g } CH_2Cl_2} \times 100 = \mathbf{73.141\% \text{ Cl}}$$

21 Molecules of Life

21.1 What is polarized light, and what is meant by the term "optical activity"?

In polarized light, the electric fields of all photons are parallel to each other. Optical activity is the ability of a substance to rotate the plane of the polarized light passing through it.

21.3 Define the following terms: "chiral molecule or compound", "chiral carbon atom", "enantiomers", "racemic mixture".

A chiral molecule is optically active and can rotate the plane of polarized light. A chiral carbon atom is one which is bonded to four different groups or atoms. This means that it is capable of stereoisomerism. Enantiomers are stereoisomers which are nonsuperimposible mirror images of each other. A racemic mixture is an equimolar mixture of two enantiomers. The racemic mixture does not rotate the plane of polarized light because the effects of the two stereoisomers cancel each other out.

21.5 Which of the following can exist as a pair of enantiomers?

a. CHBrClF b. HClC=CClH c. CCl_2F_2 d. $CH_3-CH(NH_2)-CO_2H$

Analysis/Plan: To exist as enantiomers, the molecule must have at least one chiral carbon (a carbon bonded to four different groups or atoms). Inspect the structures and determine which fit this classification.

Work:

a. The carbon is bonded to four different atoms (H, Br, Cl and F). The compound **can exist** as a pair of enantiomers.

b. The carbons are only bonded to three different atoms or groups (H, Cl and CClH). The compound **cannot** exist as a pair of enantiomers.

c. The carbon is only bonded to only two different atoms (F and Cl). It **cannot exist as a** pair of enantiomers.

d. The central carbon is bonded to four different groups or atoms (H, CH_3, NH_2 and COOH). The compound **can** exist as a pair of enantiomers.

21.7 How many chiral carbons are there in 2-bromobutane? How many optical isomers of this molecule are there?

Analysis: The number of chiral carbons is determined by the number of carbons bonded to four different atoms or groups. The maximum number of optical isomers is 2^n, where n is the number of chiral carbons.

Plan: Use the nomenclature rules to draw the structure of 2-bromobutane. Determine the number of chiral carbons from the structure and use the 2^n rule.

Work:

```
   H  Br H  H
   |  |  |  |
H—C—C—C—C—H
   |  |  |  |
   H  H  H  H
```

2-bromobutane

Only the "2" carbon to which the Br is bonded has four different groups or atoms (H, Br, CH_3 and CH_2CH_3) attached. There is **one chiral carbon**. There can be 2^1 = **2 optical isomers**.

21.9 How many chiral carbons are there in the cyclic form of glucose? How many optical isomers of this molecule are there?

Analysis: The number of chiral carbons is determined by the number of carbons bonded to four different atoms or groups. The maximum number of optical isomers is 2^n, where n is the number of chiral carbons.

Plan: The cyclic structures of glucose are shown in Figure 21.12 in the text and the α- form is presented below. Determine the number of chiral carbons from the structure and use the 2^n rule.

Work:

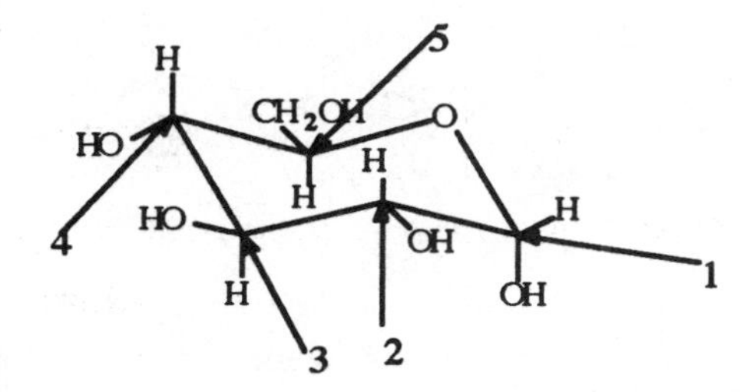

Inspection of the structure indicates that there are **5 chiral carbons**. There can be 2^5 = **32 optical isomers**.

21.11 Sketch the structures of the enantiomers of alanine.

Analysis/Plan: The structure of alanine is shown in Figure 21.1 in the text. The enantiomers have two groups interchanged to make the nonsuperimposable mirror images.

Work:

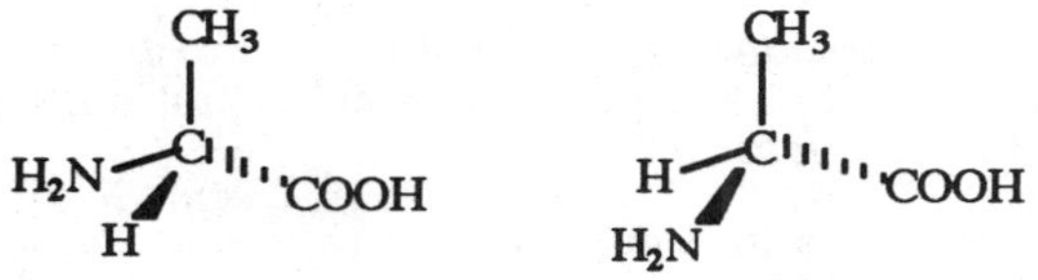

CH_3 | H_2N—C····COOH, H (wedge)

CH_3 | H—C····COOH, H_2N (wedge)

21.13 What type(s) of interaction can occur between the side-chains of glutamic acid and serine?

Glutamic acid has a carboxylic acid (-COOH) on the side-chain, while serine has an alcohol (-OH) group. See the structures in Figure 21.1 in the text. In addition to hydrogen bonding and dipole/dipole interactions, these can undergo a condensation reaction, forming the ester and releasing water as a byproduct.

21.15 Draw the structure of a typical α-amino acid. What do you think the structure of a β-amino acid might be?

On an α-amino acid, the amine group is on the first carbon next to the carboxylic acid group. In a β-amino acid, the amine group is on the second carbon from the carboxylic acid.

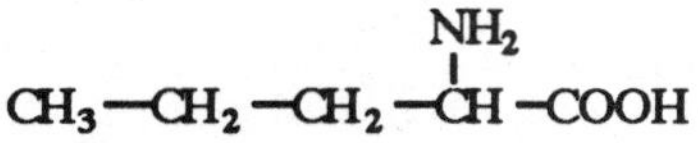

$CH_3—CH_2—CH_2—CH(NH_2)—COOH$

α-amino acid

$CH_3—CH_2—CH(NH_2)—CH_2—COOH$

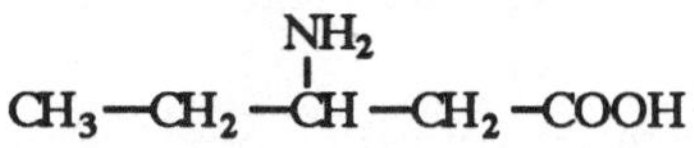

β-amino acid

21.17 Draw the structural formulas of the two dipeptides that can be formed from alanine and serine.

Analysis/Plan: Look up the structures of the two amino acids in Figure 21.1 in the text. In one dipeptide, the serine provides the NH_2 for the amide (peptide) bond and in the other dipeptide, it provides the carboxylic acid for the amide bond.

Work:

$$H_2N-CH(CH_3)-C(=O)-OH$$
alanine

$$H_2N-CH(CH_2OH)-C(=O)-OH$$
serine

$$H_2N-CH(CH_3)-C(=O)-N(H)-CH(CH_2OH)-C(=O)-OH$$
ala-ser

$$H_2N-CH(CH_2OH)-C(=O)-N(H)-CH(CH_3)-C(=O)-OH$$
ser-ala

21.19 How many different tripeptides can be formed from aspartic acid, serine, and tryptophan? Write their primary structures in the abbreviated form (asp-ser-trp, for instance).

Analysis/Plan: Shift the groups until all possible combinations have been made.
Work:
Possible combinations are: **asp-ser-trp, asp-trp-ser, ser-asp-trp, ser-trp-asp, trp-ser-asp**, and **trp-asp-ser**. This is **six** different tripeptides.

21.21 Consider lysine, an amino acid having a basic side-chain. Draw a structural formula indicating the form you expect this substance to have in strongly acid solution, say when the pH is 3 or less. What is its structure in basic solution?

Analysis/Plan: Look up the structure of lysine in Figure 21.1 in the text. In acidic solution, both amino groups will be protonated. In basic solution, the acid group will be deprotonated.
Work:

$$H_2N-CH((CH_2)_4NH_2)-C(=O)-OH$$
lysine

$$^{+}H_3N-CH((CH_2)_4NH_3^{+})-C(=O)-OH$$
in acid

$$H_2N-CH((CH_2)_4NH_2)-C(=O)-O^{-}$$
in base

21.23 What is an essential amino acid? How many are there?

An essential amino acid cannot be synthesized by human beings, but, instead, must be ingested in the diet. There are eight essential amino acids.

21.25 Name some functions performed by proteins in living organisms.

Proteins transport oxygen, serve as connective tissues, catalyze reactions, provide structural support for tissues, etc.

21.27 What forces are responsible for creating or maintaining the 1° structure of proteins? The 2° structure? The 3° structure?

The 1° structure is the sequence of amino acids and is held in place by the covalent bonds formed when the amino group condenses with the carboxylic acid group to form the peptide bond. The 2° structure is the way that the peptide bonds interact to make the characteristic structure (typically an α–helix or a β-pleated sheet) and is maintained by hydrogen bonding between the peptide bonds. The 3° structure is the way in which the secondary structure is twisted and folded to make a three-dimensional array. It is held in place by additional hydrogen bonding between subchains, dipole/dipole interactions, electrostatic interactions and by disulfide (S-S) bonds.

21.29 Distinguish between *aldose* and *ketose.*

Both aldoses and hexoses are simple sugars (polyhydroxy aldehydes and ketones). In aldoses, the carbonyl group is an aldehyde (R-CHO), while, in ketoses, the carbonyl group is a ketone (R-CO-R').

21.31 Why are carbohydrates generally soluble in water, but not in non-polar organic solvents?

Because carbohydrates are rich in alcohol groups, they are quite polar. They are soluble in water, a very polar solvent with which they can hydrogen bond. Organic solvents are often nonpolar and do not provide the intermolecular attractions necessary for high solubility.

21.33 What is the difference between the α and β forms of a carbohydrate?

When a cyclic sugar molecule is formed, a carbon atom which was originally bonded to three different groups is now bonded to a fourth group (the ring closing oxygen). This new chiral carbon is capable of stereoisomerism, one form of which is called the α form and the other the β. See Figure 21.12 in the text for the α and β forms of glucose.

21.35 Look at Figure 21.15, and explain why amylose, (a component of starch) is sometimes described as a polymer of maltose rather than glucose.

Maltose is composed of two α-glucose molecules and starch is a polymer of α-glucose molecules. Because all glycosidic linkages are α, a polymer of maltose would be identical to a polymer of α–glucose.

21.37 How is the structure of starch different from that of cellulose?

In starch, all glycoside linkages are α-1,4, while in cellulose, they are all β-1,4.

21.39 List the components of the two types of nucleic acid.

DNA is a polymer in which each monomer unit is composed of (1) the sugar deoxyribose; (2) a phosphate; and (3) a nitrogen base (adenine, thymine, guanine and cytosine). RNA is a polymer in which each monomer unit is composed of (1) the sugar ribose; (2) a phosphate; and (3) a nitrogen base (adenine, uracil, guanine and cytosine). In both nucleic acids, the "backbone" is an alternating sequence of the phosphate and the sugar, held by an ester linkage. The nitrogen bases are bonded to the sugar.

21.41 Distinguish between the 1° and 2° structure of DNA.

The 1° structure is the sequence of nitrogen bases on the sugar-phosphate backbone. The 2° structure is the α-helix in which hydrogen bonding between the bases holds the two strands of DNA together.

21.43 Suppose one strand of a DNA double helix has a segment with the base sequence -GGACA-. What must be the sequence on the opposite, complementary strand?

In DNA, cytosine is always paired with guanine and adenine is paired with thymine. The complementary strand would be **-CCTGT-**.

21.45 How are lipids defined?

A lipid is any biological molecule which is soluble in nonpolar solvents such as carbon tetrachloride.

21.47 What is meant by the terms "*saturated*" and "*polyunsaturated*" when applied to fats and oils?

In a saturated fat, there are no carbon-carbon double bonds. In a polyunsaturated substance, there are at least two carbon-carbon double bonds. Because more double bonds lower the melting point of the fat, a polyunsaturated compound is more likely to be an oil, which means that the fat is liquid at room temperature.

21.49 What is a triglyceride? What classes of naturally-occurring substances are triglycerides?

A triglyceride is an ester produced by the condensation reaction of three fatty acid molecules with one molecule of glycerol (1,2,3-propanetriol). Fats, oils and waxes are triglycerides.

21.51 Why are steroids classified as lipids even though they are not triglycerides?

They are classified as lipids because they are biological molecules which are soluble in nonpolar solvents.

21.53 What is meant by the term "hydrophobic"? Give some examples of hydrophobic substances.

Hydrophobic means "water-hating". Hydrophobic substances are nonpolar substances which will not dissolve in water, which is a polar molecule. Fats, oils, steroids, carbon tetrachloride, benzene and all hydrocarbons are hydrophobic.

21.55 What distinguishes a phosphoglyceride from a fat or oil?

In a phosphoglyceride, a phosphate group, rather than a fatty acid, is esterified to one of the alcohol groups on the glycerol molecule. Therefore, in a phosphoglyceride there are only two fatty acid esters, while in a fat or oil there are three.

21.57 What products would be expected from the saponification of one mole of a phosphoglyceride?

Saponification hydrolyzes all the ester bonds. Products are one mole of glycerol, one mole of phosphate ester and two moles of fatty acid.

21.59 What is the difference between *exothermic* and *exergonic*? Is it possible for a reaction to be both exothermic and endergonic?

A process is exothermic if $\Delta H < 0$. It is exergonic if $\Delta G < 0$. $\Delta G = \Delta H - T\Delta S$. A process can be both exothermic and endergonic if the $T\Delta S$ term is negative and of greater magnitude than the ΔH term. This will happen if ΔH is small and negative and if ΔS is large and negative or if ΔS is negative and the temperature is sufficiently high.

21.61 How many moles of ATP must be converted to ADP in order to drive a reaction whose free energy change is +962 kJ per mole of reaction?

Analysis/Plan: The text (Equation 21.2) indicates that the conversion of ATP to ADP is exergonic with $\Delta G = -33$ kJ mol^{-1}. Divide the energy required by that released in the conversion reaction to determine the number of moles of ATP needed.

Work:

$$? \text{ mol ATP} = \frac{962 \text{ kJ}}{33 \text{ kJ mol}^{-1} \text{ ATP}} = \textbf{29 mol ATP}$$

21.63 A second phosphate group can be lost from ADP, forming adenosine monophosphate (AMP). The reaction ADP + H_2O ⟶ AMP + P_i is exergonic by about 30 kJ. How many moles of ATP must be converted to AMP in order to drive a reaction that requires 672 kJ, if the process is 62.2% efficient?

Analysis/Plan: We can use Hess's Law to determine the amount of energy release in the conversion ATP ⟶ AMP. Divide the energy required but that released in the conversion and use the factor-label method with the efficiency factor to determine the number of moles of ATP required.

Work:

Find the energy released in the ATP ⟶ AMP conversion process.

ATP + H_2O ⟶ ADP + P_i	$\Delta G = -33$ kJ (Equation 21.2 in the text)
ADP + H_2O ⟶ AMP + P_i	$\Delta G = -30$ kJ
ATP + 2 H_2O ⟶ AMP + 2 P_i	$\Delta G = -33 + (-30) = -63$ kJ

Determine the number of moles of ATP needed, dividing and using the factor-label method with the efficiency factor.

$$? \text{ mol ATP} = \frac{672 \text{ kJ}}{63 \text{ kJ mol}^{-1} \text{ ATP}} \times \frac{100}{62.2} = \textbf{17 mol ATP}$$

22 The Main-Group Elements

22.1 How many neutrons are there in each of the stable isotopes of boron?

Analysis/Plan: The text (Section 22.1) indicates that the two stable isotopes of boron are B-10 and B-11. The number which follows the symbol is the mass number which is equal to the sum of the number of protons and neutrons in the nucleus. The atomic number gives the number of protons and is equal to 5. The number of neutrons can be determined by subtraction. Review Chapter 2 in the text if necessary.

Work:

B-10 has 10 - 5 = **5 neutrons**.
B-11 has 11 - 5 = **6 neutrons**.

22.3 What reducing agent is used in the preparation of elemental boron from its ore?

Magnesium is used to prepare boron from its ore. The reaction requires heat and is:

$B_2O_3(s) + 3\ Mg(s) \longrightarrow 2\ B(s) + 3\ MgO(s)$.

22.5 Boron compounds display two unusual types of bonding. Describe them, and give examples.

In many compounds, such as the boron halides (BF_3, BCl_3, etc.), boron has only six electrons in its valence shell. This arrangement which gives boron a formal charge of zero and places a zero formal charge on the more electronegative halogens. In compounds with boron and hydrogen (boranes), there are some bonds in which three atoms, rather than two, share a pair of electrons.

22.7 What is the empirical formula of the binary compound of boron and chlorine? What is the molecular shape of this and other boron halides?

The empirical formula of the boron/chlorine compound is BCl_3. All boron halides have a trigonal planar molecular shape.

22.9 The acid ionization constant of boric acid is 7.3×10^{-10}. Calculate the pH of a 0.050 M solution.

Analysis:

Target: The pH of a solution of weak acid. (? pH) =

Knowns: The molarity and the K_a of the acid.

Relationships: The structured approach to equilibrium problems is appropriate. (Review Chapter 13 if necessary.) pH = $-\log [H_3O^+]$.

Plan: Use the structured approach to equilibrium problems to determine the $[H_3O^+]$ and then use the formula to determine the pH.

Work:

Balanced	H_3BO_3 +	H_2O	$\rightleftharpoons$	H_3O^+	+	$H_2BO_3^-$
Initial	0.050			0		0
Change	- x			+ x		+ x
Equilibrium	0.050 - x			x		x

Insert the values into the K_a expression.

$$K_a = \frac{[H_3O^+][H_2BO_3^-]}{[H_3BO_3]} = \frac{(x)(x)}{(0.050 - x)} = 7.3 \times 10^{-10}$$

Solve algebraically. Because K_a is very small, assume $0.050 - x \approx 0.050$.

$\frac{x^2}{0.050} = 7.3 \times 10^{-10}$; therefore, $x^2 = 0.050(7.3 \times 10^{-10}) = 3.6 \times 10^{-11}$; $x = \sqrt{3.6 \times 10^{-11}}$

$x = [H_3O^+] = 6.0 \times 10^{-6}$ M; pH = -log $[H_3O^+]$ = -log (6.0×10^{-6}) = -(-5.22) = **5.22**

Check: $\frac{6.0 \times 10^{-6}}{0.050} = 0.00012$; so the approximation that $0.050 - x \approx 0.050$ is valid.

22.11 Give the name and formula of a boron compound that (a) reacts with water to give HBr, (b) is a good reducing agent, (c) is a useful abrasive, and (d) results from the strong heating of boric acid.

a. BBr_3 (boron tribromide) reacts with water to give HBr. The reaction is:

$BBr_3(\ell) + 3\ H_2O(\ell) \longrightarrow H_3BO_3(aq) + 3\ HBr(aq)$.

b. $NaBH_4$ (sodium borohydride) is a good reducing agent. We saw it used for the reduction of aldehydes and ketones to alcohols in Chapter 20.

c. BN (boron nitride) is a good abrasive.

d. B_2O_3 (diboron trioxide, more commonly known as boric oxide) is one of the decomposition products of boric acid, H_3BO_3. The reaction is: $2\ H_3BO_3(s) \longrightarrow B_2O_3(s) + 3\ H_2O(g)$

22.13 Carbon has two stable isotopes. Identify them, and give their approximate natural abundances.

Carbon is about 99% C-12 and about 1% C-13.

22.15 Which allotrope of carbon is useful as a lubricant? Explain.

Graphite is a good lubricant, because it is organized in layers. The weak forces between the separate planes allow the layers to slide easily against each other.

22.17 Give the name and formula of a compound that reacts with water to produce methane. Write a balanced equation for the reaction.

Beryllium carbide, Be_2C reacts with water to produce methane. The reaction is:

$Be_2C(s) + 4\ H_2O(\ell) \longrightarrow 2\ Be(OH)_2(s) + CH_4(g)$

22.19 Give the name and formula of three oxides of carbon. Are these molecular or ionic compounds? Which of them reacts with metals to form adducts?

The three oxides are CO (carbon monoxide), CO_2 (carbon dioxide) and C_3O_2 (carbon suboxide). These are all molecular compounds. CO reacts with metals to form carbonyl compounds such as $Fe(CO)_5$.

22.21 For each of the following species, draw the Lewis structure and indicate the hybridization of the carbon atom.

a. HCO_3^- b. HCN c. HCO_2^-

Analysis:

Target: The Lewis structure and the hybridization of an atom.
Knowns: The formula of the molecule or ion.
Relationships: The hybridization is related to the geometry of the species through the known orientation of orbitals in hybrid sets. This information is summarized in Table

9.1 in the text. The geometry of the species is related to the Lewis structure through VSEPR theory.

Plan: We need to know the Lewis structure and the appropriate relationships.

Step 3. Identify the hybrid orbital set having the proper orientation.
Step 2. Use VSEPR to determine the shape around the central atom.
Step 1. Draw the Lewis structure of the species.

Work:

a. **Step 1.** C is the central atom. The total number of valence electrons is 1(4 C) + 1(1 H) + 3(6 O) + 1(1 charge) = 24. One of the resonant structures is shown on the right.
Step 2. With 3 electron regions about carbon, the electron geometry is trigonal planar.
Step 3. The hybridization is ***sp^2***.

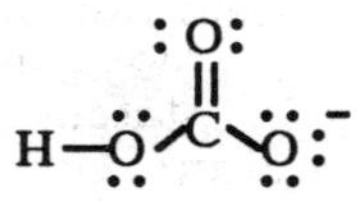

b. **Step 1.** C is the central atom. The total number of valence electrons is 1(4 C) + 1(1 H) + 1(5 N) + 1(1 charge) = 10. The structure is shown on the right.
Step 2. With 2 electron regions, the electron geometry is linear.
Step 3. The hybridization is ***sp***.

H—C≡N:

c. **Step 1.** C is the central atom. The total number of valence electrons is 1(4 C) + 1(1 H) + 2(6 O) + 1(1 charge) = 18. One of the resonant structures is shown on the right.
Step 2. With 3 electron regions, the electronic geometry is trigonal planar.
Step 3. The hybridization is ***sp^2***.

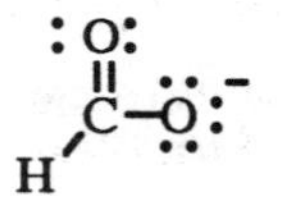

22.23 Rank the following in order of increasing acid strength: H_2CO_3, HCO_2H, HCO_3^-.

Analysis/Plan: These are all weak acids. K_a is a measure of acid strength. An acid with a higher K_a is a stronger acid. Look the appropriate K_a values up in Appendix **J** and compare them.

Work:

HCO_3^- (K_a = 4.0 x 10^{-11}) is a weaker acid than H_2CO_3 (K_a = 4.0 x 10^{-7}) which is a weaker acid than HCO_2H (K_a = 1.8 x 10^{-4}). **HCO_3^- < H_2CO_3 < HCO_2H**

22.25 Would you expect the pH of water dripping from a stalactite in a limestone cavern to be less than 7, approximately 7, or greater than 7? Explain.

The water in the cave should be saturated with calcium carbonate ($CaCO_3$) from the limestone of which many caves are composed. Carbonate ion is a base (K_b = 2.5 x 10^{-4} from K_w/K_a. The K_a of HCO_3^- was given in Problem 22.23.) The solution should be slightly basic because of hydrolysis of this base.

22.27 What carbon compound is the monomer of one of the bases in nucleic acids?

HCN is the monomer of adenine. The polymerization reaction is below.

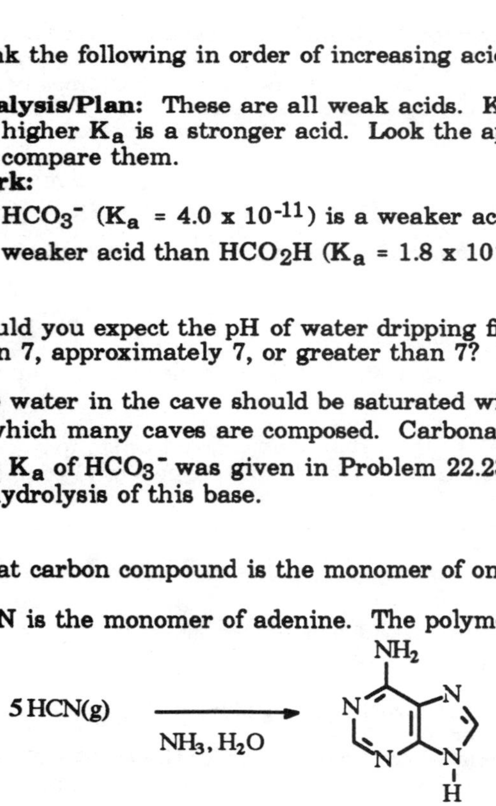

$$5\,HCN(g) \xrightarrow{NH_3,\ H_2O} \text{adenine}$$

22.29 How many liters (STP) of acetylene (ethyne) result from the reaction of 100 g CaC_2 with excess water?

Analysis:

Target: The volume of ethyne (C_2H_2) at STP. (? L C_2H_2) =

Knowns: The mass of CaC_2. The balanced chemical equation is given in the text and is:

$CaC_2(s) + 2\ H_2O(\ell) \longrightarrow Ca(OH)_2(s) + C_2H_2(g)$. The molar masses of the substances are implicit knowns.

Relationships: The balanced chemical equation gives mole ratios of all species. The molar mass provides the equivalence statement g $\leftrightarrow$ mol. At STP, the equivalence statement 1 mol(g) $\leftrightarrow$ 22.414 L is true.

Plan: Use the standard stoichiometric procedure (SSP) discussed in the text, with the appropriate conversion factors.

$$g\ CaC_2 \longrightarrow mol\ CaC_2 \longrightarrow mol\ \ C_2H_2 \longrightarrow L\ \ C_2H_2$$

Work: The SSP can be treated as a single 3 step conversion process.

Molar mass CaC_2 = 1(40.078 Ca) + 2(12.011C) = 64.100 g mol^{-1}

$$?\ L\ C_2H_2 = 100\ g\ CaC_2 \times \frac{1\ mol\ CaC_2}{64.100\ g\ CaC_2} \times \frac{1\ mol\ C_2H_2}{1\ mol\ CaC_2} \times \frac{22.414\ L\ C_2H_2}{1\ mol\ C_2H_2} = \mathbf{35.0\ L\ C_2H_2}$$

22.31 Suggest a reason why no silanes with more than six catenated Si atoms are known.

The Si-Si bond is weak compared to a C-C bond. This means that the molecules are not as stable and therefore long chain length cannot be sustained.

22.33 What is a silicone?

Silicones are long-chain polymers of the form -$[SiR_2O-]_n$. The presence of oxygen in the chain confers a stability to the ring, which is not seen in chains containing only Si-Si bonds.

22.35 What structural feature is common to the silicate minerals? How do aluminosilicates differ from ordinary silicates?

All silicate minerals have a tetrahedral SiO_4 unit. In aluminosilicates, some of the Si sites are replaced by the aluminum atoms (which have a similar radius). Other metal ions are found in silicate minerals, however, their atoms are found outside of the tetrahedron rather than replacing a silicon atom in the tetrahedral matrix.

22.37 What accounts for the optical activity of quartz?

Quartz (empirical formula SiO_2) is composed of long -Si-O-Si-O- chains which are helically arranged to form the quartz crystal. For a given crystal, light will be rotated one direction or the other, as the helices within a crystal are either all right-handed or all left-handed.

22.39 How much SiO_2 is produced when 10.0 liters (STP) of silane reacts with an excess of water?

Analysis:

Target: The mass of SiO_2 formed. (? g SiO_2) =

Knowns: The mass of silane, SiH_4, The balanced chemical equation is given in the text and is: $SiH_4(g) + 2\ H_2O(\ell) \longrightarrow SiO_2(s) + 4\ H_2(g)$. The molar masses of the substances are implicit knowns.

Relationships: The balanced chemical equation gives mole ratios of all species. The molar masses provide the equivalence statements g $\leftrightarrow$ mol. At STP, the equivalence statement 1 mol(g) $\leftrightarrow$ 22.4 L is true.

Plan: Use the standard stoichiometric procedure (SSP) discussed in the text, with the appropriate conversion factors.

L SiH_4 ⟶ mol SiH_4 ⟶ mol SiO_2 ⟶ g SiO_2

Work: The SSP can be treated as a single 3 step conversion process.

Molar mass SiO_2 = 1(28.0855 Si) + 2(15.9994 O) = 60.0843 g mol^{-1}

$$? \text{ g } SiO_2 = 10.0 \text{ L } SiH_4 \times \frac{1 \text{ mol } SiH_4}{22.4 \text{ L } SiH_4} \times \frac{1 \text{ mol } SiO_2}{1 \text{ mol } SiH_4} \times \frac{60.0843 \text{ g } SiO_2}{1 \text{ mol } SiO_2} = \mathbf{26.8\ SiO_2}$$

22.41 What are the correct names of the compounds GeH_4, SnH_4, and PbH_4.

These compounds are germane (GeH_4), stannane (SnH_4) and plumbane (PbH_4).

22.43 Both tin and lead form many ionic compounds such as $PbSO_4$ and $Sn(NO_3)_2$, but there are no known compounds of Si^{+2}. Account for this difference in terms of periodic trends.

The ease of ionization and, thus, an increase in metallic character increases as atomic number increases within a group. Lead and tin are more metallic than silicon and are more likely to enter into the ionic compounds characteristic of metals.

22.45 Write an equation for the reaction you would expect to occur between plumbane and water.

Because plumbane is in the same family as silane, a reaction similar to that of silane with water would be expected. The reaction is:

$PbH_4 + 2\ H_2O(\ell) \longrightarrow PbO_2(s) + 4\ H_2(g)$.

22.47 One of the two stable isotopes of nitrogen has a natural abundance of 0.37%. How many neutrons are contained in the nucleus of this isotope?

Analysis/Plan: See the discussion in Problem 22.1 Determine the isotope by reading the text and apply the rules.

Work:

The isotope is N-15 (Section 22.4). N has 7 protons, so has 15 - 7 = **8 neutrons**.

22.49 With what industrial process is the name Fritz Haber associated? Write the equation(s) for this process.

The Haber process is for the synthesis of ammonia from its constituent elements.

$$N_2(g) + 3H_2(g) \xrightarrow[\text{Fe}]{500 \text{ atm, } 450^\circ\text{C}} 2\ NH_3(g)$$

22.51 Give the formulas and names of, and the hybridization of the nitrogen atom(s), in the two hydrides of nitrogen.

Analysis:

Target: The formula, name and hybridization of an nitrogen in the nitrogen hydrides.

Knowns: The nitrogen hydrides can be looked up in Section 22.4 and are ammonia (NH_3) and hydrazine (N_2H_4).

Relationships: The hybridization is related to the geometry of the species through the known orientation of orbitals in hybrid sets. This information is summarized in Table 9.1 in the text. The geometry of the species is related to the Lewis structure through VSEPR theory.

Plan: We need to know the Lewis structure and the appropriate relationships.

Step 3. Identify the hybrid orbital set having the proper orientation.
Step 2. Use VSEPR to determine the shape around the central atom.
Step 1. Draw the Lewis structure of the species.

Work:

Step 1. In NH_3, N is the central atom. The total number of valence electrons is 1(5 N) + 3(1 H) = 8. The structure is shown on the right.

Step 2. With 4 electron regions, the electron geometry is tetrahedral.

Step 3. The hybridization is sp^3.

```
  ••
H–N–H
  |
  H
```

Step 1. In N_2H_4, the total number of valence electrons is 2(5 N) + 4(1 H) = 14. The structure is shown on the right.

Step 2. With 4 electron regions about each N atom, the electron geometry is tetrahedral.

Step 3. The hybridization about each N atom is sp^3.

```
  •• ••
H–N–N–H
  |  |
  H  H
```

22.53 Draw the dot structures of the following nitrogen-containing ions. Describe the geometry of the triatomic ions.

a. N_3^- (azide ion) b. CN^- c. NCO^- d. CNO^-

Analysis:

Target: The Lewis structure and the geometry of an ion.
Knowns: The formula of the ion.
Relationships: The geometry of the species is related to the Lewis structure through VSEPR theory.

Plan: We need to know the Lewis structure and the appropriate relationships.
Step 2. Use VSEPR to determine the shape around the central atom.
Step 1. Draw the Lewis structure of the species.

Work:

a. **Step 1.** The total number of valence electrons is 3(5 N) + 1(1 charge) = 16. The structure is shown on the right.
Step 2. With 2 electron regions about the central atom, the ion geometry is **linear**.

$:\ddot{N}=N=\ddot{N}:^-$

b. **Step 1.** The total number of valence electrons is 1(4 C) + 1(5 N) + 1(1 charge) = 10. The structure is shown on the right.

$:C\equiv N:^-$

c. **Step 1.** The total number of valence electrons is 1(4 C) + 1(5 N) + 1(6 O) + 1(1 charge) = 16. The structure is shown on the right.
Step 2. With 2 electron regions about the central atom, the ion geometry is **linear**.

$:\ddot{N}=C=\ddot{O}:^-$

d. **Step 1.** The total number of valence electrons is 1(4 C) + 1(5 N) + 1(6 O) + 1(1 charge) = 16. The structure is shown on the right.
Step 2. With 2 electron regions about the central atom, the ion geometry is **linear**.

$:\ddot{C}=N=\ddot{O}:^-$

22.55 What is an endothermic compound, and which of the nitrogen oxides fit this description? Is either CO or CO_2 endothermic?

An endothermic compound is one which has a positive heat of formation. Inspection of Appendix G indicates that **all** of the nitrogen oxides are endothermic compounds. The heats of formation of the carbon oxides are negative; therefore, they are exothermic compounds.

22.57 What are the two oxoacids of nitrogen? Are they strong or weak? Are they stable in aqueous solution? Which oxides are the anhydrides of these acids?

The two oxoacids of nitrogen are nitrous acid (HNO_2) and nitric acid (HNO_3). Nitrous acid is a weak acid which is unstable in aqueous solution. Its anhydride is dinitrogen trioxide (N_2O_3). The reaction is $N_2O_3(g) + H_2O(\ell) \longrightarrow 2\ HNO_2(aq)$. Nitric acid is a strong acid, which is not stable but decomposes fairly slowly. The decomposition is responsible for the brown color concentrated HNO_3 solutions develop on long term storage. The anhydride of nitric acid is N_2O_5. The reaction is $N_2O_5(g) + H_2O(\ell) \longrightarrow 2\ HNO_3(aq)$.

22.59 Like many nitrogen-containing compounds, hydrazine is a base: in acid solution it exists primarily as the $N_2H_5^+$ ion. Hydrazine is also a reducing agent. Write a balanced equation for the redox reaction of hydrazine with IO_3^- in acid solution. The products of the reaction include $N_2(g)$ and $I_2(s)$.

Analysis/Plan: Follow the steps detailed in Section 18.3 of the book for the ion-electron method. They are, for acidic solutions:

1. Write the appropriate half-reactions.
2. Balance all atoms other than H and O.
3. Add H_2O to the side deficient in O to balance O.
4. Add H^+ to the side deficient in H to balance H. This completes the mass balance.
5. Add electrons to the more positive (less negative) side to balance the charge.
6. Multiply each half-reaction by factors chosen to make the number of electrons lost equal to the number of electrons gained. Add the equations, deleting like terms.

Work:

1. Write the appropriate half-reactions.

   $N_2H_5^+ \longrightarrow N_2 \qquad\qquad IO_3^- \longrightarrow I_2$

2. Balance the atoms other than O and H.

   $N_2H_5^+ \longrightarrow N_2 \qquad\qquad 2\ IO_3^- \longrightarrow I_2$

3. Add H_2O to the side deficient in O to balance O.

   $N_2H_5^+ \longrightarrow N_2 \qquad\qquad 2\ IO_3^- \longrightarrow I_2 + 6\ H_2O$

4. Add H^+ to the side deficient in H to balance H.

   $N_2H_5^+ \longrightarrow N_2 + 5\ H^+ \qquad\qquad 12\ H^+ + 2\ IO_3^- \longrightarrow I_2 + 6\ H_2O$

5. Add electrons to the more positive (less negative) side to balance the charge.

   $N_2H_5^+ \longrightarrow N_2 + 5\ H^+ + 4\ e^- \qquad\qquad 10\ e^- + 12\ H^+ + 2\ IO_3^- \longrightarrow I_2 + 6\ H_2O$

6. The oxidation half-reaction has 4 e^-, and the reduction half-reaction has 10 e^-. the lowest common multiple is 20. Multiply the nitrogen half-reaction by 5 and the iodine half-reaction by 2 and add the equations.

   $5[N_2H_5^+ \longrightarrow N_2 + 5\ H^+ + 4\ e^-]$

   $2[10\ e^- + 12\ H^+ + 2\ IO_3^- \longrightarrow I_2 + 6\ H_2O]$

   $5\ N_2H_5^+ + 20\ e^- + 24\ H^+ + 4\ IO_3^- \longrightarrow 5\ N_2 + 25\ H^+ + 20\ e^- + 2\ I_2 + 12\ H_2O$

   Electrons and H^+ are common to both sides. The net ionic equation is:

   $5\ N_2H_5^+(aq) + 4\ IO_3^-(aq) \longrightarrow 5\ N_2(g) + H^+(aq) + 2\ I_2(s) + 12\ H_2O(\ell)$

22.61 In the allotropic form called "white phosphorus", the element occurs as P_4 molecules. Describe the geometry and hybridization in P_4.

According to the text (Section 22.4), the four phosphorus atoms are at the vertices of a tetrahedron. The unnumbered figure shows that each phosphorus atom has three bonds and one lone pair. With four electron regions, the phosphorus is sp^3 hybridized. The Lewis structure is shown on the right.

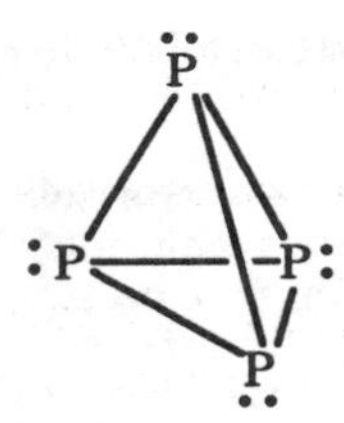

22.63 Write equations for the reaction of (a) the spontaneous combustion of white phosphorus and (b) the reaction of the combustion product with water.

a. Combustion is a reaction with oxygen. $\mathbf{P_4(s) + 5\ O_2(g) \rightarrow P_4O_{10}(s)}$.

b. The product is an acidic anhydride. When reacted with water it yields phosphoric acid.

$$\mathbf{P_4O_{10}(s) + 6\ H_2O(\ell) \rightarrow 4\ H_3PO_4(aq)}$$

22.65 Both H_3PO_2 and H_3PO_3, and their anions, are good reducing agents. In particular, $\mathscr{E}°$ for the half-reaction $H_3PO_4 + 2\ H^+ + 2\ e^- \rightarrow H_3PO_3(aq) + H_2O$ is -0.276 at 25 °C. Write the equation for the reaction that takes place when, under standard conditions of concentration, $H_3PO_3(g)$ is mixed with (a) $Pb^{2+}(aq)$ and (b) $Zn^{2+}(aq)$.

Analysis/Plan: The spontaneous process must have $\mathscr{E}°_{cell} > 0$. Look up the appropriate $\mathscr{E}°_{red}$ in Appendix I. The half-reaction with the more positive $\mathscr{E}°_{red}$ will be the reduction and function as the cathode. The half-reaction with the more negative $\mathscr{E}°_{red}$ will reverse, becoming the oxidation, and will function as the anode.

Work:

a. The two appropriate half-reactions and SRP's are:

$Pb^{2+}(aq) + 2\ e^- \rightarrow Pb(s)$ $\quad \mathscr{E}°_{red} = -0.126$ V

$H_3PO_4 + 2\ H^+ + 2\ e^- \rightarrow H_3PO_3(aq) + H_2O$ $\quad \mathscr{E}°_{red} = -0.276$

The phosphorus reaction reverses and becomes the oxidation. Neither reaction needs to be multiplied because 2 electrons are involved in each half-reaction.

anode: $H_3PO_3(aq) + H_2O \rightarrow H_3PO_4 + 2\ H^+ + 2\ e^-$ $\quad \mathscr{E}°_{ox} = -(-0.763V) = +0.276V$

cathode: $Pb^{2+}(aq) + 2\ e^- \rightarrow Pb(s)$ $\quad \mathscr{E}°_{red} = -0.126$ V

net: $\mathbf{H_3PO_3(aq) + H_2O(\ell) + Pb^{2+}(aq) \rightarrow H_3PO_4(aq) + 2\ H^+(aq) + Pb(s)}$

b. The two appropriate half-reactions and SRP's are:

$Zn^{2+}(aq) + 2\ e^- \rightarrow Zn(s)$ $\quad \mathscr{E}°_{red} = -0.763$ V

$H_3PO_4 + 2\ H^+ + 2\ e^- \rightarrow H_3PO_3(aq) + H_2O$ $\quad \mathscr{E}°_{red} = -0.276$

The zinc reaction reverses and becomes the oxidation. Neither reaction needs to be multiplied because 2 electrons are involved in each half-reaction.

anode: $Zn(s) \rightarrow Zn^{2+}(aq) + 2\ e^-$ $\quad \mathscr{E}°_{ox} = -(-0.763V) = +0.763$ V

cathode: $H_3PO_4 + 2\ H^+ + 2\ e^- \rightarrow H_3PO_3(aq) + H_2O$ $\quad \mathscr{E}°_{red} = -0.276V$

net: $Zn(s) + H_3PO_4(aq) + 2\ H^+(aq) \rightarrow Zn^{2+}(aq) + H_3PO_3(aq) + H_2O(\ell)$

Because this would be the spontaneous process of the cell formed from these two half-reactions, there would be **no reaction** if Zn^{2+} and H_3PO_3 were mixed.

22.67 Which element(s) in Group 5A exist(s) as tetratomic molecules?

Phosphorus (P_4), arsenic (As_4) and antimony (Sb_4) all exist as tetratomic molecules.

22.69 What are the two major sources of sulfur used in the United States?

Sulfur can be mined from deposits that are essentially pure elemental sulfur. In addition sulfur can be recovered from the oxide stack gases from coal and natural gas combustion and from smelting of sulfide ores via the Claus process. The reaction for the Claus process is:

$SO_2(g) + 2\,H_2S(g) \longrightarrow 3\,S(\ell) + 2\,H_2O(g)$.

22.71 What anions are formed by sulfur in its binary ionic compounds?

The most common ion of sulfur found in binary compounds is the S^{2-} ion. The disulfide ion (S_2^{2-}), which is analogous to the peroxide ion (O_2^{2-}), and the polysulfide S_n^{2-} (where n is an integer greater than 2) also are found.

22.73 Complete and balance the following equations.

a. $K_2S + H_2O \longrightarrow$ b. $K_2S + HCl(aq) \longrightarrow$ c. $SF_4 + H_2O \longrightarrow$

a. The sulfide ion is a fairly strong base and hydrolyzes in water. The reaction is:

$\mathbf{K_2S(s) + H_2O(\ell) \longrightarrow 2\,K^+(aq) + HS^-(aq) + OH^-(aq)}$.

b. The reaction of the sulfide ion (a base) with the strong acid HCl produces $H_2S(g)$.

The reaction is: $\mathbf{K_2S(s) + 2\,HCl(aq) \longrightarrow H_2S(g) + 2\,KCl(aq)}$.

c. SF_4 is a reactive gas which hydrolyzes rapidly to yield SO_2 and HF. The reaction is:

$\mathbf{SF_4(g) + 2\,H_2O(\ell) \longrightarrow SO_2(g) + 4\,HF(g)}$.

22.75 There are two common oxides of sulfur, and both are anhydrides of sulfur oxoacids. Name the oxides, and the corresponding oxoacids, and write the equation for the reactions of the oxides with water.

The two common oxides of sulfur are SO_2 (sulfur dioxide) and SO_3 (sulfur trioxide). Sulfur dioxide is the anhydride of sulfurous acid (H_2SO_3). The reaction of sulfur dioxide is:

$\mathbf{SO_2(g) + H_2O(\ell) \longrightarrow H_2SO_3(aq)}$.

Sulfur trioxide is the anhydride of sulfuric acid (H_2SO_4). The reaction is:

$\mathbf{SO_3(g) + H_2O(\ell) \longrightarrow H_2SO_4(aq)}$.

22.77 How much SO_2 is produced from the combustion of 1.00 ton (2000 lbs) of coal containing 5.2% (by mass) sulfur?

Analysis:

Target: The mass of SO_2 formed. (? g SO_2) = and ? (lb) SO_2.=

Knowns: The mass of coal and the percent sulfur in the coal. The balanced chemical equation for combustion is given in the text (Equation 22-63) and is:

$S(s) + O_2(g) \longrightarrow SO_2(g)$. The molar masses of the substances are implicit knowns.

Relationships: The balanced chemical equation gives mole ratios of all species. The molar masses provide the equivalence statements g $\leftrightarrow$ mol. The percent provides the factor $\frac{5.2\text{ g sulfur}}{100\text{ g coal}}$. The equivalence statement 1 lb $\leftrightarrow$ 453.59 g is also needed.

Plan: Determine the mass of sulfur in the coal. Use the standard stoichiometric procedure (SSP) discussed in the text, with the appropriate conversion factors. Use the factor-label method to make mass conversions as necessary.

lb coal ⟶ g coal ⟶ g S ⟶ mol S ⟶ mol SO_2 ⟶ g SO_2 ⟶ lb SO_2

Work: The SSP can be treated as a single multi-step conversion process.
Molar Mass S = 1(32.06 S) = 32.06 g mol^{-1}
Molar Mass SO_2 = 1(32.06 S) + 2 (15.9994 O) = 64.06 g mol^{-1}

$$? \text{ g } SO_2 = 2000 \text{ lb coal} \times \frac{453.59 \text{ g coal}}{1 \text{ lb coal}} \times \frac{5.2 \text{ g S}}{100 \text{ g coal}} \times \frac{1 \text{ mol S}}{32.06 \text{ g S}} \times \frac{1 \text{ mol } SO_2}{1 \text{ mol S}} \times$$

$$\frac{64.06 \text{ g } SO_2}{1 \text{ mol } SO_2} = \mathbf{9.4 \times 10^4 \text{ g } SO_2}$$

$$? \text{ lb } SO_2 = 9.4 \times 10^4 \text{ g } SO_2 \times \frac{1 \text{ lb } SO_2}{453.59 \text{ g } SO_2} = \mathbf{2.1 \times 10^2 \text{ lb } SO_2}$$

22.79 Name the following compounds.
a. K_2SO_4 b. $KHSO_4$ c. K_2SO_3 d. $KHSO_3$

The nomenclature rules (Chapter 2) should be reviewed.

a. K_2SO_4 is **potassium sulfate**, derived from sulfuric acid (H_2SO_4).
b. $KHSO_4$ is **potassium hydrogen sulfate** or **potassium bisulfate**.
c. K_2SO_3 is **potassium sulfite**, derived from sulfurous acid (H_2SO_3).
d. $KHSO_3$ is **potassium hydrogen sulfite** or **potassium bisulfite**.

22.81 Describe the geometry of, and give the hybridization of the central atom in each of the following.
a. sulfate ion b. thiosulfate ion

Analysis:
Target: The geometry and the hybridization of an atom.
Knowns: The name of the ion. The formula of the ion is an implicit known.
Relationships: The hybridization is related to the geometry of the species through the known orientation of orbitals in hybrid sets. This information is summarized in Table 9.1 in the text. The geometry of the species is related to the Lewis structure through VSEPR theory.

Plan: We need to know the Lewis structure and the appropriate relationships.
Step 4. Identify the hybrid orbital set having the proper orientation.
Step 3. Use VSEPR to determine the shape around the central atom.
Step 2. Draw the Lewis structure of the species.
Step 1. Determine the formula of the ion from its name.

Work:
a. **Step 1.** A sulfate ion is SO_4^{2-}.
Step 2. S is the central atom. The total number of valence electrons is 1(6 S) + 4(6 O) + 1(2 charge) = 32. The structure is shown on the right.
Step 3. With 4 electron regions about sulfur, the electron geometry is **tetrahedral**.
Step 4. The hybridization is sp^3.

:Ö: 2−
:Ö–S–Ö:
:Ö:

b. **Step 1.** A thiosulfate ion is $S_2O_3^{2-}$.
Step 2. One S is the central atom. The total number of valence electrons is 2(6 S) + 3(6 O) + 1(2 charge) = 32. The structure is shown on the right.
Step 3. With 4 electron regions about the central sulfur, the electron geometry is **tetrahedral**.
Step 4. The hybridization is sp^3.

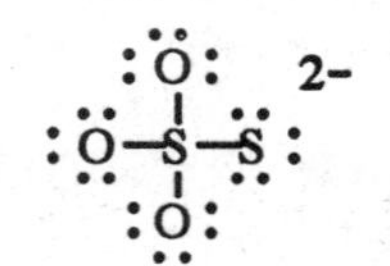

22.83 In alkaline solution, sulfite salts react with permanganate salts to produce a precipitate of manganese dioxide. Write a balanced net ionic reaction for this process.

Analysis/Plan: Follow the steps detailed in Section 18.3 of the book and repeated in answer to Problem 22.59 for the ion-electron method. Add one additional step to convert the result to a basic solution.

7. Convert to basic by adding a number of OH^- equal to the number of H^+ to both sides (water is formed) and again deleting like terms.

Work:

1. Write the appropriate half-reactions.
$SO_3^{2-} \rightarrow SO_4^{2-}$ $\qquad$ $MnO_4^- \rightarrow MnO_2$
2. Balance the atoms other than O and H.
$SO_3^{2-} \rightarrow SO_4^{2-}$ $\qquad$ $MnO_4^- \rightarrow MnO_2$
3. Add H_2O to the side deficient in O to balance O.
$H_2O + SO_3^{2-} \rightarrow SO_4^{2-}$ $\qquad$ $MnO_4^- \rightarrow MnO_2 + 2\ H_2O$
4. Add H^+ to the side deficient in H to balance H.
$H_2O + SO_3^{2-} \rightarrow SO_4^{2-} + 2\ H^+$ $\qquad$ $4\ H^+ + MnO_4^- \rightarrow MnO_2 + 2\ H_2O$
5. Add electrons to the more positive (less negative) side to balance the charge.
$H_2O + SO_3^{2-} \rightarrow SO_4^{2-} + 2\ H^+ + 2\ e^-$ $\qquad$ $3\ e^- + 4\ H^+ + MnO_4^- \rightarrow MnO_2 + 2\ H_2O$
6. The oxidation half-reaction has 2 e^-, and the reduction half-reaction has 3 e^-. the lowest common multiple is 6. Multiply the sulfur half-reaction by 3 and the manganese half-reaction by 2 and add the equations.
$3[H_2O + SO_3^{2-} \rightarrow SO_4^{2-} + 2\ H^+ + 2\ e^-]$
$2[3\ e^- + 4\ H^+ + MnO_4^- \rightarrow MnO_2 + 2\ H_2O]$
$3\ H_2O + 3\ SO_3^{2-} + 6\ e^- + 8\ H^+ + 2\ MnO_4^- \rightarrow 3\ SO_4^{2-} + 6\ H^+ + 6\ e^- + 2\ MnO_2 + 4\ H_2O$

Electrons, water and H^+ are common to both sides. Deleting like terms gives:
$3\ SO_3^{2-} + 2\ H^+ + 2\ MnO_4^- \rightarrow 3\ SO_4^{2-} + 2\ MnO_2 + H_2O$

7. Add 2 OH^- to each side and then subtract 1 H_2O from each side. This yields the net ionic equation:
$\mathbf{3\ SO_3^{2-}(aq) + H_2O(\ell) + 2\ MnO_4^-(aq) \rightarrow 3\ SO_4^{2-}(aq) + 2\ MnO_2(s) + 2\ OH^-(aq)}$

22.85 Write an equation for the reaction of gaseous SO_3 with concentrated $H_2SO_4(aq)$. Name the product.

This is equation 22.64 in the text. $\mathbf{SO_3(g) + H_2SO_4(aq) \rightarrow H_2S_2O_7(aq)}$.
The product is named **pyrosulfuric acid**.

22.87 What differences in structure exist among sulfuric, selenic, and telluric acids?

Sulfuric acid (H_2SO_4) and selenic acid (H_2SeO_4) are tetrahedral in structure. Telluric acid (H_6TeO_6) is nearly octahedral in structure as expected from its other formula representation, $Te(OH)_6$.

22.89 Are the metalloid elements found in the main groups or among the transition elements, or both? Which are the metalloid elements?

The metalloid elements (B, Si, Ge, As, Sb, Te, and At) are adjacent to the "stair-step" line on the periodic table. They are all main group elements.

22.91 Which would you expect to have the higher boiling point, H_2Se or H_2Te?

Both these compounds are polar, and are composed of nonmetals in the same group. Because H_2Te has a higher molar mass, it has a higher boiling point.

22.93 If 20.0 g of white phosphorus burns completely in air, and the combustion product is dissolved in 750 mL water, what is the molarity of the solution?

Analysis:

Target: The molarity of the combustion product dissolved in water. (? mol L^{-1}) =

Knowns: The initial mass of white phosphorus and the volume of solution. The appropriate reactions were shown in Problem 22.63. The molar mass is an implicit known.

Relationships: The balanced chemical equation gives mole ratios of all species. The molar mass provides the equivalence statement g ↔ mol. The equivalence statement 1 mL ↔ 10^{-3} L is also needed. Molarity is defined as $\frac{\text{mol solute}}{\text{1 L solution}}$

Plan: Determine the overall equation for the process. Use the standard stoichiometric procedure (SSP) discussed in the text, with the appropriate conversion factors to determine the number of moles of H_3PO_4. Determine the molarity from its definition, using the factor-label method to make the necessary volume conversion.

g P_4 ⟶ mol P_4 ⟶ mol H_3PO_4

Work:

Determine the overall equation for the process. The two pertinent equations are:

$$P_4(s) + 5\ O_2(g) \longrightarrow P_4O_{10}(s)$$

$$P_4O_{10}(s) + 6\ H_2O(\ell) \longrightarrow 4\ H_3PO_4$$

Adding the equations and deleting like terms yields:

$$P_4(s) + 5\ O_2(g) + 6\ H_2O(\ell) \longrightarrow 4\ H_3PO_4(aq)$$

Now, the SSP can be treated as a single two step conversion process.

Molar Mass P_4 = 4(30.97376 P) = 123.895 g mol^{-1}

$$\text{? mol } H_3PO_4 = 20.0\text{ g } P_4 \times \frac{1\text{ mol } P_4}{123.895\text{ g } P_4} \times \frac{4\text{ mol } H_3PO_4}{1\text{ mol } P_4} = 0.646\text{ mol } H_3PO_4$$

Use the definition of molarity with the appropriate volume conversion.

$$\text{? M } H_3PO_4 = \frac{0.646\text{ mol } H_3PO_4}{750\text{ mL}} \times \frac{1\text{ mL}}{10^{-3}\text{ L}} = \mathbf{0.861\text{ M } H_3PO_4}$$

23 The Transition Elements

23.1 What features of electron configuration distinguish the transition elements from metals in the A Groups of the periodic table?

All of the transition elements except Zn, Cd, and Hg have a partially filled *d* subshell in the neutral atom or in one or more of their common oxidation states.

23.3 What is the lanthanide contraction?

In the lanthanides (4*f*) elements, electrons are being added to an inner subshell. As a consequence, the elements following La in the *d* block have approximately the same radius as those elements directly above them on the periodic table. This is in contrast to the normal tendency for elements to be at least 10% larger than the element directly above them.

23.5 Which elements in the second and third transition series have electron configurations that depart from the *aufbau* prediction?

To determine this, the actual electron configuration must be compared with that predicted from the *aufbau* rule. The actual electron configurations are shown in the text (Figure 7.13). The elements which differ are Nb, Mo, Ru, Rh, Pd, and Ag, from the second transition series, and Pt and Au, from the third transition series.

23.7 Which of the actinides exist(s) naturally in other than trace quantities in the earth's crust?

Only uranium ($_{92}U$) and thorium ($_{90}Th$) exist in other than trace quantities on earth.

23.9 Some chemists classify Zn, Cd, and Hg with the main-group elements rather than with the transition elements. What is the justification for this?

The justification is that these three elements have completely filled *d* subshells in their neutral atoms and in their common monatomic ions.

23.11 Write the electron configurations for the following elements.

a. titanium b. nickel c. ruthenium
d. molybdenum e. yttrium f. gold

Analysis/Plan: Determine the total number of electrons, equal to the atomic number, in each of these elements. Follow the normal *aufbau* filling pattern as discussed in Chapter 7, unless the element is one of the exceptions detailed in Problem 23.5.

Work:

a. Ti with 22 e^- follows the normal *aufbau* filling pattern. The electron configuration is $[Ar]4s^2 3d^2$.

b. Ni with 28 e^- follows the normal *aufbau* filling pattern. The electron configuration is $[Ar]4s^2 3d^8$.

c. Ru with 44 e^- is an exception to the normal *aufbau* filling pattern. The electron configuration is $[Kr]5s^1 4d^7$.

d. Mo with 42 e^- is an exception to the normal *aufbau* filling pattern. The electron configuration is $[Ar]5s^1 4d^5$.

e. Y with 39 e^- follows the normal *aufbau* filling pattern. The electron configuration is $[Kr]5s^2 4d^1$.

f. Au with 79 e^- is an exception to the normal *aufbau* filling pattern. The electron configuration is $[Xe]6s^1 4f^{14} 5d^{10}$.

23.13 Identify the oxidation state of the transition element in each of the following species.

a. TiO_3^{2-} b. $AgNO_3$ c. $TiCl_4$ d. $K_3Fe(CN)_6$

Analysis/Plan: All oxidation-number problems should be approached through a *stepwise* application of the rules detailed in Table 18.1 of the text. Charges for polyatomic ions must be memorized to assign oxidation numbers in compounds with more than one element not covered by the rules. Rule 1 states that the sum of the oxidation numbers equals 0 in a compound and equals the ion charge in a polyatomic ion.

Work:

a. Rule 5 states that the oxidation number of O is almost always -2. Ti is not listed as one of the exceptions; therefore, the oxidation number of O is -2.

Atom Type	Number of Atoms	Oxidation Number	Total charge
Ti	1	x	x
O	3	-2	-6

This is an ion, so: x - 6 = -2; x = +4. The oxidation number of Ti is +4.

b. Ag is in a compound with the nitrate ion, NO_3^-. Because this is a compound, x - 1 = 0; and, x = 1. The oxidation number of Ag is +1.

c. Rule 3 states that the oxidation number of a halogen is almost always -1. Titanium is not listed as one of the exceptions; therefore, the oxidation number of Cl is -1.

Atom Type	Number of Atoms	Oxidation Number	Total charge
Ti	1	x	x
Cl	4	-1	-4

This is a compound, so: x - 4 = 0; x = +4. The oxidation number of Ti is +4.

d. Rule 1 states that the oxidation number of a 1A metal in compounds is always +1. The cyanide ion (CN^-) has a charge of -1

Atom or Ion Type	Number of Atoms or Ions	Oxidation Number	Total charge
K	3	+1	+3
Fe	1	x	+x
CN	6	-1	-6

This is a compound, so: 3 + x + (-6) = 0; x = +3. The oxidation number of Fe is +3.

23.15 Which +2 ion among the fourth-period transition elements has the highest standard reduction potential?

Inspection of the table of standard reduction potentials (Appendix I) reveals that $Cu^{2+}(aq)$, with $\mathcal{E}°_{red}$ = 0.336 V has the highest SRP of the metals in the 4th period.

23.17 What is the highest oxidation number commonly found among the elements of the first transition series? Which element(s) exhibit(s) this oxidation state? Give examples.

Mn exhibits the highest oxidation number among the elements in the first transition series. It has an oxidation number of +7 in permanganates (MnO_4^- = permanganate ion).

23.19 Which of the metals Sc through Zn is most abundant in the earth's crust? Which has the second greatest abundance?

Iron (Fe, 4.7%) is the most abundant of these elements in the earth's crust. Titanium (Ti, 0.58%) is the second most abundant of these elements.

23.21 With some exceptions, the melting point of the fourth-period transition elements rises from Sc to Cr, then falls. Explain this trend.

As we progress across the *d* block, the number of unpaired *d* electrons increases in accordance with Hund's rule to a maximum of 5 in the middle of the group. The number of unpaired electrons then begins to decrease as electrons are paired within orbitals. A greater number of unpaired electrons gives rise to greater cohesive forces, which in turn cause melting points to be higher.

23.23 Choose any two elements from the first transition series and describe how the metal is obtained from the ore.

See Section 23.3 in the text.

23.25 Define the terms "Lewis acid" and "Lewis base."

A Lewis acid is a species with an incompletely filled valence shell. It is capable of filling its valence shell with an electron pair provided by the base. A Lewis base is a species that has one or more lone pairs of electrons. A base can share a pair with an acid to form a coordinate covalent bond and fill the acid's valence shell.

23.27 Distinguish between coordination *compound* and coordination *complex.*

In both cases, a metal ion is bonded via coordinate covalent bonds to several ligands (either neutral or charged). A coordination compound is a neutral species. A coordination complex is a charged species which would be associated with an oppositely charged ion to form a neutral compound.

23.29 Define coordination number.

Coordination number is the number of ligands bonded to the metal in the coordination compound or the coordination complex.

23.31 What is a chelate? How is a chelate different from an ordinary coordination complex?

A chelate is a specific form of coordination compound or complex, in which the ligands bonded to the metal are polydentate (capable of bonding at more than one position).

23.33 Can the ammonium ion, NH_4^+, serve as a ligand in a coordination complex?

The ammonium ion cannot serve as a ligand because it has no lone pairs of electrons to donate to the metal. Instead, the ammonium ion is the result of coordinate covalent bond formation between an H^+ ion (the acid) and NH_3 (the base).

23.35 Classify each of the following ligands as monodentate, bidentate, etc.

a. NO b. pyridine, C_5H_5N c. oxalate, $C_2O_4^{2-}$ d. iodo

Some of these common ligands are shown or noted in Table 23.3 in the text. NO (a) and I^- (d) are **monodentate** ligands. Oxalate (c) is a **bidentate** ligand because of its geometry. Pyridine (b) whose structure is shown on the next page is also **monodentate**.

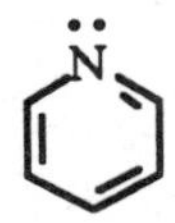

23.37 The structure of *cyclam*, an unsaturated compound related to porphine, is shown below.

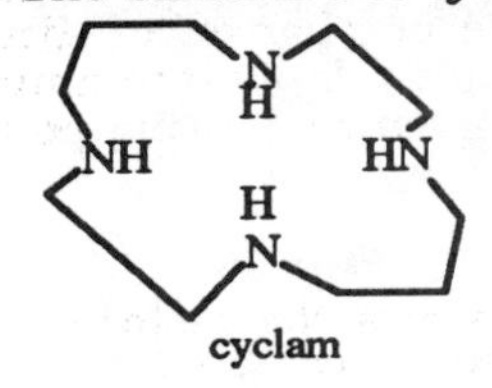

How many coordinate bonds can cyclam form with a metal ion?

Recognize that each N atom also has a lone pair of electrons, not expressly shown in the Lewis structure above. With those four lone pairs, cyclam can make 4 coordinate bonds.

23.39 Determine the oxidation state of the transition metal in each of the following species.
a. $[V(en)_3]^{3+}$ b. $[Ir(NH_3)_4Cl_2]^+$ c. $[PtCl_4]^{2-}$ d. $[Cr(NH_3)_5SO_4]^{2+}$

Analysis:
Target: The oxidation number of a metal.
Knowns: The formula of the species and the charge on the ligand.
Relationships: The charge on the species must equal the sum of the charge of the metal and the charge of all ligands.
Plan: Add up the charge of the ligands and subtract this total from the charge of the species to determine the charge of the metal.
Work:
a. Table 23.3 indicates that en is neutral, so its charge is 0. The total ligand charge is 3(0) = 0. Ion charge - ligand charge is 3 - 0 = +3. The oxidation number of V is **+3**.

b. NH_3 is neutral and Cl is -1. The total ligand charge is 4(0) + 2(-1) = -2. Ion charge - ligand charge is +1 - (-2) = +3. The oxidation number of Ir is **+3**.

c. Cl is -1. The total ligand charge is 4(-1) = -4. Ion charge - ligand charge is -2 - (-4) = +2. The oxidation number of Pt is **+2**.

d. NH_3 is neutral and SO_4 is -2. The total ligand charge is 5(0) + 1(-2) = -2. Ion charge - ligand charge is +1 - (-2) = +3. The oxidation number of Cr is **+3**.

23.41 Name each of the following complex ions. (Ox is the abbreviation for $C_2O_4^{2-}$, the oxalate ion.)
a. $[Fe(H_2O)_2(CN)_4]^-$ b. $[Co(NH_3)_2(en)Cl_2]^+$ c. $[Ni(NH_3)_2(ox)_2]^{2-}$
d. $[Zn(OH)_4]^{2-}$

Analysis:
Target: The name of the complex ion.
Knowns: The formula of the species and the charge on the ligand.
Relationships: The charge on the species must equal the sum of the charge of the metal and the charge of all ligands. Table 23.4 in the text details the rules for naming coordination compounds.
Plan: Add up the charge of the ligands and subtract this total from the charge of the species to determine the charge of the metal ion. Name the complex according to the listed rules.

Work:

a. H_2O (aquo) is neutral and CN (cyano) is -1. The total ligand charge is 2(0) + 4(-1) = -4. Ion charge - ligand charge is -1 - (-4) = +3. The oxidation number of Fe is +3. This is **diaquotetracyanoferrate(III)**.

b. NH_3 (ammine) is neutral, en (ethylenediamine) is neutral, and Cl (chloro) is -1. The total ligand charge is 2(0) + 1(0) + 2(-1) = -2. Ion charge - ligand charge is +1 - (-2) = +3. The oxidation number of Co is +3. This is **diamminedichloroethylenediaminecobalt(III)**.

c. NH_3 (ammine) is neutral and ox (oxalato) is -2. The total ligand charge is 2(0) + 2(-2) = -4. Ion charge - ligand charge is -2 - (-4) = +2. The oxidation number of Ni is +2. This is **diamminedioxalatonickelate(II)**.

d. OH (hydroxo) is -1. The total ligand charge is 4(-1) = -4. Ion charge - ligand charge is -2 - (-4) = +2. The oxidation number of Zn is +2. This is **tetrahydroxozincate(II)**.

23.43 Determine the oxidation state of the transition metal in each of the following compounds.
a. $K[Au(CN)_2]$ b. $Rb_4[Ni(ox)_3]$ c. $[Cr(en)_2(NO)_2]Cl_2$
d. $[Ni(en)_3]Br_2$

Analysis:

Target: The oxidation number of a metal.
Knowns: The formula of the species and the charges on the ligand and other ions.
Relationships: Because these are all compounds, the sum of the charge of the metal and the charge of all ligands and other ions must equal 0.

Plan: Add up the charge of the ligands and known ions and subtract this total from 0 to determine the charge of the metal.

Work:

a. K is +1 and CN is -1. The total ligand and ion charge is 1(+1) + 2(-1) = -1. Zero - total ion and ligand charge is 0 - (-1) = +1. The oxidation number of Au is **+1**.

b. Rb is +1 and ox is -2. The total ligand and ion charge is 4(+1) + 3(-2) = -2. Zero - total ion and ligand charge is 0 - (-2) = +2. The oxidation number of Ni is **+2**.

c. NO and en are neutral and Cl is -1 The total ligand and ion charge is 2(0) + 2(0) + 2(-1) = -2. Zero - total ion and ligand charge is 0 - (-2) = +2. The oxidation number of Cr is **+2**.

d. En is neutral and Br is -1. The total ligand and ion charge is 3(0) + 2(-1) = -2. Zero - total ion and ligand charge is 0 - (-2) = +2. The oxidation number of Ni is **+2**.

23.45 Name each of the following compounds. (Phen is the abbreviation for 1,10-phenanthroline, whose structure is given in Problem 23.38.)
a. $[Rh(CO)_5Cl]Cl_2$ b. $[Ru(phen)_3]Cl_2$ c. $Na_2[Fe(EDTA)]$
d. $Pt(NH_3)_2Br_2$

Analysis:

Target: The name of a coordination compound.
Knowns: The formula of the species and the charges on the ligand and other ions.
Relationships: Because these are all compounds, the sum of the charge of the metal and the charge of all ligands and other ions must equal 0. Table 23.4 in the text details the rules for naming coordination compounds.

Plan: Add up the charge of the ligands and known ions and subtract this total from 0 to determine the oxidation number of the metal. Name the compound using the listed rules.

Work:

a. CO (carbonyl) is neutral and Cl (chloro) is -1. The total ligand and ion charge is 5(0) + 3(-1) = -3. Zero - total ion and ligand charge is 0 - (-3) = +3. The oxidation number of Rh is +3.
The compound is **pentacarbonylchlororhenium(III) chloride**.

b. Phen (1,10-phenanthroline) is neutral and Cl (chloro) is -1. The total ligand and ion charge is 3(0) + 2(-1) = -2. Zero - total ion and ligand charge is 0 - (-2) = +2. The oxidation number of Ru is +2.
The compound is **tris(1,10 phenanthroline)ruthenium(II) chloride**.

c. Na is +1 and EDTA (ethylenediaminetetracetato) is -4. The total ligand and ion charge is 2(+1) + 1(-4) = -2.
Zero - total ion and ligand charge is 0 - (-2) = +2. The oxidation number of Fe is +2.
The compound is **sodium ethylenediaminetetracetatoferrate(II)**.

d. NH_3 (ammine) is neutral and Br (bromo) is -1. The total ligand and ion charge is 2(0) + 2(-1) = -2. Zero - total ion and ligand charge is 0 - (-2) = +2. The oxidation number of Pt is +2.
The compound is **diammineplatinum(II) bromide**.

23.47 Write the empirical formula of each of the following complex ions. Do not omit the ion charge from the formula.

a. tetraaquodihydroxoaluminum(III)
b. hexachloroferrate(II)
c. triamminebromozinc(II)
d. tris(ethylenediamine)nickel(II)

Analysis:

Target: The empirical formula of a complex ion.
Knowns: The name of the species and the charges on the ligands. The oxidation number of the metal is implicitly known from its name.
Relationships: The sum of the oxidation numbers must equal the ion charge. The rules for nomenclature and the order in which the species must be written is given in Table 23.4 in the text.

Plan: Determine the charge and then write the components in order. The order is metal, then the ligands in alphabetical order.

Work:

a. The appropriate groups are H_2O (aquo), OH^- (hydroxo) and Al^{3+} (aluminum(III)). The sum of the component charges is 4(0) + 2(-1) + 1(+3) = +1. The ion is **$[Al(H_2O)_4(OH)_2]^+$**.

b. The appropriate groups are Cl^- (chloro) and Fe^{2+} (ferrate(II)). The sum of the component charges is 6(-1) + 1(+2) = -4. The ion is **$[FeCl_6]^{4-}$**.

c. The appropriate groups are NH_3 (ammine), Br^- (bromo) and Zn^{2+} (zinc(II)). The sum of the component charges is 3(0) + 1(-1) + 1(+2) = +1. The ion is **$[Zn(NH_3)_3Br]^+$**.

d. The appropriate groups are en (ethylenediamine) and Ni^{2+} (nickel(II)). The sum of the component charges is 3(0) + 1(+2) = +2. The ion is **$[Ni(en)_3]^{2+}$**.

23.49 Write the empirical formula of each of the following compounds.

a. iron(III) hexacyanoferrate(II)
b. lithium diamminetetrabromochromate(II)
c. chlorotrinitroplatinum(II) chloride
d. calcium triaquotrichloronickelate(II)

Analysis:

Target: The empirical formula of a compound.
Knowns: The name of the species and the charges on the ligands. The oxidation number of the metal is implicitly known from its name.

Relationships: The sum of the oxidation numbers must equal the ion charge. The rules for nomenclature and the order in which the species must be written is given in Table 23.4 in the text.

Plan: Determine the charge of each ion. Choose subscripts to make the species neutral and then write the components in order. The cation is written first and the anion last. Within a complex ion, the order is metal, then the ligands in alphabetical order.

Work:

a. The appropriate groups are Fe^{3+} (iron(III)), CN^- (cyano) and Fe^{2+} (ferrate(II)). The sum of the component charges in the ion is 6(-1) + 1(+2) = -4. The lowest common multiple of 4 and 3 is 12. The compound is $\mathbf{Fe_4[Fe(CN)_6]_3}$.

b. The appropriate groups are Li^+ (lithium), NH_3 (ammine), Br^- (bromo), and Cr^{2+} (chromate(II)). The sum of the component charges in the ion is 2(0) + 4(-1) + 1(+2) = -2. The lowest common multiple of 1 and 2 is 2. The compound is $\mathbf{Li_2[Cr(NH_3)_2Br_4]}$.

c. The appropriate groups are Cl^- (chloro and chloride), NO_2 (nitro), and Pt^{2+} (platinum(II)). The sum of the component charges in the ion is 1(-1) + 3(0) + 1(+2) = +1. The lowest common multiple of 1 and 1 is 1. The compound is $\mathbf{[PtCl(NO_3)_3]Cl}$.

d. The appropriate groups are Ca^{2+} (calcium), H_2O(aquo), Cl^- (chloro) and Ni^{2+} (nickelate(II)). The sum of the component charges in the ion is 3(0) + 3(-1) + 1(+2) = -1 The lowest common multiple of 1 and 2 is 2. The compound is $\mathbf{Ca[Ni(H_2O)_3Cl_3]_2}$.

23.51 Distinguish between structural isomerism and stereoisomerism.

In structural isomers, atoms are bonded with a different connectivity and would be named differently. In stereoisomers, the atoms are bonded with the same connectivity, however, some structural feature limits rotation around bonds and causes the structures to differ in their spatial arrangement.

23.53 Let W, X, Y, and Z represent monodentate ligands in a square planar complex. Which of the following exist(s) as a pair of geometrical isomers: MWXYZ, MX_2YZ, MX_2Y_2, MXY_3?

Anytime 4 different groups are bonded to a central atom, geometrical isomerism is possible; therefore, **MWXYZ** can have geometrical isomers.

```
  X——Y      Z——X      X——Z      Y——Z      Z——Y      Y——X
 / M /     / M /     / M /     / M /     / M /     / M /
W——Z      W——Y      W——Y      W——X      W——X      W——Z
```

Note that the first three represent geometrical isomers and the last three are stereoisomers (specifically enantiomers of those three.)

In $\mathbf{MX_2YZ}$ and $\mathbf{MX_2Y_2}$, the X atoms can be either next to each other or across from each other, so this is capable of geometrical isomerism.

```
  X——Z      Y——X          X——Y      Y——X
 / M /     / M /         / M /     / M /
X——Y      X——Z          X——Y      X——Y
```

In MXY_3, any switching of atoms still leaves X next to Y and two Y's opposite each other. This is not capable of geometrical isomerism.

23.55 In which of the following is optical isomerism possible: $M(en)Cl_2$, $M(en)_3$, M(en)FClBrI, or $M(en)_2FCl$?

In **$M(en)Cl_2$**, en (ethylenediamine) is bidentate and occupies two corners of the square plane. Any shift leaves, the Cl atoms next to each other. This is not capable of geometric isomerism. With **$M(en)_3$**, the structure is octahedral. The following optical isomers are among those possible.

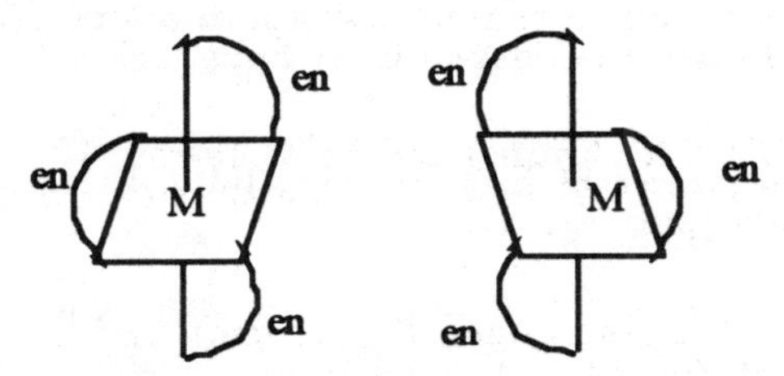

Similarly, **$M(en)FClBrI$** and **$M(en)_2FCl$** can have geometric isomers. These can be drawn by shifting the positions of the halogens relative to the ethylenediamine. Some examples of geometric isomers of each are shown below.

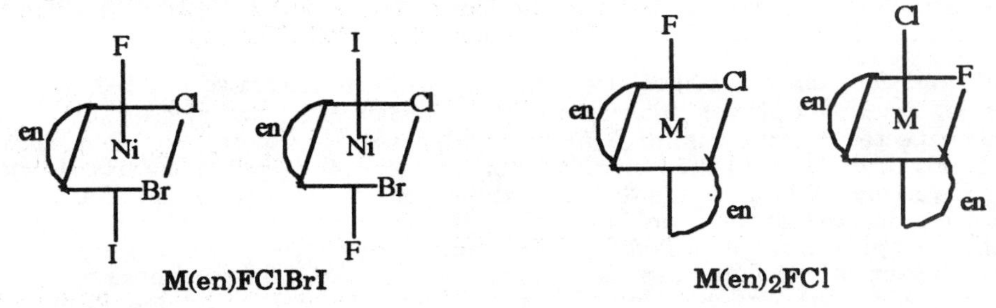

M(en)FClBrI M(en)$_2$FCl

23.57 Draw the structures of four isomers of the compound $Ni(NH_3)_2BrCl(en)$.

Analysis/Plan: Draw one basic structure and then shift the position of the groups relative to each other, checking to see that no rotation makes the structures superimposable.

Work: There are just four isomers possible.

23.59 Name and briefly describe each of the three theories that are used to describe the bonding in coordination complexes.

Valence bond theory treats the metal-ligand bonds as simple coordinate covalent bonds. The shape of the complex is explained in terms of the metal's hybrid orbitals. While this approach is useful for predicting magnetic properties, it does not successfully predict the color of the complex. Molecular orbital theory constructs bonding, nonbonding and antibonding orbitals and extends them over the entire complex. Crystal field theory discusses complex bonding in terms of an electrostatic attraction between the metal cation and the dipoles or point charges of the ligands. This theory accounts for both the magnetic properties and the colors of coordination compounds and complexes.

23.61 Compounds of Sc^{3+} are colorless, and compounds of Ti^{3+} are highly colored. Account for this difference.

The electron configuration of Sc^{3+} is [Ar], while that of Ti^{3+} is $[Ar]3d^1$. It is generally the *d* electrons that are involved in the absorption of visible light that leads to color. Sc^{3+} with no *d* electrons is colorless, while Ti^{3+}, with 1 *d* electron is colored.

23.63 For halogen ligands, crystal field splitting increases with electronegativity. Arrange the halide ions in order of increasing crystal field splitting.

Electronegativity increases from bottom to top within a group. The electronegativities and, therefore, the crystal field splitting increases in the order: $I^- < Br^- < Cl^- < F^-$.

23.65 In mystery stories, victims of cyanide poisoning are often described as having a blue color to the skin. Suggest a reason for this based on the similarity in size and shape between O_2 and CN^-.

Both CN^- and O_2 have 12 valence electrons and are formed from second period elements of similar size. Because of this similarity and the presence of lone electron pairs on each species, cyanide is expected to have similar ligand capabilities to that of oxygen. If it, rather than oxygen, bonds to the iron in the hemoglobin of the blood, the cyanide interferes with the action of the hemoglobin molecule to the detriment of the affected person. In addition, the electrostatic properties of the cyanide ion are different from that of oxygen. A cyanide-hemoglobin complex is not expected to be the same color as an oxygen-hemoglobin complex. The color differential would be most obvious in tissues (like the lips) where many blood vessels are close to the surface. Note that veins, which are returning deoxygenated blood to the heart, also appear blue.

23.67 For each of the following complexes, determine the oxidation number of the metal, whether it is a high- or low-spin case, and predict the number of unpaired electrons.

a. $[VF_6]^-$ b. $[V(CN)_6]^-$ c. $[VF_6]^{2-}$ d. $[V(CN)_6]^{2-}$

Analysis:

Target: The oxidation number of a metal, whether it is high-spin or low spin and the number of unpaired electrons.

Knowns: The formula of the species and the charges on the ligand and other ions.

Relationships: Because these are all ions, the sum of the charge of the metal and the charge of all ligands and other ions must equal the ion charge. Section 23.9 presents the spectrochemical series showing which ligands are weak field (high-spin) ligands and which are strong field (low-spin) ligands. The number of valence electrons from the metal is known from its configuration and charge.

Plan: The following series of steps is required.

1. Add up the charge of the ligands and known ions and subtract this total from the ion charge to determine the charge of the metal.
2. From the charge of the metal and the electron configuration of the neutral atom determine the ion configuration.
3. Use the spectrochemical series to differentiate between the high-spin and low-spin cases. Complete the orbital diagram to determine the magnetic properties.

Work:

a. 1. F is -1. The total ligand charge is 6(-1) = -6. Ion charge - ligand charge is -1 - (-6) = +5. The oxidation number of V is +5.

2. The electron configuration of V is $[Ar]4s^23d^3$. Removing 5 electrons gives an ion configuration of [Ar].

3. F is a weak field ligand so this is a **high-spin** state. There are no electrons outside of the noble gas core, so there are **0 unpaired** electrons.

b. **1.** CN is -1. The total ligand charge is 6(-1) = -6. Ion charge - ligand charge is -1 - (-6) = +5. The oxidation number of V is **+5**.
2. The electron configuration of V is [Ar]$4s^2 3d^3$. Removing 5 electrons gives an ion configuration of [Ar].
3. CN^- is a strong field ligand so this is a **low-spin** state. There are no electrons outside of the noble gas core, so there are **0 unpaired** electrons.

c. **1.** F is -1. The total ligand charge is 6(-1) = -6. Ion charge - ligand charge is -2 - (-6) = +4. The oxidation number of V is **+4**.
2. The electron configuration of V is [Ar]$4s^2 3d^3$. Removing 4 electrons gives an ion configuration of [Ar]$3d^1$.
3. F is a weak field ligand so this is a **high-spin** state. There is 1 electron outside of the noble gas core, so there is **1 unpaired** electron.

d. **1.** CN is -1. The total ligand charge is 6(-1) = -6. Ion charge - ligand charge is -2 - (-6) = +4. The oxidation number of V is **+4**.
2. The electron configuration of V is [Ar]$4s^2 3d^3$. Removing 4 electrons gives an ion configuration of [Ar]$3d^1$.
3. CN^- is a strong field ligand so this is a **low-spin** state. There is 1 electron outside of the noble gas core, so there is **1 unpaired** electron.

23.69 Predict the number of unpaired electrons in each of the following complexes.
a. tetrafluoromanganese(IV) b. hexafluorotungstate(V)
c. potassium hexachlororhenate(IV)

Analysis:
Target: The number of unpaired electrons in a complex.
Knowns: The name of the complex which gives the metal charge and the charges on the ligand and other ions.
Relationships: Section 23.9 presents the spectrochemical series showing which ligands are weak field (high-spin) ligands and which are strong field (low-spin) ligands. The number of valence electrons from the metal is known from its configuration and charge.
Plan: The following series of steps is required.
1. From the charge of the metal and the electron configuration of the neutral atom determine the ion configuration.
2. Use the spectrochemical series to differentiate between the high-spin and low-spin cases. Complete the orbital diagram to determine the magnetic properties.
Work:
a. 1. The electron configuration of Mn is [Ar]$4s^2 3d^5$. Removing 4 electrons gives an ion configuration of [Ar]$3d^3$.
2. F^- is a strong field ligand so this is a low-spin state. The orbital diagram is:

[ ] [ ]
[↑] [↑] [↑] There are **3 unpaired electrons**.

b. 1. The electron configuration of W is [Xe]$6s^2 4f^{14} 5d^4$. Removing 5 electrons gives an ion configuration of [Xe]$4f^{14} 5d^1$.
2. F^- is a strong field ligand so this is a low-spin state. The orbital diagram is:

[ ] [ ]
[↑] [ ] [ ] There is **1 unpaired electron**.

c. 1. The electron configuration of Rh is [Kr]$5s^1 4d^8$. Removing 4 electrons gives an ion configuration of [Kr]$4d^5$.
2. Br^- is not shown on the spectrochemical series but it is less electronegative than Cl so it is a weak field ligand. This is a high-spin state. The orbital diagram is:

[↑] [↑] [↑] [↑] [↑] There are **5 unpaired electrons.**

23.71 Is it possible for an "inert" complex to react? Explain.

In an inert complex, the rate of ligand exchange is slow, however, it is capable of reacting to some extent.

23.73 Calculate the pH of a 0.00100 M solution of chromium trichloride, assuming that all chloride is present as counterion, not ligand. K_a for $[Cr(H_2O)_6^{3+}]$ is 1×10^{-4}. Compare this value to the pH of an acetic acid solution of the same molarity.

Analysis:

Target: To determine and compare the pH of solutions of two weak acids. (? pH) =

Knowns: The molarity of the acid. The K_a of the chromium complex is given. The K_a of acetic acid is 1.8×10^{-5} (Appendix J).

Relationships: The structured approach to equilibrium problems is appropriate. (Review Chapter 13 if necessary.) pH = -log $[H_3O^+]$.

Plan: Use the structured approach to equilibrium problems to determine the $[H_3O^+]$ and then use the formula to determine the pH.

Work:

First determine the pH of the $[Cr(H_2O)_6^{3+}]$ solution.

Balanced	$[Cr(H_2O)_6^{3+}]$ + H_2O	$\rightleftharpoons$	H_3O^+ +	$[Cr(H_2O)_5(OH)^{2+}]$
Initial	0.00100		0	0
Change	- x		+ x	+ x
Equilibrium	0.00100 - x		x	x

Insert the values into the K_a expression.

$$K_a = \frac{[H_3O^+][[Cr(H_2O)_5(OH)^{2+}]]}{[Cr(H_2O)_6^{3+}]} = \frac{(x)(x)}{(0.00100 - x)} = 1 \times 10^{-4}$$

Solve algebraically. Because the molarity is very small and the K_a is fairly high, the quadratic formula must be used. From Problem 16.7, $x = \dfrac{-K_a + \sqrt{K_a^2 + 4(K_a)(C)}}{2}$, where x is the $[H_3O^+]$, K_a is the ionization constant of the acid and C is the initial molarity of the acid.

$$x = [H_3O^+] = \frac{(-1 \times 10^{-4}) + \sqrt{(1 \times 10^{-4})^2 + 4(0.00100)(1 \times 10^{-4})}}{2} = 3 \times 10^{-4}\ M$$

$$pH = -\log[H_3O^+] = -\log(3 \times 10^{-4}) = \mathbf{3.6}$$

Now determine the pH of the CH_3COOH solution.

Balanced	CH_3COOH + H_2O	$\rightleftharpoons$	H_3O^+ +	CH_3COO^-
Initial	0.00100		0	0
Change	- x		+ x	+ x
Equilibrium	0.00100 - x		x	x

Insert the values into the K_a expression.

$$K_a = \frac{[H_3O^+][CH_3COO^-]}{[CH_3COOH]} = \frac{(x)(x)}{(0.00100 - x)} = 1.8 \times 10^{-5}$$

Solve algebraically. Because the molarity is very small and the K_a is fairly high, the quadratic formula must be used. From Problem 16.7, $x = \frac{-K_a + \sqrt{K_a^2 + 4(K_a)(C)}}{2}$, where x is the $[H_3O^+]$, K_a is the ionization constant of the acid and C is the initial molarity of the acid.

$$x = [H_3O^+] = \frac{(-1.8 \times 10^{-5}) + \sqrt{(1.8 \times 10^{-5})^2 + 4(0.00100)(1.8 \times 10^{-5})}}{2} = 1.3 \times 10^{-4}\ M$$

$$pH = -\log[H_3O^+] = -\log(1.3 \times 10^{-4}) = \mathbf{3.90}$$

The pH of the acetic acid solution is higher than that of the hydrated Cr^{3+} ion because acetic acid is weaker acid than is $[Cr(H_2O)_6{}^{3+}]$.

23.75 Calculate the ratio of the concentration of $[Fe(CN)_6]^{3-}$ to that of $[Fe(H_2O)_6]^{3+}$ in a solution that is 0.050 M in CN^-.

Analysis:
Target: To determine a concentration ratio.
Knowns: The molarity of the ligand. The K_f of the complex ion can be looked up in Table 23.7 and is equal to 8×10^{43}.
Relationships: The equilibrium constant expression can be written from the balanced chemical equation.
Plan: Write the balanced equation and the equilibrium constant expression. Substitute the molarity of CN^- into the expression and solve for the desired concentration ratio.
Work:
The balanced equation is:

$$[Fe(H_2O)_6{}^{3-}](aq) + 6\ CN^-(aq) \rightleftharpoons [Fe(CN)_6{}^{3-}](aq) + 6\ H_2O(\ell).$$

The formation constant expression is shown below.

$$K_f = \frac{[[Fe(CN)_6{}^{3-}]]}{[[Fe(H_2O)_6{}^{3-}]][CN^-]^6} = 8 \times 10^{43} = \frac{[[Fe(CN)_6{}^{3-}]]}{[[Fe(H_2O)_6{}^{3-}]][0.050]^6}$$

Rearranging and evaluating the expression yields the following answer.

$$\frac{[[Fe(CN)_6{}^{3-}]]}{[[Fe(H_2O)_6{}^{3-}]]} = (8 \times 10^{43})(0.050)^6 = \mathbf{1 \times 10^{36}}$$

23.77 A chemist wishes to remove from solution as much $Ni^{2+}(aq)$ as possible, to avoid possible interfering effects with a planned laboratory procedure. She has available 0.075 M solutions of ethylenediamine and sodium calcium EDTA. Which should she use? Explain your choice.

Analysis/Plan: From Table 23.7, the K_f of the $Ni(en)_3{}^{2+}$ complex is 1×10^{18}, while that of the $Ni(EDTA)^{2-}$ complex is 4×10^{18}. The chemist should choose the combination that gives the lowest ratio of Ni^{2+} to the complex. Write balanced chemical equations for the dissociation of both complexes and solve for the $\frac{[Ni^{2+}]}{[Ni_{complex}]}$ ratio as in Problem 23.75. Remember that when we reverse a reaction $K_{reverse} = \frac{1}{K_{forward}}$. For simplicity the Ni^{2+} will be shown unhydrated.
Work:
For the ethylenediamine complex, the dissociation reaction is:

$$[Ni(en)_3^{2+}](aq) \rightleftharpoons Ni^{2+}(aq) + 3\ en(aq)$$

The dissociation constant expression is shown below.

$$K_d = \frac{[Ni^{2+}][en]^3}{[[Ni(en)_3^{2+}]]} = \frac{1}{1 \times 10^{18}} = \frac{[Ni^{2+}][0.075]^3]}{[[Ni(en)_3^{2+}]]}$$

Rearranging and evaluating the expression yields the following answer.

$$\frac{[Ni^{2+}]}{[[Ni(en)_3^{2+}]]} = \frac{1}{1 \times 10^{18}} \times \frac{1}{(0.075)^3} = 2 \times 10^{-15}$$

For the EDTA complex, the dissociation reaction is:

$$[Ni(EDTA)^{2-}](aq) \rightleftharpoons Ni^{2+}(aq) + EDTA(aq)$$

The dissociation constant expression is shown below.

$$K_d = \frac{[Ni^{2+}][EDTA]}{[[Ni(EDTA)^{2-}]]} = \frac{1}{4 \times 10^{18}} = \frac{[Ni^{2+}][0.075]}{[[Ni(EDTA)^{2-}]]}$$

Rearranging and evaluating the expression yields the following answer.

$$\frac{[Ni^{2+}]}{[[Ni(EDTA)^{2-}]]} = \frac{1}{4 \times 10^{18}} \times \frac{1}{(0.075)} = 3 \times 10^{-18}$$

Because the ratio of free nickel to complex is much lower in the case of EDTA than in the case of ethylenediamine, EDTA should be used to minimize the concentration of nickel(II) ion in solution.

23.79 When KCN(aq) is added to a solution of $ZnSO_4$, a white precipitate forms. The precipitate dissolves when more KCN is added. Explain.

Because CN^- is the conjugate base of the very weak acid HCN ($K_a = 4.0 \times 10^{-10}$) it is a fairly strong base and will produce a basic solution when dissolved in water. The white precipitate is the $Zn(OH)_2$ formed by the Zn^{2+} reacting with hydroxide produced by the cyanide hydrolysis. The precipitation reaction is:

$$Zn^{2+}(aq) + 2\ OH^-(aq) \rightleftharpoons Zn(OH)_2(s).$$

As additional KCN is added, the complex ion $Zn(CN)_4^{2-}$ forms, dissolving the $Zn(CN)_2$ precipitate. The complexation reaction is:

$$Zn(OH)_2(s) + 4\ CN^-(aq) \rightleftharpoons Zn(CN)_4^-(aq) + 2\ OH^-(aq).$$

24 The Chemistry of Nuclear Processes

24.1 Use your own words to describe what is meant by "radioactive".

A radioactive species is a species which can spontaneously produce alpha (α), beta (β) or gamma (γ) radiation.

24.3 What is a nuclide? What is the difference between a nuclide and a nucleus?

A nuclide is a specific atom (isotope) of an element characterized by its mass number and, therefore, distinguished by the number of protons **and** neutrons in its nucleus. The nucleus is the central core of an atom and is composed of the protons (which determine which element it is) and the neutrons.

24.5 What is meant by the terms "parent" and "daughter" nuclides?

A parent nuclide is an atom that undergoes decay to form a daughter nuclide.

24.7 For each of α-, β- and γ- decay, and positron emission, tell whether there is an increase, decrease, or no change in (a) atomic number, (b) mass number, and (c) *n/p* ratio.

An alpha particle is ${}^{4}_{2}\alpha$. Alpha radiation (a) **decreases the atomic number**; (b) **decreases the mass number**; but (c) **increases the *n/p* ratio** because there are usually fewer protons than neutrons in the atoms undergoing alpha decay.

A beta particle is ${}^{0}_{-1}\beta$. Beta radiation (a) **increases the atomic number**; (b) **does not change the mass number**; but (c) **decreases the *n/p* ratio** because a neutron is converted to a proton.

A gamma ray is ${}^{0}_{0}\gamma$. Gamma radiation (a) **does not change the atomic number**; (b) **does not change the mass number**; and (c) **does not change the *n/p* ratio** because there is no change in the relative numbers of neutrons and protons.

A positron is ${}^{0}_{1}\beta$. Positron radiation (a) **decreases the atomic number**; (b) **does not change the mass number**; but (c) **increases the *n/p* ratio** because a proton is converted to a neutron.

24.9 Why is a nucleus unstable if it contains too few neutrons?

Protons, with their positive charges, repulse each other. It is thought that the presence of neutrons weakens the repulsive force between protons by separating them from each other. If the repulsive forces can be minimized, the nucleus will be more stable.

24.11 Explain the significance of the stability belt and how it may be used to predict the most likely mode of disintegration of a nucleus.

In stable isotopes, the n/p ratio increases from 1/1 to about 1/1.5 as atomic number increases from 1 to 83. Nuclides which are not stable tend to decay by the method which will bring their *n/p* ratio into the band of stability which is normal for stable isotopes.

24.13 How many stable nuclides are there? Assuming that 2000 different nuclides have been characterized (actually the number is somewhat larger), what percentage of them are radioactive?

Section 24.2 states that there are 279 naturally occurring stable isotopes. Therefore, there are about $\approx$ 2000 - 279 = $\approx$ 1721 unstable nuclides. $\frac{1721}{2000}$ x 100 $\approx$ **86%** of the isotopes are unstable.

24.15 How is it possible that an atomic nucleus can have less mass than the sum of the masses of its parts? Isn't this a violation of the law of conservation of mass?

When a nucleus is formed from neutrons and protons, some of the mass is converted into binding energy to hold the nucleus together. This is a violation of the law of conservation of mass, but is a manifestation of the more general law of conservation of mass and energy. The more specific, separate laws of conservation of mass and of energy are valid for chemical and physical changes, not nuclear processes.

24.17 Figure 24.3 (binding energy per nucleon) contains no point corresponding to hydrogen, ${}^{1}_{1}H$. Explain this omission.

This isotope of hydrogen only has a proton in its nucleus. Unless there is more than one nuclear particle to be held together, there can be no binding energy.

24.19 Why does nuclear fusion require high temperature?

When nuclei are fused, two separate positive species bind together. In order to overcome their mutual repulsion, the temperature must be very high so that the nuclei have sufficient energy to overcome the repulsions, contact each other and fuse.

24.21 What is the difference between a chain reaction and a branched chain reaction?

In a chain reaction, one neutron is produced for every nuclide that fissions and can in turn induce fission in one more nuclide. This keeps the rate of the reaction relatively constant. In contrast, in a branched chain reaction, more than one neutron is produced for every nuclide that fissions. Each neutron can cause another nuclide to undergo fission. As the number of neutrons continues to increase with each fission event, the process accelerates as fission progresses, often leading to an explosion if left unchecked.

24.23 Why is it that nuclear transformations can be induced by capture of slow-moving neutrons, but slow-moving protons are ineffective?

Slow-moving neutrons encounter no electrostatic resistance from the nucleus of the target atom, so can interact with it even though the energy is relatively low. Slow-moving protons are electrostatically repulsed by the positive target nucleus and are prevented from reacting with it.

24.25 On what physical principle is isotopic enrichment of uranium based?

Uranium isotopes to be separated are first converted to the gaseous compound UF_6. They can then be separated based on their rates of gaseous diffusion according to Graham's law which states that the rate of diffusion is inversely proportional to the square root of the molar mass.

24.27 The energy released in fusion of deuterium is 6.9 x 10^{11} J mol^{-1}, while that released in fission of ${}^{235}U$ is 2 x 10^{13} J mol^{-1}. Which process produces the greater energy per gram of fuel?

Analysis:

Target: The energy per gram of a fission process versus a fusion process. (? J g^{-1}) =

Knowns: Deuterium is $^{2}_{1}He$. The energy per mole of each isotope is given. The approximate molar mass of each isotope is implicit from its mass number.
Relationship: Molar mass provides the equivalence statement g ↔ mol.
Plan: Use the factor-label method to convert energy per mole to energy per gram for each and compare the two numbers.
Work:
Molar Mass $^{235}U \approx 235 \text{ g mol}^{-1}$

$$? \text{ J g}^{-1} = \frac{2 \times 10^{13} \text{ J}}{1 \text{ mol } ^{235}U} \times \frac{1 \text{ mol } ^{235}U}{235 \text{ g } ^{235}U} = 9 \times 10^{10} \text{ J g}^{-1}$$

Molar Mass $^{2}H \approx 2.0 \text{ g mol}^{-1}$

$$? \text{ J g}^{-1} = \frac{6.9 \times 10^{13} \text{ J}}{1 \text{ mol } ^{2}H} \times \frac{1 \text{ mol } ^{2}H}{2.0 \text{ g } ^{2}H} = 3.4 \times 10^{13} \text{ J g}^{-1}$$

The energy per gram of the deuterium fusion reaction is much greater.

24.29 What is a decay series? Why are there no significant accumulations of intermediate nuclides of a series?

A decay series is a sequence of spontaneous decays beginning with one parent isotope and ending with a stable nuclide. Inspection of the three uranium decay series presented in the text in Figure 24.6, indicates that, in each case, the half-life of the parent uranium atom is much longer than that of any of the daughters. Because all subsequent decays happen much faster than does the first step, the intermediate products decay rather than building up. In other words, the first decay is the slow or rate determining step of the kinetic process.

24.31 All of the daughter nuclides in the decay series beginning with ^{238}U have mass numbers that fit the expression $4n + 2$, where n is an integer. Explain why this is so.

The number 238 is equal to 4(59) + 2. Because each decay is either an alpha emission which changes the mass number by 4 or a beta emission which does not change the mass number, the mass number can only change by 4. After a certain number, say x, of alpha decays, the mass number of the product is 4(59 - x) + 2. This fits the general format where n = 59 - x.

24.33 What are the immediate effects of ionizing radiation on matter?

Ionizing radiation is so-called because it ionizes matter. Ions and free radicals (species with unpaired electrons) are formed.

24.35 What characteristics of radioactivity are commonly measured, and what are the common units of each? What are the SI units for these quantities?

Energy is measured in joules (J, SI) or electron volts (eV, not SI). Activity is measured in bequerels (Bq, SI) and curies (Ci, not SI). Penetrating power is measured in meters (m, SI). Absorbed dose is measure in grays (Gy, SI) and rads (rad, not SI). The sievert (Sv, SI) and the rem (not SI) measure the equivalent dose of radiation. The unit for extent of ionization when passing through air is the röntgen (not SI).

24.37 Without referring to the text, name as many non-medical uses of radioactivity as you can. Reread Section 24.8 and see if you missed any. Do you know of any not mentioned in the text?

Consult the text and compare with your own list.

24.39 Write balanced nuclear equations for the following processes.

a. α-decay of $^{232}_{94}Pu$ b. β-decay of tritium $(^{3}_{1}H)$

c. γ-decay of $^{60}_{27}Co$ d. positron decay of $^{170}_{72}Hf$

Analysis:

Target: A balanced nuclear equation.

Knowns: The reactant nuclide and the nature of the emission. The atomic number (the subscript in a nuclide designation) determines the element.

Relationships: The mass numbers and atomic numbers must be conserved. This means that the sum of the superscripts on the two sides must be equal and the sum of the subscripts on the two sides must also be equal.

Plan: Fill in the mass number and atomic number of the emission. Use algebra to determine the mass number and the atomic number of the missing species. Use the atomic number to determine the symbol of the element.

Work:

a. $^{232}_{94}Pu \longrightarrow ^{4}_{2}\alpha + ^{y}_{z}X$

Superscripts must balance, so 232 = 4 + y; y = 228.
Subscripts must balance, so 94 = 2 + z; z = 92.

The element with atomic number 92 = uranium, U. The product is $^{228}_{92}\mathbf{U}$.

b. $^{3}_{1}H \longrightarrow ^{0}_{-1}\beta + ^{y}_{z}X$

Superscripts must balance, so 3 = 0 + y; y = 3.
Subscripts must balance, so 1 = (-1) + z; z = 2.

The element with atomic number 2 = helium, He. The product is $^{3}_{2}\mathbf{He}$.

c. $^{60}_{27}Co \longrightarrow ^{0}_{0}\gamma + ^{y}_{z}X$

Superscripts must balance, so 60 = 0 + y; y = 60.
Subscripts must balance, so 27 = 0 + z; z = 27.

The element with atomic number 27 = cobalt, Co. The product is $^{60}_{27}\mathbf{Co}$.

d. $^{170}_{72}Hf \longrightarrow ^{0}_{1}\beta + ^{y}_{z}X$

Superscripts must balance, so 170= 0 + y; y = 170.
Subscripts must balance, so 72 = 1 + z; z = 71.

The element with atomic number 71 = lutetium, Lu. The product is $^{170}_{71}\mathbf{Lu}$.

24.41 Write a balanced equation that describes neutron capture by each of the following stable nuclides.

a. $^{98}_{42}Mo$ b. $^{142}_{58}Ce$

Analysis:

Target: A balanced nuclear equation.

Knowns: The reactant nuclide and the nature of the process. The atomic number (the subscript in a nuclide designation) determines the element.

Relationships: The mass numbers and atomic numbers must be conserved. This means that the sum of the superscripts on the two sides must be equal and the sum of the subscripts on the two sides must also be equal.

Plan: Use algebra to determine the mass number and the atomic number of the missing species. Use the atomic number to determine the symbol of the element.

Work:

a. $^{98}_{42}Mo + ^{1}_{0}n \longrightarrow ^{y}_{z}X$

Superscripts must balance, so 98 + 1 = y; y = 99.

Subscripts must balance, so 42 + 0 = z; z = 42.

The element with atomic number 42 = molybdenum, Mo. The product is $^{99}_{42}\mathbf{Mo}$.

b. $^{142}_{58}Ce + ^{1}_{0}n \longrightarrow ^{y}_{z}X$

Superscripts must balance, so 142 + 1 = y; y = 143.
Subscripts must balance, so 58 + 0 = z; z = 58.

The element with atomic number 58 = cerium, Ce. The product is $^{143}_{58}\mathbf{Ce}$.

24.43 Each of the following is produced (by neutron capture) from a stable nuclide in neutron activation analysis. Write a balanced equation that describes the process.

a. $^{24}_{11}Na$ b. $^{121}_{50}Sn$

Analysis:

Target: A balanced nuclear equation.
Knowns: The reactant nuclide and the nature of the process. The atomic number (the subscript in a nuclide designation) determines the element.
Relationships: The mass numbers and atomic numbers must be conserved. This means that the sum of the superscripts on the two sides must be equal and the sum of the subscripts on the two sides must also be equal.

Plan: Use algebra to determine the mass number and the atomic number of the missing species. Use the atomic number to determine the symbol of the element.

Work:

a. $^{y}_{z}X + ^{1}_{0}n \longrightarrow ^{24}_{11}Na$

Superscripts must balance, so y + 1 = 24; y = 23.
Subscripts must balance, so z + 0 = 11; z = 11.

The element with atomic number 11 = sodium, Na. The reactant is $^{23}_{11}\mathbf{Na}$.

b. $^{y}_{z}X + ^{1}_{0}n \longrightarrow ^{121}_{50}Sn$

Superscripts must balance, so y + 1 = 121; y = 120.
Subscripts must balance, so z + 0 = 50; z = 50.

The element with atomic number 50 = tin, Sn. The product is $^{120}_{50}\mathbf{Sn}$.

24.45 Supply the missing information to each of the following equations.

a. $^{53}_{24}Cr + ^{4}_{2}He \longrightarrow ^{1}_{0}n + ?$ b. $^{59}_{27}Co + ^{1}_{0}n \longrightarrow ^{56}_{25}Mn + ?$

c. $^{197}_{75}Re + \beta \longrightarrow ?$ d. $^{243}_{95}Am + ^{1}_{0}n \longrightarrow ^{244}_{?}Cm + ? + \gamma$

e. $^{35}_{17}Cl + ? \longrightarrow ^{32}_{16}S + ^{4}_{2}He$

Analysis:

Target: A balanced nuclear equation.
Knowns: The reactant nuclide and the nature of the process. The atomic number (the subscript in a nuclide designation) determines the element.
Relationships: The mass numbers and atomic numbers must be conserved. This means that the sum of the superscripts on the two sides must be equal and the sum of the subscripts on the two sides must also be equal.

Plan: Fill in the mass number and atomic number of the species for which symbols are given. Use algebra to determine the mass number and the atomic number of the missing species. Use the atomic number to determine the symbol of the element.

Work:

a. $^{53}_{24}Cr + ^{4}_{2}He \longrightarrow ^{1}_{0}n + ^{y}_{z}X$

Superscripts must balance, so 53 + 4 = 1 + y; y = 56.
Subscripts must balance, so 24 + 2 = 0 + z; z = 26.

The element with atomic number 26 = iron, Fe. The product is $^{56}_{26}Fe$.

b. $^{59}_{27}Co + ^{1}_{0}n \rightarrow ^{56}_{25}Mn + ^{y}_{z}X$
Superscripts must balance, so 59 + 1 = 56 + y; y = 4.
Subscripts must balance, so 27 + 0 = 25 + z; z = 2.

The element with atomic number 2 = helium, He or α. The product is $^{4}_{2}\alpha$ or $^{4}_{2}He$.

c. $^{197}_{75}Re + ^{0}_{-1}\beta \rightarrow ^{y}_{z}X$
Superscripts must balance, so 197 + 0 = y; y = 197.
Subscripts must balance, so 75 + (-1) = z; z = 74.

The element with atomic number 74 = tungsten, W. The product is $^{197}_{74}W$.

d. The atomic number of Cm is **96**.
$^{243}_{95}Am + ^{1}_{0}n \rightarrow ^{244}_{96}Cm + ^{y}_{z}X + ^{0}_{0}\gamma$
Superscripts must balance, so 243 + 1 = 244 + y + 0; y = 0.
Subscripts must balance, so 95 + 0 = 96 + z + 0; z = -1.

The species with atomic number -1 = beta, β. The product is $^{0}_{-1}\beta$.

e. $^{35}_{17}Cl + ^{y}_{z}X \rightarrow ^{32}_{16}S + ^{4}_{2}He$
Superscripts must balance, so 35 + y = 32 + 4; y = 1.
Subscripts must balance, so 17 + z = 16 + 2; z = 1.

The species with atomic number 1 = hydrogen, H or proton, *p*. The reactant is $^{1}_{1}H$ or $^{1}_{1}p$.

24.47 Predict whether electron or positron emission is the more likely mode of decay of the following radioisotopes.

a. $^{17}_{9}F$ b. $^{56}_{28}Ni$ c. $^{8}_{3}Li$ d. $^{49}_{20}Ca$

Analysis:

Target: The type of emission.

Known: The symbol of the nuclide, which gives the mass number and the atomic number.

Relationships: The atomic number is equal to the number of protons and the mass number minus the atomic number is the number of neutrons. The decay mode is suggested by the *n/p* ratio.

Plan: Determine the *n/p* ratio and then consult Figure 24.1 in the text. If the *n/p* ratio is higher than that for the element in the band of stability, beta emission occurs. If the *n/p* ratio is lower than that for a stable species, positron emission occurs.

Work:

a. $^{17}_{9}F$ has 9 protons and 17 - 9 = 8 neutrons. The *n/p* ratio is 8/9 = 0.89. For F the *n/p* ratio should be slightly greater than 1. This should decay by **positron** (β^{+}) emission.

b. $^{56}_{28}Ni$ has 28 protons and 56 - 28 = 28 neutrons. The *n/p* ratio is 28/28 = 1 For Ni the n/p ratio should be slightly greater than 1. This should decay by **positron** (β^{+}) emission.

c. $^{8}_{3}Li$ has 3 protons and 8 - 3 = 5 neutrons. The *n/p* ratio is 5/3 = 1.7. For Li the *n/p* ratio should be close to 1. This should decay by **electron** (β^{-}) emission.

d. $^{49}_{20}Ca$ has 20 protons and 49 - 20 = 29 neutrons. The n/p ratio is 29/20 = 1.4. For Ca the n/p ratio should be slightly greater than 1. This should decay by **electron** (β^-) emission

24.49 Which of the following pairs of nuclides do you expect to be more stable? Give your reason in each case. (This problem can be answered without calculating binding energy.)
a. ^{60}Ni, ^{66}Ni b. ^{70}Se, ^{74}Se c. ^{114}Ag, ^{114}Cd

Analysis:
Target: A prediction of relative stability.
Known: The symbols of the nuclides, which give the mass number. The atomic numbers are implicit knowns.
Relationships: The atomic number is equal to the number of protons and the mass number minus the atomic number is the number of neutrons.
Plan: Determine the number of neutrons and then consult Figure 24.1 in the text to see which is more stable. If the ratios are about the same, decide based on the number of nuclear particles. (An even number of particles is more stable.)
Work:
a. Ni has 28 protons. ^{60}Ni has 60 - 28 = 32 neutrons. ^{66}Ni has 66 - 28 = 38 neutrons. Figure 24.1 indicates that stable isotopes of Ni have fewer than 35 neutrons, so ^{60}**Ni** should be more stable.

b. Se has 34 protons. ^{70}Se has 70 - 34 = 36 neutrons. ^{74}Se has 74 - 34 = 40 neutrons. Figure 24.1 indicates that stable isotopes of Se have more than 40 neutrons, so ^{74}**Se**, which has 40 neutrons should be more stable.

c. ^{114}Ag has 47 protons and 114 - 47 = 67 neutrons. ^{114}Cd has 48 protons and 114 - 48 = 66 neutrons. From Figure 24.1, it appears that ^{114}Cd is a stable isotope of cadmium, but ^{114}Ag is not a stable isotope of silver. Even without the figure, ^{114}**Cd** would be expected to be more stable because it has an even number of both protons and neutrons.

24.51 What daughter nuclide remains after uranium-234 undergoes four successive α-decays?

Analysis/Plan: The net reaction would be $^{234}_{92}U \longrightarrow 4\,^{4}_{2}\alpha + ^{y}_{z}X$. Use algebra to determine the mass number and atomic number of the daughter. Determine the element by using its atomic number and the Periodic Table.
Work:
Superscripts must balance, so 234 = 4(4)+ y; y = 218.
Subscripts must balance, so 92 = 4(2) + z; z = 84.
The element with atomic number 84 = polonium, Po. The product is $^{218}_{84}$**Po**.

24.53 One electronvolt is defined as the kinetic energy acquired by an electron when accelerated through a potential difference of 1 volt. Using the conversion factor 1 eV ↔ 1.6022×10^{-19} J and the relationship $KE = \frac{mv^2}{2}$, calculate the velocity of (a) an electron with an energy of 1.00 eV, and (b) an electron with an energy of 1.00 keV. [*Hint:* In SI, 1 J = 1 kg m^2 s^{-2}.]

Analysis/Plan: Use the given relationship and rearrange it to solve for v, using the factor-label method to convert energy units. From the end cover of the text, the mass of an electron is 9.109390×10^{-31} kg. The equivalence statement 1 keV ↔ 10^3 eV is needed.
Work:
For both parts, $KE = \frac{mv^2}{2}$. Therefore, $2KE = mv^2$; $\frac{2KE}{m} = v^2$ and $v = \sqrt{\frac{2KE}{m}}$

a. $v = \sqrt{\frac{2KE}{m}} = \sqrt{\frac{2(1.00 \text{ eV}) \times \frac{1.6022 \times 10^{-19} \text{ J}}{\text{eV}} \times \frac{1 \text{ kg m}^2 \text{ s}^{-2}}{\text{J}}}{(9.109390 \times 10^{-31} \text{ kg})}} = \mathbf{5.93 \times 10^5 \text{ m s}^{-1}}$

b. $v = \sqrt{\frac{2KE}{m}} = \sqrt{\frac{2(1.00 \text{ keV}) \times \frac{10^3 \text{ eV}}{1 \text{ keV}} \times \frac{1.6022 \times 10^{-19} \text{ J}}{\text{eV}} \times \frac{1 \text{ kg m}^2 \text{ s}^{-2}}{\text{J}}}{(9.109390 \times 10^{-31} \text{ kg})}} = \mathbf{1.88 \times 10^7 \text{ m s}^{-1}}$

24.55 Calculate the velocity of a proton whose kinetic energy is (a) 1.00 eV, and (b) 5.00 MeV. (Do not be misled by the word "*electron*volt." The eV is an energy unit, like the joule or calorie, and may be used to describe the energy of protons and baseballs as well as electrons.)

Analysis/Plan: Use the given relationship for v, derived in answer to Problem 24.53, and insert the appropriate data for the proton. From the end cover of the text, the mass of a proton is 1.672623×10^{-27} kg. The equivalence statement 1 MeV ↔ 10^6 eV is needed.

Work:

a. $v = \sqrt{\frac{2KE}{m}} = \sqrt{\frac{2(1.00 \text{ eV}) \times \frac{1.6022 \times 10^{-19} \text{ J}}{\text{eV}} \times \frac{1 \text{ kg m}^2 \text{ s}^{-2}}{\text{J}}}{(1.672623 \times 10^{-27} \text{ kg})}} = \mathbf{1.38 \times 10^4 \text{ m s}^{-1}}$

b. $v = \sqrt{\frac{2KE}{m}} = \sqrt{\frac{2(5.00 \text{ MeV}) \times \frac{10^6 \text{ eV}}{1 \text{ MeV}} \times \frac{1.6022 \times 10^{-19} \text{ J}}{\text{eV}} \times \frac{1 \text{ kg m}^2 \text{ s}^{-2}}{\text{J}}}{(1.672623 \times 10^{-27} \text{ kg})}} = \mathbf{3.09 \times 10^7 \text{ m s}^{-1}}$

24.57 Use the results of Problem 24.53 and the de Broglie relationship (Section 6.4) to calculate the wavelength of (a) a 1.00 eV electron, and (b) a 1.00 keV electron.

Analysis:

Target: A wavelength. (? pm) = or (? nm) =

Knowns: The velocity (found in Problem 24.53) and mass of the electron. Planck's constant, h, is equal to 6.626076×10^{-34} J s.

Relationships: The de Broglie equation for matter waves is $\lambda = \frac{h}{mv}$. The necessary equivalence statements are: 1 pm ↔ 10^{-12} m, 1 nm ↔ 10^{-9} m and 1 J ↔ 1 kg m^2 s^{-2}.

Plan: Use the de Broglie equation and the factor-label method to make necessary conversions. Express the wavelength in an appropriate unit.

Work:

a. $? \text{ nm} = \lambda = \frac{(6.626076 \times 10^{-34} \text{ J s}) \times \left(\frac{\frac{1 \text{ kg m}^2}{\text{s}^2}}{\text{J}}\right) \times \frac{1 \text{ nm}}{10^{-9} \text{ m}}}{(9.109390 \times 10^{-31} \text{ kg})(5.93 \times 10^5 \text{ m s}^{-1})} = \mathbf{1.23 \text{ nm}}$

b. $? \text{ pm} = \lambda = \frac{(6.626076 \times 10^{-34} \text{ J s}) \times \left(\frac{\frac{1 \text{ kg m}^2}{\text{s}^2}}{\text{J}}\right) \times \frac{1 \text{ pm}}{10^{-12} \text{ m}}}{(9.109390 \times 10^{-31} \text{ kg})(1.88 \times 10^7 \text{ m s}^{-1})} = \mathbf{38.7 \text{ pm}}$

24.59 Calculate the mass defect of the radioisotope ^{14}O (a positron emitter), whose atomic mass is 14.00860 amu. Express your answer in amu, then convert to binding energy and binding energy per nucleon in J nucleon^{-1}.

Analysis:

Target: The mass defect and binding energy of ^{14}O.

Known: The mass of the atom and the nuclide symbol. The molar masses of the nuclear particles are implicit knowns.

Relationship: The mass defect is given by the nucleon mass - nuclear mass. The nucleon mass is the sum of the masses of the protons and neutrons. The nuclear mass is the nuclide mass - the electron mass. Avogadro's number 6.022137×10^{23} atoms ↔ 1 mol is needed. The equivalence statement 1 amu ↔ 1.492419×10^{-10} J from the end cover is also useful.

Plan: Determine the number of nuclear particles from the mass number and the symbol. Find the total mass of the nuclear particles and subtract the nuclide mass from it to get the mass defect. Convert that mass to joules using the factor-label method and then divide by the number of nucleons.

Work:

^{14}O has 8 protons, 8 electrons and 14 - 8 = 6 neutrons.
Molar Mass proton = 1.007276 amu   Molar Mass neutron = 1.008665 amu
Molar Mass electron = 0.00054858 amu

Nucleon mass = proton mass + neutron mass
Nucleon mass = 8(1.007276 amu) + 6(1.008665 amu) = 14.110198 amu

Nuclear mass = nuclide mass - electron mass
Nuclear mass = 14.00860 amu - 8(0.00054858 amu) = 14.00421 amu

Mass defect = 14.110198 amu - 14.00421 amu = **0.10599 amu** per atom

The binding energy is:

$$\frac{0.10599 \text{ amu}}{\text{atom}} \times \frac{6.022137 \times 10^{23} \text{ atom}}{\text{mol}} \times \frac{1.492419 \times 10^{-10} \text{ J}}{\text{amu}} = \mathbf{9.6256 \times 10^{12} \text{ J mol}^{-1}}$$

Binding energy per nucleon is:

$$\frac{\frac{0.10599 \text{ amu}}{\text{atom}} \times \frac{1.492419 \times 10^{-10} \text{ J}}{\text{amu}}}{14 \text{ nucleons atom}^{-1}} = \mathbf{1.1298 \times 10^{-12} \text{ J nucleon}^{-1}}$$

24.61 Calculate the nuclear binding energy of ^{71}Ga, whose atomic mass is 70.924700 amu.

Analysis:

Target: The mass defect and binding energy of ^{71}Ga.

Known: The mass of the atom and the nuclide symbol. The molar masses of the nuclear particles are implicit knowns.

Relationship: The mass defect is given by the nucleon mass - nuclear mass. The nucleon mass is the sum of the protons and neutrons. The nuclear mass is the nuclide mass - the electron mass. Avogadro's number 6.022137×10^{23} atoms ↔ 1 mol is needed. The equivalence statement 1 amu ↔ 1.492419×10^{-10} J from the end cover is also useful.

Plan: Determine the number of nuclear particles from the mass number and the symbol. Find the total mass of the nuclear particles and subtract the nuclide mass from it to get the mass defect. Convert that mass to joules using the factor-label method and then divide by the number of nucleons.

Work:

^{71}Ga has 31 protons, 31 electrons and 71 - 31 = 40 neutrons.
Molar Mass proton = 1.007276 amu   Molar Mass neutron = 1.008665 amu
Molar Mass electron = 0.00054858 amu

Nucleon mass = proton mass + neutron mass
Nucleon mass = 31(1.007276 amu) + 40(1.008665 amu) = 71.57216 amu
Nuclear mass = nuclide mass - electron mass
Nuclear mass = 70.924700 amu - 31(0.00054858 amu) = 70.90769 amu

Mass defect = 71.57216 amu - 70.90769 amu = 0.66447 amu per atom

The binding energy is:

$$\frac{0.66447 \text{ amu}}{\text{atom}} \times \frac{6.022137 \times 10^{23} \text{ atom}}{\text{mol}} \times \frac{1.492419 \times 10^{-10} \text{ J}}{\text{amu}} = \mathbf{5.9719 \times 10^{13} \text{ J mol}^{-1}}$$

Binding energy per nucleon is:

$$\frac{\dfrac{0.66447 \text{ amu}}{\text{atom}} \times \dfrac{1.492419 \times 10^{-10} \text{ J}}{\text{amu}}}{71 \text{ nucleons atom}^{-1}} = \mathbf{1.3967 \times 10^{-12} \text{ J nucleon}^{-1}}$$

24.63 The atomic mass of ^{24}Mg is 23.985042 amu. Calculate its nuclear binding energy per nucleon, and express the result in MeV.

Analysis:

Target: The mass defect and binding energy of ^{24}Mg.
Known: The mass of the atom and the nuclide symbol. The molar masses of the nuclear particles are implicit knowns.
Relationship: The mass defect is given by the nucleon mass - nuclear mass. The nucleon mass is the sum of the protons and neutrons. The nuclear mass is the nuclide mass - the electron mass. The equivalence statement 1 amu ↔ 931.4943 MeV is needed.

Plan: Determine the number of nuclear particles from the mass number and the symbol. Find the total mass of the nuclear particles and subtract the nuclide mass from it to get the mass defect. Convert that mass to MeV using the factor-label method and then divide by the number of nucleons.

Work:

^{24}Mg has 12 protons, 12 electrons and 24 - 12 = 12 electrons.
Molar Mass proton = 1.007276 amu Molar Mass neutron = 1.008665 amu
Molar Mass electron = 0.00054858 amu

Nucleon mass = proton mass + neutron mass
Nucleon mass = 12(1.007276 amu) + 12(1.008665 amu) = 24.19129 amu

Nuclear mass = nuclide mass - electron mass
Nuclear mass = 23.985042 amu - 12(0.00054858 amu) = 23.978459 amu

Mass defect = 24.19129 amu - 23.978459 amu = 0.21283 amu per atom

The binding energy is:

$$\frac{0.21283 \text{ amu}}{\text{atom}} \times \frac{1 \text{ atom}}{24 \text{ nucleons}} \times \frac{931.4943 \text{ MeV}}{\text{amu}} = \mathbf{8.2605 \text{ MeV nucleon}^{-1}}$$

24.65 Some people prefer to express binding energy in joules per *gram*. Almost all stable nuclei have binding energies in the range 1.2 - 1.5 x 10^{-12} J $nucleon^{-1}$. Express this range in J g^{-1}.

Analysis/Plan: This can be solved by the factor-label method, using the following equivalence statements: 1 amu ↔ 1.660540 x 10^{-27} kg, 1 kg ↔ 10^3 g, and 1 nucleon ~ 1.01 amu.

Work:

$$\frac{1.2 \times 10^{-12}\ \text{J}}{\text{nucleon}} \times \frac{1\ \text{nucleon}}{1.01\ \text{amu}} \times \frac{1\ \text{amu}}{1.660540 \times 10^{-27}\ \text{kg}} \times \frac{1\ \text{kg}}{10^3\ \text{g}} = \mathbf{7.2 \times 10^{11}\ J\ g^{-1}}$$

$$\frac{1.5 \times 10^{-12}\ \text{J}}{\text{nucleon}} \times \frac{1\ \text{nucleon}}{1.01\ \text{amu}} \times \frac{1\ \text{amu}}{1.660540 \times 10^{-27}\ \text{kg}} \times \frac{1\ \text{kg}}{10^3\ \text{g}} = \mathbf{8.9 \times 10^{11}\ J\ g^{-1}}$$

24.67 The energy released in a nuclear transformation is the difference between the total mass of the products and the total mass of the reactants. Calculate the energy released in the α-decay of ${}^{235}_{92}U$. Atomic masses (in amu) are : ${}^{235}_{92}U$, 235.04394, ${}^{231}_{90}Th$, 231.036298; ${}^{4}_{2}He$, 4.00260.

Analysis/Plan: First write the balanced equation for the α-decay. The change in mass can be determined by subtracting the mass of reactant from the total mass of the products. The factor-label method can then be used to convert the mass change in amu into energy in MeV and kJ mol^{-1}. The equivalence statements 1 amu ↔ 931.4943 MeV, 1 eV ↔ 96.48531 kJ mol^{-1} and 1 MeV ↔ 10^6 eV are appropriate for the conversion.

Work:

The balanced reaction is: ${}^{235}_{92}U \longrightarrow {}^{231}_{90}Th + {}^{4}_{2}He$.

The change in mass is found first.

? amu = 231.036298 amu + 4.00260 amu - 235.04394 amu = -0.00504 amu.

The energy released is then determined by the factor-label method.

$$?\ \text{MeV} = \frac{-0.00504\ \text{amu}}{\text{atom U}} \times \frac{931.4943\ \text{MeV}}{\text{amu}} = \mathbf{-4.70\ MeV\ atom^{-1}\ U}$$

$$?\ \text{kJ mol}^{-1} = -4.70\ \text{MeV} \times \frac{10^6\ \text{eV}}{\text{MeV}} \times \frac{96.48531\ \text{kJ mol}^{-1}}{\text{eV}} = \mathbf{-4.53 \times 10^8\ kJ\ mol^{-1}\ U}$$

Remember that the negative sign indicates that energy is released in the process.

24.69 Calculate the energy released when 1.00 mole of deuterium undergoes fusion according to the overall reaction

$$3\ {}^{2}_{1}H \longrightarrow {}^{4}_{2}He + {}^{1}_{1}H + {}^{1}_{0}n$$

Neutron mass is 1.008665 amu; relevant atomic masses in amu are: ${}^{1}H$, 1.007825; ${}^{2}H$, 2.0140; ${}^{4}He$, 4.00260.

Analysis/Plan: The change in mass can be determined by subtracting the mass of reactant from the total mass of the products. The factor-label method can then be used to convert the mass change in amu into kJ mol^{-1}. The equivalence statements 1 amu ↔ 1.492419 x 10^{-10} J and 6.022137 x 10^{23} atoms ↔ 1 mol are appropriate for the conversion. The coefficients in the equation give the mole ratios.

Work:

The change in mass is found first, from the mass of products - the mass of reactants.

? amu = 4.00260 amu + 1.007825 amu + 1.008665 - 3(2.0140) amu = -0.0229 amu.

The energy released is then determined by the factor-label method.

$$?\ \text{J mol}^{-1} = \frac{-0.0229\ \text{amu}}{3\ \text{atoms}\ {}^{2}_{1}H} \times \frac{1.492419 \times 10^{-10}\ \text{J}}{\text{amu}} \times \frac{6.022137 \times 10^{23}\ \text{atoms}}{1\ \text{mol}} =$$

$$\mathbf{-6.86 \times 10^{11}\ J\ mol^{-1}}$$

24.71 A γ-ray source used in food preservation has an activity of 2500 Ci. Express this in becquerels.

Analysis/Plan: Use the factor-label method for this conversion. The equivalence statement required is 1 Ci ↔ 3.7 x 10^{10} Bq.
Work:

$$? \text{ Bq} = 2500 \text{ Ci} \times \frac{3.7 \times 10^{10} \text{ Bq}}{1 \text{ Ci}} = \mathbf{9.3 \times 10^{13} \text{ Bq}}$$

24.73 How many decay events occur in one hour in a 5.0 mCi source?

Analysis/Plan: A decay event is a disintegration. Use the factor-label method for this conversion. The equivalence statements required are 1 mCi ↔ 10^{-3} Ci, 1 Ci ↔ 1 disintegration s^{-1}, 60 s ↔ 1 min and 60 min ↔ 1 hour. Assume that one hour has 3 significant digits.
Work:

$$? \text{ disintegration} = 1.00 \text{ hr} \times \frac{60 \text{ min}}{1 \text{ hr}} \times \frac{60 \text{ s}}{1 \text{ min}} \times \frac{3.7 \times 10^{10} \text{ disintegration s}^{-1}}{\text{Ci}} \times \frac{10^{-3} \text{ Ci}}{1 \text{ mCi}} \times$$

$$5.0 \text{ mCi} = \mathbf{6.7 \times 10^{11} \text{ disintegrations}}$$

24.75 A scintillation counter registers a total of 22,400 counts from a radioactive sample during a period of 10 minutes. Assuming that the counter detects all of the emitted radiation, what is the activity of the sample?

Analysis/Plan: Activity is defined in terms of disintegrations per second (becquerels). The number of counts is equal to the number of disintegrations. Use the factor-label method to convert counts per 10 minutes to becquerels using the equivalence statements 1 min ↔ 60 s, and 1 Bq ↔ 1 count s^{-1}.
Work:

$$? \text{ Bq} = \frac{22400 \text{ counts}}{10 \text{ min}} \times \frac{1 \text{ min}}{60 \text{ s}} \times \frac{1 \text{ Bq}}{1 \text{ count s}^{-1}} = \mathbf{37 \text{ Bq}}$$

24.77 A 1.5 kg sample exposed to radiation receives a total of 33 J. What is the absorbed dose in rads?

Analysis/Plan: Use the factor-label method to convert to rads, using the equivalence statement 1 rad ↔ 0.01 J kg^{-1}.
Work:

$$? \text{ rad} = \frac{33 \text{ J}}{1.5 \text{ kg}} \times \frac{1 \text{ rad}}{0.01 \text{ J kg}^{-1}} = \mathbf{2.2 \times 10^{3} \text{ rad}} \text{ (or } \mathbf{2.2 \text{ krad}}\text{)}$$

24.79 A sample receives an absorbed dose of 0.050 mrads of α-radiation. What is the equivalent dose in rems?

Analysis/Plan: Use the factor-label method and the definition that rad(Q) = rem, where Q is an experimentally determined quality factor that equals 20 for α-radiation. The equivalence statement
1 mrad ↔ 10^{-3} rad is needed.
Work:

$$? \text{ rem} = 0.050 \text{ mrad} \times \frac{10^{-3} \text{ rad}}{1 \text{ mrad}} \times 20 = \mathbf{1.0 \times 10^{-3} \text{rem}}$$

24.81 The rate constant for the decay of ^{78}Br, a positron emitter, is 5.0 x 10^{-3} s^{-1}. What is its half-life?

Analysis/Plan: Use the relationship $t_{1/2} = \frac{0.693}{k}$, where $t_{1/2}$ is the half-life and k is the rate constant. Use the equivalence statement 60 s ↔ 1 min to convert to a more manageable form for the time unit.

Work:

$$? \text{ min} = \frac{0.693}{5.0 \times 10^{-3}\ s^{-1}} \times \frac{1 \text{ min}}{60 \text{ s}} = \mathbf{2.3 \text{ min}}$$

24.83 What is the decay rate constant of ^{72}Se, which undergoes electron capture with a half-life of 9.7 d?

Analysis/Plan: Use the relationship $k = \frac{0.693}{t_{1/2}}$, where $t_{1/2}$ is the half-life and k is the rate constant. Rate constants are typically reported in s^{-1}, so the equivalence statements 24 hr ↔ 1 d, 1 hr ↔ 60 min and 1 min ↔ 60 s should be used with the factor-label method to convert d^{-1} to s^{-1}.

Work:

$$? \ s^{-1} = \frac{0.693}{9.7 \text{ d}} \times \frac{1 \text{ d}}{24 \text{ hr}} \times \frac{1 \text{ hr}}{60 \text{ min}} \times \frac{1 \text{ min}}{60 \text{ s}} = \mathbf{8.3 \times 10^{-7}\ s^{-1}}$$

24.85 Thorium-220 emits α-particles with a rate constant of $2.7 \times 10^{-6}\ s^{-1}$. What is the activity, in disintegrations per second, of one mole of this isotope?

Analysis/Plan: Each α emitted signifies a disintegration. Use the factor-label method with the equivalence statements 1 mol ↔ 6.022137×10^{23} atoms and 1 α ↔ 1 Th atom.

Work:

$$? \text{ disintegrations } s^{-1} = 1.00 \text{ mol Th} \times \frac{6.022137 \times 10^{23} \text{ atoms Th}}{1 \text{ mol Th}} \times \frac{1\ \alpha}{1 \text{ atom Th}} \times$$

$$2.7 \times 10^{-6}\ s^{-1} = \mathbf{1.6 \times 10^{18}\ \alpha\ s^{-1}} \text{ or } \mathbf{1.6 \times 10^{18} \text{ Bq}}$$

24.87 Cesium-137, a radioisotope used in cancer therapy, decays by β-emission to ^{137}Ba with a half-life of 33 y. (The product nucleus is formed in an excited state, and immediately emits a γ-ray and drops to the ground state, so that this is a "double-barreled" weapon in medicine.) Use the half-life to determine the activity (disintegrations per second) of 1.0 g ^{137}Cs.

Analysis/Plan: This is a two step problem. First, use the relationship $k = \frac{0.693}{t_{1/2}}$, where $t_{1/2}$ is the half-life and k is the rate constant. Rate constants are typically reported in s^{-1}, so the equivalence statements 1 y ↔ 365 d, 24 hr ↔ 1 d, 1 hr ↔ 60 min and 1 min ↔ s should be used with the factor-label method to convert y^{-1} to s^{-1}. Second, determine the number of atoms decaying, using the molar mass, which gives the equivalence statement g ↔ mol, and Avogadro's number (1 mol ↔ 6.022137×10^{23} atoms). The activity is given by the number of atoms decaying per second.

Work:

First, determine the rate constant, k.

$$? \ s^{-1} = \frac{0.693}{33 \text{ y}} \times \frac{1 \text{ y}}{365 \text{ d}} \times \frac{1 \text{ d}}{24 \text{ hr}} \times \frac{1 \text{ hr}}{60 \text{ min}} \times \frac{1 \text{ min}}{60 \text{ s}} = 6.7 \times 10^{-10}\ s^{-1}$$

Next, determine the number of atoms of Cs in the 1.0 g sample.

Molar Mass $^{137}Cs \approx 137 \text{ g mol}^{-1}$

$$? \text{ atoms} = 1.0 \text{ g Cs} \times \frac{1 \text{ mol Cs}}{137 \text{ g Cs}} \times \frac{6.022137 \times 10^{23} \text{ atoms Cs}}{1 \text{ mol Cs}} = 4.4 \times 10^{21} \text{ atoms Cs}$$

Finally find the activity which is expressed in atoms s^{-1} (Bq).

$$? \text{ atoms } s^{-1} = (4.4 \times 10^{21} \text{ atoms}) \times (6.7 \times 10^{-10}\ s^{-1}) = \mathbf{2.9 \times 10^{12} \text{ atoms } s^{-1} \text{ (Bq)}}$$

24.89 One of the lingering products of atmospheric testing of atomic weapons in the 1950s is ^{90}Sr. It is among the more harmful constituents of fallout because its chemical similarity to calcium causes it to be concentrated in dairy products, and, ultimately, in the bones and teeth of consumers. Given that its half-life is 25 years, what percentage of the original activity remains now, after 35 years?

Analysis:

Target: The percent of an isotope remaining. (? %) =

Knowns: The half-life of the isotope and the time of decay.

Relationships: $\ln\left(\frac{N}{N_o}\right) = -kt$, where N is the number of atoms of Sr remaining, N_o is the original number of atoms, k is the rate constant and t is the elapsed time, and $k = \frac{0.693}{t_{1/2}}$. The percent is given by $\frac{N}{N_o} \times 100$. The activity is directly proportional to the amount remaining.

Plan: Combine the first two equations to make a direct relationship between the half-life and $\frac{N}{N_o}$. Insert the known values and solve for $\frac{N}{N_o}$, then use the definition of percent.

Work:

Combining the first two relationships gives $\ln\left(\frac{N}{N_o}\right) = -\frac{0.693t}{t_{1/2}}$.

Insert the known values and solve.

$$\ln\left(\frac{N}{N_o}\right) = -\frac{(0.693)(35\ y)}{25\ y} = -9.70, \text{ therefore, } \frac{N}{N_o} = e^{-9.70} = 0.38$$

$$? \%\text{ remaining} = \frac{N}{N_o} \times 100 = 0.38 \times 100 = \mathbf{38\%}$$

24.91 How long a time must elapse before the activity of ^{91}Kr, a waste product of uranium fission, is reduced to 0.1% of its original level? The half-life is 10 s.

Analysis:

Target: The age of the sample. (? s) =

Knowns: The half-life of the isotope and the percent remaining.

Relationships: From Problem 24.89, $\ln\left(\frac{N}{N_o}\right) = -\frac{0.693t}{t_{1/2}}$, where N is the number of atoms of Kr remaining, N_o is the original number of atoms, t is the elapsed time, and $t_{1/2}$ is the half-life. If the activity is reduced to 0.1%, then

$$\frac{N}{N_o} \times 100 = 0.1\%$$

Plan: Determine $\frac{N}{N_o}$ from the percent, then insert the known values into the formula and solve for the time.

Work:

$$\frac{N}{N_o} = \frac{0.1\%}{100} = 0.001$$

Now, use the formula and solve for t.

$$\ln\left(\frac{N}{N_o}\right) = -\frac{(0.693)(t)}{t_{1/2}} = \ln(0.001) = -\frac{0.693t}{10\ s}$$

$$?\ s = t = -\frac{\ln(0.001)(10\ s)}{(0.693)} = \mathbf{1 \times 10^2\ s}$$

24.93 Rubidium-87 decays by β-emission (to ^{87}Sr, a stable nuclide) with a half-life of 4.9×10^{10} y. Some fossils of single-celled organisms were found in which 4.0% of the originally-present ^{87}Rb had converted to ^{87}Sr. Estimate the age of the fossils.

Analysis:

Target: The age of the sample. (? y) =

Knowns: The half-life of the isotope and the percent reacted.

Relationships: From Problem 24.89, $\ln\left(\frac{N}{N_o}\right) = -\frac{0.693t}{t_{1/2}}$, where N is the number of atoms of Rb remaining, N_o is the original number of atoms, t is the elapsed time, and $t_{1/2}$ is the half-life. If 4.0% of the ^{87}Rb has converted to ^{87}Sr, then 100% - 4.0 % = 96.0% is left and $\frac{N}{N_o} \times 100 = 96.0\%$.

Plan: Determine $\frac{N}{N_o}$ from the percent, then insert the known values into the formula and solve for the time.

Work:

$$\frac{N}{N_o} = \frac{96.0\%}{100} = 0.960$$

Now, use the formula and solve for t.

$$\ln\left(\frac{N}{N_o}\right) = -\frac{(0.693)(t)}{t_{1/2}} = \ln(0.960) = -\frac{0.693t}{4.9 \times 10^{10}\ y}$$

$$?\ y = t = -\frac{\ln(0.960)(4.9 \times 10^{10}\ y)}{(0.693)} = \mathbf{2.9 \times 10^9\ y}$$

24.95 The ultimate stable product of ^{238}U decay is ^{206}Pb. A certain sample of pitchblende ore was found to contain ^{238}U and ^{206}Pb in the ratio 68.3 atoms ^{238}U : 31.7 atoms ^{206}Pb. Assuming that all the ^{206}Pb arose from uranium decay, and that no U or Pb has been lost by weathering, how old is the rock? The half-life of ^{238}U is 4.48×10^9 years.

Analysis:

Target: The age of the sample. (? y) =

Knowns: The half-life of the isotope and the elemental analysis of the sample.

Relationships: From Problem 24.89, $\ln\left(\frac{N}{N_o}\right) = -\frac{0.693t}{t_{1/2}}$, where N is the number of atoms of U remaining, N_o is the original number of atoms, t is the elapsed time, and $t_{1/2}$ is the half-life. All the uranium which decayed has been converted to lead.

Plan: Determine $\frac{N}{N_o}$ from the number of atoms of U and Pb present in the sample. Insert the known values into the formula and solve for the time.

Work:

If all Pb came from U, then initially there were 68.3 U + 31.7 Pb = 100 U atoms. Therefore, N = 68.3 U atoms and N_o = 100 U atoms.

Now, use the formula and solve for t.

$$\ln\left(\frac{N}{N_o}\right) = -\frac{(0.693)(t)}{t_{1/2}} = \ln\left(\frac{68.3}{100}\right) = \ln(0.683) = -\frac{0.693t}{4.48 \times 10^9\ y}$$

$$? \text{ y} = t = -\frac{\ln(0.683)(4.48 \times 10^9 \text{ y})}{(0.693)} = \mathbf{2.46 \times 10^9 \text{ y}}$$

24.97 One of the problems associated with the $^{238}U/^{206}Pb$ dating method is that uranium-containing minerals frequently contain natural lead in addition to lead formed by uranium decay, so that some of the ^{206}Pb in the samples does *not* arise from uranium. However, one lead isotope, ^{204}Pb, does not arise from decay of any heavier isotope, so its presence can be used to determine the amount of lead that was present when the rock was formed. Isotopic abundances in Pb are such that there are 19.2 ^{206}Pb atoms for each ^{204}Pb atom in natural lead. Calculate the age of a rock sample containing these isotopes in the ratio 1 atom ^{204}Pb : 22.1 atoms ^{206}Pb : 3.8 atoms ^{238}U. [*Hint:* Determine x, the number of "extra" or decay-product ^{206}Pb atoms for each ^{204}Pb atom. Then the fraction of ^{238}U atoms that has decayed is $\frac{3.8}{3.8 + x}$.]

Analysis:

Target: The age of the sample. (? y) =

Knowns: The half-life of the isotope, the elemental analysis of the sample and the ratio of ^{204}Pb to ^{206}Pb in natural samples.

Relationships: From Problem 24.89, $\ln\left(\frac{N}{N_o}\right) = -\frac{0.693t}{t_{1/2}}$, where N is the number of atoms of U remaining, N_o is the original number of atoms, t is the elapsed time, and $t_{1/2}$ is the half-life.

Plan: Use the hint to find $\frac{N}{N_o}$ from the number of atoms of U and both isotopes of Pb present in the sample. Insert the known values into the formula and solve for the time.

Work:

If the isotope ratios hold, then, if there is 1 ^{204}Pb atom in the sample, there must have been:

$$1 \text{ atom } ^{204}Pb \times \frac{19.2 \ ^{206}Pb \text{ atoms}}{1 \ ^{204}Pb \text{ atom}} = 19.2 \ ^{206}Pb \text{ atoms in the original sample.}$$

There are now 22.1 ^{206}Pb atoms, so 22.1 - 19.2 = 2.9 ^{206}Pb atoms came from decay of ^{238}U. The original sample of U must have had 3.8 ^{238}U + 2.9 ^{206}Pb = 6.7 atoms ^{238}U.

Now, use the formula and solve for t.

$$\ln\left(\frac{N}{N_o}\right) = -\frac{(0.693)(t)}{t_{1/2}} = \ln\left(\frac{3.8}{6.7}\right) = \ln(0.57) = -\frac{0.693t}{4.48 \times 10^9 \text{ y}}$$

$$? \text{ y} = t = -\frac{\ln(0.57)(4.48 \times 10^9 \text{ y})}{(0.693)} = \mathbf{3.7 \times 10^9 \text{ y}}$$